CONSERVATION MEDICINE

ECOLOGICAL HEALTH IN PRACTICE

Edited by

A. Alonso Aguirre

Richard S. Ostfeld

Gary M. Tabor

Carol House

Mary C. Pearl

OXFORD
UNIVERSITY PRESS

2002

OXFORD
UNIVERSITY PRESS

Oxford   New York
Auckland   Bangkok   Buenos Aires   Cape Town   Chennai
Dar es Salaam   Delhi   Hong Kong   Istanbul   Karachi   Kolkata
Kuala Lumpur   Madrid   Melbourne   Mexico City   Mumbai   Nairobi
São Paulo   Shanghai   Singapore   Taipei   Tokyo   Toronto

and an associated company in Berlin

Published by Oxford University Press, Inc.
198 Madison Avenue, New York, New York 10016

www.oup.com

Oxford is a registered trademark of Oxford University Press

Library of Congress Cataloging-in-Publication Data
Conservation medicine : ecological health in practice / edited by A. Alonso Aguirre . . . [et al.].
    p. : cm.
  Includes bibliographical references and index.
  ISBN 978-0-19-515093-3
    1. Environmental health. 2. Ecology. I. Aguirre, A. Alonso.
  [DNLM: 1. Environmental Health. 2. Conservation of Natural Resources. 3. Ecology.
  4. Environmental Medicine. WA 30 C755 2002]
  RA566.C667 2002
  615.9'02—dc21   2001036851

9 8 7 6 5

Printed in the United States of America
on recycled, acid free paper

This book is dedicated to the scientists and practitioners
in the field, on the ground, and around the globe—
the new generation of ecological care givers.

Foreword: Giant Moths and Doing No Harm

Michael Soulé

In the future, the Legend of the Great Dying will be recited to the children of the Third Planet:

It happened thusly. First, there was the Great Explosion in human numbers and in technological prowess. In 200 Earth years, all the wild places were degraded or destroyed. Next, the chemicals and gases released by agriculture and industry impaired the health of the surviving species and changed the climate. The Great Heat then occurred, as did the Second Great Flood. Simultaneously, thousands of species of plants and animals were transported across natural barriers and became invasive species in their new surroundings; this was known as the Great Mixing. Near the end of that era there were many new plagues—the Great Sickness—that ravaged the weakened, unprepared human beings and other species. After that, the survivors left.

The birth of conservation medicine is timely. It is an essential response to the emergence of new diseases and the physiological threats to human beings and millions of other species caused by industry, agriculture, and commerce. Even while the life expectancy of people from the richest northern countries is increasing, the defenses against the relentless old diseases, such as malaria and tuberculosis, are weakening. Moreover, there are indications that disease is becoming a major cause of ecological simplification, including extinction. The most shocking harbinger is the global decline in amphibians caused by habitat loss, pollution, and a virulent fungus. Conservation medicine addresses the two-way exchange of pathogens between human society and self-regulated nature, and it calls for interdisciplinary research that might lead to new solutions. For this we can only applaud the emergence of conservation medicine.

The ultimate cause of the Great Dying is thought to have been the failure of humanity to restrain its fecund industry. Prophet David Ehrenfeld, in his book The Arrogance of Humanism, *quoted René Dubos: "Developing countertechnologies to correct the new kinds of damage constantly being created by technological innovations is a policy of despair."*

Our descendents learned too late the simple, karmic law of ecology: all is inter-dependent and all is interconnected. As bad philosophers continued to debate whether human beings were part of nature or its butcher, a spiral of dreadful causation erased this illusory dualism, and it became evident that the destiny of humanity on Earth was to be both victim and executioner of creation. At the end all earthly beings became joined in an intimate, slow dance of death.

The new discipline arrives just in time. New diseases are emerging; pathogens and parasites are spreading more rapidly and infecting new plant and animal hosts. The widespread misuse of antibiotics, insecticides, fertilizers, and other anthro-pogenically amplified and created chemicals is contributing to the decline of phys-iological vigor and demographic vital rates in multitudes of species.

Conservation medicine is the right medicine. We need more expertise, more knowledge, new cures, interdisciplinary research. Conservation medicine can draw more attention to the problems, attract funding for analysis and new therapies, educate citizens and decision makers, and help activists lobby for more effective regulations.

In time, conservation medicine must define itself; it is not the same as envi-ronmental medicine—a purely anthropocentric discipline—but pressure exists to move in that direction. Most human beings equate value (goodness) with human welfare. Granted, this human-centered bias will make it easier for conservation medicine to carve out a niche and raise funds. And the more it emphasizes the health of human beings, the more money can be raised. This may be good or bad for nature, depending on the interests of the grantees. But money virtually always corrupts, as it has in the field of sustainable development, where multimillion dollar projects have led neither to sustainable economic development nor to ef-fective protection of nature. For the sake of saving nature, conservation medicine will also need to clarify whether its salvational mission is ecocentric or human-itarian. For institutional reasons, it cannot remain on the fence, even though it is desirable to erase disciplinary borders.

An early warning of the Great Dying was the global disappearance of most amphibians, soon followed by reptiles and most freshwater organisms. Causes were discovered but cures were never implemented because human greed and expansion trumped other values. Human beings systematically exterminated any species that stood in the way of profit and comfort. For example, the tsetse fly, a vector for sleeping sickness in Africa, was eradicated to create more settlement areas for people, even though the project also eliminated most of the remaining wildlands on the continent.

A caveat: It would be morally wrong for advocates of conservation medicine to become agents of human hegemony. Even though nature and human beings are inseparable ecologically, it does not follow that everything that benefits human beings economically and medically is also good for nature. Malaria causes much human suffering, but the draining of wetlands to prevent mosquito reproduction is not the ethical equivalent of a drug that suppresses the parasite in our bodies. The concern about such ethical issues is what distinguishes conservation medicine from environmental medicine.

Skeptics began to greet the birth of each new conservation-related discipline with suspicion. A new field meant an emerging crisis and that a new threat to nature had reached a threshold demanding an organized response. While the living world was shattering around them, scientists invented new defensive tactics that grew into new professions, new government bureaucracies, new journals, and new professional organizations.

In 1942, prophet Aldo Leopold wrote, "In 1909, when I first saw the West, there were grizzlies in every major mountain mass, but you could travel for months without meeting a conservation officer. Today there is some kind of conservation officer 'behind every bush.' Yet as wildlife bureaus grow, our most magnificent mammal retreats steadily toward the Canadian border."

A cautionary tale: When I was growing up in San Diego, my passion for nature was inspired by beautiful lepidoptera such as giant moths in the silkworm family, including the ceanothus silk moth and the polyphemus moth. I remember being enthralled by their unexpected nocturnal appearances at the porch light. Now, these great creatures have all but disappeared in much of the United States, depriving children of an opportunity to experience the mystery of contact with "the other." What happened?

A century ago, U.S. government scientists were looking for a biological way to control gypsy moths, a serious forest defoliator. They eventually discovered a fly (*Compsilura concinnata*) from Europe that parasitizes the caterpillars of many insects, including gypsy moths. Now we know that the severe declines in these glorious silk moths are likely to have been caused by the fly—a cure that turned out to be a scourge. It is an old story: good intentions, combined with ecological ignorance, are a recipe for iatrogenic disaster.

Conservation medicine holds the promise that by integrating our knowledge of pathogens and physiology with our knowledge of ecology, society can avoid disasters like *Compsilura* in the future and begin to evolve toward a more mature relationship with living nature.

Preface

Traditional approaches to the development of health strategies and environmental protection offer limited solutions to increasingly complex challenges. We are starting to see a connection between ecosystem health, human health, and animal health with the greater understanding of global climate change, and the wide-ranging influence of human effects on the planet. One out of five human health concerns has an environmental cause. With an ever-increasing human population putting pressure on the planet's resources, and increasing globalization of the world's economy, the ecology of the earth is experiencing dramatic changes, manifested by novel and potentially catastrophic health consequences.

This book provides a broad survey of the intersection of ecology and health sciences as it applies to achieving a more sustainable future for our species and others. It examines ecological health issues from various standpoints, including the emergence, reemergence, and resurgence of infectious diseases; the increasing biological effects of toxic chemicals and hazardous substances; and the health implications of ecological alterations, such as habitat fragmentation and degradation, loss of biodiversity and ecosystem services, and global climate and atmospheric changes. The thesis of this book is that health connects all species on the planet. Ecology teaches us about the interdependence of species, and health is a fundamental aspect of this tenet.

Our book is intended to be a primer for ecological and health practitioners and their students who seek to define and achieve ecological health as part of a more sustainable way for humans to live on a planet of finite resources. The definition of health is changing with time as our awareness of the multifaceted effects on health come to light. Environmental influences on health are more apparent. Conservation medicine has evolved to provide a framework for creating solutions to the various inextricably connected signs of environmental distress and dysfunction. It is the solution-oriented practice of achieving ecological health. While ecosystem health literature describes the problem, conservation medicine describes the potential solutions. In this volume, we discuss issues and problems that are relevant to policy makers and planners in the environmental and health fields. We hope this book will become the standard reference for human and veterinary medicine, public health, conservation biology, and ecology students interested the fields of conservation medicine and ecosystem health.

Conservation Medicine: Ecological Health in Practice is an outcome of an international conference held in the spring of 1999 at the White Oak Plantation and Conservation Center in Yulee, Florida. The conference brought together a broad spectrum of scientists and practitioners in health and environmental fields, including conservation biology, ecology, epidemiology, human health, public health, toxicology, and veterinary health. The meeting forged links between communities that focus on climate change, toxic and persistent pollutants, emerging infectious diseases, conservation biology, and ecosystem health. In addition, the conference launched a special Symposium on Conservation Medicine at the June 2000 annual meeting of the Society for Conservation Biology in Missoula, Montana, sponsored by the Consortium for Conservation Medicine.

The chapters in this book represent the first paints on the canvas of defining and refining the field of conservation medicine. The perspectives presented are an invitation to expand and enhance our ability to promote ecological health on a planetary scale. You are invited to take that next step in fine-tuning our work and building bridges to other necessary fields of allied human endeavor. In the end, we hope a masterpiece of collected human knowledge will provide our species with the solutions necessary to distance our planet from its present environmental predicament. We wish and hope for your success.

A.A.A.
G.M.T.

Acknowledgments

The Center for Conservation Medicine, an ecological health collaborative of Tufts University School of Veterinary Medicine, the Wildlife Trust, and the Harvard Medical School Center for Health and Global Environment, provided the institutional and intellectual milieu for discussions and research in the areas of ecological health and conservation medicine. In 2000, the Center for Conservation Medicine expanded into the Consortium for Conservation Medicine, with the explicit goal of increasing collaboration with other like-minded institutions—collaboration is essential in finding solutions to complex ecological health problems.

We are deeply grateful to the New York Community Trust for their continuing support and vision throughout the development of the field of conservation medicine. We are particularly indebted to NYCT for specific support toward the publication of this book.

This book is the result of a collaborative approach to science and conservation, supported by all the risk-taking foundations and funders who helped us along the way: V. Kann Rasmussen Foundation, National Fish and Wildlife Foundation, Oak Foundation, Geraldine R. Dodge Foundation, Howard Gilman Foundation and the White Oak Conservation Center, Jessie B. Cox Charitable Trust, Educational Foundation of America, Homeland Foundation, Rockefeller Financial Services, Morris Animal Foundation, Scholl Foundation, Acadia Foundation, Weeden Foundation, and the Fanwood Foundation. Research support for several of the authors included the Nathan Cummings Foundation, Richard Ivey Foundation, National Science Foundation, the National Institutes of Health, the National Institute for Environmental Health Sciences, the National Center for Biotechnology, the U.S. Environmental Protection Agency, Global Change Research Program, and the U.S. Military HIV Research Program. We most humbly thank them all.

The essential heavy-lifting role in enabling the editors and authors to reach completion was played by Cynthia Barakatt, program officer for the Center for Conservation Medicine.

The book is a product of many people, and our thanks go to Phil Kosch, Sheila Moffat, Shelley Rodman, Carolyn Corsiglia, Heidi Weiskel, Gretchen Kaufman, and George Saperstein from Tufts; Fred Koontz, Robyn Cashwell, Joanne Gullifer, and Marguerite Cunning from the Wildlife Trust; Rita Chan and Sheryl Barnes from the Center for Health and the Global Environment; and B. Zimmerman.

Contents

II. Monitoring Ecological Health

III: Ecological Health and Humans

Contributors

A. Alonso Aguirre, D.V.M., M.S., Ph.D.
Director for Conservation Medicine
Wildlife Trust
Palisades, New York
aguirre@wildlifetrust.org

Anthony Allchurch, B.V.Sc., M.R.C.V.S.
Jersey Zoo
Durrell Wildlife Conservation Trust
Jersey, Channel Islands, United Kindom
aallchurch@durrell.org

Christopher M. Anjema, B.S., B.E.D.,
    M.E.S., M.D.
Professor
Department of Ophthalmology
Ivey Institute of Ophthalmology
Faculty of Medicine and Dentistry
The University of Western Ontario
London, Ontario, Canada

John R. Bend, Ph.D.
Professor
Departments of Paediatrics, and Pharma-
    cology and Toxicology
Faculty of Medicine and Dentistry
The University of Western Ontario
London, Ontario, Canada
jack.bend@med.uwo.ca

Roberto Bertollini, Ph.D.
Director
European Centre for Environment
    and Health
World Health Organization, Rome
    Division
Rome, Italy
rbe@who.it

Roman Biek
Wildlife Biology Program
School of Forestry
University of Montana
Missoula, Montana
rbiek@selway.umt.edu

JoAnn M. Burkholder, Ph.D.
Department of Botany
North Carolina State University
Raleigh, North Carolina
JoAnn_Burkholder@ncsu.edu

Eric Chivian, M.D.
Director
Center for Health and the Global
    Environment
Harvard Medical School
Boston, Massachusetts
eric_chivian@hms.harvard.edu

Edward E. Clark Jr.
Executive Director
The Wildlife Center of Virginia
Waynesboro, Virginia
eclark@wildlifecenter.org

Christine M. Clarke, Ph.D.
Professor
Department of Pathology
University of Washington School
   of Medicine
Seattle, Washington

Theo Colborn, Ph.D.
World Wildlife Fund
Washington, D.C.
colborn@wwfus.org

Michael Cranfield, D.V.M.
Director
Mountain Gorilla Veterinary Project
c/o Baltimore Zoo
Druid Hill Park
Baltimore, Maryland
mrcranfi@bcpl.net

Andrew A. Cunningham, B.V.M.S., Ph.D.,
   M.R.C.V.S.
Head
Wildlife Epidemiology
Institute of Zoology
Zoological Society of London
London, United Kindom
Andrew.Cunningham@ioz.ac.uk

Peter Daszak, Ph.D.
Executive Director
Consortium for Conservation Medicine
c/o Wildlife Trust
Palisades, New York
daszak@aol.com

James G. Else, D.V.M., M.S.
Associate Professor/Director

Center for Conservation Medicine
Tufts University School of Veterinary
   Medicine
North Grafton, Massachusetts
jim.else@tufts.edu

Paul R. Epstein, M.D., M.P.H.
Associate Director
Center for Health and the Global
   Environment
Harvard Medical School
Boston, Massachussetts
paul_epstein@hms.harvard.edu

Lynne Gaffikin, Ph.D.
Department of International Health
Division of Health Systems
School of Hygiene and Public Health
Johns Hopkins University
Baltimore Maryland

Edward J. Gentz, M.S., D.V.M., Dipl.
   A.C.Z.M.
Director of Veterinary Services
The Wildlife Center of Virginia
Waynesboro, Virginia
NGentz@cabq.gov

Colin M. Gillin, M.S., D.V.M.
Assistant Research Professor
Center for Conservation Medicine
Tufts University School of Veterinary
   Medicine
Rocky Mountain Field Office
Wilson, Wyoming
GillinCM@aol.com

Thaddeus K. Graczyk, M.Sc., Ph.D.
Associate Research Professor
Departments of Molecular Microbiology
   and Immunology, and
   of Environmental Health Sciences
School of Hygiene and Public Health
Johns Hopkins University
Baltimore, Maryland
tgraczyk@jhsph.edu

Carol House, Ph.D.
Formerly: U.S. Department of Agriculture

Animal and Plant Health Inspection Services
National Veterinary Services Laboratories
Foreign Animal Disease Diagnostic Laboratory
Greenport, New York
Currently: 29745 Main Road
Cutchogue, New York
Jhousefish@aol.com

James A. House, D.V.M.
Formerly: U.S. Department of Agriculture
Animal and Plant Health Inspection Services
National Veterinary Services Laboratories
Foreign Animal Disease Diagnostic Laboratory
Greenport, New York
Currently: 29745 Main Road
Cutchogue, New York
Jhousefish@aol.com

John Howard, M.D., F.R.C.P.C.
Professor of Medicine and Paediatrics
Faculty of Medicine and Dentistry
The University of Western Ontario
London, Ontario, Canada

Kathleen E. Hunt, Ph.D.
Center for Conservation Biology
Department of Zoology
Seattle, Washington
huntk@u.washington.edu

David A. Jessup D.V.M., M.P.V.M., Dipl.
A.C.Z.M.
California Department of Fish and Game
Marine Wildlife Veterinary Care and Research Center
Santa Cruz, California
djessup@OSPR.DFG.CA.GOV

Douglas L. Jones, Ph.D., F.A.C.C.
Professor, Departments of Physiology and Medicine
Faculty of Medicine and Dentistry
The University of Western Ontario
London, Ontario, Canada

William B. Karesh, D.V.M.
Director
Field Veterinary Program
Wildlife Conservation Society
Bronx, New York
wkaresh@wcs.org

James J. Kay, Ph.D.
Department of Environmental Studies
University of Waterloo
Waterloo, Canada

Felicia Keesing, Ph.D.
Department of Biology
Bard College
Annandale, New York
keesing@bard.edu

Michael D. Kock, M.R.C.V.S., M.P.V.M.
19 Oak Street
P.O. Box 106
Greyton, South Africa
mdkock@kingsley.co.za

Robert Lannigan, M.D., F.R.C.P.C.
Professor of Microbiology and
Immunology
Faculty of Medicine and Dentistry
The University of Western Ontario
London, Ontario, Canada

Robert McMurtry, M.D., F.R.C.S.C.,
F.A.C.S.
G.D.W. Cameron Visiting Chair, Health
Canada (Ottawa, Ontario)
Professor of Surgery and Former Dean
Faculty of Medicine and Dentistry
The University of Western Ontario
London, Ontario, Canada

Gary K. Meffe, Ph.D.
Editor
Conservation Biology
Department of Wildlife Ecology and Conservation
University of Florida
Gainesville, Florida
meffe@gnv.ifas.ufl.edu

David H. Molyneux, M.A., Ph.D., D.Sc., F.I.Biol.
Professor of Tropical Health Sciences
School of Tropical Medicine
University of Liverpool
Liverpool, United Kingdom
fahy@liverpool.ac.uk

Gary R. Mullins, Ph.D.
Agricultural Economist
PAN Livestock Services
Reading, United Kingdom
mullins@info.bw

Linda Munson, D.V.M., Ph.D.
Department of Veterinary Pathology,
    Microbiology, and Immunology
School of Veterinary Medicine
University of California
Davis, California
lmunson@ucdavis.edu

Tamsyn P. Murray, M.Sc.
Health and Ecosystem Management
Centro Internacional de Agricultura
    Tropical
Centro Eco-Regional
Pucallpa, Peru
tamsynmurray@netscape.net
t.murray@cgiar.org

Todd M. O'Hara, D.V.M., Ph.D., Dipl.
    A.B.V.T.
Research Biologist
Department of Wildlife Management
North Slope Borough
Barrow, Alaska
tohara@co.north-slope.ak.us

Richard S. Ostfeld, Ph.D.
Scientist
Institute of Ecosystem Studies
Millbrook, New York
Rostfeld@ecostudies.org

Jonathan A. Patz, M.D., M.P.H.
Assistant Professor and Director
Program on Health Effects of Global En-
    vironmental Change

Department of Environmental Health Sci-
    ences
Johns Hopkins School of Public Health
Baltimore, Maryland
jpatz@jhsph.edu

Mary C. Pearl, Ph.D.
Executive Director
Wildlife Trust
Palisades, New York
pearl@wildlifetrust.org

Jeremy S. Perkins, Ph.D.
Department of Environmental Sciences
University of Botswana
Gaborone, Botswana

Mark A. Pokras, M.S., D.V.M.
Professor/Director
Wildlife Clinic
Tufts School of Veterinary Medicine
North Grafton, Massachussetts
Mark.Pokras@tufts.edu

Mary Poss, D.V.M., Ph.D.
Division of Biological Sciences
University of Montana
Missoula, Montana
mposs@selway.umt.edu

Trent Preszler, Ph.D.
Fogarty International Center
National Institutes of Health
Bethesda, Maryland

Ernesto F. Ráez-Luna
Institute for Resources and Environment
University of British Columbia
Vancouver, British Columbia, Canada
doft@hotmail.com

David J. Rapport, Ph.D., F.L.S.
Professor
Department of Rural Planning and
    Development
College Faculty of Environmental Design
    and Rural Development
University of Guelph
Guelph, Ontario, Canada
drapport@oac.uoguelph.ca

Jamie K. Reaser, Ph.D.
U.S. Department of State
OES/ETC, Room 4333
Washington, D.C. 20520
Sprgpeeper@aol.com

J. Michael Reed, Ph.D.
Department of Biology
Tufts University
Medford, Massachusetts
mreed@emerald.tufts.edu

Allen Rodrigo, Ph.D.
Computational and Evolutionary Biology
  Laboratory
School of Biological Sciences
University of Auckland
Auckland, New Zealand
a.rodrigo@auckland.ac.nz

Matthew Rooney
Colorado State University
Veterinary Teaching Hospital
Fort Collins, Colorado

Joshua P. Rosenthal, Ph.D.
Program Director for Biodiversity,
  Ecology, and Informatics
Fogarty International Center
National Institutes of Health
Bethesda, Maryland
joshua_rosenthal@nih.gov

Eric M. Schauber, Ph.D.
Scientist
Institute of Ecosystem Studies
Millbrook, New York
schaubere@ecostudies.org

Kenneth A. Schmidt, Ph.D.
Department of Biology
Williams College
Williamstown, Massachusetts
kenneth.a.schmidt@williams.edu

Jonathan Sleeman, Vet.M.B., M.R.C.V.S.
Director of Veterinary Services
Wildlife Center of Virginia
Waynesboro, Virginia
jsleeman@wildlifecenter.org

Colin L. Soskolne, Ph.D., F.A.C.E.
Professor of Epidemiology
Faculty of Medicine and Dentistry
Department of Public Health Sciences
University of Alberta
Edmonton, Alberta, Canada
colin.soskolne@ualberta.ca

Michael E. Soulé, Ph.D.
Science Director
The Wildlands Project
Hotchkiss, Colorado
soule@co.tds.net

Terry R. Spraker, D.V.M., Ph.D., Dipl.
  A.C.V.P.
Colorado State University
Veterinary Diagnostic Laboratory
Fort Collins, Colorado
tspraker@vth.colostate.edu

Sara Sullivan, A.B.
Center for Health and the Global Environ-
  ment
Harvard Medical School
Boston, Massachusetts

Gary M. Tabor, V.M.D., M.E.S.
Wilburforce Foundation
Yellowstone to Yukon Program Office
Bozeman, Montana
Gary@wilburforce.org

Sam R. Telford III, D.Sc.
Lecturer in Tropical Public Health
Harvard School of Public Health
Boston, Massachusetts
stelford@hsph.harvard.edu

David Waltner-Toews, Ph.D.
Department of Population Medicine
University of Guelph
Guelph, Ontario, Canada
dwaltner@uoguelph.ca

Samuel K. Wasser, Ph.D.
Endowed Chair in Conservation Biology
Center for Conservation Biology
Department of Zoology

Seattle, Washington
wassers@u.washington.edu

Mary E. Wilson, M.D.
Associate Professor of Medicine,
  Harvard Medical School
Associate Professor of Population and In-
  ternational Health,
Harvard School of Public Health

Chief of Infectious Diseases, Mount Au-
  burn Hospital
Cambridge, Massachusetts
mary_wilson@harvard.edu

Nathan D. Wolfe
Department of Environmental Health Sci-
  ences
Johns Hopkins School of Public Health
Baltimore, Maryland

Part I

Ecological Health
and Change

1

Introduction

James G. Else

Mark A. Pokras

A primary goal of conservation medicine is the pursuit of ecological health or, by extension, the health of ecosystems and their inhabitants. Ecological health is, and will remain, in a continuous state of flux. As our environment continues to change, so will the disease patterns and their effects on the health of human and animal populations.

These types of interactions by their very nature will remain unpredictable and poorly understood. When evaluating the health of an ecosystem, there is a need to tease out normal background noise to look for long-term trends and changes, rather than short-term occurrences that may or may not be significant. One needs to draw a composite picture from an array of perceived trends that as a whole may indicate the health of the environment and its occupants.

This book is a compilation of chapters defining the potential boundaries of the field of conservation medicine. These chapters describe the wide array of issues presently involving health and the environment. On one end of the spectrum is a degraded biosphere being consumed by insatiable human needs, and on the other, the health effects on human and other species as a result of this consumptive behavior. The decline in ecological health is happening on all scales from the global to the local. The chapters in this book present effects in terrestrial, marine, and freshwater environments in the developed and developing worlds. Attention is also given to changes in the practice of human, veterinary, and ecosystem health care. To understand the scope of conservation medicine, a framework for the exploration and practice of the field is presented.

In part I, Tabor (chap. 2) offers an in-depth discussion of the concept of conservation medicine with an invitation to the reader to join in the development of this practice and both refine its parameters and extend its scope to additional allied fields. The definition is developed in the context of related disciplines by Ostfeld et al. (chap. 3), who present conservation medicine as a "crisis discipline" as defined by Soulé. Epstein (chap. 4) examines the contributions of social parameters, ecological change, and climate trends to species dynamics in controlling disease and suggests guidelines for selecting biological indicators for improving disease surveillance and ecosystem monitoring. Daszak and Cunningham (chap. 5) and House and colleagues (chap.9) highlight the emerging and resurgent infectious diseases as one manifestation of ecological health. Colborn (chap. 6) presents two studies illustrating the damage done to both human and animal endocrine systems by chemical contamination: one, on ingestion of tainted fish, and the second, on exposure to a fungicide.

The monitoring of ecological health requires both interpretation and extrapolation, and the synthesis of scientific information so that it can be presented in a usable form to managers, policy makers, and the public. This synthesis is important because political decisions often have to be made before a sufficient scientific base has been established.

Part II focuses on the monitoring and assessment of these complex interactions, using a range of ecological, biomedical, and behavioral indicators. In order to monitor ecological health, and to interpret perceived trends, we need to have at our disposal a set of tools that span a wide range of ecologic, biologic, and demographic measures. Consequently, this section is not about monitoring techniques per se, but rather the incredibly wide spectrum of parameters that the conservation medicine practitioner should take into account in making a diagnosis. The challenge is to synthesize properly what we already know, while incorporating an ever-increasing range of monitoring indices and assessment techniques. It takes a well-coordinated and truly multidisciplinary team effort to accomplish these tasks—a recurring theme that cuts across each of the chapters in this section.

Since the mid twentieth century we have experienced unprecedented environmental change that has affected the health of both marine and terrestrial ecosystems. Until very recently the capacity of the ocean seemed limitless—but now we are beginning to see the effects of generations of dumping and extractions. Aguirre and colleagues (chap. 7) provide evidence of the extent to which we are undermining the health of the ocean, while Munson and Karesh (chap. 8) turn our attention to terrestrial ecosystems, stressing the importance of wildlife health and viable animal populations to a healthy ecosystem. House and colleagues (chap. 9) discuss the significance of the emergence and increasing incidence of a number of marine microbial diseases. Fundamental to this is the understanding of the ecology of infectious agents and toxic compounds. The authors discuss the value of long-term disease monitoring, particularly when faced with an epidemic or the need to make wildlife management or policy decisions.

Poss and colleagues (chap. 10) examine the role of viruses both in shaping host population structure and as a tool to monitor host population dynamics. The authors underscore the importance of accurate laboratory isolation and characterization of suspected disease agents when studying disease outbreaks and the emergence of new disease syndromes, and how genetic typing can help elucidate disease etiology. It is necessary, however, to distinguish between the presence of a disease agent and actual disease with these increasingly sensitive diagnostic techniques. Wasser and colleagues (chap. 11) review noninvasive monitoring procedures that can be used to track physiological parameters of selected wildlife species, and in the study of disease. Reed (chap. 12) touches upon the ways behavior can contribute to population persistence, risk, and recovery and can be used as a tool to evaluate ecosystem health. Recognizing the limitations of some of the popular bioindicators of ecosystem health, including cost and the need for long-term data sets, Reed points out how particular behaviors can be used as indicators of ecosystem health and how knowledge of behavior can contribute to the design of more effective monitoring techniques. It is clear that the authors in this section recognize the need to take a holistic approach to monitoring and assessment that transcends traditional scientific or medical disciplines. This is a basic tenet of conservation medicine and should lead to the identification of new patterns or connections that perhaps previously went unnoticed.

The impact of ecological change on human health is presented in part III. Each chapter acknowledges the common truth that the expansion of our human population and its ever more consumptive lifestyles are increasingly causing conflicts between the interests of humans, domestic animals, and wildlife. These include the crowding of more and more free-ranging species onto less and less habitat, declining quality of air and water, accelerated rates of climate change, increased spread of disease and disease vectors, introduction of nonnative species, and a host of other threats. Patz and Wolfe (chap. 13) describe the effects of some of the natural and human-induced changes in the environment on the incidence, proliferation, and emergence of disease. Chivian and Sullivan (chap. 14) present examples of medications found in nature that may have already disappeared, or will in the near future, as a result of species loss. The authors illustrate the use of endangered animals as research models and subjects as well as infectious diseases affected by disruption of ecosystem and biodiversity equilibria.

Molyneux (chap. 15) discusses the impact of deforestation and reforestation on vectors and infectious diseases in humans. Biodiversity and disease risk using Lyme disease as a case study is addressed (chap. 16) demonstrating that by fragmenting the eastern seaboard forests into fragments below a minimum critical size, we lose the predators that keep the primary Lyme disease carriers (white-footed mice and chipmunks) in check, thereby providing more carriers of the disease. Graczyk (chap. 17) describes the impact of zoonoses on public health, wildlife conservation, and agriculture. Nutrient loading and water contamination as a result of intensive agricultural practices have led to potentially harmful algal blooms (Burkeholder, chap. 18).

Earlier sections of this volume have focused on documenting how widespread and serious ecological health issues are becoming and on techniques used to monitor such environmental perturbations. Part IV focuses on integrative approaches that are aimed at effecting solutions. The various applications presented reflect the diversity of skills, knowledge and approaches that are required if people are to implement creative, workable solutions to the complex problems facing wildlife and natural ecosystems. Controversial wildlife disease issues in protected areas are discussed by Gillin et al. (chap. 19); whereas Kock et al. (chap. 20) overview the conservation issues in the rural livelihood of Botswanan communities. The role of zoos as conservation entities is also discussed (chap. 21). Cranfield et al. (chap. 22) address the human health impacts on the severely endangered mountain gorilla populations in the Impenetrable and Virunga forests. These endangered primates face an uncertain future due to their small population size, their limited habitat and their susceptibility to disease. A scenario that can be applied to many dwindling animal species. Murray and colleagues (chap. 23) take a human health perspective, recognizing that the determinants of health occur from the individual through the landscape level, overlaying ecological change onto human behavior patterns, resource utilization, and health. Telford (chap. 24) concludes this section with a detailed exposition on the causes of tick-borne diseases (especially Lyme disease), the process by which the deer tick became zoonotic, and the environmental changes contributing to the emergence of deer tick infestations and pathogens.

The key to changing the way we approach conservation may well lie in education, an important theme in part V, which addresses the future challenges for conservation medicine. Several authors point out the need for more and better training at every level, but they are not seeking more narrowly trained specialists or more facts. Rather, they seek to break down the traditional professional barriers, encourage better communication, and develop models for multidisciplinary team-oriented problem solving. Rosenthal and Prezler (chap. 25) emphasize the role that educational and research facilities can play in reaching the public. Rapport and colleagues (chap. 26) highlight the need for more holistic, cross-disciplinary training at every level. Of course, proponents of conservation medicine would advocate for such educational efforts to take place in many more ways, from elementary environmental education to postgraduate seminars. We must certainly be attentive to including both human and veterinary medical education in this effort . . . but we cannot stop there.

Realizing that no one discipline has a lock on the truth, we must learn to put our professional egos aside and encourage the creation of multidisciplinary working groups early in the life of a project. Biologists, medical scientists, imaging experts, and economists must all be able to share the table and respect the knowledge of their peers. Like the blind men examining the elephant, we can see the magnitude and shape of an issue only by sharing our observations. Another advantage of this approach is that it may allow us to become more proactive in developing conservation policies (chap. 28) rather than reactive as traditionally observed in most conservation efforts. We spend the vast majority of our time, effort, and resources attempting to repair or redress damage that is already been done. But how do we develop the foresight that will allow us to act in a preventive manner?

Reaser and colleagues (chap. 29) convincingly show that without a healthy environment and a curb on human population growth, social and economic pressures over clean water, clean air, natural resource utilization, economic development, and a host of other issues will become increasingly significant threats to global harmony and security. So the need to educate social scientists (lawyers, political scientists, economists, the military, ad infinitum) early in their training is critical. But as several authors so ably point out, this new educational model must do more than have biologists talking to biologists and economists talking to economists. We must build the programs in which students and faculty from different disciplines, speaking different professional dialects, interact on a regular basis and develop better methods of group problem solving.

First we must recognize the limits to our knowledge. In reading this book, one is struck by the problems presented in chapter after chapter. Whether the author is working in the laboratory or in the field, it is apparent that one consistent limiting factor is lack of knowledge. This ranges from the biomedical—lack of "normal" endocrine and serum chemistry values for most species—to the ecological—food habits or population dynamics unknown for a particular species—to such global issues as climate change. Again and again, the lack of definitive data is shown to impair modeling and hamstring decision making and policy initiatives. It is apparent that the gathering and dissemination of such information should be identified as a priority for research and monitoring efforts on all scales.

So we have concluded almost where we started. Conservation medicine is defined by action, by effective implementation. The efforts of this book are not so much to define a new discipline as to demonstrate new ways in which existing disciplines must tear down walls and work together in a new kind of team if we are to achieve the conservation goals we hold dear.

2

Defining Conservation Medicine

Gary M. Tabor

In 1827, Charles Darwin, founder of modern evolutionary biology, decided to leave medical school at the University of Edinburgh to pursue studies in religion and natural history at Cambridge. This was a period in medical training when natural history and zoology were considered integral disciplines of the medical arts. Although Darwin never completed his medical training, the combination of his exposure to multiple disciplines and his worldly experience aboard the *HMS Beagle* led to his penning *On the Origin of Species.* Perhaps the power and depth of his observations that led to his groundbreaking work were based in part on his diverse training and intellectual pursuits. Like Darwin, health practitioners in his era and before were routinely trained in both medicine and natural history; these were disciplines that were once closely aligned.

Modern-day environmental threats necessitate renewing the link between medicine and natural history—more specifically, ecology—once again. Drawing from the past as we go forward, here we present a primer on the new field of conservation medicine. In essence, conservation medicine is founded upon reconnected disciplines long separated by time and tradition. In our intellectual evolution of specialization, our society has lost a range of problem-solving skills based on interdisciplinary observation and problem definition. The connection of natural history and medicine is an element known to Darwin but foreign in today's thinking.

In the nineteenth century, much of the world Darwin explored was relatively pristine, and natural processes were still intact. Today, such a journey would reveal a very different reality. The continuum of current environmental problems due in large part to human effects is a theme repeated often in this book and elsewhere. The global loss of biodiversity; the destruction, degradation, and fragmentation of natural habitat; the increase of pervasive chemical pollutants in air, water, and soil; and the change in climate and depletion of stratospheric ozone are all working in concert to undermine the ecological processes of the planet. In a sense, natural selection is giving way to an insidious global-scale artificial selection—a human imprint on life, as Darwin perhaps could never have imagined.

The emerging field of conservation medicine is as much a historic concept as it is a new idea. The term "conservation medicine" itself was first introduced by Koch (1996) to describe the broad ecological context of health. By bringing the disciplines of health and ecology together, conservation medicine represents an attempt to examine the world in an inclusive way. Health effects ripple throughout the web of life. Health connects all species. The interaction of species is inextricably linked to the ecological processes that govern life.

Conservation biology and conservation medicine share the common aim of trying to achieve ecological health. Conservation medicine studies the multiple two-way interactions between pathogens and disease on the one hand, and between species and ecosystems on the other. It focuses on the study of the ecological context of health and the remediation of ecological health problems. In response to the growing health implications of environmental degradation, conservation medicine has emerged as a new interdisciplinary field to address the complex interrelationship between health and ecological concerns. As such, conservation medicine's purview includes examining the linkages among (a) changes in habitat structure and land use; (b) emergence and reemergence of infectious agents, parasites, and environmental contaminants; and (c) maintenance of biodiversity and ecosystem functions as they sustain the health of plant and animal communities including humans. For example, conservation medicine is concerned with the effects of disease on rare or endangered species and on the functioning of ecosystems. It is also concerned with the effects of changes in species diversity or rarity on disease maintenance and transmission (Tabor et al. 2001).

The dynamic balance that we term "health" is viewed on a series of widely varying spatial scales by many disciplines, including human and public health, epidemiology, veterinary medicine, toxicology, ecology, and conservation biology. Conservation medicine represents an approach that bridges these disciplines to examine health of individuals, of groups of individuals, and of the landscapes in which they live as an indivisible continuum (Meffe 1999; Pokras et al. 1999; Norris 2001; Tabor et al. 2001).

By reaching out to multiple disciplines, conservation medicine provides new skills, new tools, and new vision to the field of both conservation biology and medicine. This includes bringing biomedical research and diagnostic resources to address conservation problems, for example, developing new noninvasive health-monitoring techniques; training veterinarians, physicians, and conservation biologists in the promotion and practice of ecological health; and establishing interdisciplinary teams of health and ecological professionals to assess and redress ecological health problems.

Conceptually, conservation medicine is at the nexus of the fields of human health, animal health, and ecosystem health (figure 2.1). And yet it is more than interdisciplinary, because it represents an integration of knowledge among disci-

CONSERVATION MEDICINE

Ecosystem Health

Animal Health   **Human Health**

plines. The term "transdisciplinary" is being advanced in the ecosystem health arena to describe this intellectual trend (Rapport 1995). Conservation medicine requires professionals from diverse disciplines to work together in addressing the complex aspects of the intersection of health and the environment. In essence, conservation medicine is practiced through collaborations beyond the individual, the institutional, and the interdisciplinary. As Albert Einstein once said, "We can't solve problems by using the same kind of thinking we used when we created them." Equally so, we need to adapt our efforts in addressing complex ecological health problems to include novel methods of human understanding and intervention.

Many disciplines are recognizing ecological health problems from the unique lens of their own intellectual perspective. What has yet to happen is the sharing of perspectives in a broader, more holistic fashion. The International Society for Ecosystem Health has taken the lead in promoting a transdisciplinary approach. Conservation medicine in practice not only maintains a goal of achieving ecological health but also endeavors to address the ecological context of health through transdiciplinary means.

In comparison to human and veterinary medicine, conservation medicine is the examination of ecological health concerns beyond the species-specific approach. Clearly, in some respects, human health holds a greater priority than does the health of other species. To date, human medicine has been limited in looking at the health connections between the human species and the environment. Human health intervention has been focused on the downstream effects of environmental impacts (e.g., health consequences of landscape changes or pollution emissions) rather than encompassing a preventive approach in looking at upstream events (e.g., prevention of massive ecological degradation or pollution prevention).

Unlike physicians in Darwin's era, today's human doctors are narrowly trained and lack the comparative health perspective that is required of veterinary practitioners. That is not to say that veterinary medicine is without limitation in perspective as well, since most veterinarians focus on the health of just eight domestic species (i.e., the cow, horse, dog, cat, goat, sheep, pig, and chicken). As Dr. Eric Chivian, Executive Director of the Center for Health and the Global Environment at Harvard Medical School, remarked in his 1999 commencement speech at Tufts University School of Veterinary Medicine, "Physicians are really specialized 'medium-to-large-sized animal' veterinarians." Conservation medicine attempts to demonstrate that health connects all species when it is viewed from an ecological context. Human health is linked to the health of other species and vice versa. And the health of all beings is connected to the ecosystems (or environments) in which they live (Karr and Chu 1995).

Human effects on the health of the planet can be characterized within four areas of environmental concern (Colborn et al. 1993, 1996; Rees 1996; Wilson 1988; Woodwell and Houghton 1990; G. Woodwell, personal commun. 2000):

1. increasing biological impoverishment, which includes the loss of biodiversity and modification of ecological processes;
2. increasing global "toxification," which includes the spread of hazardous wastes and toxic substances and the impact of endocrine-disrupting hormones;
3. global climate change and ozone depletion; and
4. increasing human ecological footprint as a result of our exponentially growing population and use of resources (an ecological footprint is an area-based calculation used in environmental economics to determine human effects on natural resources).

The discrete and cumulative effects of these human-induced global changes not only have diminished the environmental capital of the planet but also have yielded an array of health concerns, including the increasing spread of infectious (e.g., malaria) and noninfectious (e.g., melanoma) diseases and the growing physiological effects on species reproductive health, developmental biology, and immune system response (e.g., endocrine disruptors) (Blaustein et al. 1994; Blaustein and Wake 1995; Colborn et al. 1996; Laurance et al. 1996; Epstein 1995; McMichael and Haines 1997; Patz et al. 1996).

The ecological effects of humans can ripple throughout ecological communities. The demise of one species, or the rise of one species at the expense of another, may establish a trophic cascade of ecological responses. When predator–prey or species competition relationships are disrupted, ecological effects may extend beyond the predator and prey or the competitors (Epstein et al. 1997; Pace et al. 1999; Ostfeld and Keesing 2000a). The spread of Lyme disease as a result of the changing ecology of white-tailed deer (*Odocoileus virginianus*) and white-footed mice (*Peromyscus leucopus*) in a landscape devoid of large predators and with diminished biodiversity is a good example of ecological magnification of disease (Ostfeld and Keesing 2000b).

Emerging and resurgent infectious diseases are one manifestation of diminishing ecological health. The common theme in conservation medicine, regardless of whether the focus is human, animal, or botanical, is the ecological backdrop to disease events. Deforestation, climate change, land-use and coastal zone management, and chemical pollution provide the medium for disease outbreaks. Another troubling aspect of infectious diseases is the growing recognition of cross-species disease transmission, such as the recent outbreaks of West Nile virus in animals and humans in the greater New York metropolitan area and now spreading outward in the eastern United States, reaching Florida. Two modalities of cross-species contact are on the rise. Physical means of contact through transportation, trade, and travel (e.g., airplane travel) are well documented (Wilson 1995). What is less apparent and perhaps more pervasive are the increasing direct interactions of humans and animals (Burkholder and Glasgow 1997). Habitat fragmentation, human settlement encroachment, ecotourism, and intensive agricultural systems are creating a new milieu for the cross-species spread of disease.

In essence, with increasing human numbers and the expanding human footprint on the planet, human and animal disease interactions become more frequent and more real. For example, mountain gorillas, an endangered primate species in Central Africa, are at risk for human diseases such as influenza, measles, and tuberculosis due to ecotourism pressures. Intensive agricultural practices resulting in increased nutrient loading and decreased water quality give rise to *Cryptosporidium* contaminations and potentially harmful algal blooms (Burkholder and Glasgow 1997).

In addition to concern about infectious diseases is the emerging focus on persistent toxic pollutants and endocrine-disrupting hormones. Not since Rachel Carson's *Silent Spring* have the public and policy makers been made so aware of the wide-ranging ramifications of a toxic chemical assault on nature by endocrine disruptors, chemical compounds that compromise endocrine, immune, reproductive, and nervous systems at lethal and sublethal levels. As the earth accumulates its own soup of synthetic chemicals, the effects of these substances on the web of life are still barely understood (Colborn et al. 1996).

Climate change is one of the most pressing ecological health concerns. By all scientific evidence to date, climate change is a reality. The health consequences are pervasive. The effects of climate change in promoting the spread of infectious disease from more tropical ranges to temperate areas is being observed in such diseases as malaria and dengue fever (Epstein 1995, 1997; Patz et al. 1997). The impact of climate change on all ecological processes is profound: increased precipitation in some regions and drought in others, increased erosion of the coastal zone with rising sea levels, and the inability of many species to adapt to the relatively rapid changes in climatic regimes, potentially resulting in mass extinctions (Epstein 1999; McMichael 1997; McMichael et al. 1999). Changes in atmospheric volatilization and deposition of pollutants are of equal concern (Blais et al. 1998; Schindler 1999).

Today, humanity as a whole uses over one third more resources and eco-services than nature can regenerate. Preliminary estimates show that the ecological footprint of today's consumption in food, forestry products, and fossil fuels alone might already exceed global carrying capacity by roughly 30%. About ¾ of the current consumption goes to the 1.1 billion people who live in affluence, while ¼ of the consumption remains for the other 4.6 billion people. By definition, each organism uses resources from the ecosystem to exist. This essential requirement is expressed as an area of the planet that annually supplies these requirements each year and defines this as the organism's ecological footprint. Ecological footprint refers what extent their consumption can be supported by their local ecological capacity (Wackernagel and Rees 1995).

In the face of these threats, conservation medicine can be viewed as a call to arms. There are no silver bullets in tackling global environmental problems. A multipronged strategy is required. At the heart of any solution-oriented agenda is the need for better problem definition. Policy scientists have long known that better solutions arise through better problem definition (Lasswell 1971; Clark 1992). By bringing disciplines together, conservation medicine can contribute to solving environmental problems by improving problem definition.

In improving problem definition, new tools for assessing and monitoring ecological health concerns are required. One possible approach is the development of some form of integrated ecological health assessment, which incorporates aspects of environmental indicator studies with specific biomedical diagnostic tools (Karr and Chu 1999; Epstein 1997). These include the development of noninvasive physiological and behavioral monitoring techniques, the adaptation of modern molecular biological and biomedical techniques, the design of population-level disease-monitoring strategies, the creation of ecosystem-based health and sentinel species surveillance approaches, and the adaptation of health-monitoring systems appropriate for situations in developing countries.

Beyond monitoring and assessment is action. This takes several forms. Improving medical and ecological education is a fundamental goal of conservation medicine. Looking back to Darwin, the natural sciences and the medical sciences have since parted ways, branching like the intellectual equivalent of an evolutionary tree. Rapport and colleagues (chap. 27) report their methods for reforming undergraduate human medical education at Western Ontario School of Medicine using the concepts of ecosystem health. The Center for Health and the Global Environment at Harvard Medical School has recently created a new curriculum in global climate change and health consequences. Veterinary medicine is undergoing similar changes, as evidenced by the creation of "Envirovet Summer Institute," a conservation medicine–oriented educational effort of the schools of veterinary medicine at the University of Illinois, University of California Davis, and Tufts University. Wildlife Trust has pioneered the transdisciplinary education of scientists in Brazil through international conservation medicine leadership courses. As a result, several veterinary colleges and biology schools have implemented conservation medicine in their curricula. In short, conservation medicine training has already begun.

There is a need for global public policy development that assists in promoting ecological health. The current view of the U.S. Department of State and other agencies focused on the long-term health and prosperity of nations is that environmental instability presents very real, very serious global security problems. So it would seem that the health and well-being of nations are also linked to ecological health. Sub-Saharan Africa is now faced with political and cultural destabilization as a result of HIV infections in 16–45-year-olds, which is decimating earning capacity in many families and regions. In New York City, West Nile encephalitis virus is presently assaulting residents in several ways: through the fear of possible infections, through the effects of massive pesticide spraying, and through the concerns that city hall and governmental health agencies are really ineffective in stopping the spread of the disease.

Perhaps the most controversial public policy issue relating to wildlife disease is brucellosis in bison (*Bison bison*) in the greater Yellowstone ecosystem. Every year, bison are killed by the Department of Livestock in the State of Montana as the animals migrate northward outside of Yellowstone National Park. The fear of bison transmitting brucellosis to domestic cattle in the face of contrary evidence yields questionable public policy. The issues relating to wildlife conservation, public policy, and disease management are complex.

As zoos and aquaria have become more involved in conservation and especially in advancing the field of veterinary contributions to endangered species recovery and wildlife reintroductions, conservation medicine can help to improve these efforts by providing an ecological framework. Clinical zoo medicine has an opportunity to reach beyond the confines of the captive environment to provide health assessment and maintenance in situ. For example, Jersey Zoo in the United Kingdom historically adopted a conservation approach early in its mission and continues to explore future opportunities for zoo involvement in conservation medicine. The Wildlife Conservation Society, under former Chief Veterinarian Emil Dolensek, initiated similar field-based conservation efforts in the 1980s. And today, the Wildlife Trust with the Center for Conservation Medicine at Tufts University School of Veterinary Medicine is providing training and ground support for conservation medicine in zoos throughout the tropics.

Finally, what is the relevance of Darwin's work and conservation medicine today? Charles Darwin helped to define the scientific underpinning of our modern understanding of speciation and biological diversity. Biodiversity conservation efforts to date have viewed health concepts in simplistic, market-oriented terms. Much has been stated about bioprospecting and the medicinal qualities of such rare species as the Pacific yew (*Taxus brevifolia*), source of the cancer-fighting drug taxol, or the rosy periwinkle (*Vinca rosea*) from Madagascar, source of the cancer-fighting drugs vincristine and vinblastine (Chivian 1997; Grifo et al. 1997). Proffering biodiversity as a medicinal commodity creates a market for the exploitation of a scarce resource. The health aspects of biodiversity are not solely found in the selling of its products and services. They lie also in the fact that all species are interconnected and interdependent—including their health.

When biodiversity is lost, species compositions are disturbed, and ecological processes are disrupted, serious health implications arise. As health and ecological professions understand the interconnections of species and the complexity of ecological health problems, conservation medicine provides a critical stepstone for the future.

References

Blais, J.M., D.W. Schindler, D.C.G. Muir, D.B. Donald, and B. Rosenberg. 1998. Accumulation of persistent organochlorine compounds in mountains of western Canada. Nature 395:585–588.

Blaustein, A.R., P.D. Hoffman, D.G. Hokit, and J.M. Kiesecker. 1994. UV repair and resistance to solar UV-B in amphibian eggs: a link to population declines. Proc Natl Acad Sci USA 91:1791–1795.

Blaustein, A.W., and D.B. Wake. 1995. The puzzle of declining amphibian populations. Sci Am 272:52–57.

Burkholder, J.M., and H.B. Glasgow Jr. 1997. *Pfiesteria piscicida* and other *Pfiesteria*-like dinoflagellates: behavior, impacts and environmental controls. Limnol Oceanogr 42:1052–1075

Chivian, E. 1997. Global environmental degradation and biodiversity loss: implications for human health. *In* Grifo, F., and J. Rosenthal, eds. Biodiversity and human health, pp. 7–38. Island Press, Washington, D.C.

Clark, T.W. 1992. Practicing natural resource management with a policy orientation. Environ Manag 16:423–433.

Colborn, T., R.S. vom Saal, and A.M. Soto. 1993. Developmental effects of endocrine-disrupting chemicals in wildlife and humans. Environ Health Perspect 101:378–384.

Colborn, T., D. Dumanoski, and J.P. Meyers. 1996. Our stolen future. Dutton, New York.

Epstein, P.R. 1995. Emerging diseases and ecosystem instabilities: new threats to public health. Am J Publ Health 85:168–172.

Epstein, P.R. 1997. Climate, ecology, and human health. Consequences 3:2–19.

Epstein, P.R. 1999. Climate and health. Science 285:347–348.

Epstein, P.R., A., Dobson, and J. Vandermeer. 1997. Biodiversity and emerging infectious diseases: integrating health and ecosystem monitoring. *In* Grifo, F., and J. Rosenthal, eds. Biodiversity and human health, pp. 60–86. Island Press, Washington, D.C.

Grifo, R., D. Newman, A.S. Fairfield, B. Bhattacharya, and J.T. Grupenhoff. 1997. The origins of prescription drugs. *In* Grifo, F., and J. Rosenthal, eds. Biodiversity and human health, pp. 131–163. Island Press, Washington, D.C.

Karr, J.R. and E.W. Chu. 1995. Ecological integrity: reclaiming lost connections. *In* Westra, L., and J. Lemons, eds. Perspectives on ecological integrity, pp. 34–48. Kluwer, Dordrecht.

Karr, J.R., and E.W. Chu. 1999. Restoring life in running waters: better biological monitoring. Island Press, Washington, D.C.

Koch, M. 1996. Wildlife, people, and development. Trop Anim Health Prod 28:68–80.

Lasswell, H.D. 1971. A preview of policy sciences. Elsevier, New York.

Laurance, W.F., K.R. McDonald, and R. Speare. 1996. Epidemic disease and the catastrophic decline of Australian rain forest frogs. Conserv Biol 10:406–413.

McMichael, A.J. 1997. Global environmental change and human health: impact assessment, population vulnerability, and research priorities. Ecosyst Health 3:200–210.

McMichael, A.J., and A. Haines. 1997. Global climate change: the potential effects on health. Br. Med. J 315:805–809.

McMichael, A.J., B. Bolin, R. Costanza, G.C. Daily, C. Folke, K. Lindahl-Kiessling, B. Lindgren, and E. Niklasson. 1999. Globalization and the sustainability of human health: an ecological perspective. Bioscience 49:205–210.

Meffe, G. 1999. Conservation medicine. Conserv Biol 13:953–954.

Norris, S. 2001. A new voice in conservation medicine. Bioscience 51:7–12.

Ostfeld, R.S. and F. Keesing. 2000a. Pulsed resources and community dynamics of consumers in terrestrial ecosystems. TREE 15:232–237.

Ostfeld, R.S. and F. Keesing. 2000b. Biodiversity and disease risk: the case of Lyme disease. Conserv Biol 14:722–728.

Pace, M.L., J.J. Cole, S.R. Carpenter and J.F. Kitchell. 1999. Trophic cascades revealed in diverse ecosystems. TREE 14:483–488.

Patz, J.A., P.R. Epstein, T.A. Burke, and J.M. Balbus. 1996. Global climate change and emerging infectious diseases. JAMA 275:217–223

Pokras, M., G.M. Tabor, M. Pearl, D. Sherman, and P. Epstein. 1999. Conservation medicine: an emerging field. *In* Raven, P., and T. Williams, eds. Nature and human society: the quest for a sustainable world, pp. 551–556. National Academy Press, Washington D.C.

Rapport, D.J. 1995. Ecosystem health: an emerging integrative science. *In* Rapport, D.J., C.L. Gaudet, and P. Calow, eds. Evaluating and monitoring the health of large scale ecosystems, pp. 5–34. Springer, Heidelberg.

Rees, W.E. 1996. Revisiting carrying capacity: area-based indicators of sustainability. Popul Environ 7:195–215.

Schindler, D.W. 1999. From acid rain to toxic snow. Ambio 28:350–355.

Tabor, G.M., R.S. Ostfeld, M. Poss, A.P. Dobson, and A.A. Aguirre. 2001. Conservation biology and the health sciences: defining the research priorities of conservation medicine. *In* Soulé, M.E., and G.H. Orians, eds. Research priorities in conservation biology, 2nd ed., pp. 155–173. Island Press, Washington, D.C.

Wackernagel, M. and W.E. Rees. 1995. Our ecological footprint; Reducing human impact on the earth. New Society Publishers, Philadelphia.

Wilson, E.O. 1988. The current state of biological diversity. *In* Wilson, E.O., ed. Biodiversity, pp. 3–19. National Academy Press, Washington, D.C.

Wilson, M.E. 1995. Travel and the emergence of infectious diseases. Emerg Infect Dis 1:39–46.

Woodwell, G.M., and R.A. Houghton. 1990. The experimental impoverishment of natural communities: effects of ionizing radiation on plant communities 1961–1976. *In* Woodwell, G.M., ed. The earth in transition: patterns and processes of biotic impoverishment, pp. 9–25. Cambridge University Press, Cambridge.

3

Conservation Medicine
The Birth of Another Crisis Discipline

Richard S. Ostfeld

Gary K. Meffe

Mary C. Pearl

The existence of an environmental crisis that includes widespread extinctions, biodiversity loss, habitat destruction, damaging pollution, and global climate change is not in doubt (Groombridge 1992; Meffe et al. 1997; Frumkin 2001). The causes of this crisis are complex and multifaceted, and the severity of its long-term consequences for the earth's biota, including human health, is difficult to predict. Scientists are increasingly recognizing that solutions to the environmental crisis will be as complex and interrelated as the factors that have led to the crisis. In 1986 one of the pioneers in the field, Michael Soulé, called conservation biology a "crisis discipline." By this he meant that action must be taken without complete knowledge, because waiting to collect full and complete data could mean inaction that would destroy the effort. Consequently, conservation biologists usually work under a high degree of uncertainty, as do ecologists in general. Conservation medicine (CM), like conservation biology before it, is also a "crisis discipline," developing in response to a web of problems scientists recognize as beyond the scope of a single health or wildlife management discipline, and requiring intervention with incomplete information (Pullin and Knight 2001).

The principal goals of CM are to develop a scientific understanding of the relationship between the environmental crisis and both human and nonhuman animal health, and to develop solutions to problems at the interface between environmental and health sciences. To accomplish these goals, it will be necessary to define this new discipline, to suggest ways in which CM is connected to other disciplines, to assess its importance, and to anticipate impediments to its strong and rapid development. The purpose of this chapter is to pursue these four actions, drawing upon the other chapters in this book.

3.1 Conservation Medicine in the Context of Related Disciplines

Conservation medicine is an emerging, multidisciplinary scientific field devoted to understanding the interactions among (1) human-induced and natural changes in climate, habitat structure, and land use; (2) pathogens, parasites, and pollutants; (3) biodiversity and health within animal communities; and (4) health of humans. Conservation medicine combines the pursuit of basic issues, such as how destruction and alteration of natural habitats influence community diversity and population size of wildlife, with the pursuit of practical issues, such as how to break the cycle of pathogen transmission between domestic stock and imperiled wildlife populations. Conservation medicine thus embraces participation by practitioners of ecology; organismal, cellular, and molecular biology; veterinary medicine; and human medicine, including epidemiology and public health. In addition, perspectives from the social and political sciences are fundamental in understanding the underlying causes of human-induced changes in climate, habitat, and land use. Although CM is a scientific discipline, it may provide the basis for political positions on the conservation and management of species or ecosystems, because CM may reveal utilitarian functions of preserving biodiversity.

Conservation medicine is unique in that it combines multiple linkages among our four goals. For example, the interaction between pathogens and (nonhuman) animal health is the realm of veterinary science, but when this interaction is influenced by habitat destruction or alteration, such study falls within the domain of CM. Similarly, the impact of human-induced climate change on species diversity within animal communities is the purview of conservation biology. But when the effects of climate change include changes in the distribution, abundance, or transmission of pathogens or pollutants (such as endocrine disruptors) within those animal communities, the focus falls within the domain of CM.

A key challenge in developing the discipline of CM is uniting, both in theory and in practice, several scientific disciplines that on first glance seem unrelated. Toward that end, topics covered in this book include veterinary care of endangered wildlife, the function of biodiversity and landscape change in zoonotic diseases, the effects of water pollution on food webs, medical practice that includes a global perspective on environmental causes of disease, and virus–host coevolution. How are these issues united? Each has both a component that is related to environmental conservation (e.g., toxics, habitat destruction, biodiversity loss, protected area management) and a component related to medicine or health (e.g., pathogens, wildlife or human disease of both infectious and noninfectious nature, disease vectors). The existence of many crucial issues at the interface between these two disciplines demands that such a discipline be developed.

How do we recognize or create research programs in CM? First a specific focus of research (or application) is identified that consists of either a medical/health issue or a conservation issue. Then, the link of this focus to the other component is developed. For example, a research target might be an imperiled wildlife population, and the link to medicine is a focus on pathogens or toxic substances that play a role in the population's conservation status. Another example is a human disease as a research target, and a link to conservation consists of the influence of biodiversity, habitat alteration, or climate change on the transmission, maintenance, or pathogenicity of the disease agent. Yet another example is a wildlife species as a target, with the link to conservation and health consisting of the impact of habitat or biodiversity loss on pathogens that in turn influence the wildlife species. These examples highlight the importance of developing specific theories and approaches that can facilitate an understanding of general principles underlying CM. Some of these theoretical issues are described at the end of this chapter. The field of CM, as with the related field of conservation biology, is broad and encompassing, and an amalgamation of many established fields. Our hope is that it will also develop emergent properties of its own.

3.2 Why Is Conservation Medicine Important?

Biomedical and conservation sciences have many traits in common, including a focus on entities that are failing to thrive but that we care about as a society or as individuals. In both biomedical and conservation sciences, it is evident that fixing the causes that underlie the failure to thrive can prevent irreparable loss or damage and can avoid the expenditure of enormous costs, both financial and ethical. Of course, understanding the causes of ill health in both humans and ecological systems presents a major challenge to scientists but is necessary in order to identify solutions.

Clearly, there are interactions between the causes of ill health of humans and nonhuman animals on the one hand, and the causes of ill health of the various ecosystems within which people and animals are embedded on the other. The causal arrows describing these interactions may point both ways. For example, diseases can erode biodiversity by decimating populations (an arrow from health to environment), and human-caused environmental change can erode both human and animal health (an arrow from environment to health). Understanding the nature and dynamics of those interactions demands rigorous scientific approaches to both the health/medicine issues and the environmental/conservation issues. The purpose of a new scientific agenda embodied in CM is to promote rigorous science and application at the interface between these two types of issues.

The status quo is a relatively high degree of specialization within both the medical and ecological arenas. Training and research programs in, for example, the epidemiology of infectious diseases, wildlife veterinary medicine, ecology, and conservation biology are robust and healthy. However, researchers with expertise in more than one of these disciplines remain relatively rare. Such a broad expertise will be necessary to tackle the fundamentally interdisciplinary nature of the problems. In a world where the global impact of human activity is apparent in the dynamics of every local ecosystem and habitat, we must recognize that health management to maintain ecosystem function is increasingly not an option but an imperative.

3.3 What Are the Impediments to the Strong Development of Conservation Medicine?

3.3.1 Technical Languages

As in any interdisciplinary field, understanding and accurately applying the technical terms from each of the specific disciplines can be problematic. The rich terminology of the biomedical and veterinary sciences poses particular difficulties for ecologists and conservation biologists. But the risk of misunderstanding or misapplication of terms exists for any of the subdisciplines of CM. For example, the term "ecosystem" has a fairly specific definition in the field of ecology as a level of ecological study that includes all the organisms in a particular area and the abiotic factors with which the organisms interact. Because this definition involves both a circumscribed area and the organisms and abiotic features that occur within that area, ecologists are often baffled when the term ecosystem is used without reference to boundaries or synonymously with biosphere. A solution to the terminology problem should include careful definition of terms for nonspecialists and avoidance of borrowing and redefining terms from other disciplines.

3.3.2 Ignoring Context

As described above, CM consists of both the ecological or conservation context of health and the health context of ecological or conservation issues. Specialists in ecology or conservation biology will often consider health issues as at least partially external to their main focus; likewise, for medical specialists, the ecological or conservation arena will constitute the external context of their main interests. For example, a health professional who is monitoring an epidemic or epizootic may be unaware of some ecological change that has facilitated rapid disease transmission. Or a wildlife biologist may be unaware of the role of some invisible pathogen in the decline of an animal population. The conservation context of health and the health context of conservation will become more accessible only when they are recognized and pursued directly.

3.3.3 Black-Boxing Other Disciplines

Perhaps more insidious than ignoring the context that is considered external to the main disciplinary focus is the use of shallow or naive perceptions of that context. Obviously, there are many cases in which it is desirable or necessary to draw metaphorical boundaries around the focus of scientific inquiry and to treat other scientific foci as externalities that can be simplified or "black-boxed." However, such an approach poses special dangers in an interdisciplinary arena. In essence, the decision to simplify externalities to the focal system must be made on the basis of prior knowledge that the interactions embedded within the black box are not relevant to the focal system. For example, an epidemiologist trying to predict how changes in global climate will influence the distribution of a mosquito-borne disease (such as malaria) may assume that the ecological interactions that determine the distribution of the mosquito can be simplified by describing a range of temperature and precipitation conditions conducive to vector persistence. This specialist might be tempted to black-box all biotic interactions affecting mosquito populations as external to a set of abiotic conditions, which can be described using remote sensing data and geographic information systems. This decision to consider biotic interactions as external or irrelevant should be made only on the basis of empirical support for such a conclusion. Similarly, a wildlife ecologist might expect that, because predators cause the obvious mortality in a population of imperiled animals, it is appropriate to assume that diseases are unimportant, or simply add stochastic "noise" to a demographic model featuring predation-caused mortality. If pathogens cause changes in susceptibility to predators, or influence growth and reproduction even subtlely, it may be inappropriate to ignore or simplify disease.

3.3.4 Not Embracing Individual- and Population-Based Perspectives in the Practice of Conservation Medicine

It is often difficult for practitioners of individual organism medicine (human and animal health professionals) to embrace a population-based understanding of disease in practice. Whereas public health specialists and population biologists consider disease in relation to populations, physicians and veterinarians are more likely to consider disease in relation to individuals. They are trained to combat pathogens and restore health. In contrast, population biologists are more likely to be observers rather than interventionists. This mismatch in scope and practice becomes problematic in building the applied field of CM. Today, health issues require that public health specialists and physicians develop for cooperative research and application across this divide as do wildlife population biologists and wildlife veterinarians.

3.3.5 Ignoring the Bias Toward Terrestrial Ecosystems

Conservation biology developed with a terrestrial, forest bias, and there are some long-neglected areas that need greater attention. These include freshwater and marine systems, many of which are experiencing serious and accelerating degradation. We should not let this bias enter the field of CM, and instead should ensure that the effects of degradation of aquatic, marine, and other areas on human and wildlife health are fully explored.

3.3.6 Dangers in the Use of Analogies

A key challenge to CM is understanding the desired states of individual health, population health, and ecosystem health, so that science can guide policy and management. A useful tool for promoting interdisciplinary understanding is the use of analogies that allow concepts and entities from one field to be described to another. But the use of analogies can undermine, rather than promote, interdisciplinary understanding. A key example of these dangers are the analogies that are sometimes drawn between the health of patients, which is the purview of the biomedical and veterinary sciences, and the health of ecosystems, which is the focus of some ecologists and conservation biologists. These analogies entail some pitfalls that can obfuscate key issues.

"Ecosystem health" is notoriously difficult to define. Even the health of individual humans or other organisms is a slippery concept. Health is a condition defined by the lack of disease, the ability to resist infectious and noninfectious insults, and the ability to quickly resolve illnesses once they occur. In defining ecosystem health, it may be convenient to draw on analogies from the health sciences. Indeed, ecosystems, and their components (e.g., populations of different species, biomass) and processes (e.g., capture and transfer of energy), can be characterized by measurable features such as stability (constancy through time in components and process rates), resistance (maintenance of current state when disturbed), and resilience (rate of return to a prior state after disturbance) that appear directly analogous to human or animal patients. Nevertheless, the focal entities of ecosystem health and medical health are vastly different, straining the use of analogies. The major difference is in the degree of integration of the component parts of organisms (organs, tissues, cells) versus ecosystems (populations of overwhelmingly large numbers of species and their abiotic environments). In the case of organisms, the component parts are highly integrated, such that localized or systemic disturbances caused by pathogens or toxic substances cause highly predictable outcomes for the organism. But in the case of ecosystems, the component parts may interact strongly or weakly with one another, in large or small networks, and may interact uniquely with their abiotic environments. Such variability in the strength and nature of interconnections compromises the ability of scientists to predict the consequences of environmental insults on ecosystems.

Just as humans and nonhuman animals have normal ranges of fluctuations in, for example, heart rate, brain activity, and blood chemistry, so too do ecological systems have normal ranges of fluctuations, although what is normal for an ecological system is vastly more difficult to define. For humans and most domesticated animals, norms are easily definable because of huge sample sizes of individuals under study, and enormous quantities of information on most or all subsystems. In contrast, for ecosystems, sample size may be as small as one (i.e., no replication at all), and often only a tiny fraction of subsystems are amenable to thorough study. Ecosystem health, therefore, should not be considered an entity that can be defined by analogy or described thoroughly and rigorously, but instead is suggested by a range of conditions that allow normal degrees of stability, resistance, and resilience to be maintained.

Similarly, the controls on flux rates often will be different for an ecosystem compared to an organism. The science of ecology has shifted in recent decades to incorporate a new paradigm, termed the "flux of nature," which has replaced key aspects of an older paradigm, termed the "balance of nature" (Pickett et al. 1992). The flux of nature paradigm emphasizes that ecological systems (1) are open to external forces; (2) often do not to seek or maintain stable-point equilibria; (3) undergo directional change (succession) that may be strongly influenced by stochastic forces; (4) are influenced by idiosyncracies of current community composition and recent history; (5) are subject to common, natural disturbances; and (6) have experienced important human influences typically for hundreds to thousands of years (Pickett and Ostfeld 1995). Humans and many other organisms, in contrast, tend to be closely internally regulated, maintain stable-point equilibria, and undergo directional change (e.g., aging) that is largely deterministic. Any focus on ecosystem health as a scientific, policy, or management goal must include a range of normal variability and must incorporate the tenets of the new paradigm in ecology (Pimm 1991; Wagner and Kay 1993; Pickett and Ostfeld 1995). Neither stasis nor vague notions of ecological "balance" should be equated with health. Monitoring of what we consider to be healthy ecological systems is crucial for determining the range of variation, resistance, and resilience, so that states of ill health can be detected. In this pursuit, analogies to human health must be employed carefully.

3.4 Conclusions

The practice of CM involves a focus on one or more of the following entities: humans; global climate; habitat destruction and alteration; biodiversity, including wildlife populations; domestic animals; and pathogens, parasites, and pollutants (figure 3.1). Furthermore, CM also focuses on explicit linkages between these entities, as represented by the arrows in figure 3.1. As a crisis discipline, the usefulness of CM ultimately will depend on its applicability to solving problems. The perspectives and scientific findings of CM provide input into (1) education, particularly curricula in veterinary and medical schools, and (2) policy and management of ecosystems, habitats, and imperiled species (figure 3.1). In terms of education, CM provides the foundation for broadening curricula to include the ecological context for diseases of humans and nonhuman animals. In terms of environmental policy and management, CM informs decisions by focusing on the interactions between biodiversity and health of humans and other animals.

Conservation medicine receives input from both the social sciences and bioinformatics (figure 3.1). Fields such as sociology, economics, and anthropology inform the practice of CM by revealing potential causes of human behavior relevant to human-induced environmental change. These disciplines also inform CM scientists about human motivations and pressures involving the use and preservation of natural resources (including wildlife), and about patterns of human travel that influence the dispersal of parasites and pathogens. Most important, the social sciences can assess whether the legislation and policies that follow from CM-based recommendations actually work. Top-down approaches to wildlife management are being replaced by adaptive management strategies that reflect uncertainty and complexity. These new management techniques can inform CM as well as conservation biology. Bioinformatics and the creation, management, and dissemination of databases relevant to wildlife, their habitats, and their diseases may be crucial to the practice of CM.

In pursuing the practice of CM, we recommend an integrated agenda that recognizes the recent birth and current early development of this field. To this

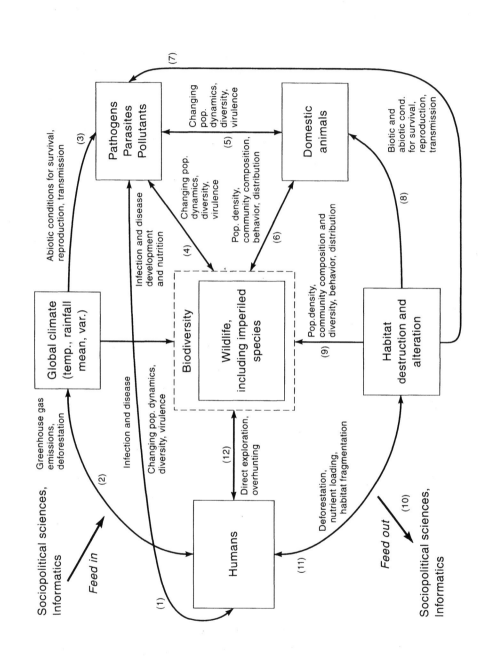

Sociopolitical sciences, Informatics

Feed in

Sociopolitical sciences, Informatics

Feed out

Global climate (temp., rainfall mean, var.)

Greenhouse gas emissions, deforestation

Pathogens Parasites Pollutants

Abiotic conditions for survival, reproduction, transmission
(3)

Infection and disease development and nutrition

Changing pop. dynamics, diversity, virulence
(4)

Changing pop. dynamics, diversity, virulence
(5)

Changing pop. dynamics, diversity, virulence

Domestic animals

Biotic and abiotic cond. for survival, reproduction, transmission
(8)

Pop. density, community composition, behavior, distribution
(6)

Biodiversity

Wildlife, including imperiled species

Pop. density, community composition and diversity, behavior, distribution
(9)

Habitat destruction and alteration

Infection and disease

Changing pop. dynamics, diversity, virulence
(2)

Direct exploration, overhunting
(12)

Humans

Deforestation, nutrient loading, habitat fragmentation
(11)

(1)

(7)

(10)

end, CM should be defined broadly and inclusively, perhaps to be narrowed and made more specific with time. We expect that exploration of a wide variety of habitats and approaches will best serve the creation of models for basic and applied research at the interface between human health and conservation issues. Some of the issues most in need of development, both in theory and practice, are biomagnification of pollutants in the environment; emerging diseases across species boundaries; building the means for measuring the relationship between biodiversity and ecosystem/human health; disease transmission in ecotones; the relationship between current trends in human transformation of the planet, such as habitat destruction and climate change, and emerging and growing health issues; and the identification of sentinel species for tracking various health issues in different types of environments. Environmental and health crises demand that the powerful and healthy disciplines of medical and veterinary science on the one hand, and ecology and conservation biology on the other, be united in a new cooperative scientific venture. This new venture will take place in a time of unprecedented environmental degradation and emerging disease, much of which is occurring in the tropics. This fact, combined with the unprecedented ease of global communication, affords us the moral imperative and the opportunity to develop CM with close international cooperation. By building the field in collaboration with practitioners and theoreticians around the world, we will have the opportunity to create a more effective and more broadly applicable conservation science.

←———————————————————————————————————————

Figure 3.1 Conceptual diagram of the relationships among entities that together comprise the new discipline of conservation medicine. Lines connect entities (within boxes) that directly interact, with arrows indicating the direction of interactions. Mechanisms associated with specific interaction arrows are described by adjacent text. Parenthetical numbers adjacent to arrows indicate the chapters from the current volume that map onto these particular interactions:
(1) Colburn, chapter 6; Burkholder, chapter 18; Ostfeld et al., chapter 16; Rosenthal and Preszler, chapter 25; Wilson, chapter 27; Reaser et al., chapter 29.
(2) Epstein, chapter 4; Patz and Wolfe, chapter 13; Chivian and Sullivan, chapter 14; Soskolne and Bertollini, chapter 28.
(3) Daszak and Cunningham, chapter 5; Patz and Wolfe, chapter 13; Chivian and Sullivan, chapter 14.
(4) Daszak and Cunningham, chapter 5; Aguirre et al., chapter 7; Munson and Karesh, chapter 8; House et al., chapter 9; Poss et al., chapter 10; Wasser et al., chapter 11; Ostfeld et al., chapter 16; Graczyk, chapter 17; Gillin et al., chapter 19; Telford, chapter 24.
(5) Graszyk, chapter 17; Allchurch, chapter 21.
(6) Daszak and Cunningham, chapter 5; Kock et al., chapter 20.
(7) Molyneux, chapter 15; Burkholder, chapter 18.
(8) Kock et al., chapter 20.
(9) Reed, chapter 12; Ostfeld et al., chapter 16; Telford, chapter 24.
(10) Rapport et al., chapter 26; Reaser et al., chapter 29.
(11) Epstein, chapter 4; Chivian and Sullivan, chapter 14; Burkholder, chapter 18.
(12) Allchurch, chapter 21; Chivian and Sullivan, chapter 14; Cranfield et al., chapter 22; Murray et al., chapter 23.

References

Frumkin, H. 2001. Beyond toxicity: human health and the natural environment. Am J Prev Med 20:234–240.

Groombridge, B., ed. 1992. Global biodiversity. Status of the earth's living resources. Chapman and Hall, London.

Meffe, G.K., C.R. Carroll, and contributors. 1997. Principles of conservation biology, 2nd ed. Sinauer, Sunderland, Mass.

Pickett, S.T.A., and R.S. Ostfeld. 1995. The shifting paradigm in ecology. *In* Knight, R.L., and S.F. Bates, eds. A new century for natural resources management, pp. 261–278. Island Press, Washington, D.C.

Pickett, S.T.A., V.T. Parker, and P. Fiedler. 1992. The new paradigm in ecology: implications for conservation biology above the species level. *In* Fiedler, P., and S. Jain, eds. Conservation biology: the theory and practice of nature conservation, preservation, and management, pp. 65–88. Chapman and Hall, New York.

Pimm, S.L. 1991. The balance of nature? University of Chicago Press, Chicago.

Pullin, A.S., and T.M. Knight. 2001. Effectiveness in conservation practice: pointers from medicine and public health. Conserv Biol 15:50–54.

Wagner, F.H., and C.E. Kay. 1993. "Natural" or "healthy" ecosystems: are U.S. national parks providing them? *In* McDonnell, M.J., and S.T.A. Pickett, eds. Humans as components of ecosystems, pp. 257–270. Springer, New York.

4

Biodiversity, Climate Change, and Emerging Infectious Diseases

Paul R. Epstein

There are multiple biological symptoms of global change, including amphibian declines on six continents, a decline in pollinators, the proliferation of harmful algal blooms along coastlines worldwide, and emerging infectious diseases across a wide taxonomic spectrum. There are also multiple social, ecological, and global factors underlying these symptoms (Levins et al. 1994; Morse 1996). There are multiple plausible mechanisms and pathways by which these social and environmental factors are altering biodiversity and influencing the emergence of infectious diseases. In general, declining biodiversity and the effects of climate change on habitat and biodiversity can decouple important biological control systems limiting emergence and spread of pests and pathogens (Walker 1994). The pathways and mechanisms include the following:

1. Monocultures and habitat simplification increase the potential for disease spread among agricultural crops and forest plants.
2. Habitat loss and human penetration into disrupted wilderness areas can bring humans in contact with previously "isolated" pathogens.
3. Declines in predators can release prey from natural biological controls, and small prey can become pests and carriers of pathogens.
4. Loss of competitors (which carry pathogens less efficiently than primary animal hosts) can remove these buffers against pathogen abundance and spread.
5. Dominance of generalists over specialists, especially among avian populations, may increase pathogen levels (in the more tolerant generalists) and increase their spread.

6. Conversely, diseases of wildlife (such as West Nile encephalitis virus) can themselves reduce animal biodiversity, further increasing vulnerability to emerging infectious diseases (Daszak et al. 2000).

The importance of diversity for control of pathogens has been highlighted by a large-scale agricultural experiment in China (Zhu et al. 2000; Wolfe 2000). This critically important study by researchers from the Yunnan Province demonstrated that planting two varieties of rice in the same field, one resistant to rice blast (a fungal disease), boosted productivity by 89% and reduced rice blast severity by 94%, obviating the use of fungicides. The implications of this study for the role of biodiversity in resisting disease in complex ecosystems deserve greater study. This chapter focuses on the effects of climate change on biodiversity and the implications of these ecological disruptions for one of nature's essential, though neglected, services: controlling the emergence, resurgence, and redistribution of infectious diseases. In this chapter are presented some examples of how declines in biodiversity, especially linked with a changing and unstable climate, may enhance susceptibility to disease emergence and spread. Special attention will be given to the hantavirus pulmonary syndrome and West Nile encephalitis virus, two diseases affecting the United States in the 1990s.

4.1 Background

Logging, land clearing for agriculture, unplanned suburban development, and use of wood for fuel are altering landscapes and community ecology of animal species. The increased intensity of extreme weather events associated with climate change (Easterling et al. 2000) can also affect habitat on short-and long-term time scales. Over the twentieth century droughts have become longer and heavy rain events (>2 inches/day) have become more frequent (Karl et al. 1995a,b). In Honduras, drought-sustained wildfires consumed 11,000 km^2 of forest during the summer preceding Hurricane Mitch (Epstein et al. 1999), widening the deforestation that magnified the flooding and devastation from the hurricane (IFRCSRCS 1999; Hellin et al. 1999). Climate models project an increase in the frequency and intensity of such extreme weather events (Easterling et al. 2000).

Climate change, with altered timing of seasons, can also affect coevolved synchronies among species, and the effects of sequences of extreme weather events associated with global warming can affect habitat and animal biodiversity. The Intergovernmental Panel on Climate Change (IPCC) Working Group II on Impacts (PCC 2001) documents over 80 studies of poleward or altitudinal shifts in plant and animal populations. Resulting alterations in predator/prey ratios, and in populations of competitors, may decouple natural biological controls over pests and pathogens.

Natural systems consist of sets of functional groups—predators, prey, competitors, nitrogen fixers, decomposers, recyclers—that may help regulate the populations of opportunistic organisms. The diversity of mechanisms and redundant (or "insurance") species (i.e., those ensuring that the function is not lost if a species is lost) can provide resilience and resistance in the face of perturbations and invasions, while the mosaics of habitat support that diversity, providing generalized defenses against the spread of opportunists. Against the background of habitat fragmentation and simplification, excessive use of toxins, and the loss of stratospheric ozone—components of global environmental change—climate change is rapidly becoming a prominent force in disrupting the relationships among species that help to prevent the proliferation of pests and pathogens. As the rate of climate change increases and the stresses increase under which sensitive biological systems are placed by long-term warming and greater weather instability, climate may play an expanding role in the emergence, resurgence, and redistribution of infectious disease in the coming decades (Epstein et al. 1997).

Synergies among climatic insults, such as extreme weather events, as well as changes in land use/land cover and sequential climatic extremes, can lead to explosions of nuisance organisms, as in the 1993 emergence of hantavirus pulmonary syndrome in the U.S. Southwest. Predator–prey relationships are key to ecosystem-level processes (Likens 1992), and losses of large predators may affect natural biological controls. Raptors, coyotes, and snakes, for example, help regulate populations of rodents—opportunistic species involved in the transmission of Lyme disease, ehrlichiosis, hantaviruses, arenaviruses, leptospirosis, and human plague. Freshwater fish, reptiles, birds, and bats help limit the abundance of mosquitoes that can carry malaria, yellow fever, dengue fever, and various forms of encephalitis. Finfish, shellfish, and marine mammals are involved in the dynamics of marine algal populations, and harmful algal blooms can be toxic or hypoxic/anoxic and can harbor bacteria (Harvell et al. 1999).

Population explosions of nuisance organisms and disease carriers may be viewed as signs of ecosystem disturbance, reduced resilience, and reduced resistance. Some species of rodents, insects, and algae represent key biological indicators, rapidly responding to environmental change—some human induced. The increasing rate of extinctions assumes additional significance in this respect: periods of mass extinctions—punctuations in evolutionary equilibrium—are followed by the emergence of new species. Will the current loss of biodiversity favor opportunistic species?

This analysis expands the classic disease epidemiological framework of host–agent–environment interactions to examine the compounding influences of social practices, ecological change, and climate trends on species dynamics in controlling disease, and suggests future directions for selecting biological indicators to improve disease surveillance and ecosystem monitoring.

4.2 Climate Change

The IPCC (2001) has concluded that the majority of climate change observed in the twentieth century can be attributed to human activites. For the past 420,000 years, as measured by the Vostok ice core in Antarctica, carbon dioxide (CO_2) concentration has remained between 180 and 280 parts per million (ppm) in the troposphere (Petit et al. 1999). Today, CO_2 is 366 ppm and the rate of change surpasses the rates observed in ice core records (IPCC 2001). Moreover, changes in multiple forcing factors (land use, stratospheric ozone, ice cover, and tropospheric greenhouse gas concentrations) have begun to alter the climate regime of the earth. These conclusions are based primarily on so-called "fingerprint" studies, where data match the projections of general circulation models driven with increased greenhouse gases. Three of the most prominent studies are

1. the warming pattern in the mid-troposphere in the southern hemisphere (Santer et al. 1996),
2. the disproportionate rise in nighttime and winter temperatures (Easterling et al. 1997), and
3. the statistical increase in extreme weather events occurring in many nations (Karl et al. 1995a,b).

More recently, detection of climate change has drawn upon long time series of multiple "paleothermometers" (Mann et al. 1998), which indicate that the twentieth century is the warmest in over 1,000 years. Data from recent years also demonstrate a significant increase in the rate at which warming is occurring. The rate of warming increased from about 1°C per 100 years until the 1980s to 3°C per 100 years during 1997 and 1998, the rate of change projected for the twenty-first century (Karl et al. 2000).

The Antarctic Peninsula has warmed some 2.5°C this century, and there is new declassified submarine data demonstrating significant shrinking of Arctic Sea ice (Krabill et al. 1999). Measuring 3.1 meters thick during the 1950s to 1970s, it thinned to 1.8 meters in the 1990s, a decrease of some 40% (Parkinson et al. 1999, Rothrock et al. 1999; Johannessen et al. 1999).

Continued reduction in polar ice will alter albedo (reflectivity), a change that can provide a positive feedback to the warming process. Permafrost melting in the tundra and boreal regions could release methane, an additional positive feedback. Deregulation of such positive feedbacks would increase the potential for abrupt changes in climate through alterations of ocean thermohaline circulation.

Overall, oceans are warming, ice is melting, atmospheric water vapor is rising, and changes in the hydrological cycle underlie the disproportionate warming and intensification of extremes (Karl et al. 1995a). A warmer atmosphere holds more moisture (6% more for every 1°C), and the disproportionate nature of the changes may be attributable in part to the intensified hydrological cycle (IPCC 1996). Increased evaporation (ocean 86%, land evapotranspiration 14%) produces more clouds and higher humidity, reducing daytime warming and retarding nighttime cooling (Karl et al. 1993). Meanwhile, warming oceans fuel storms and alter ocean and air currents (e.g., the jet stream), affecting weather patterns globally. Large systems may be shifting due to land mass warming. The Asian monsoons appear to be intensifying and losing their previous association with El Niño events, due to warming of the Eurasian landmass (Rajagopalan et al. 1999).

Easterling and colleagues (1997) report that since 1950, maximum temperatures have risen at a rate of 0.88°C per 100 years while minimum temperatures (TMINs) increased at a rate of 1.86°C per 100 years. In both hemispheres TMINs increased abruptly in the late 1970s. There are several implications of these changes for infectious diseases. In particular, the disproportionate rise in TMINs accompanying climate change favors insect overwintering and activity, and pathogen maturation rates.

At high elevations, the overall trends regarding summit glacial retreat, the upward displacement of plants, vector-borne diseases (VBDs) or their vectors, and upward shift of the freezing isotherm show remarkable internal consistency, and there is consistency between model projections and the ongoing changes (Epstein et al. 1998). Such emerging patterns consistent with model projections may be considered biological "fingerprints" of climate change and suggest that the climate regime has already affected the geographic distribution of biota on the earth.

High-resolution ice core records indicate that periods of increased variance from norms often precede phase-state changes or rapid climate change events, such as changes of average global temperatures of 2–3°C within just a few years. Since 1976, variability in the climate system has increased (Karl et al. 1995a.b; Trenberth 2000), and several small stepwise adjustments appear to have reset the climate system (CLIVAR 1992; Trenberth 2000). Pacific Ocean temperatures warmed significantly in 1976, warmed further in 1990, and cooled in 2000. Cold upwelling in the Pacific in 2000 could portend a multidecade correction that stores accumulating heat at intermediate ocean layers. Meanwhile, two decades of warming in the North Atlantic have melted North Pole and Greenland ice, and this anthropogenic influence may be contributing to the cold water (affecting windstorms) ranging from Labrador across to Europe (Hurrell et al. 2001).

4.3 Climate Change and Biological Responses

Several aspects of climate change are particularly important to the responses of biological systems. First, global warming is not uniform. Warming is occurring disproportionately at high latitudes, just above the earth's surface. Parts of Antarctica, for example, have already warmed almost 5°F this century. Since 1950 spring surfaces earlier in the northern hemisphere and fall appears later. While as yet unstudied, alterations the timing of seasons has the potential of leading to herbivore emergence out of sequence with insect predators.

Disproportionate warming during winter and nighttime (Easterling et al. 1997) plays particularly strong roles in altering biological processes. Elevated humidity (boosting the heat index) and lack of nighttime relief during heat waves both contribute to human mortality (T. Karl, NCDC, personal commun. 1998, in reference to 1995 Chicago heat wave mortalities). These conditions also affect arthropods. Warm winters facilitate overwintering, thus northern migration of ticks that carry Lyme disease (Tälleklint and Jaenson 1998; Lindgren et al. 2000), for example—the most important VBD in the United States. Agricultural zones are shifting northward—but not as swiftly as are key pests, pathogens, and weeds (Rosenzweig et al. 1998,) that, in today's climate, consume 52% of the growing and stored crops worldwide (Pimentel 1991, 1995).

An accelerated hydrological (water) cycle—ocean warming (Bindoff and Church 1992, Parilla et al. 1994; Levitus et al. 2000), sea ice melting, and rising atmospheric water vapor—is demanding significant adjustments from biological systems. Climate-related poleward shifts in marine fauna and flora movement were first documented in Monterey Bay in California (Barry et al. 1995), with further findings documented by the IPCC (2001).

4.4 Emerging Infectious Diseases

According to the World Health Organization, 30 new diseases have emerged since the mid 1970s. Included are HIV/AIDS, Ebola, Lyme disease, legionnaires' disease, toxic *Escherichia coli*, a new hantavirus, a new strain of cholera, and a rash of rapidly evolving antibiotic-resistant organisms (CDC 1993). In addition, there has been a resurgence and redistribution of old diseases on a global scale. Diseases such as malaria and dengue ("breakbone") fever carried by mosquitoes are among those undergoing resurgence and redistribution.

Throughout this century public health researchers have understood that climate circumscribes the distribution of VBDs, while weather affects the timing and intensity of outbreaks (Gill 1920,1921; Dobson and Carper 1993). Some arthropods are exquisitely sensitive to climate. Paleoclimatic data with beetle fossils, for example, demonstrate rapid distributional shifts with past climate changes (Elias 1994). Their distribution—using the mutual climatic range method to map fossilized species assemblages—is particularly sensitive to changes in TMINs. Dramatic shifts in beetle distributions accompanied the beginning and the end of an abrupt cool period called the Younger Dryas (a drop of 2°C within decades) near the end of the last glacial maximum.

From an historical perspective, changes in herbivore distributions may have been key to climate control during the carboniferous period (Retallack 1997): by developing and coevolving multiple means of defending against herbivory, woody terrestrial plants may have thrived, drawing down atmospheric carbon and helping to cool the biosphere (Retallack 1997).

Regarding human health, a growing number of investigators propose that VBDs (e.g., involving insects and snails as carriers) could shift their range in response to climate change (Leaf 1989; Shope 1991; Patz et al. 1996; McMichael et al. 1996; Haines et al. 2000). Models, incorporating vectorial capacity (temperature-dependent insect reproductive and biting rates, and microorganism reproductive rates), uniformly indicate the potential for spread of the geographic areas that could sustain VBD transmission to higher elevations and higher latitudes under global warming ($2 \times CO_2$) scenarios (Martens et al. 1997; Matsuoka and Kai 1994; Martin and Lefebvre 1995; Rogers and Randolph 2000). Statistical modeling by Rogers and Randolph (2000) projects less of a change than earlier dynamic models. But it is significant that all the models use average temperatures, rather than the changes in winter temperatures that are occurring even more rapidly in northern latitudes than in the tropics (e.g., TMINs are rising ~4°C/century in boreal areas, vs. ~2°C/century [1.86°C since 1970] overall; see IPCC 2001). Thus, the current models may be underestimating the potential biological responses.

With earlier arrival of summer and later arrival of fall, the transmission season of VBDs may also be extended. This is another parameter to be monitored. In terms of latitudinal shifts, studies in the United States indicate a potential for the northern movement of mosquito-borne encephalitides (e.g., western equine encephalitis and St. Louis encephalitis) within the continental United States and Canada (Reisen et al. 1993; Reeves et al. 1994). Simulations of the changes in malaria virulence due to global temperature increases during the past several decades show patterns strikingly similar to those found in the double-CO_2 simulations (Poveda et al. 2000).

Weeds, mice and rats, mosquitoes, and microorganisms are opportunistic species; they grow rapidly, are small in body size, produce huge broods, and have good dispersal mechanisms (thus high r, the intrinsic rate of increase). Small mammals, of course, consume insects; thus, the interplay and emergence of pests are complex. Extreme weather events (floods and droughts) can be conducive to insect explosions (Epstein 1999; Kovats et al. 1999) and rodents fleeing inundated burrows (Epstein 1997). Overall, opportunists and r-strategists are good colonizers and achieve dominance in disturbed environments (and in weakened hosts).

The transmission of eastern equine encephalitis involves the "generalist" species starlings (*Sturnus vulgaris*) and American robins (*Turdus migratorius*) in the bird–mosquito–human–equine life cycle (A. Spielman, personal commun. 1999). It is possible that hardy generalists might be better able to acquire, maintain, and circulate pathogens than those with isolated ranges and restricted diets. Generalist birds may prove more tolerant of—and therefore superior carriers of—viruses (e.g., West Nile virus), bacteria (e.g., *E. coli*), and protozoa (e.g., *Cryptosporidium*). This area requires further study, as habitat loss, simplification, and fragmentation have the potential to harm specialists with restricted ranges and diets.

4.5 Sequential Extremes and Surprises

Temporal sequences of extreme weather (e.g., droughts capped by intense rains and thunderstorms) can generate synergistic effects, helping to precipitate population explosions of pests (Epstein and Chikwenhere 1994). During hantavirus emergence in the U.S. Southwest, land-use changes and prolonged drought may have reduced rodent predators, while heavy 1993 rains (providing piñon nuts and grasshoppers) precipitated a tenfold increase in rodent populations, and a "new" disease (hantavirus pulmonary syndrome) was spread through rodent feces and urine (CDC 1993; Levins et al. 1993; Engelthaler et al. 1999).

A study of Canadian snowshoe hares (*Lepus americanus*) depicts such a synergy: reduced predation causes a twofold population increase; new food sources increase hare populations threefold. But both together convey a tenfold population explosion (Krebs et al. 1995; Martin and Turkington 1995; Stenseth 1995). Studying synergies among stresses—some anthropogenic, others natural—may prove important in understanding population explosions of opportunistic populations.

4.6 West Nile Encephalitis Virus in New York

In September 1999 at least 59 people fell ill and seven died from mosquito-borne encephalitis in the New York region. How West Nile encephalitis virus was introduced into the United States remains a mystery, but the conditions favoring diseases that cycle among birds, urban mosquitoes, and humans have been associated with warm winters followed by summer droughts (Monath and Tsai 1987). A disease with similar ecological dynamics—St. Louis encephalitis (SLE)—first occurred in the United States during the "dust bowl" in 1933, and of the 25 urban outbreaks since then, most have been associated with drought. For West Nile encephalitis virus, the outbreaks in Romania (1996), New York City (1999) and Israel (2000) were all related to drought. Climatic changes projected for U.S. regions include warmer winters and more hot, dry summers.

How drought and heat might affect the cycle of such diseases is a matter for conjecture. Mild winters and dry summers favor breeding of city-dwelling mosquitoes (*Culex pipiens*), while predators of mosquitoes (amphibians, lacewings, and ladybugs) can decline with drought. Birds may congregate around shrinking water sites, encouraging viral circulation among birds and mosquitoes, while heat amplifies viral maturation. In New York torrential rains in late August unleashed a new crop of mosquitoes, including *Aedes* spp., that possibly provided an additional "bridge" vector to humans (Epstein 2000). The same downpour on 26 August drove farm waste into an underground aquifer supplying the upstate Washington County fair. Over 1,000 became ill with toxic *E. coli*, many suffered from hemolytic uremic syndrome, and several children died.

West Nile encephalitis virus took an unusual toll on birds in New York. Either the West Nile encephalitis virus may have recently increased in virulence or the birds in the United States were immunologically naive. In either case, the unexpected outbreak of a mosquito-borne disease in New York City also serves as a reminder that pathogens anywhere on the globe—and the social and environmental conditions that contribute to those changes—can affect us all.

Emerging infectious diseases are not only affecting humans. Peter Daszak and colleagues (2000) present an excellent review of zoonoses and diseases affecting livestock and wildlife. Pests, diseases, and weeds of crops (and perennial plants), also responsive to ecological and climatic factors (Dahlstein and Garcia 1989; Sutherst 1990; Davis and Zabinski 1992); Rosenzweig and Hillel 1998; Coakley et al. 1999), appear to be spreading (Anderson and Morales 1995).

Marine-related diseases are also apparently increasing (Harvell et al. 1999), along with eutrophication, removal of wetland and watershed habitats that filter nutrients, and warming and extreme weather events that can flush organisms, nutrients, and toxins into coastal marine ecosystems.

Do current land-use practices, overuse of chemicals, and habitat fragmentation (Kruess and Tschamtke 1994) increase the chances for such "nasty synergies" (Peters and Lovejoy 1992)? A disturbance in one factor can be destabilizing; multiple perturbations can affect the resistance and the resilience of a system.

4.7 Conclusions

The resurgence of infectious diseases among humans, wildlife, livestock, crops, forests, and marine life in the final quarter of the twentieth century may be viewed as a primary symptom integrating the manifold aspects of global environmental and social change. Contemporaneous changes in greenhouse gas concentrations, ozone levels, the cryosphere, ocean temperatures, land use, and land cover threaten the stability of our epoch, the Holocene—a remarkable 10,000-year period that has followed the retreat of ice sheets from temperate zones. The effects of deforestation and climatic volatility are a particularly potent combination creating conditions conducive to disease emergence and spread. Given the rate of changes in local and global conditions, we may expect more synergies and new surprises.

The aggregate of air pollution from burning fossil fuels and felling forests provides a relentless destabilizing force on the earth's heat budget. Examining the full life cycle effects of fossil fuels also exposes layers of damages due to exploration, extraction, refining, transport, and combustion. For combustion, acid precipitation harms forests and aquatic organisms, while the direct human health effects of air pollution include pulmonary and cardiovascular disease. Returning CO_2 to the atmosphere through combustion reverses the very biological process by which plants initially drew down atmospheric carbon and generated oxygen and ozone, helping to cool and shield the planet sufficiently to support animal life.

Together, these chemical and physical changes—compounded by large-scale land-use and land-cover changes—have begun to affect biological systems. Just as we have underestimated the rate of climate change, we have apparently underestimated the sensitivity of biological systems to small changes in temperature and the instability in climate accompanying those changes. The public and policy makers must be increasingly concerned with the biological consequences and the societal costs associated with climate change.

References

Anderson, P.K., and F.J. Morales. 1993. The emergence of new plant diseases: the case of insect-transmitted plant viruses. *In* Wilson, M.E., R. Levins, and A. Spielman, eds. Disease in evolution: global changes and emergence of infectious diseases, pp. 181–194. New York Academy of Sciences, New York.

Barry, J.P., C.H. Baxter, R.D. Sagarin, and S.E. Gilman. 1995. Climate-related, long-term faunal changes in a California rocky intertidal community. Science 267:672–675.

Bindoff, N.L., and J.A. Church. 1992. Warming of the water column in the southwest Pacific. Nature 357:9–62.

Centers for Disease Control and Prevention (CDC). 1993. Hantavirus infection—southwestern United States: interim recommendations for risk reduction. Morbid Mortal Wkly Rep, 42:1–13.

Climate Variability and Predictability (CLIVAR). 1992. A study of climate variability and predictability. World Climate Research Program, World Health Organization, Geneva.

Coakley, S.M., H. Scherm, and S. Chakraborty. 1999. Climate change and plant disease management. Ann Rev Phytopathol 37:399–426.

Dahlstein, D.L., and R. Garcia, eds. 1989. Eradication of exotic pests: analysis with case histories. Yale University Press, New Haven, Conn.

Daszak P., A.A. Cunnningham, and A.D. Hyatt. 2000. Emerging infectious diseases of wildlife—threats to biodiversity and human health. Science 287:443–449.

Davis, M.B., and C. Zabinski. 1992. Changes in geographical range resulting from green-house warming: effects on biodiversity in forests. *In* Peters, R.L., and T.E. Lovejoy, eds. Global warming and biological diversity, pp. 297–308. Yale University Press, New Haven, Conn.

Dobson, A., and R. Carper. 1993. Climate change and human health: biodiversity. Lancet 342:1096–1031.

Easterling, D.R., B. Horton, P.D. Jones, T.C. Peterson, T.R. Karl, D.E. Parker, M.J. Salinger, V. Razuvayev, N. Plummer, P. Jamason, and C.K. Folland. 1997. Maximum and minimum temperature trends for the globe. Science 277:363–367.

Easterling, D.R., G.A. Meehl, C. Parmesan, S.A. Changnon, T.R. Karl, and L.O. Mearns. 2000. Climate extremes: observations, modeling, and impacts. Science 289:2068–2074.

Elias, S.A. 1994. Quaternary insects and their environments. Smithsonian Institution Press, Washington, D.C.

Engelthaler, D.M., D.G. Mosley, J.E. Cheek, C.E. Levy, K.K. Komatsu, P. Ettestad, T. Davis, D.T. Tanda, L. Miller, J.W. Frampton, R. Porter, and R.T. Bryan. 1999. Climatic and environmental patterns associated with hantavirus pulmonary syndrome, Four Corners region, United States. Emerg Infect Dis 5:87–94.

Epstein, P.R. 1995. Emerging diseases and ecosystem instabilities: new threats to public health. Am J Publ Health 85:168–172.

Epstein, P.R. 1999. Climate and health. Science 285:347–348.

Epstein, P.R. 2000. Is global warming harmful to health? Sci Am 283:50–57.

Epstein, P.R., and G. Chikwenhere. 1994. Biodiversity questions. Science 265:1510–1511.

Epstein, P.R., A. Dobson, and J. Vandermeer. 1997. Biodiversity and emerging infectious diseases: integrating health and ecosystem monitoring. *In* Grifo, F., and J. Rosenthal, eds. Biodiversity and human health, pp. 60–86. Island Press, Washington, D.C.

Epstein, P.R., H.F. Diaz, S. Elias, G. Grabherr, N.E. Graham, W.J.M. Martens, E. Mosley-Thompson, and J. Susskind. 1998. Biological and physical signs of climate change: focus on mosquito-borne disease. Bull Am Meteorol Soc 78:409–417.

Gill, C.A. 1920. The relationship between malaria and rainfall. Ind J Med Res 7:618–632.

Gill, C.A. 1921. The role of meteorology and malaria. Ind J Med Res, 8:633–693.

Haines, A., A.J. McMichael, and P.R. Epstein. 2000. Global climate change and health. Can Med Assoc J. 163:729–734.

Harvell, C.D., K. Kim, J.M. Burkholder, R.R. Colwell, P.R. Epstein, J. Grimes, E.E. Hofmann, E. Lipp, A.D.M.E. Osterhaus, R. Overstreet, J.W. Porter, G.W. Smith, and G. Vasta. 1999. Diseases in the ocean: emerging pathogens, climate links, and anthropogenic factors. Science 285:1505–1510.

Hellin, J., H. Haigh, and F. Marks. 1999. Rainfall characteristics of Hurricane Mitch. Nature 399:316.

Hurrell, J., Y. Kushnir, and M. Visbeck. 2001. The North Atlantic oscillation. Science 291:603–604.

Intergovernmental Panel on Climate Change (IPCC). 1996. Climate change '95: the science of climate change. *In* Houghton, J.T., L.G. Meiro Filho, B.A. Callander, N. Harris,

A. Kattenberg, and K. Maskell, eds. Contribution of Working Group I to the Second Assessment Report of the IPCC, chap. 3, p. 149, and chap. 7, pp. 370–374. Cambridge University Press, Cambridge.

Intergovernmental Panel on Climate Change (IPCC). 2001. IPCC Third Assessment Report: Climate Change 2001. Website: http//www.ipcc.ch/, Accessed, October 3, 2001.

International Federation of Red Cross and Red Crescent Societies (IFRCRCS). 1999. World disasters report 1999. Oxford University Press, New York.

Johannessen, O.M., E.V. Shalina, and M.W. Miles. 1999. Satellite evidence for an Arctic Sea ice cover in transformation. Science 286:1937–1939.

Karl, T.R., P.D. Jones, R.W. Knight, G. Kukla, N. Plummer, V. Razuvayev, K.P. Gallo, J. Lindsay, R.J. Charlson, and T.C. Peterson. 1993. A new perspective on recent global warming: asymmetric trends of daily maximum and minimum temperature. Bull Am Meteorol Soc 74:1007–1023.

Karl, T.R., R.W. Knight, D.R. Easterling, and R.G. Quayle. 1995a. Trends in U.S. climate during the twentieth century. Consequences 1:3–12.

Karl, T.R., R.W. Knight, and N. Plummer. 1995b. Trends in high-frequency climate variability in the twentieth century. Nature 377:217–220.

Karl, T.R., R.W. Knight, and B. Baker. 2000. The record breaking global temperatures of 1997 and 1998: evidence for an increase in the rate of global warming? J Geophys Res Lett 27:719–722.

Kovats, R.S., M. Bouma, and A. Haines. 1999. El Niño and health. World Health Organization, Geneva.

Krabill, W., E. Frederick, S. Manizade, C. Martin, J. Sonntag, R. Swift, R. Thomas, W. Wright, and J. Yungel. 1999. Rapid thinning of parts of the southern Greenland ice. Sheet Sci 283:1522–1524.

Krebs, C.J., S. Boutin, R. Boonstra, et al. 1995. Impact of food and predation on the snowshoe hare cycle. Science 269:1112–1115.

Kruess, A., and T. Tschamtke. 1994. Habitat fragmentation, species loss and biological control. Science 264:1581–1584.

Leaf, A. 1989. Potential health effects of global climate and environmental changes. N Engl J Med 321:1577–1583.

Levins, R., P.R. Epstein, M.E. Wilson et al., 1993. Hantavirus disease emerging. Lancet 342:1292.

Levins, R., T. Auerbach, U. Brinkmann, et al. 1994. The emergence of new diseases. Am Sci 82:52–60.

Levitus, S., J.I. Antonov, T.P. Boyer, and C. Stephens. 2000. Warming of the world oceans. Science 287:2225–2229.

Likens, G.E. 1992. The ecosystem approach: its use and abuse. Ecology Institute, Oldendorf Luhe, Germany.

Lindgren, E., L. Tälleklint, and T. Polfeld. 2000. Impact of climatic change on the northern latitude limit and population density of the disease-transmitting European tick *Ixodes ricans*. Environ Health Perspect 108:119–123.

Mann, M.B., R.S. Bradley, and M.K. Hughs. 1998. Global-scale temperature patterns and climate forcing over the past six centuries. Nature 392:779–787.

Martens, W.J.M., T.H. Jetten, and D. Focks. 1997. Sensitivity of malaria, schistosomiasis and dengue to global warming. Climatic Change 35:145–156.

Martin, K., and R. Turkington. 1995. Impact of food and predation on the snowshoe hare cycle. Science 269:1112–1115.

Martin, P.H., and M.G. Lefebvre. 1995. Malaria and climate: sensitivity of malaria potential transmission to climate. Ambio 24:200–209.

Matsuoka, Y., and K. Kai. 1994. An estimation of climatic change effects on malaria. J Global Environ Engin 1:1–15.

McMichael, A.J., A. Haines, R. Slooff, S. Kovats, et al., eds. 1996. Climate change and human health. World Health Organization, World Meteorological Organization, United Nations Environmental Program, Geneva, Switzerland.

Monath, T.P., and T.F. Tsai. 1987. St. Louis encephalitis: lessons from the last decade. Am J Trop Med Hygiene 37:405–595.

Morse, S.S. 1996. Factors in the emergence of infectious diseases. Emerg Infect Dis 1:7–15.

Parkinson, C.L., D.J. Cavalieri, P. Gloersen, H.J. Zwally, and J.C. Comiso. 1999. Spatial distribution of trends and seasonality in the hemispheric sea ice covers. J Geophys Res 104:20827–20835.

Parrilla, G., A. Lavin, H. Bryden, M. Garcia, and R. Millard, 1994. Rising temperatures in the sub-tropical North Atlantic ocean over the past 35 years. Nature 369:48–51.

Patz, J.A., P.R. Epstein, T.A. Burke, and J.M. Balbus. 1996. Global climate change and emerging infectious diseases. JAMA 275:217–223.

Peters, R.L., and T.E. Lovejoy (eds.), 1992. Global warming and biodiversity. Yale University Press, New Haven, Conn.

Petit, J.R., J. Jouze, D. Raynaud, N.I. Barkov, J.M. Barnola, I. Basile, M. Bender, J. Chapellaz, M. Davis, G. Delaygue, M. Delmotte, V.M. Kotlyakov, M. Legrand, V.Y. Lipenkov, C. Lorius, L. Peplin, C. Ritz, E. Saltzman, and M. Stievenard. 1999. Climate and atmospheric history of the past 420,000 years from the Vostok ice core, Antarctica. Nature 399:429–436.

Pimentel D. 1991. Handbook of pest management in agriculture, 2nd ed. U.S. Department of Agriculture, Washington, D.C.

Pimentel, D. 1995. Pest management, food security, and the environment: history and current status. Paper presented at the IFPRI workshop: Pest Management, Food Security, and the Environment: The Future to 2020, May 10–11, Washington, D.C.

Poveda, G., N.E. Graham, P.R. Epstein, W. Rojas, M.L. Quiñonez, I.D. Vélez, and W.J.M. Martens. 2000. Climate and ENSO variability associated with vector-borne diseases in Colombia. *In* Diaz, H.F., and V. Markgraf, eds., El Niño and the southern oscillation: multiscale variability and global and regional impacts, pp. 183–204. Cambridge University Press, Cambridge.

Rajagopalan, B., K.K. Kumar, and M.A. Cane. 1999. On the weakening relationship between the Indian monsoon and ENSO. Science 284:2156–2159.

Reeves, W.C., J.L. Hardy, W.K. Reisen, and M.M. Milby. 1994. Potential effect of global warming on mosquito-borne arboviruses. J Med Entomol 31:323–332.

Reisen, W.K., R.P. Meyer, S.B. Preser, and J.L. Hardy. 1993. Effect of temperature on the transmission of western equine encephalomyelitis and St. Louis encephalitis viruses by *Culex tarsalis* (Diptera: Culicadae). J Med Entomol 30:51–160.

Retallack, G.J. 1997. Early forest soils and their role in Devonian global change. Science 276:583–585.

Rogers, D.J., and S.E. Randolph. 2000. The global spread of malaria in a future, warmer world. Science 289:1763–1766.

Rosenzweig, C., and D. Hillel. 1998. Climate change and the global harvest. Oxford University Press, New York.

Rothrock, D.A., Y. Yu, and G.A. Marykut. 1999. Thinning of the Arctic Sea ice cover. J Geophys Res Lett 26:3469–3472.

Santer, B.D., K.E. Taylor, T.M.L. Wigley, T.C. Johns, P.D. Jones, D.J. Karoly, J.F.B. Mitchell, A.H. Oort, J.E. Penner, V. Ramaswamy, M.D. Schwarzkopf, R.J. Stouffer, and S. Tett. 1996. A search for human influences on the thermal structure of the atmosphere. Nature 382:39–46.

Shope, R. 1991. Global climate change and infectious disease. Environ Health Perspect 96:171–174.

Stenseth, N.H. 1995. Snowshoe hare populations squeezed from below and above. Science 269:1061–1062.

Sutherst, R.W. 1990. Impact of climate change on pests and diseases in Australasia. Search 21:230–232.

Tälleklint, L., and T.G.T. Jaenson. 1998. Increasing geographical distribution and density of *Ixodes ricinus* (Acari: Ixodidae) in central and northern Sweden. J Med Entomol 35:521–526.

Trenberth, K.E. 2000. The extreme weather events of 1997 and 1998. Consequences 5:3–15.

Walker, B.H. 1994. Landscape to regional-scale responses of terrestrial ecosystems to global change. Ambio 23:67–73.

Wolfe, M.S. 2000. Crop strength through diversity. Nature 406:681–682.

Zhu, Y., H. Chen, J. Fan, et al. 2000. Genetic diversity and disease control in rice. Nature 406:718–722.

5

Emerging Infectious Diseases
A Key Role for Conservation Medicine

Peter Daszak
Andrew A. Cunningham

5.1 Emerging Infectious Diseases

Emerging infectious diseases (EIDs) are diseases that have recently increased in incidence, have recently expanded in geographic or host range, are newly recognized, or are caused by newly evolved pathogens (Lederberg et al. 1992; Morse 1993; Daszak et al. 2000a). The term "emerging" was first used widely for human diseases in the 1970s (Sencer 1971; Moore and Robbs 1979), but achieved notoriety in the late 1980s following the discovery of a group of highly pathogenic diseases that included AIDS, hantavirus pulmonary syndrome, Legionnaire's disease, toxic-shock syndrome, multidrug-resistant bacterial infections, and Lyme disease (Krause 1992). The number of human EIDs continues to expand, with new pathogens discovered at an alarming rate and a number of older diseases resurging in incidence. These latter "reemerging diseases" include those caused by drug-resistant pathogens (e.g., tuberculosis in Europe and malaria throughout the tropics) and new threats from wildlife reservoirs (e.g., rabies in the eastern United States). Emerging infectious diseases have a high impact on human health, with some infections causing significant morbidity and mortality panglobally (e.g., drug-resistant malaria, AIDS) or on national or regional scales (e.g., dengue fever, Japanese encephalitis). Other EIDs cause more limited mortality but are a particular threat due to their extremely high case fatality rates and lack of available vaccines or therapy (e.g., Ebola and Marburg virus disease, Hendra virus disease, hantavirus pulmonary syndrome, Nipah virus disease).

Recent work has broadened the scope of EID research. By using the same criteria that define EIDs of humans, EIDs affecting domestic animals (Mahy and Brown 2000), crops (Rybicki and Pietersen 1999), wildlife (Daszak et al. 2000a), and wild plants (Anderson and May 1986; Real 1996) have been identified. These have a significant impact on the global economy by reducing agricultural production and cause biodiversity loss on national, regional, and global scales. Domestic animal diseases such as foot and mouth disease, Newcastle disease, and bluetongue, which are enzootic in many regions, can form a significant economic threat as EIDs to countries currently free of infection. The recent emergence of citrus canker, caused by the bacterium *Xylella citri*, in Florida has resulted in considerable economic and political fallout following the instigation of outbreak control measures. Globally, crop EIDs cause significant human mortality and economic loss (Real 1996; Rybicki and Pietersen 1999). In North America, pathogens are responsible for around 12% of the annual loss in crop production, a figure likely to be much higher in developing countries (Rosensweig et al. 2000). Forestry pathogens such as *Phytophthora cinamomi* have had a substantial impact on tropical forest conservation in Australia (Wills 1992). In the United States, chestnut blight caused by the fungal pathogen *Cryphonectria/Endothia parasitica* (Castello et al. 1995) and dogwood anthracnose (Carr and Banas 2000) have completely changed the structure of forests in the southeastern United States and along the length of the Appalachian Mountains. The recent discovery of a phytophthoran disease of oaks in California (Anonymous 2000) demonstrates the continual threat disease emergence poses to wild plant biodiversity.

Marked recent increases in the incidence and impact of wildlife EIDs have received particular attention (Daszak et al. 2000a). These include outbreaks caused by canine distemper virus and newly discovered morbilliviruses in marine mammals (Harvell et al. 1999); an epizootic of blindness in kangaroos caused by a novel orbivirus (Hooper et al. 1999); canine distemper virus disease in black-footed ferrets and African lions and with rabies in African wild dogs (Thorne and Williams 1988; Roelke-Parker et al. 1996; Cleaveland and Dye 1995; Alexander et al. 1996); mycoplasmal conjunctivitis and bacterial infections of passerine birds (Kirkwood 1998; Fischer et al. 1997); a newly discovered panzootic mycosis and outbreaks of ranavirus disease in amphibians (Cunningham et al. 1996; Berger et al. 1998; Daszak et al. 1999); disease outbreaks in marine fish, such as pilchard herpesvirus disease (Whittington et al. 1997), and in freshwater fish, such as whirling disease of salmonids (Hoffman 1970), infectious salmon anemia (Murray et al. 2002), and epizootic hematapoietic necrosis virus (EHNV) of rainbow trout and red-fin perch in southern Australia (Whittington et al. 1996); and a range of diseases in marine invertebrates (Harvell et al. 1999), among many others. Wildlife EIDs can have a particularly strong effect on small, fragmented populations and are responsible for a series of important extinction events on local and global scales (McCallum and Dobson 1995; Daszak and Cunningham 1999; Woodroffe 1999).

5.2 Environmental Change, Evolution, and the Emergence of Disease

Efforts to understand the emerging disease phenomenon have focused on identifying common factors that drive emergence. For human EIDs these are almost entirely changes in human population demography and behavior (Lederberg et al., 1992), with other large-scale anthropogenic environmental changes (e.g., climate change) also playing a role (Epstein 1999). These essentially ecological changes (Schrag and Wiener 1995) alter the dynamic equilibrium of the host–parasite relationship, increasing disease transmission rates within the human population or between zoonotic reservoir hosts and humans. They act within a background of microbial diversity, selecting for the dominance of pathogen strains adapted to the new environmental conditions. This pattern of emergence is exemplified by drug-resistant microbial strains, which have increased prevalence significantly in the last few decades (Cohen 1992). These strains are present at low prevalence in populations never previously exposed to antibiotic use (Gilliver et al. 1999). Emergence reflects a within-population change in strain prevalence, driven by changes in antibiotic use and hospital treatment regimes, rather than very recent evolution of novel strains. It follows that the most common source of emerging diseases is not newly evolved pathogens but rather long-present, often previously undetected pathogens driven to emerge by anthropogenic environmental change. The agents of Lyme disease, Ebola virus disease, Nipah virus disease, and hantavirus pulmonary syndrome all emerged in this way from reservoir hosts with which they have probably coevolved over many millennia (Telford et al. 1997; Monroe et al. 1999; Klenk 1999; Chua et al. 2000). Evolution plays a more important role in the emergence of pathogens that are thought to have recently evolved from animal microbes, following transmission to humans. However, changes in human ecology still underpin emergence consequent to these crossover events. The viruses causing AIDS (HIV-1 and HIV-2) are thought to have crossed over at least seven times from their chimpanzee and sooty mangabey ancestors, respectively, but only emerged as pandemics following increased volume of trade and transport of fresh-killed bushmeat and increasing human population size and mobility (Gao et al. 1992, 1999; Hahn et al. 2000).

An Institute of Medicine committee, convened in 1991 to review the human EID problem, identified six major factors leading to EID emergence: (1) recent increases in international travel and commerce, (2) changes in human demographics and behavior, (3) advances in technology and industry (particularly food and livestock industries and health care), (4) increased economic development and land use, (5) the breakdown of public health measures, and (6) microbial adaptation and change (Lederberg et al. 1992). These factors reflect increasing globalization of trade and human population movement during the latter half of the twentieth century (McMichael et al. 1999). Intensive livestock husbandry and centralized food processing have led to a range of EIDs of domestic animals, some with zoonotic potential (Lederberg et al. 1992; Mahy and Brown 2000). For example, globalization of the processed meat trade and the recent centralization of slaughter houses in the United Kingdom is thought to have led to the epizootic of foot and mouth disease in 2001. Technological advances in antibiotic chemotherapy to counter these threats have themselves driven disease emergence, for example, the rise of antibiotic-resitant strains (Cohen 1992). It has also been proposed that development of sulfonamide antibiotics and the virtual eradication of *Salmonella pullorum* in domestic poultry stocks revealed a vacant microbial niche and led to the emergence of *S. typhimurium* and *S. enteritidis* (Bäumler et al. 2000). Changes in livestock feed production and centralization of food processing and distribution have led to the emergence of pathogenic bacterial and viral infections in humans (Beuchat and Ryu 1997; Tauxe 1997), and novel transmissible spongiform encephalopathies (bovine spongiform encephalitis [BSE] or "mad cow disease" in cattle and variant Creutzfeld-Jakob disease [vCJD] in humans) (TSO 2000). Globalization of trade in nonanimal products can also foster disease emergence, for example, via the unintentional international transport of mosquito vectors (Lederberg et al. 1992).

Human population migrations associated with urbanization, war, poverty, famine, or globalization of the workforce allow the movement of infected individuals into previously unexposed populations. Such demographic changes are thought to have driven the emergence of dengue hemorrhagic fever in rapidly developing tropical cities (Lederberg et al. 1992), the early spread of AIDS in central Africa (Hahn et al. 2000), and the recent spread of multidrug-resistant tuberculosis (Berkelman et al. 1994) and may be implicated in the recent emergence of West Nile virus disease in the United States (Lanciotti et al. 1999). One of the results of exponential human population increase is a gradual encroachment into wildlife habitat (usually associated with deforestation) and increased contact with wildlife reservoirs of known or unknown zoonotic pathogens (the "zoonotic pool" of Morse 1993). Encroachment into wildlife habitat may have driven the emergence of Lyme disease, hantavirus pulmonary syndrome, and Nipah virus disease, within a background of increased deer populations, climatic factors, and intensified pig production, respectively (Barbour and Fish 1993; Mills et al. 1999; Chua et al. 2000). The emergence of these diseases demonstrates complex interactions between environment, biodiversity, and anthropogenic change in altering host–parasite transmission. In developed countries, the term "urban sprawl" has replaced "encroachment." This accelerated human spread into previously agricultural or wildlife habitat brings humans into contact with a range of wildlife reservoirs of zoonotic disease, including tick-borne encephalitides and plague.

Striking similarities exist between the factors that drive disease emergence in human, domestic animal, wildlife, and plant populations. Disease emergence across these groups is essentially driven by different presentations of the same anthropogenic forms of environmental change (Daszak et al. 2000a, 2001). Each of the causal categories listed by the Institute of Medicine (Lederberg et al. 1992) for human EIDs has a direct parallel that drives emergence in these other populations. For example, international travel by humans is paralleled by unprecedented international transport in agricultural products, live animals, plants, topsoil, ballast water, and other materials. Agricultural changes have caused disease emergence across the board: the feeding of cattle with mammal-derived protein led to the emergence of BSE and vCJD in Europe (TSO 2000), the increasing dependence on monoculture in arable farming leads to heightened risk of disease emergence in crop plants (Zhu et al. 2000), and the expansion of open range cattle and sheep ranching in the Rocky Mountains of the United States has led to the emergence of pasteurellosis in Rocky Mountain bighorn sheep (Jessup et al. 1991). Human population expansion into wildlife habitat drives zoonotic disease emergence, but may also drive wildlife EIDs by fragmenting populations, altering species complements, reducing biodiversity, and increasing exposure to harmful chemical substances (Daszak et al. 2001; Ross et al. 2000). The complex nature of disease emergence may result in a double effect on biodiversity conservation. For example, anthropogenic habitat fragmentation causes direct biodiversity loss and may also lead to wildlife EID emergence, which itself can cause biodiversity loss. There is increasing evidence that biodiversity loss may result in disease emergence: Ostfeld and Keesing (2000) hypothesize that regional biodiversity differences may explain variations in Lyme disease incidence in the United States. They cite a "dilution effect" whereby increased biodiversity of less competent vectors lessens the risk of infection to final hosts (Schmidt and Ostfeld 2001). This may act as a model for understanding potential mechanisms by which biodiversity loss may lead to disease emergence.

5.3 Pathogen Pollution

The most commonly cited cause of disease emergence in wild animals and plants is the anthropogenic introduction of pathogens (with or without their host) to new geographic regions, a process that has been termed "pathogen pollution" (Daszak et al. 2000a). Indeed, recent quantitative analysis suggests pathogen pollution may account for 60% of the emerging diseases of wildlife (Dobson and Foufopoulos 2001). Pathogen pollution is a previously underestimated form of environmental change that is largely a product of the globalized traffic in domestic animals, wildlife, plants, their products, and other materials. The origins of pathogen pollution probably lie early in human history: anthropogenic introduction of disease has been implicated in a series of Pleistocene megafaunal extinctions that followed human migration into North America (MacPhee and Marx 1997). This spate of rapid population declines may represent the earliest anthropogenic virgin-soil disease outbreaks in immunologically naive populations ("hyperdiseases"; MacPhee and Marx 1997). Similar events have been a by-product of European colonization of the planet throughout the last 900 years, with devastating consequences for native American peoples, who were immunologically naive to smallpox, measles, and other pathogens, and to European colonists with no prior exposure to yellow fever and malaria (Crosby 1986; Dobson and Carper 1996). The recent emergence of West Nile encephalitis virus in the United States and foot and mouth disease in the United Kingdom demonstrates that, despite heightened surveillance, diseases continue to move across international boundaries and into new geographic areas.

The introduction of human, domestic animal, and crop diseases into areas where they have been historically absent or recently eradicated threatens global health and national economies. In this chapter, we concentrate on the effect of pathogen pollution on wild animals and plants, due to its effect on biodiversity conservation. Pathogen pollution parallels chemical pollution in its global occurrence, rapid dissemination, often hidden nature, and high impact on wildlife and plant communities, even in pristine locales. For example, *Salmonella* of potentially human origin and antibodies to a poultry pathogen, infectious bursal disease virus, have been reported in Antarctic penguins (Gardner et al. 1997; Olsen et al. 1996). Outbreaks of the domestic dog disease canine distemper have been reported in Siberian and crabeater seals (Mamaev et al. 1995; Bengston et al. 1991). Introduced diseases are capable of moving rapidly through susceptible host communities, via host–host contact, aerosol transmission, movement of vectors, or the transport of fomites. For example, rinderpest, a directly transmitted morbilliviral disease of hoofstock introduced into the Horn of Africa in 1889, had reached the Cape by 1897 (Plowright 1982). Whirling disease of salmonid fish, caused by a myxozoan parasite, spread almost globally during the twentieth century via the anthropogenic movement of infected fish (Hoffman 1970). In cointroduction events (where host and pathogen are introduced), disease emergence may occur in endemic populations remote from the point of introduction and prior to direct contact with the invading host. For example, during European colonization of Australia, aboriginal people making contact with Europeans for the first time occasionally showed evidence of smallpox, transmitted via other aboriginal people, moving ahead of the European invasion (Crosby 1986). Because stochastic processes are involved in disease transmission, emergence may also lag significantly behind the introduction event, making prediction difficult. For example, an epizootic of mycoplasmal conjunctivitis first emerged in the eastern U.S. house finch population in 1994, despite this host population having been introduced into the region over 50 years earlier (Dhondt et al. 1998). It remains unknown whether this pathogen was cointroduced with the host.

Pathogen pollution is essentially the disease equivalent of biological introductions and invasions, which are themselves increasingly recognized threats to biodiversity. Indeed, pathogen pollution often acts hand-in-hand with biological invasions or introductions (e.g., whirling disease introduction into the United States). In these cointroduction cases, pathogen pollution may have a heightened ecological impact by conferring competitive advantage to the invading host—a phenomenon known as "apparent competition" (Hudson and Greenman 1998). The introduction of North American gray squirrels into the United Kingdom has been followed by a gradual decline in the U.K. red squirrel population. It appears that a parapox virus infection, to which gray squirrels are relatively resistant, may have enabled them to outcompete the more susceptible endemic species (Sainsbury et al. 2000).

Pathogen pollution can have a particularly strong effect where domestic animals are the introduced host species. Here, introduced pathogens may continually invade endemic wildlife populations by a process known as "disease spillover." The introduced host acts as a reservoir species, artificially raising overall host population size or density and allowing pathogen transmission even when the endemic host population has been reduced below the threshold density (the host density below which a pathogen is normally unable to maintain transmission). In this way, spillover of rabies and canine distemper from the large domestic dog population in the Serengeti enabled local extinction of the fragmented populations of African wild dogs in this region (Cleaveland and Dye, 1995). In some cases, the introduction of uninfected hosts may also drive disease emergence. For example, brushtail possums introduced into New Zealand have become new reservoirs for bovine tuberculosis, which has significantly increased in incidence since their introduction (Coleman 1988). Pathogen pollution has similar profound long-term effects on ecosystems as biological introductions. Introduced diseases may significantly alter community structure. For example, outbreaks of anthrax in impala in Tanzania led to establishment of even-aged acacia stands, mirroring earlier events caused by the rinderpest panzootic that have altered the landscape for a century (Dobson and Crawley 1994). As introduced diseases become enzootic, ripple effects may linger throughout ecosystems. Plague, caused by the bacterium *Yersinia pestis*, introduced into the United States during the last great pandemic in the 1860s, persists in prairie dog hosts in the western United States, where it causes periodic epizootics with mortality rates often 99% or higher (Cully et al. 1997). One of these catastrophic die-offs (in conjunction with distemper) precipitated the near-extinction of the black-footed ferret (Thorne and Williams 1988) following the loss of its staple (prairie dog) prey.

The mechanics of pathogen pollution are largely those of globalization of trade, agriculture, hunting, and recreational use of wildlife habitat. International transport of live cattle, for example, led to the emergence of brucellosis in the United States and rinderpest in Africa. The soil-borne plant pathogen *Phytophthora cinamomi* has been widely disseminated in western Australia along forestry management tracks, presumably transported by forestry vehicles (Wills 1992). The importation of frozen mixed by-catch to feed farmed tuna has been implicated in the emergence of pilchard herpesvirus disease in Australia. Translocation of rabies-infected raccoons to bolster hunting stocks in Middle Atlantic United States in the 1980s (Rupprecht et al. 1995) led to a new rabies threat to humans in that country. Deliberate introduction of pathogens also occurs in biocontrol programs, such as the introduction of myxomatosis and rabbit hemorrhagic disease to reduce rabbit populations in Europe and Australia. Fish stocking and transport of fomites for recreational fishing are thought to have led to the emergence of EHNV in Australia (Whittington et al. 1996). Similarly, movement of fish between marine farms led to the recent emergence of infectious salmon anemia in the United Kingdom (Murray et al. 2002). The scale of trade and transport in produce is often alarming, even in relatively pristine, protected areas. A year-long survey of products imported into Puerto Ayora, Galapagos Islands (Whelan 1995), revealed numerous potential vectors for biological and disease introduction, including large consignments of unknown goods marked "various," 65,000 kg of fruit and vegetables, and 79 sacks of soil!

5.4 The Crossroads of Conservation and Public Health

Humans share geography, habitat, and pathogens with a range of domestic and wild animals. Anthropogenic environmental changes act across this host–parasite ecological continuum to drive disease emergence (Daszak et al. 2000a). It is not surprising, therefore, that the majority of human EIDs are caused by zoonotic pathogens. Taylor and Woodhouse (2000) showed that 73% of human EIDs are caused by zoonotic pathogens, a figure that may be higher if we include those pathogens that have recently jumped from animal reservoirs. Similarly, most of the EIDs listed by the Institute of Medicine (Lederberg et al. 1992) are zoonotic. Zoonotic emergence from domestic animals has its origins in early human history (Dobson and Carper 1996). Measles virus, a recently evolved paramyxovirus, is closely related to rinderpest virus, a cattle pathogen, and most likely evolved from the latter as a close human–cattle relationship developed during the formation of early civilizations (Griffin and Bellini 1996). Pandemic influenza usually occurs following recombination of bird and porcine virus strains in pigs (Alexander and Brown 2000). Thus, the domestication of pigs and poultry was probably a key event in the emergence of this still-significant zoonotic threat to human health. Recent emerging zoonoses have generally been caused by pathogens that rarely cause disease in their wildlife reservoir hosts (e.g., Lyme disease) and are often rarely transmitted to humans, requiring domestic animal amplifier hosts (e.g., Hendra virus and Nipah virus diseases); others result from rare encounters with reservoir hosts (e.g., hypothesized Ebola virus reservoirs). More rarely, zoonotic EIDs affect both humans and wildlife (e.g., West Nile encephalitis).

The existence of multiple hosts and diverse patterns of host–parasite life cycle allows great complexity in the interaction of disease and environment. For Lyme disease, a tick-borne disease of humans that infects white-footed mice and deer, cycles in the population density of the introduced gypsy moth drive cycles in oak mast production, rodent populations, tick density, and therefore disease risk (Jones et al. 1998). Lyme disease emergence was probably originally driven by refor-estation (and subsequent deer population increase) following movement of arable farming away from the northeastern United States. Conversely, deforestation and subsequent loss of rodent biodiversity and increase in reservoir species density has driven the emergence of Venezuelan hemorrhagic fever in South America (Doyle et al. 1998).

As humans encroach into more remote habitats, new zoonoses and reservoir hosts have emerged. Human monkeypox first emerged during the 1980s from chimpanzees in Zaire (Mutombo et al. 1983). The reservoir hosts of the Ebola and Marburg viruses remain unknown, and although outbreaks in human beings are rare and sporadic, infection with these viruses often results in exceptionally high mortality rates (Klenk 1999). Fruit bat reservoirs have been implicated by experimental infection studies and by the presence of bat roosts at sites of out-breaks in geographically disparate regions (Monath 1999). Fruit bats are increas-ingly important as reservoirs for zoonotic pathogens in Southeast Asia and Aus-tralia. Three viruses pathogenic to humans have recently been described from these regions, and their presence in fruit bats has been strongly implicated by serology or confirmed by virus isolation: Menangle virus, a rubulavirus that causes a mild, febrile illness in humans and abortion in pigs (MacKenzie 1999), and Hendra and Nipah viruses (paramyxoviruses closely related to each other), which cause human mortality and emerged via horse and pig amplifier hosts, respectively (Murray et al. 1995; Chua et al. 1999). Isolation of Nipah virus from fruit bats has heightened concern that these wildlife hosts may act as reservoirs for many more potentially zoonotic infections than are currently recognized: a relatively small survey of frugivorous bats in one Malaysian locality revealed Nipah virus and another new rubulavirus (Tioman virus) of unknown zoonotic potential (Enserink 2000). Considering the diversity of fruit bats throughout the Indian and Pacific oceans and their high endemicity, further work is likely to reveal other potential human pathogens. It will also reveal increasing complexity within emergence, due to the complex interplay of different driving causes. For example, K.B. Chua and colleagues (2001, pers.comm.) suggested that climate change, via an extreme El Niño event and anthropogenic forest burning, may have altered fruit bat migration and blocked fruit production, causing Nipah virus to migrate to the piggery at which the index case was identified.

The emergence of zoonoses has a significant economic and conservation effects. The Nipah virus outbreak in Malaysia has resulted in the slaughter of over a million pigs and losses of around 36,000 jobs and US$120 million in exports. Treatment of zoonoses is costly, and where infection leads to more chronic problems, the economic burden may be far greater (e.g., Lyme disease, which is thought to cost as much as $500 million per year in the United States). Threats to conservation involve both disease-mediated mortality of reservoir and sympatric species (e.g., avian mortality due to West Nile virus) and mortality via human-mediated control measures (e.g., eradication of prairie dog reservoirs of plague). Zoonotic threats may be particularly important in conservation programs that involve direct human contact with wildlife (Hill et al. 1998), and the instigation of protective measures can substantially increase the cost of these programs.

5.5 Emerging Diseases Threaten the Conservation of Biodiversity

Wildlife disease emergence results in biodiversity loss due to increased mortality and decreased fecundity (including a number of local extinction events and global [species] extinctions) (Daszak et al. 2000a). Just as pandemic AIDS and reemerging malaria have caused significant human mortality, outbreaks of wildlife EIDs, such as amphibian chytridiomycosis, pilchard herpesvirus disease, and marine mammal morbillivirus infections, have severely depleted host populations. Localized outbreaks of human EIDs causing limited mortality (e.g., due to Ebola virus) are paralleled by a plethora of small-scale outbreaks of wildlife and plant EIDs. However, for wildlife EIDs, mortality is usually significantly underestimated due to rapid scavenging or decay of carcasses and lack of surveillance. In some cases, disease emergence has resulted in local (population) extinction—the complete removal of discrete populations, often over large areas. Amphibian chytridiomycosis, for example, has caused a number of multiple-species local extinctions of montane rainforest amphibians in Central America and Australia (Daszak et al. 1999). Some amphibian populations have failed to recover, and no evidence of successful recolonization has been reported, despite repeated surveys (Lips 1999). Similarly, a number of plant EIDs have caused the effective removal of species from an ecosystem. Throughout the Appalachians of the eastern United States, forest community structure has been radically changed by the introduction of the chestnut blight pathogen (*Cryphonectria/Endothia parasitica*) at the beginning of the twentieth century. Although chestnuts still persist in some areas, significant growth of this once codominant overstorey species is prevented by the continued presence of the pathogen (Castello et al. 1995).

Emerging diseases have also been implicated in a growing number of global (species) extinctions, such as the disappearance of certain Hawaiian bird species following introduction of avian malaria, avian pox, and mosquito vectors into Hawaii, and the extinction of the thylacine in Tasmania following a proposed epizootic of distemperlike illness possibly cointroduced with hunting dogs (Daszak and Cunningham 1999; Woodroffe 1999). Pathogen pollution appears to play an important role in this phenomenon, particularly where reservoir hosts are present or where endemic populations are immunologically naive to the introduced pathogen. Recently, the first definitively proven case of species extinction due to infectious disease has been documented. Cunningham and Daszak (1998) described disseminated microsporidiosis caused by a *Steinhausia* sp. parasite as the cause of death of the last remaining group of the Polynesian tree snail, *Partula turgida*, in an ex situ captive breeding program. Extinction by infection is not readily predicted by host–parasite ecological theory (Anderson and May 1986). Diseases that cause high mortality rapidly reduce populations to a level at which transmission cannot occur, while the increasing proportion of immune individuals reduces the proportion of susceptibles. This is the threshold density effect, which a parasite must circumvent if it is to cause extinction. Unusual ecological circumstances, such as the presence of a reservoir host or a pathogen that can survive outside the host in the environment, are therefore required for extinction by infection to occur. Following Woodroffe (1999), a number of potential mechanisms of extinction by infection can be envisaged (see also McCallum and Dobson 1995; Daszak et al. 2001): (1) Deterministic extinction in which a disease kills host individuals more rapidly than the population can expand. This is an important risk for small, fragmented populations or low-fecundity species. (2) Extinction by "hyperdisease" is an extreme form of deterministic extinction wherein an extremely virulent pathogen (one that rapidly causes death and results in a very high case fatality rate) is introduced into an immunologically naive population. After rapid depopulation, the relict breeding group is either too small or too dispersed to sustain recovery. MacPhee and Marx (1997) hypothesized that Pleistocene megafaunal extinctions occurred following a hyperdisease panzootic from which low-fecundity species (i.e., those with large body size) were unable to recover. (3) Extinction from stochastic factors is possible after a virulent epizootic disease dramatically reduces population size. (4) Extinction directly due to epizootic diseases is possible in the presence of reservoir hosts (e.g., spillover events). Here, the threshold effect is circumvented by the presence of a continual source of infection in a reservoir host (e.g., a domestic animal). This is a particular threat if the endemic population is small and the reservoir population large—a situation that led to local extinction of African wild dog populations in the Serengeti due to domestic dog rabies and canine distemper (Cleaveland and Dye 1995). (5) Extinction following infection by pathogens with unusual life cycles that confer naturally low-threshold densities has been hypothesized for amphibian chytridiomycosis, caused by a fungal pathogen that causes aclinical infections in larvae and can survive saprobically in the laboratory (Daszak et al. 1999). (6) Extinction due to indirect effects of disease outbreak can occur. For example, extinction of the large blue butterfly in the United Kingdom occurred after introduced myxomatosis removed rabbit populations, which reduced grazing pressure, and changed habitat for the ant species that nurtures butterfly larvae (Dobson and May 1986). Similarly, a host-specific eelgrass limpet, *Lottia alveus*, became extinct early in the twentieth century following an outbreak of eelgrass wasting disease, which removed its habitat throughout the eastern United States (Daszak and Cunningham 1999).

Wildlife EIDs raise a number of novel, complex conservation problems, primarily due to conflicts between conservation of biodiversity and actual or perceived threats to public health or agriculture. For example, where wildlife species harbor zoonotic pathogens, or are suspected as reservoirs, they are often willfully eradicated. In the United Kingdom badgers are exterminated due to their potential status as bovine tuberculosis reservoirs, and in the western United States prairie dog towns are removed in part due to the public health threat of plague emergence. Although a lyssavirus fatal to humans has been recently described in Australian fruit bats, vaccination, education, and disease surveillance programs have allayed public fears of zoonotic emergence, and the once-proposed eradication of urban bat colonies has not occurred. Zoonoses harbored by captive wildlife kept as pets have emerged as a public health threat. For example, reptile-associated salmonellosis is responsible for a small number of deaths annually in the United States and has also been reported in Europe (Friedman et al. 1998). Ad hoc conservation measures adopted by the public may also lead to disease emergence. For example, it has been estimated that 15,000 tons of peanuts are fed annually to wild birds from backyard feeders in the United Kingdom alone (Moss and Cottridge 1998). Such actions may result in unexpected consequences such as the rise in salmonellosis (Kirkwood 1998) and mycoplasmal conjunctivitis (Fischer et al. 1997) in wild birds that frequent backyard feeding stations and the occurrence of mycoplasmosis in red-footed tortoises following the release of infected pet tortoises into the wild (Jacobson 1993). Ecotourism, adopted by many conservation programs as an economically viable method for promoting wildlife conservation, has led to the emergence of human pathogens as a threat to great apes (Homsy 1999; Ferber 2000). Conservation programs that involve translocation or captive breeding involve an intrinsic risk of disease emergence by fostering pathogen transmission between allopatric species (Viggers et al. 1993; Cunningham 1996). For example, scrapielike spongiform encephalopathies have emerged in a range of zoo animals in Europe fed BSE-contaminated feed (Kirkwood and Cunningham 1994; Bons et al. 1999); toxoplasmosis has caused mass mortality in new-world primates following exposure to the causative organism in captivity (Cunningham et al. 1992; Epiphanio et al. 2000); and captive co-maintenance of African and Asian elephants has led to a fatal disease in the latter caused by a herpesvirus benign to the former (Richman et al. 1999). Where animals are released into the wild, the potential effects of diseases acquired in captivity are heightened. Measures have been put in place to reduce these risks (IUCN 1998; Cunningham 1996), but compliance is often poor (Daszak et al. 2000a). Conservation programs that do not translocate animals may also present a disease threat to wildlife. For example, Ferrer and Hiraldo (1995) reported a markedly higher prevalence of staphylococcal infections in wild Spanish imperial eagle nestlings handled without gloves compared to those handled with gloves. Protection of the southern elephant seal in regions of the United States where natural predators have been eradicated has led to the rapid expansion of populations. These now vie with human and domestic animals for valuable coastal land, leading to a perceived disease risk and a complex conservation dilemma (Brownell et al. 2000; Daszak et al. 2000b).

Ethical issues regarding parasite conservation (Harris 1983; Dobson and May 1986; Windsor 1990; Stork and Lyall 1993; Gompper and Williams 1998) raise unique problems, because pathogens are the only group containing species for which coordinated global eradication programs exist (e.g., the WHO programs to eradicate smallpox, polio, and river blindness). Conservation of parasites is hindered further by their largely negative public image and the general consensus that their biodiversity is of less value than that of their hosts. However, this field has become more than an interesting academic concept, since a number of parasites in imminent risk of extinction can be identified. For example, the conservation status of the parasite that caused the extinction of the snail *Partula turgida* is unknown (Cunningham and Daszak 1998). Examples of parasite extinction have also been reported. Stork and Lyall (1993) cite the extinction of the passenger pigeon lice *Columbicola extinctus* and *Campanulotes defectus* along with their host, as representing a much larger cohort of extinct (or "coextinct") parasite fauna. Gompper and Williams (1998) reported an eimerian parasite of the black-footed ferret that probably became extinct when the last wild population was brought into captivity. These authors suggest practical methods of parasite conservation such as use of alternative hosts.

Parasite conservation has been largely overlooked by conservationists, despite obvious practical advantages to the conservation of hosts (Cunningham 1996). In captive breeding and release programs, parasite conservation may benefit hosts, in the short term by maintaining integrity of host immunity to pathogens still present in wild populations, and in the long term due to the role parasites play in host population biology and evolution (e.g., evolution of sex, maintenance of diversity, and regulation of populations). Furthermore, to many people, parasites have intrinsic beauty, either due to their appearance (e.g., striking arrangements of hooks on the scolex of some cestode species) or due to their exquisitely complex life cycles that demonstrate the harmony of coevolution (e.g., the synchronized life cycles of desert toads and their parasites). Thus, there is no ethical reason not to conserve parasite biodiversity on an equal footing with host biodiversity.

5.6 Rising to the Challenge

Emerging infectious diseases represent a key threat to public health and the conservation of biodiversity. They emerge almost entirely as a result of recent anthropogenic changes, shifts in human demography and technology, habitat alterations, and global climate change. The scale and intensity of impact of wildlife and plant EIDs act as indicators of the level of anthropogenic change on the global ecosystem, that is, of ecosystem health (Epstein 1998; Harvell et al. 1999; Daszak et al. 2000a). For these reasons, infectious disease emergence encapsulates the breadth and ethos of the conservation medicine field, drawing on aspects of conservation biology, veterinary medicine, and public health, within the framework of human environmental change and ecosystem health. So far, the threat of EIDs to biodiversity conservation has been greatly underestimated to the extent of largely being ignored. This state of affairs needs to rapidly change if the conservation of wildlife and their habitats is to be an achievable goal. This is perhaps the greatest challenge facing those in the field of conservation medicine, a challenge that is possibly best met by focusing on each of the following three goals.

5.6.1 Broadening Our View of Conservation

Disease emergence has been largely ignored as a threat to conservation. The complexity of disease ecology is reflected in growing, complex conservation problems—threats to small, fragmented populations of endangered species, translocation of exotic pathogens within conservation programs, and the contrasting needs of humans and wildlife that clash via zoonotic disease emergence. A new approach to conservation is required that considers the disease consequences of each step in the conservation process. One step toward this will be the quantification of the threat EIDs pose to conservation efforts. New approaches to captive breeding programs will require increased surveillance, adherence to guidelines for disease screening, and consideration of the conservation of host–parasite assemblages. While the value of pathogen biodiversity surveys to human health is clear, concerted efforts to assess parasite biodiversity are required. A recent analysis of monogenean parasites in Australian teleost fish suggests that 90% remain undescribed, and at current rates of study, 900 years will be required to describe all extant species (Whittington 1998), future "coextinction" events notwithstanding. To successfully promote this enlightened approach to disease and parasites will require new approaches to public education that explain the risks of pathogen pollution, the disease threats involved in habitat alterations, and the benefits of parasite conservation.

5.6.2 Integration within Conservation Medicine

Emerging infectious diseases represent a unique, complex challenge that requires unprecedented integration between disciplines in research and outbreak investigation. Because most human EIDs are zoonoses, EID research already involves collaboration between medics, veterinarians, wildlife biologists, ecologists, epidemiologists, anthropologists, sociologists, and economists. However, significant gaps remain. Elucidating the role of the environment and anthropogenic changes to habitat in disease emergence requires further research. Cutting-edge disciplines from conservation biology and ecology (ecology of fragmented habitats, global change biology, biological invasions) must be fused with the recent advances in host–parasite ecology and evolution (modeling of disease invasion, "apparent competition," evolution of virulence) and our understanding of public health issues and the economic costs of disease and prevention.

Addressing this need for integration will require scientists to step outside the security of their narrow fields and tackle complex issues in related disciplines. For medics and public health researchers, this will entail an understanding of ecological processes, conservation, and anthropogenic effects on wildlife. For ecologists and conservation biologists, this means understanding the depth of the impact parasites and disease have at individual, population, and community levels and taking their role in the environment far more seriously. For veterinarians, this means understanding the linkage between environment, ecology, and disease within populations.

Research goals should include efforts to understand the scope of pathogen pollution and the mechanisms that drive it. Much has been hypothesized about the role of climate change in human disease emergence, but little regarding wildlife diseases. Chemical pollution has been implicated as a factor in a number of disease outbreaks, but there is little empirical work on the effects of toxicants on immune function or on population-level disease incidence and severity. Work on current threats of disease to biodiversity via local and global extinction events is crucial to understanding wildlife EID effects. Previous extinction events where disease has been hypothesized as an underlying or proximate cause should be reexamined using the latest pathological, microbiological, and disease-modeling techniques. Surveys of pathogen biodiversity would be relatively cheap and particularly cost-effective compared to the economic impact of disease emergence.

Conservation medicine can learn from the ground-level approach of conservation programs run by nongovernmental organizations and develop new ways to actively counter wildlife EIDs and to pass information to the public that diseases are a significant threat to biodiversity. At a practical level, we support the proposals of Reaser and colleagues (chap. 29) and urge the formation of interdisciplinary task forces comprising veterinarians, microbiologists, ecologists, modelers, public health experts, the medical profession, and conservation biologists. These teams would coordinate rapid-response outbreak investigation programs that deal with disease emergence on national and global scales. These could follow the format of the very successful teams set up by the Centers for Disease Control and Prevention in the United States and the Commonwealth Scientific and Industrial Research Organization in Australia to investigate human EID outbreaks. The National Wildlife Health Center in Madison, Wisconsin, may be able to play a key role in coordinating such teams, but it is clear that global solutions are required. Surveillance programs and pathogen biodiversity surveys should be expanded nationally and internationally. Justifying the cost of these will require a more focused assessment of the probably far higher cost of managing disease emergence after the fact.

5.6.3 Closing the Funding Gap

The broader definitions of EIDs laid out in this chapter are covered by different branches of research funding in Europe and the United States. In the United States, the National Science Foundation oversees funding for research on ecology, biology, and the environment but specifically does not fund veterinary or medical research; the National Institutes of Health oversee medical research funding, and the U.S. Department of Agriculture funds work on domestic animal and crop diseases. No specific government programs exist to fund wildlife disease research (apart from hunted or managed wildlife). Similar funding gaps exist in the United Kingdom. Huge disparities exist between the funding of medical and wildlife diseases in the United States despite the obvious zoonotic connections.

To close these gaps, we should stress the zoonotic connection between human health and the emergence of disease in wildlife. Anthropogenic environmental changes cause the emergence of disease in wildlife and plants that directly threaten human health and well-being. Disease-induced biodiversity loss also affects humans. A solid analysis of the economic impact of pathogen pollution (in the same way economists have put a dollar value on biological invasions; Pimentel et al. 2000) will strengthen this argument. The significant interest in biological invasions should be used as a platform for promoting awareness of the key role of disease invasion and pathogen pollution in biodiversity loss.

The development of the emerging disease concept was partly a political movement within science in response to two scientific and sociopolitical problems: a series of high-profile disease outbreaks in the United States and Europe during the 1980s and a reduction in antimicrobial surveillance programs within the U.S. public health laboratory system. In some ways, the EID movement can be seen as a concerted, very successful attempt to increase the funding for infectious disease research and surveillance (Daszak 2000). The developing interest in EIDs has significantly strengthened funding for infectious disease surveillance and research and formed a new approach to disease outbreak investigation. But wildlife EIDs have largely been left behind. There is a need for a determined approach to follow the recent high-profile descriptions of a number of wildlife EIDs, such as marine mammal morbillivirus outbreaks, coral diseases, rabies and canine distemper in African wild dogs, amphibian chytridiomycosis, and West Nile encephalitis in the United States.

5.7 Conclusions

In this chapter we have highlighted a specific group of diseases that form a significant threat to public health and the conservation of biodiversity. Emerging infectious diseases form only part of the broad problem that conservation medicine addresses. However, due to their unique origins in recent anthropogenic environmental changes and the complex conservation issues that result, EIDs challenge us in a novel way. The borders between human populations, domestic animals and plants, and wild ecosystems grow more blurred each year as human populations increase, expand, and encroach. We need to recognize that, because of these blurred borders, disease emergence in wild animals and plants is a direct result of our own actions, which in turn threatens our own health, economy, and well-being via zoonotic diseases, crop diseases, and the loss of biodiversity. Although we only partly understand the linkage between environmental damage and disease emergence and between disease emergence and extinction, it is increasingly clear that diseases have been underreported and their effects underestimated in wild animals and plants. Extinction by infection, hypothesized for a number of historic cases, has finally been demonstrated with the passing of the snail *Partula turgida*, albeit in unusual ecological circumstances. What is required now is a reassessment of previous extinction events where disease has been implicated, and increasing attention to ongoing extinctions, such as those caused by panzootic amphibian chytridiomycosis and disease introduction into endangered, fragmented populations of African wild dogs and great apes. Understanding these newly broadened definitions of emerging diseases will require coordinated integration of public health, wildlife disease research, and conservation biology—a key goal of conservation medicine (Meffe 1999). By integrating recent key advances in host–parasite ecology and the effects of habitat fragmentation, invasive species, and environmental change on ecosystems, conservation medicine may catalyze a crucial step toward understanding this important phenomenon.

References

Alexander, D.J. and I.H. Brown. 2000. Recent zoonoses caused by influenza A viruses. Rev Sci Tech 19:197–225.

Alexander, K.A., P.W. Kat, L.A. Munson, A. Kalake, and M.J.G. Appel. 1996. Canine distemper-related mortality among wild dogs (*Lycaon pictus*) in Chobe National Park, Botswana. J Zoo Wildl Med 27:426–427.

Anderson, R.M., and R.M. May. 1986. The invasion, persistence and spread of infectious diseases within animal and plant communities. Philos Trans R Soc Lond B 314:533–570.

Anonymous. 2000. Culprit named in "sudden oak death." Science 289:859.

Barbour, A.G., and D. Fish. 1993. The biological and social phenomenon of Lyme disease. Science 260:1610–1616.

Bäumler, A.J., B.M. Hargis, and R.M. Tsolis. 2000. Tracing the origins of *Salmonella* outbreaks. Science 287:50–52.

Bengston, J.L., P. Boveng, U. Franzen, P. Have, M.-P. Heide-Jorgensen, and T.J. Harkionen. 1991. Antibodies to canine distemper virus in antarctic seals. Mar Mamm Sci 7:85–87.

Berger L., R. Speare, P. Daszak, D.E. Green, A.A. Cunningham, C.L. Goggin, R. Slocombe, M.A. Ragan, A.D. Hyatt, K.R. McDonald, H.B. Hines, K.R. Lips, G. Marantelli, and H. Parkes. 1998. Chytridiomycosis causes amphibian mortality associated with population declines in the rainforests of Australia and Central America. Proc Natl Acad Sci USA 95:9031–9036.

Berkelman, R.L., R.T. Bryan, M.T. Osterholm, J.W. LeDuc, and J.M. Hughes. 1994. Infectious disease surveillance: a crumbling foundation. Science 264:368–370.

Beuchat, L.R., and J.-H. Ryu. 1997. Produce handling and processing practices. Emerg Infect Dis 3:459–465.

Bons, N., N. Mestre-Frances, P. Belli, F. Cathala, D.C. Gajdusek, and P. Brown. 1999. Natural and experimental oral infection of nonhuman primates by bovine spongiform encephalopathy agents. Proc Natl Acad Sci USA 96:4046–4051.

Brownell, R.L., B.E. Curry, W. Van Bonn, and S.H. Ridgway. 2000. Conservation conundrum. Science 288:2319.

Carr, D.E., and L.E. Banas. 2000. Dogwood anthracnose (*Discula destructiva*): effects of and consequences for host (*Cornus florida*) demography. Am Midland Nat 143:169–177.

Castello, J.D., D.J. Leopold, and P.J. Smallidge. 1995. Pathogens, patterns, and processes in forest ecosystems. Bioscience 45:16–24.

Chua, K.B., K.J. Goh, KT. Wong, A. Kamarulzaman, P.S.K. Tan, T.G. Ksiazek, S.R. Zaki, G. Paul, S.K. Lam, and C.T. Tan. 1999. Fatal encephalitis due to Nipah virus among pig-farmers in Malaysia. Lancet 354:1257–1259.

Chua, K.B., W.J. Bellini, P.A. Rota, B.H. Harcourt, A. Tamin, S.K. Lam, T.G. Ksiazek, P.E. Rollin, S.R. Zaki, W.-J. Shieh, C.S. Goldsmith, D.J. Gubler, J.T. Roehrig, B. Eaton, A.R. Gould, J. Olson, H. Field, P. Daniels, A.E. Ling, C.J. Peters, L.J. Anderson, and B.W.J. Mahy. 2000. Nipah virus: a recently emergent deadly paramyxovirus. Science 288:1432–1435.

Cleaveland, S., and C. Dye. 1995. Maintenance of a microparasite infecting several host species: rabies in the Serengeti. Parasitology 111:S33–S47.

Cohen, M.L. 1992. Epidemiology of drug resistance: implications for a post-antimicrobial era. Science 257:1050–1055.

Coleman, J.D. 1988. Distribution, prevalence, and epidemiology of bovine tuberculosis in brushtail possums, *Trichosurus vulpecula*, in the Hohonu Range, New Zealand. Aust Wildl Res 5:651–663.

Crosby, A.W. 1986. Ecological imperialism. The biological expansion of Europe, 900–1900. Cambridge University Press, New York.

Cully, J.F., A.M. Barnes, T.J. Quan, and G. Maupin. 1997. Dynamics of plague in a Gunnison's prairie dog colony complex from New Mexico. J Wildl Dis 33:706–719.

Cunningham, A.A. 1996. Disease risks of wildlife translocations. Conserv Biol 10:349–353.

Cunningham, A.A., and P. Daszak. 1998. Extinction of a species of land snail due to infection with a microsporidian parasite. Conserv Biol 12:1139–1141.

Cunningham, A.A., D. Buxton, and K.M. Thomson. 1992. An epidemic of toxoplasmosis in a captive colony of squirrel monkeys (*Saimiri sciureus*) J Comp Pathol 107:207–209.

Cunningham, A.A., T.E.S. Langton, P.M. Bennett, J.F. Lewin, S.E.N. Drury, R.E. Gough, and S.K. Macgregor. 1996. Pathological and microbiological findings from incidents of unusual mortality of the common frog (*Rana temporaria*). Philos Trans R Soc Lond B 351:1529–1557.

Daszak, P. 2000. Emerging infectious diseases: bridging the gap between humans and wildlife. Scientist 14:14.

Daszak, P., and A.A. Cunningham. 1999. Extinction by infection. TREE 14:279.

Daszak, P., L. Berger, A.A. Cunningham, A.D. Hyatt, D.E. Green, and R. Speare. 1999. Emerging infectious diseases and amphibian population declines. Emerg Infect Dis 5:735–748.

Daszak, P., A.A. Cunningham, and A.D. Hyatt. 2000a. Emerging infectious diseases of wildlife—threats to biodiversity and human health. Science 287:443–449.

Daszak, P., A.A. Cunningham, and A.D. Hyatt. 2000b. Conservation conundrum, response. Science 288:2320.

Daszak, P., A.A. Cunningham, and A.D. Hyatt. 2001. Anthropogenic environmental change and the emergence of infectious diseases in wildlife. Acta Trop 78:103–116.

Dhondt, A.A., D.L. Tessaglia, and R.L. Slothower. 1998. Epidemic mycoplasmal conjunctivitis in house finches from eastern North America. J Wildl Dis 34:265–280.

Dobson, A.P., and E.R. Carper. 1996. Infectious diseases and human population history. Bioscience 46:115–126.

Dobson, A., and M. Crawley. 1994. Pathogens and the structure of plant communities. TREE 9:393–398.

Dobson, A.P., and J. Foufopoulos. 2001. Emerging infectious pathogens of wildlife. Philos Trans R Soc Lond B, 356:1001–1012.

Dobson, A.P., and R.M. May. 1986. Disease and conservation. *In* Soulé, M. ed. Conservation biology: the science of scarcity and diversity, pp. 345–365. Sinauer, Sunderland, Mass.

Doyle, T.J., R.T. Bryan, and C.J. Peters. 1998. Viral hemorrhagic fevers and hantavirus infections in the Americas. Infect Dis Clin N Am 12:95–110.

Enserink, M. 2000. Malaysian researchers trace Nipah virus outbreak to bats. Science 289:518–519.

Epiphanio, S., M.A.B.V. Guimaraes, D.L. Fedullo, S.H.R. Correa, and J.L. Catao-Dias. 2000. Toxoplasmosis in golden-headed lion tamarins (*Leontopithecus chrysomelas*) and emperor marmosets (*Sanguinus imperator*) in captivity. J Zoo Wildl Med 31:231–235.

Epstein, P.R. 1998. Marine ecosystems: emerging diseases as indicators of change. Harvard Medical School, Boston.

Epstein, P.R. 1999. Climate and health. Science 285:347–348.

Ferber, D. 2000. Human diseases threaten great apes. Science 289:1277–1278.

Ferrer, M., and F. Hiraldo. 1995. Human-associated staphylococcal infection in Spanish imperial eagles. J Wildl Dis 31:534–536.

Fischer, J.R., D.E. Stallknecht, M.P. Luttrell, A.A. Dhondt, and K.A. Converse. 1997. Mycoplasmal conjunctivitis in wild songbirds: the spread of a new contagious disease in a mobile host population. Emerg Infect Dis 3:69–72.

Friedman, C.R., C. Torigian, P.J. Shillam, R.E. Hoffman, D. Heltzel, J.L. Beebe, G. Malcolm, W.E. DeWitt, L. Hutwagner, and P.M. Griffin. 1998. An outbreak of salmonellosis among children attending a reptile exhibit at a zoo. J Pediatr 132:802–807.

Gao, F., L. Yue, A.T. White, P.G. Pappas, J. Barchue, A.P. Hanson, B.M. Greene, P.M. Sharp, G.M. Shaw, and B.H. Hahn. 1992. Human infection by genetically diverse SIVsm-related HIV-2 in West Africa. Nature 358:495–499.

Gao, F., E. Bailes, D.L. Robertson, Y. Chen, C.M. Rodenburg, S.F. Michael, L.B. Cummins,

L.O. Arthur, M. Peeters, G.M. Shaw, P.M. Sharp, and B.H. Hahn. 1999. Origin of HIV-1 in the chimpanzee *Pan troglodytes.* Nature 397:436–441.

Gardner, H., K. Kerry, M. Riddle, S. Brouwer, and L. Gleeson. 1997. Poultry virus infection in Antarctic penguins. Nature 387:245.

Gilliver, M.A., M. Bennett, M. Begon, S.M Hazel, and A. Hart. 1999. Enterobacteria—antibiotic resistance found in wild rodents. Nature 401:233–234.

Gompper, M.E. and E.S. Williams. 1998. Parasite conservation and the black-footed ferret recovery program. Conserv Biol 12:730–732.

Griffin, D.E. and W.J. Bellini. 1996. Measles virus. *In* Fields, B.N. D.M. Knipe, P.M. Howley, R.M. Chanock, J.L. Melnick, T.P. Monath, B. Roizman, and S.E. Sraus, eds. Fields' virology, pp. 1267–1312. Lippincott-Raven, Philadelphia.

Hahn, B.H., G.M. Shaw, K.M. de Cock, and P.M. Sharp. 2000. AIDS as a zoonosis: scientific and public health implications. Science 287:607–614.

Harris, E.A. 1983. Helminths in danger: the forgotten few. Parasitology 87:R29.

Harvell, C.D., K. Kim, J.M. Burkholder, R.R. Colwell, P.R. Epstein, D.J. Grimes, E.E. Hofmann, E.K. Lipp, A.D.M.E., Osterhaus, R.M. Overstreet, J.W. Porter, G.W. Smith, and G.R. Vasta. 1999. Emerging marine diseases—climate links and anthropogenic factors. Science 285:1505–1510.

Hill, D.J., R.L. Langley, and W.M. Morrow. 1998. Occupational injuries and illnesses reported by zoo veterinarians in the United States. J Zoo Wildl Med 29:371–385.

Hoffman, G.L. 1970. Intercontinental and transcontinental dissemination and transfaunation of fish parasites with emphasis on whirling disease (*Myxosoma cerebralis*). *In* Snieszko, S.F. ed. Symposium on the diseases of fishes and shellfishes, pp. 69–81 American Fisheries Society, Washington, D.C.

Homsy, J. 1999. Ape tourism and human diseases: how close should we get? Consultancy report for the International Gorilla Conservation Programme. Fauna and Flora International, Cambridge.

Hooper, P.T., R.A. Lunt, A.R. Gould, A.D. Hyatt, G.M. Russell, J.A. Kattenbelt, S.D. Blacksell, L.A. Reddacliff, P.D. Kirkland, R.J. Davis, P.J.K. Durham, A.L. Bishop, and J. Waddington. 1999. Epidemic of blindness in kangaroos—evidence of a viral aetiology. Austr Vet J 77:529–536.

Hudson, P., and J. Greenman. 1998. Competition mediated by parasites: biological and theoretical progress. TREE 13:387–390.

International Union for the Conservation of Nature (IUCN). 1998. Guidelines for re-introductions. Prepared by the IUCN/SSC Re-introduction Specialist Group. IUCN, Gland, Switzerland.

Jacobson, E.R. 1993. Implications of infectious diseases for captive propagation and introduction programs of threatened/endangered reptiles. J Zoo Wildl Med 24:245–255.

Jessup, D.A., W.M. Boyce, and R.K. Clarke. 1991. Diseases shared by wild, exotic and domestic sheep. *In* Renecker, L.A., and R.J. Hudson, eds. Wildlife production: conservation and sustainable development, pp. 438–445. University of Alaska, Fairbanks.

Jones, C.G., R.S. Ostfeld, M.P. Richard, E.M. Schauber and J.O. Wolff. 1998. Chain reactions linking acorns to gypsy moth outbreaks and Lyme disease risk. Science 279:1023–1026.

Kirkwood, J.K. 1998. Population density and infectious disease at bird tables. Vet Rec 142:468.

Kirkwood, J.K., and A.A. Cunningham. 1994. Epidemiological observations on spongiform encephalopathies in captive wild animals in the British Isles. Vet Rec 135:296–303.

Klenk, H.-D., ed. 1999. Marburg and Ebola viruses. Current Topics in Microbiology and Immunology, No. 235. Springer, Berlin.

Krause, R.M. 1992. The origins of plagues: old and new. Science 257:1073–1078.

Lanciotti, R.S., J.T., Roehrig, V. Deubel, J. Smith, M. Parker, K. Steele, B. Crise, K.E. Volpe, M.B. Crabtree, J.H. Scherret, R.A. Hall, J.S. MacKenzie, C.B. Cropp, B. Panigrahy, E. Ostlund, B. Schmitt, M. Malkinson, C. Banet, J. Weissman, N. Komar, H.M. Savage, W. Stone, T. McNamara, and D.J. Gubler. 1999. Origin of the West Nile virus responsible for an outbreak of encephalitis in the northeastern United States. Science 286:2333–2337.

Lederberg, J., R.E. Shope, and S.C. Oakes, Jr. 1992. Emerging infections: microbial threats to health in the United States. Institute of Medicine, National Academy Press, Washington, D.C.

Lips, K.R. 1999. Mass mortality and population declines of anurans at an upland site in western Panama. Conserv Biol 13:117–125.

MacKenzie, J.S. 1999. Emerging viral diseases: an Australian perspective. Emerg Infect Dis 5:1–8.

MacPhee, R.D. and P.A. Marx. 1997. The 40,000 year plague: humans, hyperdisease, and first-contact extinctions. *In* Goodman, S. M., and B.D. Patterson, eds. Natural change and human impact in Madagascar, pp. 169–217. Smithsonian Institution Press, Washington, D.C.

Mahy, B.W.J. and C.C. Brown. 2000. Emerging zoonoses: crossing the species barrier. Rev Sci Tech OIE 19:33–40.

Mamaev L.V., N.N. Denikina, S.I. Belikov, V.E. Volchkov, I.K.G. Visser, M. Fleming, C. Kai, T.C. Harder, B. Liess, A.D.M.E. Osterhaus, and T. Barrett. 1995. Characterization of morbilliviruses isolated from Lake Baikal seals (*Phoca sibirica*). Vet Microbiol 44: 251–259.

McCallum, H. and A. Dobson. 1995. Detecting disease and parasite threats to endangered species and ecosystems. TREE 10:190–194.

McMichael, A.J., B. Bolin, R. Costanza, G.C. Daily, C. Folke, K. Lindahl-Kiessling, E. Lindgren, and B. Niklasson. 1999. Globalization and the sustainability of human health. Bioscience 49:205–210.

Meffe, G.K. 1999. Conservation medicine. Conserv Biol 13:953–954.

Mills, J.N, T.G. Ksiazek, C.J. Peters, and J.E. Childs. 1999. Long-term studies of hantavirus reservoir populations in the southwestern United States: a synthesis. Emerg Infect Dis 5:135–142.

Monath, T.P. 1999. Ecology of Marburg and Ebola viruses: speculations and directions for future research. J Infect Dis 179(suppl 1):S127–S138.

Monroe, M.C., S.P. Morzunov, A.M. Johnson, M.D. Bowen, H. Artsob, T. Yates, C.J. Peters, P.E. Rollin, T.G. Ksiazek, and S.T. Nichol. 1999. Genetic diversity and distribution of *Peromyscus*-borne hantaviruses in North America. Emerg Infect Dis 5: 75–86.

Moore, S.W., and J.V. Robbs. 1979. Acute appendicitis in the Zulu—an emerging disease? S Afr Med J 55:700.

Morse, S.S., ed. 1993. Emerging viruses. Oxford University Press, New York.

Moss, S., and D. Cottridge. 1998. Attracting birds to your garden. New Holland, London.

Murray, A. G., R. J. Smith, and R. M. Stagg. 2002. Shipping and the spread of infectious salmon anemia in Scottish aquaculture. Emerg Infect Dis 8:1–5.

Murray, K., P. Selleck, P. Hooper, A. Hyatt, A. Gould, L. Gleeson, H. Westbury, L., Hiley, L. Selvey, B. Rodwell, and P. Ketterer. 1995. A morbillivirus that caused fatal disease in horses and humans. *Science* 268:94–97.

Mutombo, M.W., I. Arita, and Z. Jezek. 1983. Human monkeypox transmitted by a chimpanzee in a tropical rain-forest area of Zaire. Lancet 2:735–737.

Olsen, B., S. Bergstrom, D.J. McCafferty. M. Sellin, and J. Wistrom. 1996. *Salmonella enteritidis* in Antarctica: zoonosis in man or humanosis in penguins? Lancet 348: 1319–1320.

Ostfeld, R.S. and F. Keesing. 2000. Biodiversity and disease risk: the case of Lyme disease. Conserv Biol 14:722–728.

Pimentel D., L. Lach, R. Zuniga, and D. Morrison. 2000. Environmental and economic costs of nonindigenous species in the United States. Bioscience 50:53–65.

Plowright, W. 1982. The effects of rinderpest and rinderpest control on wildlife in Africa. Symp Zool Soc Lond 50:1–28.

Real, L.A. 1996. Sustainability and the ecology of infectious disease. Bioscience 46:88–97.

Richman, L.K., R.J. Montali, R.L. Garber, M.A. Kennedy, J. Lehnhardt, T. Hildebrandt, D. Schmitt, D. Hardy, D.J. Alcendor, and G.S. Hayward. 1999. Novel endotheliotropic herpesviruses fatal for Asian and African elephants. Science 283:1171–1176.

Roelke-Parker, M.E., L. Munson, C. Packer, R. Kock, S. Cleaveland, M. Carpenter, S.J. O'Brien, A. Pospischil, R. Hofmann-Lehmann, H. Lutz, G.L.M. Mwamengele, M.N. Mgasa, G.A. Machange, B.A. Summers, and M.J.G. Appel. 1996. A canine distemper virus epidemic in Serengeti lions (*Panthero leo*). Nature 379:441–445.

Rosenzweig, C., A. Iglesias, X.B. Yang, P.R. Epstein, and E. Chivian. 2000. Climate change and U.S. agriculture: the impacts of warming and extreme weather events on productivity, plant diseases, and pests. Consultancy Report. Center for Health and the Global Environment, Harvard Medical School, Boston.

Ross, P.S., J.G. Vos, L.S. Birnbaum, and A.D.M.E. Osterhaus. 2000. PCBs are a health risk for humans and wildlife. Science 289:1878–1879.

Rupprecht, C.E., J.S. Smith, M. Fekadu, and J.E. Childs. 1995. The ascension of wildlife rabies: a cause for public health concern or intervention? Emerg Infect Dis 1:107–114.

Rybicki, E.P., and G. Pietersen. 1999. Plant virus disease problems in the developing world. Adv Virus Res 53:127–175.

Sainsbury, A.W., P. Nettleton, J. Gilray, and J. Gurnell. 2000. Grey squirrels have high seroprevalence to a parapoxvirus associated with deaths in red squirrels. Anim Conserv 3:229–233.

Schmidt, K.A. and R.S. Ostfeld. 2001. Biodiversity and the dilution effect in disease ecology. Ecology. 82:609–619.

Schrag, S.J., and P. Wiener. 1995. Emerging infectious diseases: what are the relative roles of ecology and evolution? TREE 10:319–324.

Sencer, D.J. 1971. Emerging diseases of man and animals. Ann Rev Microbiol 25:465–486.

Stork, N.E., and C.H.C. Lyall. 1993. Extinction or "co-extinction" rates? Nature 366:307.

Tauxe, R.V. 1997. Emerging foodborne diseases: an evolving public health challenge. Emerg Infect Dis 3:425–434.

Taylor, L.H., and M.E.J. Woodhouse. 2000. Zoonoses and the risk of disease emergence. *In* Proceedings of the international conference on emerging infectious diseases, poster 122. CDC, Atlanta, Georgia.

Telford, S.R., P.M. Armstrong, P. Katavolos, I. Foppa, A.S.O. Garcia, M.L. Wilson, and A. Spielman. 1997. A new tick-borne encephalitis-like virus infecting New England deer ticks, *Ixodes dammini*. Emerg Infect Dis 3:165–170.

Thorne, E.T., and E.S. Williams. 1988. Disease and endangered species: the black-footed ferret as a recent example. Conserv Biol 2:66–74.

Trading Standards Office (TSO) 2000. The BSE inquiry: the inquiry into BSE and variant CJD in the United Kingdom. Stationary Office, Norwich, U.K.

Viggers, K.L., D.B. Lindenmayer, and D.M. Spratt. 1993. The importance of disease in reintroduction programmes. Wildl Res 20:687–698.

Whelan, P.M. 1995. An inspection and quarantine system for the Galapagos Islands: why and how. *In* Proceeding of the symposium on invasive species on oceanic islands (6 April 1995), pp. 23–26. Royal Society, London.

Whittington, I.D. 1998. Diversity "down under": monogeneans in the Atipodes (Australia) with a prediction of monogenean biodiversity worldwide. Int J Parasitol 28:1481–1493.

Whittington, R.J., C. Kearns, A.D. Hyatt, S Hengstberger, and T. Rutzou. 1996. Spread of epizootic haematopoietic necrosis virus (EHNV) in redfin perch (*Perca fluviatilis*) in Southern Australia. Austral Vet J 73:112–114.

Whittington, R.J., J.B. Jones, P.M. Hine, and A.D. Hyatt. 1997. Epizootic mortality in the pilchard *Sardinops sagax neopilchardus* in Australia and New Zealand in 1995. I. Pathology and epizootiology. Dis Aquat Org 28:1–16.

Wills, R.T. 1992. The ecological impact of *Phytophthora cinnamomi* in the Stirling Range National Park, Western Australia. Austral J Ecol 17:145–159.

Windsor, D.A. 1990. Heavenly hosts. Nature 348:104.

Woodroffe, R. 1999. Managing disease threats to wild mammals. Anim Conserv 2:185–193.

Zhu, Y.Y., H.R. Chen, J.H. Fan, Y.Y. Wang, Y. Li, J.B. Chen, J.X. Fan, S.S. Yang, L.P. Hu, H. Leung, T.W. Mew, P.S. Teng, Z.H. Wang, and C.C. Mundt. 2000. Genetic diversity and disease control in rice. Nature 406:718–722.

6

Effects of Endocrine Disruptors on Human and Wildlife Health

Theo Colborn

This chapter briefly describes the endocrine system and the role it plays in survival and quality of life for both wildlife and humans. It provides evidence that products and chemical by-products of modern technology can pose a threat to the future integrity of many species on the earth, including humans. Two studies are presented in which reported untoward health effects in wildlife led to studies to determine whether the same effects were being expressed in humans. Based on insight gained from the two studies, the need for well-directed, long-term research that involves wildlife, human, and laboratory collaborations is discussed.

6.1 Background

The endocrine system comprises the sex organs (mammary glands, ovaries, uterus, testes, and prostate), brain (which includes the pituitary and the hippocampus), thyroid, adrenals, thymus, and pancreas, to mention the more familiar organs. It operates through the use of chemical messengers called hormones, produced by the organs. This system controls the metabolism of fat, response to stress, how the brain develops, daily function, maturation, and senescence. It controls how the immune system functions to protect animals from infection, and most important, it controls sexual development and the ability to reproduce. Historically, the endocrine system assured the integrity and survival of animals—until humans began to produce large quantities of synthetic chemicals that interfere with the hormones and other chemical messengers that control development.

The essential role of the endocrine system in development makes it especially vulnerable to chemicals of this nature (Colborn and Clement 1992).

The concentrations of a number of chemicals that interfere with the endocrine system are now at or above levels in the environment where they can compromise the health and potential of a sizable proportion of wildlife and humans born today (Johnson et al. 1999). When scientists began to recognize the unique and insidious effects of this group of chemicals, they named them "endocrine disruptors" to distinguish their activity from the traditionally understood activity of toxic chemicals such as acute toxicity, DNA damage, gross and obvious effects on development, and outright mortality (Bern et al. 1992). The effects of endocrine disruptors are easily overlooked because the exposure that causes the damage generally takes place before the individual is hatched or born. For humans, this could be any time during the 266 days of gestation. In addition, the effects are often missed because prenatal exposure can lead to changes in the developing individual that, in many cases, are not visible at birth but are expressed many years later as the individual ages (Colborn and Clement 1992). Among wildlife, the exposed individuals may not reach sexual maturity (Leatherland 1998; Government of Canada 1991) or they may develop poorly, fail to thrive, and die prematurely (Colborn 1991; Cook et al. 1995). In the case of mammals, the mother shares the chemicals in her blood with her unborn embryo and fetus. After the offspring is born, because of hormonally driven changes in her physiology, the mother transfers much higher concentrations of endocrine disruptors to the newborn through her breast milk than she did through her blood (Jensen and Slorach 1990; Béland 1996).

Embryologists and endocrinologists agree that prenatal development affects an individual's future more than any period after birth. During gestation, infinitesimally small concentrations of naturally produced hormones control how cells develop to become the reproductive, endocrine, and immune systems and how the brain is wired (Haddow et al. 1999). For example, both mouse and human embryos develop in the presence of 10^{-13} gm (0.1 parts per trillion, or ppt) of free estradiol (vom Saal et al. 1992; Welshons et al. 1999). Thyroid hormones in humans are present in parts per trillion and parts per billion (ppb) (Brucker-Davis 1998, Haddow et al. 1999). There is now ample evidence that exposure to widely dispersed synthetic chemicals during development can restrict an individual's potential and have an effect at the population level (Jacobson and Jacobson 1996; Stewart et al. 2000a, b, Brouwer et al. 1995, Garry et al. 1996; Hose et al. 1989, Jobling et al. 1998).

6.2 Exposure

Every human or wild animal born today will already have some endocrine disruptors in his or her body. These high-volume chemicals are globally distributed by human activity (commerce and trade) and then further dispersed via air and water. They range from industrial chemicals such as plastics and plastics components to pesticides, including herbicides, fungicides, and insecticides. As we move into the era of globalization, more and more of these chemicals will be dispersed in the environment. For example, the plastics industry grew steadily at 12% per year in the United States between 1980 and 2000, and more recently, as much as 40% per year in developing countries (Society of the Plastics Industry 1996). Plastics components that are known endocrine disruptors include plastic monomers and plastic additives, such as fire retardants, softeners, and fluidizers. Many of these compounds are long-lived and have been accumulating in the environment for over 50 years (Wams 1987).

In addition to industrial chemicals, 1.3 billion pounds (590 million kg) of active pesticide ingredients were produced in 1994–95 in the United States alone (Short and Colborn 1999). Some plastic components that are known endocrine disruptors are included as inert pesticide ingredients, contributing even more to their wide dispersal (Short and Colborn 1999). Over 60% of all herbicides produced are known endocrine disruptors (Short and Colborn 1999). Pesticides are tested for their carcinogenicity, obvious acute neurotoxicity, lethality, and skin and eye irritation, but no screens or assays exist to test them even cursorily for endocrine disruption (EPA 1998). This also applies to industrial chemicals.

Compounding the problem of exposure, the global environment is dusted with the residues of many persistent, lipophilic, organochlorine chemicals (OCs), such as polychlorinated biphenyls (PCBs), dichlorodiphenyltrichloroethane (DDT), dieldrin, toxaphene, and chlordane, whose legacy will remain for hundreds of years. There are 209 different PCBs, 177 toxaphenes, 135 furans, and 75 dioxins (tetrachlorodibenzo-p dioxins, or TCDDs), based on the number and position of chlorine substitutions on the parent aromatic benzene ring. These are called congeners. PCBs are further segregated as homologs, such as mono-, di-, tri, tetra-, through octa-chlorinated biphenyls. Isomers belong to the same homologous class. The toxic effects of PCB congeners with no chlorine substitution in the ortho position are the same as 2,3,7,8-TCDD with reduced potency. These particular congeners are not readily metabolized by fish and therefore accumulate in fish tissue. Toxicity equivalency factors (TEQs) have been determined for many PCBs and other OCs based on their potency compared with dioxin. Their sum is then used to determine the toxicity resulting from environmental exposure.

The implications of widespread exposure of persistent OCs were first signaled in the late 1960s and 1970s, when gulls on the Great Lakes and Pacific coast started to experience obvious behavioral changes and difficulty reproducing (Hunt and Hunt 1977, Conover and Hunt 1984; Fry and Toone 1981; Fox et al. 1976). The birds were carrying elevated levels of OCs that had biomagnified in the food webs of their respective aquatic systems. A number of OCs were regulated in the early 1970s, resulting in a marked decline in their concentrations in the tissue of Great Lakes fishes and birds (Gilbertson et al. 1998). But since 1985, there has been no further significant decline, and the most recent health evaluation for Great Lakes wildlife reveals that the same developmental problems exist in the same regions where they were first reported in the 1960s and 1970s (Giesy et al. 1994, Gilbertson et al. 1998). Reductions in dieldrin and DDT abated acute mortality and eggshell thinning, respectively, but the more insidious developmental damage in the young animals still persists (Bowerman 1993). Animals and humans do not have to spend time in the farm belt, in industrialized areas, or next to a Superfund site to be exposed to endocrine disruptors. For example, pesticides applied at 40° N latitude move in ocean and air currents and are later found from 60° to 70° N latitude (International Air Quality Advisory Board 1988). The triazine and acetanilid herbicides are regularly found in drinking water supplies in every part of the United States and almost always in surface water (Short and Colborn 1999). Atrazine and other pesticides are found in fog, snow, rain, and melting ice in the Chukchi and Bering Seas (Chernyak et al. 1996). The highest concentrations are found at the air-water interface in the top 100 μm of the sea surface microlayer and in sea fog. Marine and terrestrial animals in several Arctic regions carry some of the highest body burdens of PCBs and other OCs in the world (Weis and Muir 1997; Norstrom et al. 1998). Humans in eastern Arctic Canada carry five to seven times more PCBs (Dewailly et al. 1993), and those in western coastal Greenland, 14 times more, than humans in lower Canada or the United States (Arctic Monitoring and Assessment Programme 1997).

Even albatrosses that feed only on the surface of the North Pacific Ocean are as biochemically hot as the birds in the dirty areas of the Great Lakes because of their burden of PCBs, dioxins, and furans (Auman et al. 1997). The black-footed albatross that flies the northeastern Pacific Ocean is exhibiting signs of developmental difficulty. Logistically, acquisition of appropriate data to evaluate this situation requires many years of intense effort in the field. Fortunately, albatrosses nest in large colonies—albeit remote and difficult to reach—and are relatively tame, which makes data gathering easier than working with animals that live and reproduce in seclusion. However, more reclusive species could disappear without anyone's knowledge because of the effects of endocrine disruption on development and reproduction (Bantle et al. 1995).

The inability to recognize endocrine disruptors effects on development was recently demonstrated when core drills in Lake Michigan revealed that dioxin levels were high enough in the 1940s and 1950s to prevent the survival of top predator fish species there (Cook et al. 1995, Cook 1999). The dioxin is attributable to the emergence of the chlor-alkali industry after World War I with the large-scale production of free chlorine that marked a turning point in the technology of catalytic chemistry. It also coincided with the extirpation of Great Lakes salmonids and other top predator species in the system. Controlled laboratory studies done in the 1990s demonstrated that only 5 picograms of dioxin in lake trout eggs will cause the embryos to develop abnormally, leading to death of the fry (Cook et al. 1995). Even today, wildlife managers still blame the crash of the economically important salmonids in the Great Lakes on overfishing and habitat destruction, and continue to restock the lakes on an annual basis with coho, chinook, and pink salmon and lake, rainbow, and brook trout (J.F. Leatherland, personal commun. 1998).

6.3 Cross Species Effects

The following studies reveal the advantages of and need for increased interdisciplinary discussion and cooperation among medical professionals and wildlife and laboratory researchers. The studies involve two frequent health anomalies reported in children born in the United States today—learning disabilities and hypospadias. In each case, the studies would probably not have been undertaken if teams of scientists working in the field and laboratory had not reported serious developmental problems in wildlife and received the attention of alert public health officials.

6.3 Study 1: The Great Lakes

Repeated reports about the damage to wildlife dependent upon fish from the Great Lakes led to concern about what might happen to the children born of women who also ate the fish from the lakes. This prompted the director of the U.S. Environmental Protection Agency's (EPA) Great Lakes, Grosse Ile Research Center to request a prospective study in 1979 with healthy mothers and infants who, like the wildlife, were at the top of the aquatic food web in Lake Michigan. It became evident that the intelligence and behavior of a sizable proportion of children born in the United States today are being affected as a result of prenatal exposure to PCBs and their cocontaminants (Jacobson and Jacobson 1996). A team of Detroit psychologists found that infants born between 1980 and 1981 to mothers who ate two to three meals per month of Lake Michigan fish, at least six years prior to their pregnancies, were behind at birth in neuromuscular maturity compared with controls (Jacobson et al. 1984). At age four, based on elevated PCB concentrations in serum, these children had difficulty with short-term memory, processed information more slowly, had reduced auditory and verbal skills, and had lower quantitative and visual discrimination memory (Jacobson and Jacobson 1993). At age 11, the children most highly exposed during gestation had a mean 6.2-point intelligence quotient (IQ) deficit and were more than a year behind their peers in word and reading comprehension (Jacobson and Jacobson 1996).

Most important, trained psychologists were able to measure neurodevelopmental delays at birth in the offspring of mothers with 1.00 parts per million (ppm) dioxin in fat during gestation. At 1.25 ppm PCB, the changes were significant (p = 0.0001) with little variation, suggesting a populationwide effect. The authors of this study point out that the average PCB fat levels hover around 1.00 ppm in the U.S. population, and women do not have to eat fish to carry this amount of PCBs in their bodies. At age four, 17 children were removed from the study because they were intractable and refused to be tested. Later, it was discovered that these were the children of the 17 women with the highest PCB concentrations in the study (Jacobson et al. 1990). At the end of the study, another child was removed because of an IQ below 70 and was considered mentally retarded. If all of these children had remained in the study, 20% of the study population would have exhibited effects of high PCB concentrations. This suggests that a sizable proportion of the general population born today is compromised. That only healthy mothers and infants were selected for this study emphasizes the robustness of the results.

In a second healthy mother–infant study that commenced 12 years later, infants of mothers who ate more than 40 pounds (18.1 kg) of Lake Ontario fish (about two to three fish meals a month) prior to their pregnancies also displayed the same neuromuscular delays at birth as the Lake Michigan infants compared with controls. When tested with an additional set of tasks, these children were found to smile and laugh less and express more fear and were hyperreactive (Stewart et al. 2000b), similar to the hyperreactivity reported in rats fed Lake Ontario fish (Daly 1993). This study, which replicated the Lake Michigan study (and included more measurable parameters), reveals that the same effects are being reported in children today as was reported over 10 years ago. The results are similar to the recent findings in Great Lakes wildlife where the adverse effects still persist (Geisy et al. 1994; Gilbertson et al. 1998).

It is important to note that in these Great Lakes studies, even though the decrement in the child was statistically significant, it took skilled psychologists and technicians to quantify the changes in the children. These children are not retarded or obviously different; nevertheless, they are not developing to their fullest potential. And certainly, the quality of their lives and their parents lives will be reduced. The implications for society as a whole are disturbing, and they beg the question, What invisible functional effects are occurring in wildlife that we will never discover?

In the Netherlands, another healthy mother–infant study of the general population was undertaken, *not necessarily looking at fish eaters*. Infants of mothers with the highest PCB-dioxin TEQs exhibited psychomotor delays at three months (Koopman-Esseboom et al. 1996). Thyroid hormone levels (Koopman-Esseboom et al. 1994) and monocyte and granulocyte numbers were inversely associated with increases in TEQs, whereas T-lymphocytes were directly associated with increased TEQs (Weisglas-Kuperus et al. 1995). Prenatal exposure in this case was measured using the contaminants in maternal plasma during the last month of pregnancy. Again, a sizable proportion of the study population was affected (Patandin et al. 1997). Research by the EPA reveals that a technical PCB mixture (Aroclor 1254) interferes, through a thyroid mechanism, with the development of the cochlea, the area of the ear that recognizes sound (Goldey and Crofton 1998; Goldey et al. 1995a, b; Herr et al. 1996). Rats prenatally exposed to the PCBs experience reduced motor coordination and do not hear low- or intermediate-frequency clicking sounds. Loss of auditory discrimination such as this in humans can create difficulty with phonics, reading, and intellectual development, similar to the problems the children displayed in the Great Lakes studies.

With respect to wildlife, the data revealing the detrimental effects of PCBs through a thyroid mechanism raise questions about the stability of populations of marine mammals that are dependent upon hearing for direction and communication. Marine mammals carry some of the highest PCB concentrations of all species (Colborn and Smolen 1996). The data also raise questions about the thyroid problems reported in Great Lakes fish and gulls. In the 1970s, Canadian wildlife biologists could not find a salmonid in the Great Lakes that did not have an enlarged thyroid gland (Leatherland 1998), or a Great Lakes herring gull with normal thyroid hormone ratios (Fox et al. 1998). (Note: the top predator fish species present in the Great Lakes today are the result of a major stocking effort by the United States and Canada.) Today, even following the significant decline of concentrations of some of the OCs (including PCBs) in the early 1980s in response to regulatory action, the thyroids in Lake Erie salmonids are so large that they are rupturing and visible (Leatherland 1998). This suggests that something other than OCs may be causing the problem.

6.3.2 Study 2: The Florida Everglades

The second study shifts geographically to Florida in 1992. Armed with the new knowledge about endocrine disruptors in the environment and abnormal gonadal development in birds and fish from the Great Lakes, a team of Florida biologists went back to Lake Apopka, where they had been trying to figure out why the alligator population was not reproducing as expected (Guillette et al. 1994). Taking a second look, they discovered male alligators with reduced hemipenises and turtles hatching only as females or intersex, that is, no males (Guillette and Crain 1996; Guillette 1998).

Only through a serendipitous meeting between the field biologist working with the alligators and the EPA's chief reproductive and developmental toxicologist did they realize that perhaps these abnormalities could be the result of a newly discovered toxicological mechanism, the antiandrogen effect. The new effect was discovered while the EPA toxicologists were trying to figure out why the fungicide vinclozolin was causing unusual penile development and loss of fertility in male rats (Gray et al. 1999a). Male pups of pregnant rats fed vinclozolin are born with femalelike anogenital distances, cleft phalluses with hypospadias, very small or no penises, vaginal sacs, undescended testicles, and permanent nipples or areolas. Because DDE, the breakdown product of DDT, was found in the alligators, the EPA tested DDE for its antiandrogen effect. Up to that time, traditional toxicological testing had missed this effect on male sexual development. The EPA team demonstrated that in utero exposure to DDE causes abnormal development of the urogenital tract in male rat pups, hypospadias, shortened ano-genital distance, and undescended testicles (Kelce et al. 1995). This startling discovery took place 54 years after DDT first went on the market. The EPA team has since added linuron (herbicide), procymidone (fungicide), methoxychlor (a widely used insecticide that replaced DDT), PCB 169, dioxin, and the plastic additives diethylhexyl-phthalate (DEHP), di-*n*-butylphthalate (DBP), and di-iso-nonylphthalate (DINP) to the list of widely used chemicals that cause hypospadias in rats (Gray et al. 1999b). They also state that there is no threshold for some of the antiandrogen effects they discovered.

As with the first study, the wildlife findings prompted concerns about human development. In response to the alligator findings, epidemiologists at the Centers for Disease Control and Prevention searched through the available health registry data on hypospadias, a reportable visible birth defect in male newborns, to determine the incidence of hypospadias in the United States (Paulozzi et al. 1997). In humans, hypospadias is a condition in which the urethra does not open at the end of the penis and may open anywhere along the shaft. They found that hypospadias had doubled in the U.S. population between 1970 and 1990, and approximately one in 125 boys was affected in 1990. The incidence of the most serious form of hypospadias, in which the urethra opens in the scrotum, has increased the most (Paulozzi et al. 1997). Hypospadias can only be triggered between days 56 and 80 of gestation in humans, when the urethra and penis begin to enlarge and lengthen simultaneously (Duckett and Baskin 1996). It is now generally accepted that hypospadias increases the odds 4.2-fold in adulthood of reduced sperm production and quality, and testicular cancer (Prener et al. 1995). Overall, testicular cancer incidence increased 51% from 1973 to 1995 in the United States (McKiernan et al. 1999). In this case, a trail can be traced from day 56 in the uterus to adulthood, when a delayed effect is expressed as cancer of a secondary sex organ. To date, there is no similar model from the uterus to the expression of breast cancer in females.

6.4 Conclusions

Table 6.1 lists reported associations between hormonally sensitive tissue in wildlife and environmental effects. Most of the effects are not immediately lethal; nevertheless, many could shorten an animal's life span. In terms of function, however, some of the effects are on reproductive success and population growth.

The recent increase of peer-reviewed reports of associations between endocrine-disrupting chemicals and damage heretofore not reported in wildlife demonstrates the need for a number of changes in how society has traditionally protected the well-being of all animals, including humans (Jobling et al. 1998; Fairchild et al. 1999). These studies, including the two studies discussed in this

Table 6.1. Associations between hormonally responsive tissue in wildlife and environmental effects

Increases	Decreases
Size of thyroid	Embryo hatchability
Size of liver	Embryo viability
Liver enzyme induction	Chick survivorship
Spontaneous abortions	Fry survivorship
Abnormal external genitalia	Egg size and number
Delayed or abnormal development of secondary sex organs	Number of animals reaching sexual maturity
Atrophied gonads	Thyroid hormone production
Hermaphrodism	Estrogen production
Sex-skewed birth defects	Testosterone production
Asymmetrical brains	Retinoids production
Asymmetrical skulls	Immune competency
Unusual behavior	

chapter, highlight the importance of creating new partnerships to investigate the overlooked effects of endocrine disruption on development and reproduction. They emphasize the urgency for more large-scale, long-term prospective studies (at least a generation, 20 years or more) of the developmental and functional health of humans, wildlife, and domestic animals. Complementary ongoing laboratory investigations must be a part of these large-scale efforts. Such collaborations require new and increased input from medical practitioners, public health authorities, medical institutions, veterinarians, and the public, all sharing ownership of the studies. From the earliest planning stages, it must be made clear that in order to distinguish endocrine disruption effects from all the other possible confounding stressors, all efforts will require a long-term commitment by all investigators involved as well as that of government funder(s). It is time to start building these sorely needed cross-discipline collaborations to direct research toward preventive measures rather than pouring vast resources into research to develop treatments and cures for damage that took place before birth and that cannot be repaired.

The wealth of information on endocrine disruption that has evolved over the past decade also highlights the need to include toxic chemicals in wildlife management algorithms for determining risk. It has been repeatedly demonstrated that very low doses of synthetic chemicals, at environmental and physiological concentrations, can have dire effects on reproduction, behavior, maturation, and survival. Managers should keep in mind the Great Lakes dioxin findings and the surprising and disturbing dioxin and furan contribution to the toxicity measured in the North Pacific albatrosses. Even vast, remote sanctuaries and protected areas are vulnerable to these stealth chemicals that invade the females and target their embryos.

In the near term, as veterinarians are called to necropsy beached and dead marine mammals, it is time to probe beyond the obvious or primary cause of death (viral, bacterial, or starvation). Monitoring schemes as described by Aguirre et al. (chap. 7) are required on a global scale to assay tissues for environmental contaminants. It is necessary to keep in mind that the thyroid plays an important role in metabolism, intelligence, and behavior and that the endocrine system is integrated with the immune system. An elegant series of pinniped studies from the Netherlands has already made connections between the immune system and the OCs (de Swart 1995; Ross 1995; Visser 1993). Perhaps it is time to collect the cochlea as well as bone samples of beached whales, while at the same time collecting samples for monitoring their contaminant loading, and to closely examine the whales for the same effects that are found in laboratory animals when treated with the same chemicals found in the whales' tissue. Examination of the bony tissue might lead to answers about the development of the whale auditory system or calcium metabolism that are affected by PCBs and other synthetic chemicals (Goldey and Crofton 1998).

As one group of scientists noted, "Populations of many long-lived species are declining, some to the verge of extinction, without society's knowledge. The presence of breeding adults and even healthy young does not necessarily reflect a healthy population. Detailed population analysis is needed to determine whether offspring have the functional capacity to survive and reproduce" (Bantle et al. 1995). As Alleva et al. (1998) concluded after reviewing the literature on endocrine disruption and its in utero effects on intelligence and behavior, "Widespread loss of this nature can change the character of human societies and destabilize wildlife populations," and society may never understand what has happened.

References

Alleva, E., J. Brock, A. Brouwer, T. Colborn, M.C., Fossi, E. Gray, L. Guillette, P. Hauser, J. Leatherland, N. MacLusky, A. Mutti, P. Palanza, S. Parmigiani, S. Porterfield, R. Santti, SA. Stein, F. vom Saal, B. Weiss. 1998. Statement from the work session on environmental endocrine-disrupting chemicals: neural, endocrine, and behavioral effects. Toxicol Indust Health 14:1–8.

Arctic Monitoring and Assessment Programme. 1997. Arctic pollution issues: a state of the Arctic environment report. Arctic Monitoring and Assessment Programme, Oslo.

Auman, H.J., J.P., Ludwig, C.L., Summer, D.A. Verbrugge, K.L. Froese, T. Colborn, and J.P. Giesy. 1997. PCBs, DDE, DDT and TCDD-eq in two species of albatross on Sand Island, Midway Atoll, North Pacific Ocean. Environ Toxicol Chem 16:498–504.

Bantle, J., W.W. Bowerman, C. Carey, T. Colborn, S. DeGuise, S. Dodson, C.F. Facemire, G. Fox, M. Fry, M. Gilbertson, K. Grasman, T. Gross, L. Guillette, C. Henny, D.S. Henshel, J.E. Hose, P.A. Klein, T.J. Kubiak, G. Lahvis, B. Palmer, C. Peterson, M. Ramsay, and D. White, 1995. Statement from the work session on environmentally induced alterations in development: a focus on wildlife. Environ Health Perspect 103(suppl 4):3–5.

Béland, P. 1996. The beluga whales of the St. Lawrence River. Sci Am, 274:74–81.

Bern, H., P. Blair, S. Brasseur, T. Colborn, G.R. Cunha, W. Davis, K.D. Dohler, G. Fox, M. Fry, E. Gray, R. Green, M. Hines, T.J. Kubiak, J. McLachlan, J.P. Myers, R.E. Peterson, P.J.H. Reijnders, A. Soto, G. Van Der Kraak, F. vom Saal, and P. Whitten, 1992. Statement from the work session on chemically-induced alterations in sexual development: the wildlife/human connection. In Colborn, T., and C. Clement, eds. Chemically induced alterations in sexual and function development: the wildlife/human connection, pp. 1–8. Scientific Publishing, Princeton, N.J.

Bowerman, W.W. IV. 1993. Regulation of bald eagle (Haliaeetus leucocephalus) productivity in the Great Lakes Basin: an ecological and toxicological approach. PhD thesis, Michigan State University.

Brouwer, A., U.G. Ahlborg, M. Van den Berg, L.S. Birnbaum, E.R. Boersma, E. Bosveld, M.S. Denison, L.E. Gray, L. Hagmar, E. Holene, M. Huisman, S.W. Jacobson, J.L. Jacobson, C. Koopman-Esseboom, J.G. Koppe, B.M. Kulig, D.C. Morse, G. Muckle, R.E. Peterson, P.J.J. Sauer, R.F. Seegal, A.E. Smits-Van Prooije, B.C.L. Touwen, N. Weisglas-Kuperus, and G. Winneke, 1995. Functional aspects of developmental toxicity of polyhalogenated aromatic hydrocarbons in experimental animals and human infants. Eur J Pharmacol 293:1–40.

Brucker-Davis, F. 1998. Effects of environmental synthetic chemicals on thyroid function. Thyroid 8:827–856.

Chernyak, S.M., C.P. Rice, and L.L. McConnell. 1996. Evidence of currently used pesticides in air, ice, fog, seawater and surface microlayer in the Bering and Chukchi Seas. Mar Poll Bull 32:410–419.

Colborn, T. 1991. Epidemiology of Great Lakes bald eagles. J Toxicol Environ Health 33: 395–453.

Colborn, T. and C.R. Clement, eds. 1992. Chemically-induced alterations in sexual and functional development: the wildlife/human connection. Princeton Scientific Publishing, Princeton, N.J.

Colborn, T. and M.J. Smolen. 1996. Epidemiological analysis of persistent organochlorine contaminants in cetaceans. Rev Environ Contam Toxicol 146:92–171.

Conover, M., and G. Hunt. 1984. Female–female pairing and sex ratios in gulls: an historical perspective. Wilson Bull 96:619–625.

Cook, P.M. 1999. Can ecotoxicology define threats to fish populations in the Great Lakes? *In* Great Lakes Science Advisory Board's meeting to assess scientific issues in relation to lakewide management plans (25–26 February 1999), pp. 10–12. International Joint Commission, Windsor, Ontario.

Cook, P.M., E.W. Zabel, and R.E. Peterson. 1995. The TCDD toxicity equivalence approach for characterizing risks for early life-stage mortality in trout. *In* Rolland, R.M., M. Gilbertson, and R.E. Peterson, eds. Proceedings from a session at the Wingspread Conference Center: Chemically induced alterations in functional development and reproduction of fishes (21–23 July 1995; Racine, Wisc.), p. 124. Society of Environmental Toxicology and Chemistry Press, Pensacola, Fla.

Daly, H.B. 1993. Laboratory rat experiments show consumption of Lake Ontario salmon causes behavioral changes: support for wildlife and human research results. J Great Lakes Res 19:784–788.

de Swart, R.L. 1995. Impaired immunity in seals exposed to bioaccumulated environmental contaminants. PhD thesis, Erasmus Universiteit, Rotterdam.

Dewailly, E., P. Ayotte, S. Bruneau, C. Laliberte, D.C.G. Muir, and R.J. Norstrom. 1993. Human exposure to polychlorinated biphenyls through the aquatic food chain in the Arctic. Vol. 14: Organohalogen compounds. Paper presented at Dioxin '93: 13th International symposium on chlorinated dioxins and related compounds, September 1993, Vienna, Austria.

Duckett, J.W., and L.S. Baskin. 1996. Hypospadias. *In* J. Gillenwater, J. Grayback, S. Howard, J. Duckett, eds. Adult and pediatric urology, 3rd ed. pp. 2549–2589. Mosby, St. Louis.

Fairchild, W.L., E.O., Swansburg, J.T. Arsenault, and S.B. Brown. 1999. Does an association between pesticide use and subsequent declines in catch of Atlantic salmon (*Salmo salar*) represent a case of endocrine disruption? Environ Health Perspect 107: 349–357.

Fox, G., S. Teeple, A. Gilman, F. Anderka, and G. Hogan. 1976. Are Lake Ontario herring gulls good parents? *In* Proceedings of the fish-eating birds of the Great Lakes and environmental contaminants symposium (2–3 December 1976), pp. 76–90. Toxic Chemical Division and Ontario Region Canadian Wildlife Service, Toronto.

Fox, G.A., S. Trudae, H. Won, and K.A. Grasman. 1998. Monitoring the elimination of persistent toxic substances from the Great Lakes; chemical and physiological evidence from adult herring gulls. Environ Monit Assess 53:147–168.

Fry, D., and M. Toone. 1981. DDT-induced feminization of gull embryos. Science 213: 922–924.

Garry, V.F., D. Schreinemachers, M.E. Harkins, and J. Griffith. 1996. Pesticide appliers, biocides, and birth defects in rural Minnesota. Environ Health Perspect 104:394–399.

Giesy, J.P., J.P. Ludwig and D.E. Tillitt. 1994. Deformities in birds of the Great Lakes region: assigning causality. Environ Sci Technol 28:128–135.

Gilbertson, M. G.A. Fox, and W.W. Bowerman, eds. 1998. Trends in levels and effects of persistent toxic substances in the Great Lakes. *In* Articles from the workshop on environmental results, hosted in Windsor, Ontario, by the Great Lakes Science Advisory Board and the International Joint Commission (12–13 September 1996), pp. 65–75. Kluwer, Dordrecht.

Goldey, E.S., and K.M. Crofton. 1998. Thyroxine replacement attenuates hypothyroxinemia, hearing loss, and motor deficits following developmental exposure to Aroclor 1254 in rats. Toxicol Sci 45:94–105.

Goldey, E.S., L.S. Kehn, C. Lau, G.L. Rehnberg, and K.M. Crofton. 1995a. Developmental exposure to polychlorinated biphenyls (Aroclor 1254) reduced circulating thyroid hormone concentrations and causes hearing deficits in rats. Toxicol Appl Pharmacol 135: 77–88.

Goldey, E.S., L.S. Kehn, G.L. Rehnberg, and K.M. Crofton. 1995b. Effects of developmental hypothyroidism on auditory and motor function in the rat. Toxicol Appl Pharmacol 135:67–76.

Government of Canada. 1991. Toxic chemicals in the Great Lakes and associated effects, vol. 2. Government of Canada, Cat. No. En 37-95/1990-1E. Ontario.

Gray, L.E., Jr., J. Ostby, E. Monosson, and W.R. Kelce. 1999a. Environmental antiandrogens: low doses of the fungicide vinclozolin alter sexual differentiation of the male rat. Toxicol Indust Health 15:48–64.

Gray L.E. Jr., C. Wolf, C. Lambright, P. Mann, M. Price, R.L. Cooper, and J. Ostby. 1999b. Administration of potentially antiandrogenic pesticides (procymidone, linuron, iprodione, chlozolinate, p,p'-DDE, and ketoconazole) and toxic substances (dibutyl- and diethylhexylphthalate, PCB 169, and ethane dimethane sulphonate) during sexual differentiation produces diverse profiles of reproductive malformations in the male rat. Toxicol Indust Health 15:94–118.

Guillette, L.J., Jr. 1998. Endocrine disrupting contaminants and alligator embryos: a lesson from wildlife? In Eisenbrand, G., H. Daniel, A.D. Dayan, P.S. Elias, W. Grunow, F.H. Kemper, F.H. Kemper, E. Loser, M. Metzler, and J. Schlatter, eds. Symposium: hormonally active agents in food, pp. 72–88. Deutsche Forschungsgemeinschaft, Wiley-VCH. Bonn, Germany.

Guillette, L.J. Jr., and D.A. Crain. 1996. Endocrine-disrupting contaminants and reproductive abnormalities in reptiles. Comments Toxicol 5:381–399.

Guillette, L.J. Jr., T.S. Gross, G.R. Masson, J.M. Matter, H.F. Percival, and A.R. Woodward. 1994. Developmental abnormalities of the gonad and abnormal sex hormone concentrations in juvenile alligators from contaminated and control lakes in Florida. Environ Health Perspect 102:680–688.

Haddow, J.E., G.E. Palomaki, W.C. Allan, J.R. Williams, G.J. Knight, J. Gagnon, C.E. O'Heir, M.L. Mitchell, R.J. Hermos, S.E. Waisbren, J.D. Faix, and R.Z. Klein. 1999. Maternal thyroid deficiency during pregnancy and subsequent neuropsychological development of the child. N Engl J Med 341:549–555.

Herr, D.W., E.S. Goldey, and K.M. Crofton. 1996. Developmental exposure to Aroclor 1254 produces low-frequency alterations in adult rat brainstem auditory evoked responses. Fundam Appl Toxicol 33:120–128.

Hose, J.E., J.N. Cross, S.G. Smith, and D. Dichl. 1989. Reproductive impairment in a fish inhabiting a contaminated coastal environment off southern California. Environ Pollut 57:139–148.

Hunt, G. and M. Hunt. 1977. Female–female pairing in western gulls (Larus occidentalis) in southern California. Science 196:1466–1467.

International Air Quality Advisory Board. 1988. Progress Report #5 to the International Joint Commission, International Air Quality Board, Downsview, Ontario.

Jacobson, J.L. and S.W. Jacobson. 1993. A 4-year follow-up study of children born to consumers of Lake Michigan fish. J Great Lakes Res 19:776–783.

Jacobson, J.L. and S.W. Jacobson, 1996. Intellectual impairment in children exposed to polychlorinated biphenyls in utero. N Engl J Med 335:783–789.

Jacobson, J.L., G.G. Fein, P.M. Schwartz, and J.K. Dowler 1984. Prenatal exposure to an environmental toxin: a test of the multiple effects model. Dev Psychol 20:523–532.

Jacobson, J.L., S.W. Jacobson, and H.E.B. Humphrey. 1990. Effects of in utero exposure to polychlorinated biphenyls and related contaminants on cognitive functioning in young children. J Pediatr 116:38–45.

Jensen, A.A., and S.A. Slorach, eds. 1990. Chemical contaminants in human milk. CRC Press, Boca Raton, Fla.

Jobling, S., M. Nolan, C.R. Tyler, G. Brighty, and J.P. Sumpter. 1998. Widespread sexual disruption in wild fish. Environ Sci Technol 32:2498–2506.

Johnson, B.L., H.E. Hicks, and C.T. De Rosa. 1999. Key environmental human health issues in the Great Lakes and St. Lawrence River basins. Environ Res 80:S2–S12.

Kelce, W., C. Stone, S. Laws, L. Grey, J.A. Kempppainen, and E. Wilson. 1995. Persistent DDT metabolite *p,p'*-DDE is a potent androgen receptor antagonist. Nature 375:581–585.

Koopman-Esseboom, C., D.C. Morse, N. Weisglas-Kuperus, I.J. Lutke-Schipholt, C.G. Van der Paauw, L.G.M.T. Tuinstra, A. Brouwer, and P.J.J. Sauer. 1994. Effects of dioxins and polychlorinated biphenyls on thyroid hormone status of pregnant women and their infants. Pediatr Res 36:468–473.

Koopman-Esseboom C., N. Weisglas-Kuperus, M.A.J. de Ridder, C. Van der Paauw, L.G.M.T. Tuinstra, and P.J.J. Sauer. 1996. Effects of polychlorinated biphenyl/dioxin exposure and feeding type on infants' mental and psychomotor development. Pediatrics 97:700–706.

Leatherland, J.F. 1998. Changes in thyroid hormone economy following consumption of environmentally contaminated Great Lakes fish. Toxicol Environ Health 14:41–57.

McKiernan, J.M., E.T. Goluboff, G.L. Liberson, R. Golden, and H. Fisch. 1999. Rising risk of testicular cancer by birth cohort in the United States from 1973 to 1995. J Urol 162:361–363.

Norstrom, R.J., S.E. Belikov, E.W. Born, G.W. Garner, B. Malone, S. Olpinski, M.A. Ramsay, S. Schliebe, I. Stirling, M.S. Stishov, M.K. Taylor, and O. Wiig. 1998. Chlorinated hydrocarbon contaminants in polar bears from eastern Russia, North America, Greenland, and Svalbard: biomonitoring of Arctic pollution. Arch Environ Contam Toxicol 35:354–367.

Patandin, S., N. Weisglas-Kuperus, M.A.J. de Ridder, C. Koopman-Esseboom, W.A. van Staveren, C.G. van der Paauw, and P.J.J. Sauer. 1997. Plasma polychlorinated biphenyl levels in Dutch preschool children either breast-fed or formula-fed during infancy. Am J Public Health 87:1711–1714.

Paulozzi, L.J., J.D. Erickson, and R.J. Jackson. 1997. Hypospadias trends in two U.S. surveillance systems. Pediatrics 100:831–834.

Prener, A., G. Engholm, and O.M. Jensen. 1995. Genital anomalies and risk for testicular cancer in Danish men. Epidemiology 7:14–19.

Ross, P.S. 1995. Seals, pollution and disease: environmental contaminant-induced immunosuppression. PhD thesis. Universiteit Utrecht.

Short, P. and T. Colborn 1999. Pesticide use in the U.S. and policy implications: a focus on herbicides. Toxicol Indust Health 15:240–275.

Society of the Plastics Industry. 1996. Facts and figures of the U.S. plastics industry. Washington, D.C.

Stewart, P., J. Pagano, D. Sargent, T. Darvill, E. Lonky, and J. Reihman. 2000a. Effects of Great Lakes fish consumption on brain PCB pattern, concentration, and progressive-ratio performance. Environ Res 82:18–32.

Stewart, P., J. Reihman, E. Lonky, T. Darvill, and J. Pagano. 2000b. Prenatal PCB exposure and Neonatal Behavioral Assessment Scale (NBAS) performance. Neurotoxicol Teratol 22:21–29.

U.S. Environmental Protection Agency (EPA). 1998. Endocrine Disruptor Screening and Testing Advisory Committee (EDSTAC) final report. Vols. 1 and 2. U.S. EPA, Washington, D.C.

Visser, I.K.G. 1993. Morbillivirus infections in seals, dolphins, and porpoises. PhD thesis, Universiteit Utrecht.

vom Saal, F.S., M.M. Montano, and M.H. Wang. 1992. Sexual differentiation in mammals. *In* Colborn, T., and C. Clement, eds. Chemically induced alterations in sexual and functional development: the wildlife/human connection, pp. 17–38. Princeton Scientific Publishing, Princeton, N.J.

Wams, T.J. 1987. Diethylhexylphthalate as an environmental contaminant: a review. Sci Total Environ 66:1–16.

Weis, I.M., and D.C.G. Muir. 1997. Geographical variation of persistent organochlorine concentrations in blubber of ringed seal (*Phoca hispida*) from the Canadian Arctic: univariate and multivariate approaches. Environ Pollut 96:321–333.

Weisglas-Kuperus, N. T.C.J. Sas, C. Koopman-Esseboom, C.W. Van Der Zwan, M.A.J. de Ridder, A. Beishuizen, H. Hooijkaas, and P.J.J. Sauer. 1995. Immunologic effects of background prenatal and postnatal exposure to dioxins and polychlorinated biphenyls in Dutch infants. Pediatr Res 38:404–410.

Welshons, W.V., S.C. Nagel, K.A. Thayer, B.M. Judy, and F.S. vom Saal. 1999. Low-dose bioactivity of xenoestrogens in animals: fetal exposure to low doses of methoxychlor and other xenoestrogens increases adult prostate size in mice. Toxicol Indust Health 15:12–25.

Part II

Monitoring Ecological Health

7

Monitoring the Health and Conservation of Marine Mammals, Sea Turtles, and Their Ecosystems

A. Alonso Aguirre
Todd M. O'Hara
Terry R. Spraker
David A. Jessup

Marine ecosystem health is a relatively new and poorly defined concept. The enormous biological diversity harbored by the oceans and the lack of taxonomic and ecological studies limit our ability to define a healthy marine ecosystem. It is generally agreed, however, that healthy ecosystems are those that have stable, or at least not declining, species abundance and diversity; do not have obvious environmental degradation, frequent pollution events, or unsustainable harvests; do not have a high frequency of emerging or reemerging diseases/intoxications with negative implications for human and wildlife health; and do not have frequent die-offs or similar stochastic events, particularly those involving "indicator" or "keystone" species.

Marine ecosystems are undergoing multiple concurrent stressors that are clearly affecting marine vertebrate health. The relatively recent, drastic accelerated transformation of coastal ecosystems in most instances has not been quantified; however, the loss of critical habitat is apparent worldwide. As coastal areas are developed, marine species tend to concentrate in smaller feeding, mating, nesting, or haul-out areas, making them more susceptible to human fisheries interactions, infectious agents, algal blooms, and environmental pollution.

The scale of the oceans makes it impossible to assess the health status of the marine environment in its entirety. Even focal subsets of marine areas such as the Hawaiian Islands or the California coast do not diminish issues of scale. Adequate scientific investigations and monitoring programs are necessary on a global scale to assess the present and future health of these fragile ecosystems. Nationwide resource agencies, universities, and nongovernmental organizations are attempting to move toward the management of ecosystems and their health while incorporating the traditional concept of individual species management.

This concept clearly applies to both marine resources currently exploited for human consumption, and those requiring protection and conservation, including marine mammals and sea turtles. Recently, mortality events involving different species of vertebrates at various trophic levels have been used as a measure of the health of the Gulf of Mexico and Atlantic Ocean.

Marine mammals, particularly indicator species that are obligate inhabitants of near-shore ecosystems, may serve as sentinels of disturbances that reflect changes in health of coastal marine environments. They represent key animal groups to monitor marine ecosystem health since most occupy higher levels in the food web and are considered top predators and/or represent important integrators of a system (i.e., productivity, sustainability). Also, they tend to bioaccumulate and bioconcentrate lipid-soluble environmental contaminants, and their condition is tied to ecosystem health. The evolutionary relatedness to some terrestrial mammalian groups (i.e., Carnivora, Bovidae) has made possible the development of diagnostic tests and immunologic assays applicable to marine species. Also the knowledge acquired from captive cetaceans and pinnipeds has allowed the design and development of safe and humane veterinary techniques for sampling and monitoring.

Recent evidence indicates that marine turtles may be particularly vulnerable to a variety of environmental insults such as high temperatures, pollutants, infectious agents, and marine biotoxins. Harmful effects include compromised physiology, impaired immune function, and an increase in disease prevalence. The high profile of marine turtles—the extensive sampling programs, and stranding networks already in place offer an effective platform for monitoring the health of marine turtle populations and, in so doing, monitoring the health of the ecosystems they inhabit.

Monitoring health of marine mammals and sea turtles has been conducted through several sources of specimens in the United States. These include monitoring of stranded marine mammals (e.g., many species of pinnipeds and dolphins) and sea turtles, monitoring of harvestable populations (e.g., native Alaskan subsistence harvests of bowhead whales and northern fur seals), and health assessment programs to monitor condition and disease of specific populations (e.g., Hawaiian monk seals, sea otters). The objective of this chapter is to integrate and describe existing monitoring programs and techniques used in surveying marine mammals and sea turtles and their ecosystems and to evaluate whether these techniques are useful for monitoring environmental and marine ecosystem health. Focusing on marine wildlife that could serve as sentinels for the quality of health in the marine environment, in this chapter we examine species that have documented health problems during stranding situations, are threatened with extinction, and are important for human health during harvest.

7.1 Monitoring of Stranded Marine Mammals

The Marine Mammal Health and Stranding Response Program (MMHSRP) was established through congressional amendment to the Marine Mammal Protection

Act in 1992, following the 1987–1988 die-off of Atlantic bottlenose dolphins (*Tursiops truncatus*). The MMHSRP was established to facilitate the dissemination of reference data on health trends in marine mammal populations, to correlate the health of marine mammals with environmental parameters, and to coordinate effective responses to unusual mortality events. One component of the program, the U.S. Stranding Network, consists of regional teams that respond to strandings of marine mammals and are equipped to collect biological information and samples that can be used to understand health, contaminant levels, population dynamics, and life histories of marine mammals. A near real-time reporting system facilitates the rapid detection of unusual mortality events. The second component is the Working Group on Unusual Marine Mammal Mortality Events. The National Marine Fisheries Service has established a national contingency plan for the investigation of such events supported by the working group (Wilkinson 1996). Several programs have been developed in conjunction with other agencies for contaminants and tissue banking, including the development of protocols for heavy metals, organochlorine contaminants, oil spills/hydrocarbons, and algal blooms.

During rehabilitation while are under veterinary care, stranded marine mammals can provide a wealth of information on the health status of a population and contribute to the development of novel techniques for the medical care of marine mammals, as well as facilitate public education. For example, following admission of a pinniped to rehabilitation, the animal receives a physical examination. Blood and fecal samples are collected for hematology, serum biochemistry, serum banking, and parasitologic evaluation. If an animal dies during rehabilitation, a complete necropsy including histopathologic and microbiologic examinations is performed, and tissues are banked for contaminant analyses. Results of these studies assist in identifying, in addition to the cause of death, health problems in the free-ranging population. A number of important pathogens that can cause epizootics in marine mammals were first identified in stranded seals, including phocine distemper virus, phocine herpesvirus, and *Leptospira interrogans* var *pomona*. The marine biotoxin domoic acid produced by the alga *Pseudonitzchia australis* was also identified in stranded California sea lions (*Zalophus californianus*), in which it caused severe seizures and hippocampal necrosis (Scholin et al. 2000). Necropsy of California sea lions dying in rehabilitation allowed the detection of metastatic carcinoma in 18% of sexually mature sea lions examined, drawing attention to a health problem in this free-ranging population (Gulland et al. 1996). Stranded marine mammals can thus be sentinels of health for wild pinniped populations, especially since diseased animals are more likely to strand and thus render the disease more likely to be detected. Because sampling is far from random, and because the sex, age, and temporal and geographic distributions of stranded animals are highly skewed, results from disease surveys must be interpreted with caution. Direct parallels in disease prevalence between stranded animals and free-ranging populations cannot be drawn; however, temporal and spatial changes may be important when considering unusual mortalities.

7.2 Monitoring of Fibropapillomatosis in Sea Turtles

Marine turtle fibropapillomatosis (FP) is by far one of the most important health problems affecting free-ranging sea turtles (George 1997). It is a condition characterized by multiple cutaneous tumor masses ranging from 0.1 to more than 30 cm in diameter that has primarily affected green turtles (*Chelonia mydas*) worldwide (Jacobson et al. 1989; Aguirre et al. 1994b). Although reported since the late 1930s in Florida, it was not until the late 1980s that it reached epizootic proportions in several sea turtle populations. The condition has been confirmed histopathologically in all other species, including loggerhead (*Caretta caretta*) (Herbst 1994), olive ridley (*Lepidochelys olivacea*) (Aguirre et al. 1999a), Kemps ridley (*L. kempii*) (Harshbarger 1991), leatherback (*Dermochelys coriacea*) (Huerta et al. 2002), and hawksbill turtles (*Eretmochelys imbricata*) (D' Amato and Moraes-Neto 2000). Flatback turtles (*Natator depressus*) in Australia were also reported with tumors by gross examination (Limpus and Miller 1994).

The disease has a circumtropical distribution and has been observed in all major oceans. Prevalence of FP varies among locations, ranging from as low as 1.4% to as high as 90%. Studies have shown that pelagic turtles recruiting to near-shore environments (about 35 cm straight carapace length) are free of the disease. After exposure to these near-shore feeding environments and interaction with other individuals, FP manifests itself with primary growths in the corner of the eyes (Aguirre et al. 1994b; Keupper-Bennett et al. 2002).

Although several viruses have been identified with the tumors, the primary etiological agent remains to be isolated and identified. Green and loggerhead turtles from various locations in Florida have been diagnosed with FP, and experimental transmission studies have demonstrated the involvement of a filterable, subcellular agent of an infectious nature. Direct experimental inoculation of turtles has consistently demonstrated a horizontal transmission, and tumors have developed at the site of inoculation in turtles free of the disease within a year of postinoculation (Herbst et al. 1995).

The potential interactions and epidemiological links regarding the causal hypotheses of FP have been the subject of discussion. Although there is convincing evidence of a viral etiology, other cofactors including parasites (Dailey and Morris 1995, Aguirre et al. 1998), genetic susceptibility, chemical carcinogens, environmental pollutants (Aguirre et al. 1994a), biotoxins (Landsberg et al. 1999), immunosuppression (Aguirre et al. 1995; Work et al. 2000), ultraviolet light, and other potential pathogens (Aguirre et al. 1994b) may play additional roles in the etiology of FP. Research in captivity suggests that immunosuppression may not be required for the manifestation of the disease. Polymerase chain reaction techniques were applied on a large number of freshly isolated tumor samples failed to detect papillomavirus. The molecular identification of one or more herpesviruses (Herbst et al. 1996, 1999, Quakenbush et al. 1998; Lackovich et al. 1999; Lu et al. 2000a), a papillomalike virus (Lu et al. 2000b), and a retrovirus (Casey et al. 1997) in Hawaiian green turtles has complicated the scenario to determine the primary cause of tumor formation.

The status of FP in Hawaii has been substantiated primarily by turtle strandings and long-term studies on manifestation of the disease in foraging pastures at Kaneohe Bay, Island of Oahu; Palaau, Island of Molokai; and the entire west coast of the Island of Hawaii, where the disease is absent (Balazs et al. 2000). The Hawaiian green turtle nesting population at French Frigate Shoals has increased since its protection 20 years ago. No population effects caused by FP have been detected in the number of nesting females, with 1997 being the highest nesting season ever recorded. The first documented case of FP for the Hawaiian Islands occurred in 1958 in Kaneohe Bay. Between 1989 and 1997, 581 green turtles were captured alive and tagged in this bay. Mild to severe FP was reported in 44% of the turtles handled, of which 17% presented oropharyngeal tumors. Annual prevalence of the disease has ranged from 42% to 65%, with no consistent trend observed. Growth rates for turtles with severe tumors were significantly lower than for turtles free of FP–1.0 versus 2.2 cm per year, straight carapace length (Balazs et al. 2000).

Over 80% of stranded turtles necropsied since 1992 have presented with FP and cardiovascular trematodes (Aguirre et al. 1998). Turtles with FP are hypoproteinemic, with an extremely low packed cell volume and severe emaciation (Aguirre and Balazs 2000; Work and Balazs 1999). Approximately 25–30% of turtles with external tumors present internal tumors diagnosed as fibromas, myxofibromas (Norton et al. 1990), and fibrosarcomas of low-grade malignancy primarily in lung, kidney, and heart (Aguirre et al. 2001). In addition, tumors are commonly identified in the nasopharyngeal region (Balazs et al. 1997).

Field observations support that prevalence of the disease has been associated with heavily polluted coastal areas, human areas of high density, agricultural runoff, and/or biotoxin-producing algae. Marine turtles can serve as excellent sentinels of ecosystem health in these near shore environments. It is possible that FP can be used as an indicator of ecosystem health, but correlations with physical and chemical characteristics of water and other factors need to be made in order for this to happen. Further research in identifying the etiologic agent and its association to other environmental variables can provide sufficient parameters to measure the health of coastal marine ecosystems, which serve not only as sources of ecotourism but also as primary feeding areas for sea turtles.

7.3 Monitoring of Harvestable Populations

7.3.1 Bowhead Whales and the Alaskan Arctic Ecosystem

The Cooperative Institute for Arctic Research (CIFAR) at the University of Alaska Fairbanks identified two areas of research priorities for the Western Arctic/Bering Sea ecosystem: the study of natural variability and the monitoring of anthropogenic influences. Both of these are components of "ecosystem health." Conservation and health preservation of the bowhead whale (*Balaena mysticetus*) are critical because it is classified as an endangered species and is an important subsistence species to many coastal villages of the Bering, Chukchi, and Beaufort seas (O'Hara et al. 1999).

The Arctic Environmental Protection Strategy (AEPS) and the Arctic Monitoring and Assessment Program (AMAP) represent an effort by the eight circumpolar Arctic countries to maintain a "healthy" arctic ecosystem (AMAP 1995). They are interested in levels and possible effects of heavy metals/minerals (Bratton et al. 1997; Krone et al. 1999), organochlorines (O'Hara et al. 1999), radionuclides (Cooper et al. 2000), and other potential stressors on Arctic marine mammals, especially the bowhead whale. Population trends, blubber indices (quantity and quality), gross examination, histologic examination, presence of lesions, serologic screening, and essential and nonessential elements are analyzed in Eskimo subsistence-harvested bowhead whales. This intensive examination of individual animals is augmented by good concurrent estimations of the population size and trends over the past two decades. Specifically, the relationships (natural variability or interactions) of the above factors (blubber indices, histology, element analyses, "health") are compared for whales harvested in the fall versus the spring, between size/age cohorts (*Ingutuk*, a young whale; subadults; adults), and considering gender/reproductive status. This systematic approach to assessing mysticete health is continually being tested and utilized and will be iteratively optimized for monitoring cetacean health using harvested whales as an indicator of overall population and ecosystem health.

Marine mammals of the Arctic are long-lived, have large lipid depots, and utilize a lipid-dependent food web. Persistent organic pollutants resulting from anthropogenic activities (e.g., pesticides) are found in tissues of bowhead whales. Bowhead whales feeding at a low trophic level tend to have a lower exposure and subsequently a low tissue burden; however, levels do accumulate with age in males (O'Hara et al. 1999). For female bowhead whales, hexachlorobenzene and total lipid levels in blubber decreased and other organochlorine (OC) compound levels did not change significantly with increasing length. Most OC contaminants have low concentrations in tissues of the bowhead whale compared to concentrations in tissues of other cetaceans, especially odontocetes (O'Shea and Brownell 1994; O'Hara et al. 1999). While no adverse effects related to these OC contaminants are predicted, investigation of low-level chronic exposure effects and a more rigorous assessment of histopathology, biomarkers, and immune status in the bowhead whale would be required to conclude "no effect" (O'Hara et al. 1999).

There is no doubt that recent human effects on ecosystems locally and globally have increased the rate at which pollutants are mobilized to and within the biosphere (Pacyna and Keeler 1995). The study of anthropogenic environmental contamination is more difficult for heavy metals such as cadmium and mercury that occur naturally. Regardless of the source, evidence has indicated that relatively high levels of nonessential metals are found in some marine mammal tissues (Wagemann et al. 1990; Muir et al. 1992). Some arctic marine mammals may have evolved proficient physiologic mechanisms to detoxify and tolerate comparatively high concentrations of at least some potentially harmful metals (Dietz et al. 1998). Both cadmium and mercury occur at apparently elevated concentrations in some marine mammal tissues as compared to some domestic and laboratory animals, and consequently they may pose a potential risk to marine mammals or human consumers (Macdonald and Bewers 1996). Concentrations of metals, mechanisms of their accumulation, means whereby they may exert toxicity, and elemental interactions with biological or ecological factors in marine mammals are poorly understood.

Tissues of bowhead whales were measured to establish "normal ranges" of essential and nonessential elements (As, Cd, Co, Cu, Pb, Mg, Mn, Hg, Mo, Se, Ag, and Zn), in a continuing study between 1983 and 1997 (Bratton et al. 1997; Krone et al. 1999). Cobalt, magnesium, manganese, and molybdenum are known to be essential for mammals and have not been found at levels of toxicological concern. Mean manganese levels were significantly greater in liver than in kidney and decrease with age indices (body length). A significant negative correlation was seen between copper and age in liver of bowheads, confirming a phenomenon previously reported in cetaceans (Wagemann et al. 1990). This decline in tissue copper levels could result from loss of copper over time, as might occur through the displacement of copper from metallothionein by zinc or cadmium. Alternatively, copper levels may simply be diluted by increased tissue mass with age or as the tissue-level requirement decreases (i.e., metabolic regulation). Cadmium is known to accumulate with age in both liver and kidney of bowhead whales (Krone et al. 1999). Highest mean concentration of arsenic occurs in blubber, which exceeds concentrations in liver about tenfold, indicating that it is probably in organic form and thus lipid soluble. Hepatic and renal arsenic in bowhead whales is 98% arsenobetaine, which is the predominant form of arsenic in marine animals. These organic arsenic compounds are considered relatively nontoxic to mammals and are apparently excreted unchanged (Bratton et al. 1997; Neff 1997). The age-dependent accumulation of some metals in certain tissues, combined with a long bowhead whale lifespan of 100 years or more, increases the probability that some metals may accumulate to high levels.

The health of the bowhead whale population (in particular, the Bering Sea population) is intimately tied to ecosystem health and the offshore and coastal industrial (i.e., oil) activities. The effects of any development or fisheries activity must be closely monitored for the input of anthropogenic substances (e.g., heavy metals) or human interactions (e.g., trauma) and environmental health. Human interactions including ship strikes and line or net entanglement, have clearly affected bowhead whales (George et al. 1994; Philo et al. 1992, 1993; George and O'Hara 1997). Oil may have acute and chronic effects on bowhead whales, including disruption or injury to the baleen, skin, eyes, respiratory system (volatile components), and gastrointestinal tract. There is much speculation and concern about the relationship of climate change in the Arctic to loss of habitat (sea ice), with possible increase in predation by killer whales and ship activity.

Hematology, biochemistry, and serologic studies in large whales are difficult due to the limitations of blood collection in live animals and the accessibility of fresh specimens. Blood can be an excellent tissue for determining health status (nutrition, stress, infectious agents) with careful interpretation, but logistic constraints have prevented this. Population level comparisons of blood chemistry have been utilized to determine nutritional insufficiencies. Since harvested or stranded bowhead whales are sampled postmortem, many of the blood indicators are not directly comparable to the living animal (Heidel et al. 1996). There are indicators, however, that are stable under these conditions. Serologic studies in bowhead whales include the determination of antibodies to marine caliciviruses (O'Hara et al. 1998; Smith et al. 1987) and other agents, and clinical chemistries (Heidel et al. 1996).

7.3.2 Pinnipeds and the Arctic/Bering Sea Ecosystem

Numerous species of pinnipeds inhabit the Arctic/Bering Sea ecosystem, including gray seals (*Halichoerus grypus*), harbor seals (*Phoca vitulina*), northern fur seals (*Callorhinus ursinus*), ringed seals (*Phoca hispida*), Steller sea lions (*Eumetopias jubatus*), and walrus (*Odobenus rosmarus*). Most marine mammal monitoring programs of the Arctic and Bering Sea are related to population status and trends, diseases, and persistent organic contaminants. Cause–effect relationships of levels of persistent contaminants and their direct link to disease processes, such as reproductive failure or decreased immunity with increased susceptibility to infections within a marine mammal population, have been extremely difficult to assess. In addition, the concentrations of these pollutants in marine mammal tissues and their long-term effects must be interpreted with caution.

The primary monitoring program for harbor seals and Steller sea lions in the northern Pacific Ocean and Bering Sea includes aerial survey population estimates that are performed every other year throughout their range. In addition, over the last 20 years approximately 1,000 Steller sea lions have been hot branded and their location and date of sighting recorded. This program has helped to determine specific ranges for these animals. Satellite radio telemetry has been used with most of these species to further delineate range and diving and feeding behaviors (Loughlin et al. 1994).

The most intensive monitoring programs have been performed on northern fur seals. Various programs monitoring these seals began as early as the 1890s (Lucas 1899). The northern fur seal has extensive recorded history because it has been harvested since the late 1700s for its valuable pelt. The population of these seals was probably five million when the Russian explorer Gerasim Loginovich Pribylov first discovered "the islands of the seals" in 1786, located in the lower half of the Bering Sea (57° 15'N/170° 20'W). Five small islands within a 20-mile radius were later named the Pribilof Islands in his honor. From the early 1800s to the early 1900s, the population went through two severe declines due to commercial harvesting of their pelts (Loughlin et al. 1994). In 1911, the Northern Fur Seal Act was passed to protect the seals from being exploited to extinction by the "sealers" working for fur companies from the United States, Russia, Canada, and Japan. The population slowly rebounded and during the 1950s was more than two million, with approximately 500,000 pups born each year. A deliberate population herd reduction was conducted during 1956–1968, which included the killing of thousands of adult females. This was followed by a decline in the population during the latter half of the 1970s through the present (York and Hartley 1981). Currently, approximately one million pups, juveniles, and adults live on the Pribilof Islands, with approximately 200,000 pups being born each year. At the present time approximately 70% of the world population of northern fur seals inhabits the Pribilof Islands (Loughlin et al. 1994).

The northern fur seal population trends are monitored by bull counts and pup production. Bull counts are performed during the second week of July every year and count the number of territorial bulls with females. Since an average number of females in each territory is estimated, an approximation of the population is made from this count (Fowler 1990). In the pup counts, a calculated number are captured in a specific predetermined area and then marked by shearing a small amount of hair from the top of the head. The marked pups are released into the rookery. One week later marked and unmarked pups are counted in these specific predetermined areas and the ratios of marked to unmarked pups are obtained. This mark-and-recapture technique is a modification of the Lincoln-Peterson index (York and Hartley 1981). As early as 1898, the program began to determine the causes of death and diseases in seals of all age cohorts within the population (Lucas 1899). The present monitoring program started in 1986. Dead pups are collected each year on a daily basis from 4 July to 10 August from rookeries on St. Paul Island, Pribilof Islands. Subadult animals are either found in the rookeries or obtained from the annual subsistence harvest of subadult males currently killed by the Aleut community. The adult animals are found dead in the haul-out areas or within rookeries. A total of 2,410 pups, 87 subadults, 96 adult females, and 29 adult males were examined at necropsy from 1986 to 1999.

Pup mortality is classified as emaciation/starvation, trauma, perinatal mortality, infections, and a specific condition characterized by multifocal necrotizing myopathy and cardiomyopathy. This new condition was noted in 1990 and 1991. It was found in 30% of 364 in 1990 and 2% of 250 pups examined at necropsy in 1991. The etiology of the condition was not determined. During the late 1890s and in the 1950s to 1970s uncinariasis caused by the hookworm *Uncinaria lucasi* was a major cause of death in northern fur seal pups (Lucas 1899; Olsen and Lyons 1965). Between the 1970s and 1990s however, uncinariasis was not important and, in fact, was found in only 15 pups with moderate burdens. One factor that may have played a role in the decline of uncinariasis was that as the population declined, the use of rookeries with a favorable microenvironment for hookworm transmission declined. In the late 1890s, it was noted that uncinariasis was found primarily on clay, sandy, and/or compacted soil types of rookeries and not on the more irregular or rocky types of rookeries (Lucas 1899). Also, in the 1960s animals were primarily collected from the Polovina rookery, which is the sandiest rookery on the island (Olsen and Lyons 1965). The rookeries studied by these earlier workers have fewer seals on them now. There is a suggestion that the number of pups dying from emaciation/starvation may be higher currently when compared to the 1960s and 1970s numbers (Keyes 1965), perhaps reflecting a poor diet quality for lactating adult seals.

The causes of mortality in the subadult male northen fur seals that were examined at necropsy included hyperthermia (41%), trauma/fall from cliff (41%), and entanglement (7%). Other observations included salmon-orange discoloration of blubber in normal animals killed in the subsistence harvest (5%), bite wounds with cellulitis (3%), neoplasia or blunt trauma (3%), and subcutaneous fibromas (2%). All animals had a relatively low level of parasitism. Causes of mortality in the adult females examined at necropsy included dystocia (9%), blunt trauma

(3%), neoplasia (3%), and miscellaneous mortality (3%). Other observations included bite wounds with cellulitis (59%).

7.4 Health Assessment Programs

7.4.1 Hawaiian Monk Seals and the Pacific Reef/Atoll Ecosystem

The Hawaiian monk seal (*Monachus schauinslandi*) is the most endangered pinniped in U.S. waters. The population has been steadily declining since the first rangewide surveys in the late 1950s and most likely will continue to decline as a result of high juvenile mortality and low reproductive recruitment. If this trend continues, it is conceivable that the Hawaiian monk seal could be on the verge of extinction within the next 20 years. Several sources of natural mortality have been described in Hawaiian monk seals, including mobbing (Hiruki et al. 1993a,b); starvation, primarily affecting young seals (Banish and Gilmartin 1992); predation by sharks, particularly tiger sharks (*Galeocerdo cuvier*) and Galapagos sharks (*Carcharhynchus galapagoensis*) (Alcorn and Kam 1986; Hiruki et al. 1993a); net entanglement (Henderson 1990); and disease and trauma (Banish and Gilmartin 1992). The importance of endoparasites as a source of mortality is unknown, although practically all seals harbor nematodes (Dailey et al. 1988). The primary causes of mortality in 65 monk seals between 1981 and 1995 identified on gross necropsy and, in some cases, histopathology included emaciation (35%), trauma (29%), infectious disease (15%), and undetermined mortality (12%). Gastrointestinal parasite burdens (69%) were abnormally high (Banish and Gilmartin 1992). The biotoxins ciguatoxin and mitotoxin have been suspected as causes of mortality in more than 50 seal carcasses found in an advanced state of decomposition at Laysan Island during a die-off in 1978 (Gilmartin et al. 1980).

More recently, rehabilitation efforts have been halted by an ocular condition of unknown etiology affecting 12 female pups brought into captivity during the summer of 1995. Seals were not released into the wild because of the risk of spreading the disease to the wild population. This blinding condition was characterized by blepharoconjunctivitis, photophobia, corneal opacities, and eventual cataract formation. To date, the etiology of this disease remains unknown despite intensive diagnostic investigations (Yantis et al. 1998; Aguirre et al. 2002).

Studies were initiated in the fall of 1996 as part of an epidemiology plan to characterize the role of health and disease in monk seal population dynamics. The plan was also established to determine if monk seals can serve as sentinels to monitor the health of reef/atoll habitats along the Hawaiian Island chain. The Hawaiian Monk Seal Epidemiology Plan incorporates specific strategies intended to enhance recovery and prevent further decline of the species and emphasizes the use of monk seals to monitor environmental health by measuring exposure to infectious agents and contaminants. The eventual intent is to develop an ecosystem health management plan as well as to attempt to recover the species. Health assessment, infectious disease surveillance, and documentation of unusual mortality and exposure to anthropogenic contaminants, spills, biotoxins, and natural disasters can serve to monitor the health of this ecosystem (Aguirre et al. 1999b).

Wild monk seals were sampled to establish normal baseline values for physical condition, hematology, biochemistry, serology, and OC contaminants. These parameters were compared for three subpopulations. The French Frigate Shoals (FFS) subpopulation, especially the juvenile component, has been declining dramatically in recent years, while the Pearl and Hermes (P&H) subpopulation has been growing at a rate of approximately 6–7% annually. Seals from both sexes at FFS were shorter, have smaller girth, and weigh less than do seals at P&H. Hematology results indicated that FFS females had higher mean white blood cell counts than do those at P&H. The mean white blood cell counts were similar among males; however, significant increases in the percentages of neutrophils and declines in the percentages of lymphocytes and eosinophils were found for FFS males. Albumin, creatinine, and Na/K ratio averages were significantly higher among both females and males on FFS. It is suggested that a subclinical disease process may be affecting monk seals at FFS, requiring further study to evaluate habitat quality (Aguirre 2000, 2001).

Serum specimens from 125 seals from FFS, P&H, and Midway Atoll (MID) were tested for selected infectious pathogens known to occur in marine mammals. Seals at these three sites were positive to antibodies against *Brucella abortus* with an overall antibody prevalence of 18% Positive titers for *Chlamydia psittaci* using the complement fixation test averaged 96% for seals tested at all sites. *Toxoplasma gondii* antibodies were identified in two adult females at FFS and MID. These zoonotic agents are known to cause abortion or reproductive impairment in terrestrial and marine mammals. All specimens were serologically negative to *Brucella canis*, *L. pomona* (plus five serovars), caliciviruses, seal influenza virus, canine distemper virus, phocine distemper virus, dolphin morbillivirus, and porpoise morbillivirus, among others. The interpretation of selected serologic results is equivocal, despite the use of standard techniques established at state and national reference laboratories for other species. Long-term studies will characterize the health of this endangered pinniped to reflect the quality of the atoll ecosystem that appears to be declining at some sites (Aguirre 2000, 2001).

7.4.2 Southern Sea Otters and the Kelp Ecosystem

The southern sea otter (*Enhydra lutris nereis*) is listed as threatened under the Endangered Species Act. Once thought to be extinct, this subspecies has increased in number and geographic distribution since the 1920s. This recovery appeared to falter during the late 1970s but resumed in the 1980s after bans moved net fisheries away from most sea otter habitats. The sea otter is a "keystone species" that strongly influences the abundance and diversity of the other species within the kelp forest ecosystem, primarily by preying upon sea urchins. Sea urchins eat the kelp stipe and holdfast and can reduce a kelp forest to a kelp barren without sea otter predation.

The sea otter is also an excellent sentinel species, as it eats approximately 25% of its body weight per day in shellfish and other invertebrates. Since many of these shellfish are used for human food, the role of the sea otter as a bioindicator has implications for human as well as marine animal health. Sea otters are also extremely susceptible to marine pollutants such as petroleum, which may foul or alter their fur's insulating properties, as shown during the 1989 *Exxon Valdez* incident (Monson et al. 2000).

Since 1996 the population has steadily declined by at least 12% annually, apparently due to increased mortality in prime-age adult animals irrespective of the El Niño and La Niña oceanic cycles. Approximately 40% of mortalities are due to infectious diseases and parasites. One of these, protozoal encephalitis caused by *Toxoplasma* spp. and *Neospora* spp., is classified as a new and emerging disease of marine mammals. Two major diseases of the southern sea otter, toxoplasmosis and coccidioidomycosis, or San Joaquin Valley fever, have human health implications and may be related to runoff or non-point-source pollution. The effects and apparently increasing frequency of these diseases may be related to a declining immune function as the animal ages.

Another major mortality factor, infestation with Acanthocephalan parasites and subsequent peritonitis, may be due to diet shifts resulting from a scarcity of preferred prey species. Sea otter population pressures, pollution, reduced ecosystem carrying capacity, or nutrient cycles may be driving these changes. Many environmental contaminants have been identified in southern sea otter tissues, and sea otters with higher pollutant loads have been shown to be more likely to die of infectious diseases than those with lower pollutant loads. In 1998, a toxic algal bloom (*Pseudonitzchia australis*) caused the death of at least one sea otter (and at least 70 sea lions). This type of algal intoxication is very rare in the Pacific Ocean and had not been reported in sea otters prior to this time (Scholin et al. 2000).

Based on the definition of ecosystem health and the clinical signs of a "sick" ecosystem, including indicator species die-offs, frequent pollution events, emerging diseases, and declining species abundance and diversity, the decline of the marine mammal species described herein may indicate declining marine ecosystem health. Conversely, these situations may reflect improved diagnostic capabilities, better cooperation, and increased monitoring and interest in marine animal mortality events. Or, some or all of these factors may be occurring simultaneously.

The development of an integrated regional monitoring program of sentinel vertebrate species as indicators of marine health will be necessary in the near future in order to predict trends of environmental health in one of the richest ecosystems in the world. Pooling these monitoring efforts into an integrated ecosystem health surveillance initiative will become a unique model to address the health of marine mammals (and sea turtles). Oil spills and depletion of fisheries are anthropogenic and usually result in severe changes to the marine ecosystem in a short period of time. Larger scale changes in prey species and in ocean nutrient or other cycles, whether due to chronic non-point-source pollution or to global warming, may cause long-term changes in the environment. These long-term changes may also be facilitating periodic events such as toxic algal blooms. As programs for monitoring mortalities continue, we should aim toward management practices that can preserve and improve the health of the ocean.

References

Aguirre, A.A. 2000. Health assessment and disease status studies of the Hawaiian monk seal. NOAA NMFS Administrative Report H-00-01. U.S. Department of Commerce, Honolulu.

Aguirre, A.A., 2001. Epidemiology of the Hawaiian monk seal are infectious agents limiting population recovery? Verh. ber. Erkrk. Zootiere 40:1–9.

Aguirre, A.A. and G.H. Balazs. 2000. Plasma biochemistry values of green turtles (*Chelonia mydas*) with and without fibropapillomas in the Hawaiian Islands. Comp Haematol Int, 10:132–137.

Aguirre, A.A., G.H. Balazs, B. Zimmerman, and F.D. Galey. 1994a. Organic contaminants and trace metals in the tissues of green turtles (*Chelonia mydas*) afflicted with fibropapillomas in the Hawaiian Islands. Mar Pollut Bull 28:109–114.

Aguirre, A.A., G.H. Balazs, B. Zimmerman, and T.R. Spraker. 1994b. Evaluation of Hawaiian green turtles (*Chelonia mydas*) for potential pathogens associated with fibropapillomas. J Wildl Dis 30:8–15.

Aguirre, A.A., G.H. Balazs, T.R. Spraker, and T.S. Gross. 1995. Adrenal and hematological responses to stress in juvenile green turtles (*Chelonia mydas*) with and without fibropapillomas. Physiol Zool 68:831–854.

Aguirre, A.A., T.R. Spraker, G.H. Balazs, and B. Zimmerman. 1998. Spirorchidiasis and fibropapillomatosis in green turtles from the Hawaiian Islands. J Wildl Dis 34:91–98.

Aguirre, A.A., T.R. Spraker, A. Chaves-Quiroz, L.A. du Toit, and G.H. Balazs. 1999a. Pathology of fibropapillomatosis in olive ridley turtles (*Lepidochelys olivacea*) in Costa Rica. J Aquat Anim Health 11:283–289.

Aguirre, A.A., J. Reif, and G.A. Antonelis. 1999. Hawaiian monk seal epidemiology plan: health and disease status studies. NOAA Technical Memorandum NMFS-SWFSC. U.S. Department of Commerce, Hawaii.

Aguirre, A.A., M. Hanson, and R. Braun. 2002. An ocular disease of unknown etiology in Hawaiian monk seals. Zoo Wildl Med (in press).

Aguirre, A. A., T. R. Spraker, R. Morris, B. Powers, and B. Zimmerman. 2001. Low-grade fibrosarcomas in green turtles (*Cheleonia mydas*) from the Hawaiian Islands. Proceedings of the 21st international sea turtle symposium. NOAA Technical Memorandum, Philadelphia, Pa. (in press).

Alcorn, D.J., and A.K.H. Kam. 1986. Fatal shark attack on a Hawaiian monk seal (*Monachus schauinslandi*). Mar Mamm Sci 3:313–315.

Arctic Monitoring and Assessment Program (AMAP). 1995. Guidelines for the AMAP Assessment. AMAP Report 95:1.

Balazs, G.H., A.A. Aguirre, and S.K.K. Murakawa. 1997. Ocurrence of oral fibropapillomas in the Hawaiian green turtle: differential disease expression. Mar Turtle Newslett 76:1–2.

Balazs, G., S.K.K. Murakawa, D.M. Ellis, and A.A. Aguirre. 2000. Manifestation of fibropapillomatosis and rates of growth of green turtles at Kaneohe Bay in the Hawaiian Islands. *In* Abreu-Grobois, F.A., R. Briseño-Dueñas, R. Márquez, and L. Sarti, comp. Proceedings of the 18th international sea turtle symposium (NOAA Tech Memo NMFS-SEFSC-436), pp. 112–113. U.S. Department of Commerce, Miami, Fla.

Banish, L.D., and W.G. Gilmartin. 1992. Pathological findings in the Hawaiian monk seal. J Wildl Dis 28:428–434.

Bratton, G.R., W. Flory, C.B. Spainhour, and E.M. Haubold. 1997. Assessment of selected heavy metals in liver, kidney, muscle, blubber, and visceral fat of Eskimo harvested bowhead whales *Balaena mysticetus* from Alaska's north coast. North Slope Borough Final Report. Department of Wildlife Management, Barrow, Alaska.

Casey, R.N., S.L. Quackenbush, T.M. Work, G.H. Balazs, P.R. Bowser, and J.W. Casey. 1997. Evidence for retrovirus infections in green turtles *Chelonia mydas* from the Hawaiian Islands. Dis Aquat Org 31:1–7.

Cooper, L.W., I.L. Larsen, T.M. O'Hara, S. Dolvin, V. Woshner, and G.F. Cota. 2000. Radionuclide contaminant burdens in arctic marine mammals harvested during subsistence hunting. Arctic 53:174–182.

D' Amato, A.F., and M. Moraes-Neto. 2000. First documentation of fibropapillomas verified by histopathology in *Eretmochelys imbricata*. Mar Turtle Newslett 89:12–13.

Dailey, M.D., and R. Morris. 1995. Relationship of parasites (Trematoda: Spirorchidae) and their eggs to the occurrence of fibropapillomas in the green turtle (*Chelonia mydas*). Can J Fish Aquat Sci 52:84–89.

Dailey, M.D., R.V. Santangelo, and W.G. Gilmartin. 1988. A coprological survey of helminth parasites of the Hawaiian monk seal from the northwestern Hawaiian Islands. Mar Mamm Sci 4:125–131.

Dietz, R., J. Nørgaard, and J.C. Hansen. 1998. Have arctic marine mammals adapted to high cadmium levels? Mar Pollut Bull 36:490–492.

Fowler, C.W. 1990. Density dependence in northern fur seals (*Callorhinus ursinus*). Mar Mamm Sci 6:171–195.

George, J.C., and T.M. O'Hara. 1997. Frequency of killer whale (*Orcinus orca*) bites, ship collisions, and entanglements based on scarring on bowhead whales (*Balaena mysticetus*). In Proceedings for the Wildlife Disease Association conference p. 102 (abstr. 90). (St. Petersburg, Fla.)

George, J.C., L.M. Philo, K. Hazard, D. Withrow, G.M. Carroll, and R. Suydam. 1994. Frequency of killer whale attacks and ship collisions based on scarring on bowhead whales of the Bering-Chukchi-Beaufort Seas stock. Arctic 47:247–255.

George, R.H. 1997. Health problems and diseases of sea turtles. In Lutz, P.L., and J.A. Musick, eds. The biology of sea turtles, pp. 363–385. CRC Press, Boca Raton, Fla.

Gilmartin, W.G., R.L. DeLong, A.W. Smith, L.A. Griner, and M.D. Dailey. 1980. An investigation into unusual mortality in the Hawaiian monk seal, *Monachus schauinslandi*. In Grigg, R.W., and R.T. Pfund., eds. Proceedings of the meeting on the status of research investigations of the northwest Hawaiian Islands UNIHI-SEAGRANT-MR-80-04, pp. 32–41. University of Hawaii, Honolulu.

Gulland, F. M. D., J. G. Trupkiewicz, T. R. Spraker, and L.J. Lowenstine. 1996. Metastatic carcinoma of probable transitional cell origin in free-living California sea lions (*Zalophus californianus*): 64 cases (1979–1994). J Wildl Dis 32:250–258.

Harshbarger, J.C. 1991. Sea turtle fibropapilloma cases in the registry of tumors in lower animals. In Balazs, G.H., and S.G. Pooley, eds. Research plan for marine turtle fibropapilloma, pp. 63–70. NOM Tech. Memo. NMFS-SWFSC-156. U.S. Department of Commerce, Honolulu.

Heidel, J.R., L.M. Philo, T.F. Albert, C.B. Andreasen, and B.V. Strang. 1996. Serum chemistry of bowhead whales (*Balaena mysticetus*). J Wildl Dis 32:75–79.

Henderson, J.R. 1990. Recent entanglements of Hawaiian monk seals in marine debris. In Proceedings of the second international conference on marine debris (2–7 April 1989, NOAA Tech Memo NOAA-TM-NMFS-SWFSC-154), pp. 540–553. U.S. Department Commerce, Honolulu.

Herbst, L.H. 1994. Fibropapillomatosis of marine turtles. Ann Rev Fish Dis 4:389–425.

Herbst, L.H., E.R. Jacobson, R. Moretti, T. Brown, J.P. Sundberg, and P.A. Klein. 1995. Experimental transmission of green turtle fibropapillomatosis using cell-free tumor extracts. Dis Aquat Org 22:1–12.

Herbst, L.H., R. Moretti, T. Brown, and P.A. Klein. 1996. Sensitivity of the transmissible green turtle fibropapillomatosis agent to chloroform and ultracentrifugation conditions. Dis Aquat Org 25:225–228.

Herbst, L.H., E.R. Jacobson, P.A. Klein, G.H. Balazs, R. Moretti, T. Brown, and J.P. Sundberg. 1999. Comparative pathology and pathogenesis of spontaneous and experimentally induced fibropapillomas of green turtles (*Chelonia mydas*). Vet Pathol 36: 551–564.

Hiruki, L.M., W.G. Gilmartin, B.L. Becker, and I. Stirling. 1993a. Wounding in Hawaiian monk seals (*Monachus schauinslandi*). Can J Zool 71:458–468.

Hiruki, L.M., I. Stirling, W.G. Gilmartin, T.C. Johanos, and B.L. Becker. 1993b. Significance of wounding to female reproductive success in Hawaiian monk seals (*Monachus schauinslandi*) at Laysan Island. Can J Zool 71:469–474.

Huerta, P., H. Pineda, A.A. Aguirre, T.R. Spraker, L. Sarti, and A. Barragan. 2002. First confirmed case of fibropapilloma in a leatherback turtle (*Dermochelys coriacea*). *In* Proceedings of the 20th international sea turtle symposium. (NOAA Tech Memo NMFS-SEFSC-). U.S. Department of Commerce. Orlando, Fla. (in press).

Jacobson, E.R., J.L. Mansell, J.P. Sundberg, L. Hajarr, M.E. Reichmann, L.M. Ehrhart, M. Walsh, and F. Murru. 1989. Cutaneous fibropapillomas of green turtles, *Chelonia mydas*. J Comp Pathol 101:39–52.

Keuper-Bennett U., P. Bennett, and G.H. Balazs. 2002. The eyes have it: manifestation of ocular tumors in the green turtle Ohana of Honokowai, West Maui. *In* Proceedings of the 21st international sea turtle symposium. NOAA Tech Memo NMFS-SEFSC. U.S. Department of Commerce, Philadelphia (in press).

Keyes, M.C. 1965. Pathology of the northern fur seal. JAVMA 147:1090–1095.

Krone, C.A., P.A. Robisch, K.L. Tilbury, J.E. Stein, E.A. Mackey, P.R. Becker, T.M. O'Hara and L.M. Philo. 1999. Elements in liver tissues of bowhead whales (*Balaena mysticetus*). Mar Mammal Sci 15:123–142.

Lackovich, J.K., D.R. Brown, B.L. Homer, R.L. Garber, D.R. Mader, R.H. Moretti, A.D. Patterson, L.H. Herbst, J. Oros, E.R. Jacobson, S.S. Curry, and P.A. Klein. 1999. Association of herpesvirus with fibropapillomatosis of the green turtle *Chelonia mydas* and the loggerhead turtle *Caretta caretta* in Florida. Dis Aquat Org 37:889–897.

Landsberg, J.H., G.H. Balazs, K.A. Steidinger, D.G. Baden, T.M. Work, and D.J. Russell. 1999. The potential role of natural tumor promoters in marine turtle fibropapillomatosis. J Aquat Anim Health 11:199–210.

Limpus, C.J., and J.D. Miller. 1994. The occurrence of cutaneous fibropapillomas in marine turtles in Queensland. *In* Proceedings of the Australian marine turtle conservation workshop (14–17 November 1994), pp. 186–188. Queensland Department of Environment and Heritage and the Australian Nature Conservation Agency, Brisbane.

Loughlin, T.R., G.A. Antonelis, J.D. Baker, A.E. York, C.W. Fowler, R.L. DeLong, and H.W. Braham. 1994. Status of the northern fur seal population in the United States during 1992. *In* Sinclair, E.H., ed. Fur seal investigations NOAA Tech Memo NMFS-AFSC-45, pp. 9–28. U.S. Department of Commerce, Anchorage, Alaska.

Lu, Y., Y. Wang, Q. Yu, A.A. Aguirre, G.H. Balazs, V.R. Nerurkar, and R. Yanagiliara. 2000a. Detection of herpesviral sequences in tissues of green turtles with fibropapilloma by polymerase chain reaction. Arch Virol 145:1–9.

Lu, Y., A.A. Aguirre, T.M. Work, G.H. Balazs, V.R. Nerurkar, and R. Yanagiliara. 2000b. Identification of a small, naked virus in tumor-like aggregates in cell lines derived from a green turtle, *Chelonia mydas*, with fibropapillomas. J Virol Meth 86:25–33.

Lucas, F.A. 1899. The causes of mortality among seals. *In* Jordan, D.S., ed. The fur seals and fur seal islands of the North Pacific Ocean, pt. 3, pp. 75–98. U.S. Government Printing Office, Washington, D.C.

Macdonald, R.W., and J.M. Bewers. 1996. Contaminants in the arctic marine environment: priorities for protection. J Mar Sci 53:537–563.

Monson, D.H., D.F. Doak, B.E. Ballachey, A. Johnson, and J.L. Bodkin. 2000. Long-term impacts of the *Exxon Valdez* oil spill on sea otters, assessed through age-dependent mortality patterns. Proc Natl Acad Sci USA 97:6562–6567.

Muir, D.C.G., R. Wagemann, B.T. Hargrave, D.J. Thomas, D.B. Peakall, and R.J. Norstrom. 1992. Arctic marine ecosystem contamination. Sci Total Environ 122:75–134.

Neff, J.M. 1997. Ecotoxicology of arsenic in the marine environment. Environ Tox Chem 16:917–927.

Norton, T.M., E.R. Jacobson, and J.P. Sundberg. 1990. Cutaneous fibropapillomas and renal myxofibroma in a green turtle, *Chelonia mydas*. J Wildl Dis 26:265–270.

O'Hara, T.M., C. House, J.A. House, R.S. Suydam, and J.C. George. 1998. Viral serologic survey of bowhead whales in Alaska. J Wildl Dis 34:39–46.

O'Hara, T.M., M.M. Krahn, D. Boyd, P.R. Becker, and L.M. Philo. 1999. Organochlorine contaminant levels in Eskimo harvested bowhead whales of arctic Alaska. J Wildl Dis 35:741–752.

Olsen, O.W., and G.T. Lyons. 1965. Life cycle of *Uncinaria lucasi* Stiles, 1901 (Nematoda: Ancylostomatidae) of the fur seals, *Callorhinus ursinus* Linn., on the Pribilof Islands, Alaska. J Parasitol 51:689–700.

O'Shea, T.J., and R.L. Brownell, Jr. 1994. Organochlorine and metal contaminants in baleen whales: a review and evaluation of conservation implications. Sci Total Environ 154:179–200.

Pacyna, J.M., and G.J. Keeler. 1995. Sources of mercury in the arctic. Water Air Soil Pollut 80:621–632.

Quackenbush, S.L., T.M. Work, G.H. Balazs, R.N. Casey, J. Rovnak, A. Chaves, L. duToit, J.D. Baines, C.R. Parrish, P.R. Bowser, and J.W. Casey. 1998. Three closely related herpesviruses are associated with fibropapillomatosis in marine turtles. Virology 246: 392–399.

Scholin, C.A., F. Gulland, G.J. Doucette, S. Benson, M. Benson, et al. 2000. Mortality of sea lions along the central California coast linked to a toxic diatom bloom. Nature 403:80–84.

Smith, A., K. Bernirschke, T.F. Albert, and J.E. Barlough. 1987. Serology and virology of the bowhead whale. J Wildl Dis 23:92–98.

Wagemann, R., R.E.A. Stewart, P. Beland, and C. Desjardins. 1990. Heavy metals and selenium in tissues of beluga whales, *Delphinapterus leucas*, from the Canadian arctic and the St. Lawrence Estuary. Can Bull Fish Aquat Sci 224:191–206.

Wilkinson, D.M. 1996. National contingency plan for response to unusual marine mammal mortality events. US Dept of Commerce, NOAA Tech. Memo. NMFS-OPR-9. National Marine Fisheries Service, Washington, DC.

Work, T., and G. Balazs. 1999. Relating tumor score to hematology in green turtles with fibropapillomatosis in Hawaii. J Wildl Dis 35:804–807.

Work, T.M., G.H. Balazs, R.A. Rameyer, S.P. Chang, and J. Berestecky. 2000. Assessing humoral and cell-mediated immune response in Hawaiian green turtles, *Chelonia mydas*. Vet Immunol Immunopathol 74:179–194.

Yantis, D., J. Dubey, R. Moeller, R. Braun, A. A. Aguirre, and C. Gardiner. 1998. Hepactic sarcocystosis in a Hawaiian monk seal (*Monachus schauinslandi*). Infect Dis/Toxicol Pathol 35:453.

York, A.E., and J.R. Hartley. 1981. Pup production following harvest of female northern fur seals. Can J Fish Aquat Sci 38:84–90.

8

Disease Monitoring for the Conservation of Terrestrial Animals

Linda Munson
William B. Karesh

errestrial environments have undergone unprecedented change over the last century, and many environments no longer resemble the ecosystems within which indigenous animals have evolved. Host–pathogen relationships have been altered, traditional sources of food have been lost, novel pathogens and toxins have been introduced, and novel forms of stress are now present from changes in land-use patterns and associated human activities. Although human impacts have occurred in all ecosystems, their effects have been greatest on terrestrial animals that directly compete with mankind for habitat. The genetic composition of many terrestrial species has also been altered, directly or indirectly, through human actions that have reduced population size or restricted gene flow between populations. Together, these genetic and environmental changes have the potential to dramatically alter the ecology of diseases in terrestrial ecosystems and the overall health of wildlife populations.

Conserving the health of terrestrial animal populations is now an integral part of the greater goal of conserving ecosystem health. Viable animal populations are essential for balanced ecosystems, and healthy, reproductively normal animals are required to maintain this population viability. How environmental changes over the last 10 decades have affected the health of terrestrial wild animals is largely unknown, because information is lacking on what diseases were present in these environments prior to these changes. Whether disease contributed to historic population declines in wild animals was rarely documented, and proof was limited by the diagnostic capabilities available at the time. It is possible that the current perceptions of "emerging diseases" only reflect our ignorance of previous epidemics. Of more concern is the possibility that new diseases are arising or old diseases are finding new host species, due to global travel and changes in ecosystems.

Understanding the role of endemic diseases in the evolution of current animal populations could be of practical use for wildlife management, but insufficient information exists to assess this role. On the other hand, historic information about the natural ecology of diseases in terrestrial environments may no longer be applicable when ecosystems have been so drastically altered. For example, animal populations may have previously recovered from epidemics through immigration from adjacent healthy populations, tactics that would now be thwarted by habitat fragmentation. It is intuitive that small, geographically isolated, and genetically depauperate populations will respond differently to infectious diseases than will larger heterogeneous populations. Therefore, the influence of disease on the health of current terrestrial animal populations may need to be defined for each unique ecosystem and species, now and in the future.

Devising conservation strategies that are practical in the current "state of the earth" will require models that include disease risks. To acquire the knowledge needed for these models, prospective consistent disease-monitoring programs that collect information on endemic diseases and indigenous microflora in all habitats for that species will be essential. During the current experiments in progress, our greatest responsibility is to collect as much data from as many individuals and sites as possible, so that future modeling can use facts, rather than the conjecture commonly used today. For conservation medicine, the collection of data involves disease monitoring.

8.1 What Should Be Monitored?

An ecosystem approach to wildlife health requires knowledge of potential pathogens in an environment, as well as the actual disease prevalences in the region. "Disease" is a disorder of body functions, systems, or organs and is not necessarily synonymous with viral or bacterial infection, infestation with parasites, or exposure to toxins. Although many toxins, microorganisms, and macroparasites (or evidence of previous infections by these organisms) can be found within animals, the occurrence of disease depends on the dose and pathogenicity of the infectious organism and the suitability and health of the host (including its immunity to that infectious agent). One species' natural microbial flora may be another species' pathogen. The current human-made constraints on terrestrial environments increase the potential for aberrant persistence or concentration of pathogens or toxins that may increase their pathogenicity in a species previously was not affected.

As new technologies increase our ability to detect exogenous compounds and agents in animals, the distinction between detection and disease becomes more critical. This is analogous to the current situation with the human genome project, where a plethora of disease-related genes have been identified, but the function of those genes in disease is largely unknown. Increasingly sensitive molecular tests now detect even more microbial infections, making assessment of the significance of these data more complicated. Increasing sensitivity of chemical detection may allow us to find trace levels of potential toxins in animal tissues, but the significance of these findings awaits our understanding of what levels of toxins cause organ dysfunction. Hence, management decisions concerning the conservation of wildlife populations will rely on knowing not only which pathogens or toxins are present in a given area, but also how they affect wildlife health.

The actual health risks posed by pathogens can be estimated by making associations between potential pathogens and disease conditions (Koch's Law or Postulates). Koch's Postulates cannot be fulfilled for most wild animal diseases, because experimental infections are not feasible. Partial fulfillment of Koch's Postulates is possible by identifying the organism in all cases with a characteristic disease. A more specific link between an organism and disease, however, can be made when the organism can be visualized within lesions that are the basis for that disease (the so-called Koch's postulate of pathologists). The availability of immunohistochemical and in situ hybridization methods to detect many pathogens in tissues has greatly facilitated this understanding. For toxins, pathologists must compare the lesions produced by a toxin in experimental species with those identified in wild animals or identify characteristic cellular responses to toxins (e.g., cytochrome P450 profiles). Because these morphologic associations are the key to determining the cause of diseases in wildlife, disease-monitoring programs in wildlife should always include a pathology component. Substantial contributions can also be made through clinical veterinary input in interpreting clinical signs, selecting appropriate diagnostic tests, and interpreting the results.

Selecting appropriate tests for disease monitoring usually is limited by budgetary constraints. During an epidemic, the selection of tests can be based on the character of the clinical signs or necropsy results. During routine monitoring, however, selection of tests is less intuitive. Ideally, all important infectious diseases known to affect that species or taxon should be evaluated. Also, levels of toxins possibly present in the ecosystem should be assessed. Given the limited resources currently available for conservation efforts worldwide, however, priorities for monitoring, research, and management actions must be considered. Proper archiving of samples during disease surveillance programs makes prioritizing decisions less critical, because additional tests can be run at later dates if needed.

When selecting tests, laboratories with particular expertise in a disease should also be selected to run the tests. For example, the American Zoo and Aquarium Association's (AZA) Canid Taxon Advisory Group sends samples for canine distemper virus and canine parvovirus tests to the New York State Diagnostic Laboratory at Cornell University where primary research was conducted on these diseases and the most sensitive and specific tests were developed. Because many North American zoo conservation programs have designated specific laboratories for disease surveillance, they now have more than 10 years of valuable consistent data on seroprevalences. In some cases, long-term collaborations with specified laboratories have made it possible to have tests run at reduced prices in exchange for coauthorship on publications. Although designating a single global lab would be ideal, it also is important to promote regional laboratory expertise in countries with in situ conservation programs for endangered species. By sharing reagents and protocols among specified laboratories and maintaining quality assurance, the consistency of results needed for effective disease monitoring can be retained. Pathology evaluations on single species should also be conducted at a central site under the supervision of a pathologist so that trends in diseases can be followed. Designation of central sites for disease information does not preclude building broader teams for consultations. As for serological and toxicological testing, selection of pathologists for species-targeted surveillance should consider prior experience and particular expertise in that species or taxon.

Besides pathology, serosurveys, and assays for infectious and toxic agents, what other tests should be run? Should hemograms, complete serum biochemistries, and complete clinical exams be done on all animals handled? Although these tests might contribute to an understanding of wildlife diseases, the blood tests are costly and field conditions often prevent rapid sample handling needed for accurate results. Also, many parameters of these profiles are greatly altered by animal handling. Therefore, cost–benefit of these tests should be considered. As the technology becomes more affordable, should genetic analyses be a routine part of disease surveillance? As more animals are translocated or placed in reserves, should stress be measured through fecal or urinary assays? Stress clearly affects infectious disease resistance and also may cause disease through metabolic and physiologic changes. Because future assessment of disease risks will likely include evaluations of inherent resistance and exogenous environmental factors, we need to begin thinking more broadly about disease surveillance and start archiving information and biomaterials for including these assessments.

8.2 Which Species Should Be Monitored?

Over the last few decades, disease investigations in conservation programs have principally focused on high-profile species that have undergone a sudden demographic crash. The extinction of wild dogs (*Lycaon pictus*) in the Masai Mara and loss of viable wild populations of the California condor (*Gymnogyps californianis*) and black-footed ferret (*Mustela nigripes*) in North America clearly demonstrated that diseases can contribute to extinctions (Kat et al. 1996; Gascoyne et al. 1993; Stringfield 1998; Thorne and Williams 1988). Disease surveys also have included species destined for translocation or reintroduction (Acton et al. 2000; Bush et al. 1993). These focused studies on high-profile animals have also been important in that they have increased public awareness of the impact of disease on animal conservation. But should only endangered species be targeted for disease monitoring? Too often a crisis, such as a catastrophic epidemic, occurs in a species without historic data on that species or ecosystem.

The ideal ecosystem approach to disease monitoring would include both endangered and other more common indigenous species whose health reflects the health of the ecosystem as a whole. Many endangered species are "bioaccumulators" at the top of the food chain, so also serve as sensitive indicators of ecosystem health. Monitoring more common species is important where access to adequate samples from endangered species may not be feasible or where an important ecosystem does not have high-profile endangered species. Other wildlife species, such as seed dispersers or flower pollinators that play important roles in maintaining habitat or ecosystem integrity, should also be included (Karesh et al. 1997). Species that add human-related value to wild areas by providing important sources of protein for local people or have important cultural meaning should also be monitored to meet long-term conservation objectives (Karesh et al. 1998). Therefore, monitoring a spectrum of terrestrial species across all major ecosystems is the ideal. But is this realistic?

Disease monitoring is not without costs, and wildlife has little value in many, rapidly urbanizing societies. What country can afford to monitor wildlife diseases when local human health needs are underfunded? Can these projects be funded by research dollars? An old adage of research, which states that funds are available for things worth doing, rarely applies to research on wildlife diseases unless there is the potential to affect human or domestic animal health. General disease surveys are criticized as not hypothesis driven and as "fishing expeditions." Tissue archives from these surveys have been considered "stamp collecting." These same surveys and archives, however, have provided the information necessary to solve complex ecosystem health issues when a crisis arose. For example, the diagnosis of canine distemper in African lions was made from good necropsy specimens, but interpreting the significance of this finding was greatly enhanced by the availability of archived samples for comparison (Roelke-Parker et al. 1996).

Many disease surveys have been financed through the donation of time and resources of individual research laboratories. To be sustainable, more permanent funding and governmental commitment must be identified to compensate scientists for their time and to encourage laboratories and institutions to pursue these topics. Commitment of resources could also be encouraged by promoting the relationship of wildlife health to human and livestock health (Osofsky et al. 2000). Wildlife health professionals need to play an active role in encouraging government agencies, universities, international nongovernmental organizations, and individuals within the private sector to commit funds and resources to studying and protecting the health of wildlife populations.

In the interim, biomaterials should be collected for future analyses, and current projects prioritized by importance by conservation or governmental organizations. If resources limit surveillance to single species, then selecting the species for monitoring should depend on the specific concerns for that ecosystem. For example, if toxins are suspected, both amphibians with their porous skin and sensitive embryologic development (Ouellet et al. 1997) and animals that bioaccumulate toxins through their status in food chain should be selected. If cattle-borne infectious diseases are of concern, indigenous ungulates and their predators should be targeted for surveillance (Karesh et al. 1995, 1997). Where there is the potential for zoonotic or anthropogenic diseases, such as with the Mountain Gorilla Project

(Kalema et al. 1998; Mudakikwa et al. 1998) and in countries with bushmeat traditions (Wolfe et al. 1998), then both humans and primates can be targeted for surveillance.

8.3 Collecting Data: Each Individual Is Important

Although the individual animal is often disregarded by the disease ecologist, disease surveillance is based on information from individual animals. Diseases are identified through clinical or pathologic evaluations of individuals; population and ecosystem trends are derived by collating this information. Therefore, any opportunity to acquire biomaterials from any animal should be exploited.

An inexpensive utilitarian approach to disease monitoring in terrestrial animals has been the opportunistic collection of biomaterials during ongoing wildlife management procedures or field studies (Karesh and Cook 1995). Whenever wild animals are handled for population census (capture-recapture) or translocations, health examinations can be performed and blood and fecal samples collected with minimal stress to most species. Animals anesthetized for any purpose can easily have blood, hair, and fecal samples taken. If wildlife veterinarians are included in translocation or census procedures, then considerably more biomedical information can be obtained through physical examinations and selected biopsy procedures. Even if resources are not immediately available for laboratory analyses of collected samples, frozen or properly preserved samples can be archived for future comparative studies.

Carcasses are the most valuable resource for acquiring information on actual disease prevalences in populations. Wildlife biologists can be alerted to the importance of submitting carcasses of all culled or deceased wildlife to a pathologist for necropsy. This practice is particularly important during an epidemic and has been instrumental in tracking important wildlife diseases in North America and Europe for decades. In addition to determining causes of death, comprehensive gross and histopathologic evaluations can also disclose subclinical disease conditions in populations that may have nutritional or genetic causes or be potentially transmittable to susceptible animals in other environments. If carcasses cannot be submitted to a pathology laboratory in a timely manner, wildlife biologists can be trained to perform the necropsy and then harvest appropriate samples for pathology and pathogen analyses by following available instructions and guidelines (e.g., Necropsy Procedures for Wild Animals, available at www.vetmed.ucdavis. edu/whc/necropsy/toc.html). During routine disease surveillance when financial resources are limited, full histopathologic and microbial analyses of carcasses are not essential, because tissues can still be preserved in formalin or frozen and then archived for future testing if needed. If performing a necropsy is not possible, then freezing the whole carcass still maintains some valuable information. Like properly stored blood, hair, and fecal samples from living animals, these formalin-fixed and frozen tissues collected from carcasses will prove an invaluable historical resource if an epidemic occurs.

8.4 How Can Disease Monitoring Data Be Applied to Conservation?

The value of long-term comprehensive disease surveillance of multiple species in an ecosystem is clearly evident at the time of major mortality events. Necropsy findings from carcasses examined during an epidemic can be contrasted with data obtained from carcasses in that ecosystem during previous surveillance, to determine if the diseases detected are the actual cause of the epidemic or a preexisting endemic disease. Results of serological and parasitological tests obtained during the epidemic can also be compared with results in that population before the epidemic. These essential comparative analyses are not possible in unsurveyed populations.

Use of disease-monitoring data for wildlife management during periods of population stability, however, should be approached with caution. As mentioned above, most "disease surveys" determine the prevalences of antibodies to viruses, protozoa, or bacteria, the prevalences of parasites, or the levels of toxins in populations. These data are then often reported out of context of their effect on animal health and disease ecology. Distinguishing between exposure/colonization and disease through comparative pathology will become increasingly important as wild populations become more intensively managed, because each decision to translocate an animal will be confounded by knowledge of its resident flora. "Zero risk" policies for animal translocations are impractical in environments teeming with microorganisms. In the future, intensive management of endangered species will also likely include interventions that both treat and prevent diseases. The overall benefit to the conservation of an ecosystem or species needs to be weighed against the risk of intervention. Justifying such measures will require a clearer association of health risk with microbial agent. The more prepared we are to answer these questions, the fewer man-made disasters will result, because no intervention is risk free. Wildlife management decisions involving disease or preventive health decisions would greatly benefit from the input of veterinarians, particularly those trained in epidemiology and risk assessment. These veterinarians have the expertise to interpret individual animal test results in the context of the disease and to assess the potential for the spread of disease through a population and its sympatric species. They also have knowledge of the efficacy and safety of medical treatment and preventive measures essential for decision making.

Disease risk assessments are becoming an integral part of many environmental impact statements for wildlife reintroduction and translocation. Risk assessments have been a component of most population and habitat viability assessments conducted by the International Union for the Conservation of Nature and Natural Resources' Conservation Breeding Specialty Group, but the value of these assessments has been limited by the lack of data on disease prevalence in most species. Objective data from properly conducted disease-monitoring programs that can be subjected to statistical analyses are critically needed to improve the accuracy of these assessments.

8.5 Who Owns the Data?

A final issue that is important to address is that of the intellectual property of data collected during disease monitoring of wild animals. Disease surveillance data are valueless if not readily accessible. It is the responsibility of those collecting disease data to report their results to government agencies or landowners who have jurisdiction over the wildlife. Additionally, surveillance information can be more widely disseminated through publication in scientific journals and government reports. Beyond the entitlement of government agencies, open access to raw data has the potential to cause conflict. Not only might those data be taken out of the context in which they were collected, but also publication of results obtained by others without permission or recognition by coauthorship might occur. Defining ethical standards for both the acquisition and dissemination of information on wildlife diseases will be essential if full cooperation is to be achieved in disease-monitoring programs in the future.

8.6 Conclusions

Wildlife health professionals clearly recognize the importance of disease surveillance and the value of understanding the ecology of infectious agents and toxic compounds in terrestrial environments. Conservation biologists are becoming increasingly aware of the impact that disease can have on their efforts to protect wildlife populations. Bridging these areas of interest could contribute to synergistic advances. Priority setting and resource allocations require effective coordination among stakeholders to ensure that resources for wildlife health efforts and resources for conservation programs can be used to achieve the most important, long-term results possible. Together, conservation professionals and wildlife health professions should be educating donor groups about the value of monitoring the health of wildlife populations in order to increase funding available for this important effort. In the meantime and with very little additional effort, field project personnel can begin the proper collection of valuable sample materials for future studies.

References

Acton, A.E., L. Munson, and W.T. Waddell. 2000. Survey of necropsy results in captive red wolves (*Canis rufus*). 1992–1996. J Zoo Wildl Med 31:2–8.
Bush, M., C.A. Beck, and R.J. Montali. 1993. Medical considerations of reintroduction. *In* Fowler, M.E., ed. Zoo and wild animal medicine: current therapy, vol. 3, pp. 24–26. Saunders, Philadelphia.
Gascoyne, S.C., S. Laurenson, S. Lelo, and M. Borner. 1993. Rabies in African wild dogs (*Lycaon pictus*) in the Serengeti region, Tanzania. J Wildl Dis 29:396–402.
Kalema, G., R.A. Kock, and E. Macfie. 1998. An outbreak of sarcoptic mange in free-ranging mountain gorillas (*Gorilla gorilla beringei*) in Bwindi Impenetrable National

Park, southwestern Uganda. In Proceedings of the AAZV/AAWV joint conference p. 438. AAZV, Omaha, Ne.

Karesh, W.B., and R.A. Cook. 1995. Applications of veterinary medicine to in situ conservation efforts. Oryx 29:244–252.

Karesh, W.B., J.A. Hart, T.B. Hart, C. House, A. Torres, E.S. Dierenfeld, W.E. Braselton, H. Puche, and R.A. Cook. 1995. Health evaluation of five sympatric duiker species (*Cephalophus* spp.). J Zoo Wildl Med 26:485–502.

Karesh, W.B., A. del Campo, W.E. Braselton, H. Puche, and R.A. Cook. 1997. Health evaluation of free-ranging and hand-reared macaws (*Ara* spp.) in Peru. J Zoo Wildl Med 28:368–377.

Karesh, W.B., M.M. Uhart, E.S. Dierenfeld, W.E. Braselton, A. Torres, C. House, H. Puche, and R.A. Cook. 1998. Health evaluations of free-ranging guanaco (*Lama guanicoe*). J Zoo Wildl Med 29:134–141.

Kat, P.W., K.A. Alexander, J.S. Smith, J.D. Richardson, and L. Munson. 1996. Rabies among African wild dogs (*Lycaon pictus*) in the Masai Mara, Kenya. J Vet Diagn Invest 8:420–426.

Mudakikwa, A.B., J. Sleeman, J.W. Foster, L.L. Meader, and S. Patton. 1998. An indicator of human impact: gastrointestinal parasites of mountain gorillas (*Gorilla gorilla beringei*) from the Virunga volcanos region, Central Africa. In Proceedings of the AAZV/AAWV joint meeting, pp. 436–437. AAZV, Omaha, Ne.

Osofsky, S.A., W.B. Karesh, and S.L. Deem. 2000. Conservation medicine: a veterinary perspective [letter]. Conserv Biol 14:336–337.

Ouellet, M., J. Bonin, J. Rodrigue, J.L. DesGranges, and S. Lair. 1997. Hindlimb deformities (ectomelia, ectodactyly) in free-living anurans from agricultural habitats. J Wildl Dis 33:95–104.

Roelke-Parker, M.E., L. Munson, C. Packer, R.A. Kock, S. Cleaveland, M. Carpenter, S.J. O'Brien, A. Pospichil, R, Hoffman-Lehmann, H. Lutz, G.L.M. Mwamengele, M.N. Mgasa, G.A. Machange, B.A. Summers, and M.J.G. Appel. 1996. A canine distemper virus epidemic in Serengeti lions (*Panthera leo*). Nature 379:441–445.

Stringfield, C.E. 1998. Medical management of free-ranging California condors. In Proceedings of the AAZV/AAWV joint meeting, pp. 422–424. AAZV, Omaha, Ne.

Thorne, E.T., and E.S. Williams. 1988. Disease and endangered species: the black-footed ferret as a recent example. Conserv Biol 2:66–74.

Wolfe, N.D., A.A. Escalante, W.B. Karesh, A. Kilbourn, A. Spielman, and A.A. Lal. 1998. Wild primate populations in emerging infectious disease research: the missing link? Emerg Infect Dis 4:149–158.

9

Emergence of Infectious Diseases in Marine Mammals

Carol House

A. Alonso Aguirre

James A. House

Emerging diseases are defined as infections that either have newly appeared or have existed and are rapidly expanding their geographic range with a corresponding increase in detection, prevalence, mortality, or morbidity (Morse 1995). Many infectious diseases have been documented as emerging or reemerging in the marine environment (table 9.1). Specific factors responsible for the emergence of a new disease or the reemergence of an old disease can be identified in most cases. This chapter identifies the factors associated with emerging microbial infectious diseases of marine mammals.

An infectious disease may emerge in a population because of changes in the properties of the disease agent, changes in the host's resistance, environmental changes causing new interactions of the host and the disease agent, or simply the host being observed more carefully. We discuss the role of each element in recently recognized diseases of marine mammals.

9.1 The Agent

Common microbial infectious disease agents of marine mammals are bacteria and viruses. Mycoplasmae, chlamydiae, fungi, prions, and other agents may contribute to an array of disease conditions but are not as well studied at this time and hence are not addressed in this chapter.

The ability of viruses or bacteria to cause disease may result from a change in the microorganism's genetic composition. This change can be categorized as a mutation (i.e., change in the nucleotide sequence) or as a reassortment/rearrangement of the nucleic acid within a segmented genome. Viable mutations are more

Table 9.1. Some emerging and reemerging disease agents recently reported in the marine environment

Disease	Species	Etiologic Agent
Acute necrotizing enteritiis	Phocids	Coronavirus
Adenovirus infection	Walrus	Walrus adenovirus W77R
Angiomatosis	Delphinids	Unknown
Aspergillosis	Sea fans; gorgonians	*Aspergillus fumigatus*
Black band disease	Scleratinian corals	Microbial consortium including *Phormidium corallyticum, Beggiatoa* spp., and *Desulfovibrio* spp.
Brevetoxicosis	Manatees	*Gymnodinium breve* brevetoxin
Brucellosis	Delphinids, phocids	*Brucella marinus; Brucella* spp.
Calcivirus infections	Marine mammals	Calciviruses (39 serotypes)
Campylobacteriosis	New Zealand sea lions	*Campylobacter*-like agent
Candidiasis	Marine mammals	*Candida albicans*
Canine distemper	Phocids	Canine distemper virus
Chlamydiosis	Sea turtles, Pinnipeds	*Chlamydia psittaci*
Coccidioidomycosis	Delphinids	*Coccidioides immitis*
Cryptosporidiasis	Sea lions; sea turtles	*Cryptosporidium* spp.
Cutaneous viral papillomatosis	Orcas, manatees	Papillomavirus
Disseminated neoplasia and germinomas	Bivalves	Unknown
Dolphin morbillivirus infection	*Stenella*	Dolphin morbillivirus
Eastern equine encephalomyelitis	Grey whale	Arbovirus
Encephalitis	Harbor seals	Herpesvirus
Encephalomyelitis/meningoencephalitis	Sea otters	*Sarcocystis neurona/Toxoplasma gondii*
Enterovirus infection	Walrus	Walrus enterovirus 7–19
Erysipelas	Delphinids	*Erysipelotrix rusiopathiae*
Fibropapillomatosis	Marine turtles	Unknown: Herpesvirus, Retrovirus, Papillomavirus??
Gastrointestinal perforations	Sea lions	*Contracaecum corderoi*
Giardiasis	Phocids	*Giardia* spp.
Helicobacteriosis	Delphinids	*Helicobacter* spp.
Hepadnaviral hepatitis	Delphinids	Hepadnavirus
Hepatic sarcocystosis	Hawaiian monk seal	*Sarcocystis*-like organism
Herpesvirus infection	Phocids; northern fur seals	Phocine herpesvirus 1 & 2; Marine herpesvirus 206
Histoplasmosis	Delphinids	*Histoplasma capsulatum*
Immunoblastic malignant lymphoma	*Stenella/Tursiops*	Unknown
Influenza	Phocids	Influenza B virus
Leptospirosis	Pinnipeds	*Leptospira* spp.
Keloidal blastomycosis/Lobomycosis/Lobo's disease	Delphinids	*Lacazia loboi*
Melioidosis	Delphinids (SE Asia, Australia)	*Burkholderia psedomallei (Pseudomonas pseudomallei)*
Meningoencephalitis	Harbor seal	*Sarcocystis neurnona*
Metastatic oral squamous cell carcinoma	Delphinids	Papillomavirus??

(continued)

Table 9.1. Continued

Disease	Species	Etiologic Agent
Monk seal morbillivirus infection	Mediterranean monk seals	Monk seal morbillivirus-WA, G; biotoxins??
Mycobacteriosis	stripped bass	*Mycobacterium* spp.
Nocardiosis	Marine mammals	*Nocardia* spp.
Opthalmic disease	Hawaiian monk seals	Unknown; herpesvirus?
Peritonitis	Sea otters	*Polymorphus* spp. and *Corynsoma* spp.
Phocine distemper	Phocids	Phocine distemper virus
Pilot whale morbillivirus infection	Long-finned pilot whale	Pilot whale morbillivirus
Placentitis	Harbor seal	*Coxiella burnetti*
Polyps and hyperplasms	Corals	Unknown
Porpoise morbillivirus infection	Porpoises	Porpoise morbillivirus
Protozoal encephalitis	Sea otters	*Neospora* spp.
Pseudomoniasis	Marine mammals	*Pseudomonas* spp.
Pulmonary squamous cell carcinoma	Delphinids	Unknown
Red-band disease	Hard star and brain corals	Cyanobacteria
Renal neoplasms	Delphinids	Unknown
Retrovirus infection	Walrus	Walrus retrovirus T2/19
Rotavirus infection	Califoirnia sea lion	California sea lion rotavirus A11R
San Joaquin Valley Fever	Sea otters	*Coccidioides immitis*
Sea lion transitional cell carcinoma	California seal lions	Herpesvirus??
Seal pox	Pinnipeds	Pox virus
St. Louis encephalitis	Orca	Arbovirus
Stress-related necrosis	Corals	High water temperatures
Toxoplasmosis	Pinnipeds, Delphinids	*Toxoplasma gondii*
Tuberculosis	Fur seals	*Mycobacterium* spp.
Vibriosis	Marine mammals	*Campylobacter* spp.
White band disease types I and II	Elkhorn and staghorn coral	Unknown

common in single-stranded RNA viruses than in double-stranded DNA viruses, as the latter have repair mechanisms to correct genetic errors. Changes in the agent's genome can result in the agent gaining the ability to infect a new species, thereby expanding its host range. The end result is an emerging disease.

The marine environment is less harsh than the terrestrial environment. Marine environments usually have moderate temperatures, physiological pH ranges, and little ultraviolet light beneath the surface of the water, all favoring microbial survival. Desiccation, heat, and ultraviolet light are major causes of microbial death in the severe terrestrial environment. Only spore-forming bacteria and viruses encased in scabs, such as poxviruses, are able to exist under hot, arid conditions.

Excellent examples of marine disease agents arising from a terrestrial ancestor are the marine morbilliviruses. Morbilliviruses are members of the family Paramyxoviridae and have a nonsegmented single-stranded RNA genome. The only morbilliviruses known prior to 1988 were human measles virus, canine distemper virus, rinderpest virus, and peste-des-petits-ruminants virus.

Recently, newly characterized morbilliviruses have been shown to be epizootic-associated pathogens in pinnipeds and cetaceans (Taubenberger et al. 1996). In 1987, a morbillivirus that appeared to be canine distemper virus (CDV) killed thousands of Baikal seals (*Phoca sibirica*) in Lake Baikal, Russia (Grachev et al. 1989). Canine distemper virus is known to infect a wide range of carnivore species. The infection of Baikal seals was caused by a wild-type CDV that was genetically related to a strain circulating in Germany and not a vaccine strain widely used in domestic dogs and farmed mink in Siberia (Mamaev et al. 1995). Dogs often accompany their masters on shoreline walks, possibly becoming the source of both CDV and, later, phocine distemper virus (PDV) infection in seals. In 1988, a mass die-off of over 17,000 harbor seals (*Phoca vitulina*) occurred off the coast of northwestern Europe, attributed to a morbillivirus later designated PDV (Kennedy et al. 1988, 1989; Osterhaus et al. 1990). The PDV was characterized as genetically related to CDV (Barrett et al. 1993).

Dolphin and porpoise morbilliviruses (DMV and PMV) were responsible for the death of bottlenose dolphins (*Tursiops truncatus*) on the Atlantic coast during 1987–88 (Geraci 1989; Lipscomb et al. 1994a; Duignan et al. 1996), and PMV alone caused the death of bottlenose dolphins during 1993–94 in the Gulf of Mexico (Lipscomb et al. 1994b, 1996). Dolphin morbillivirus was first isolated in Europe from striped dolphins (*Stenella longirostris*) along the coast of Spain in 1990 (Domingo et al. 1990, 1992), and also caused the death of Mediterranean striped dolphins (*Stenella coeruleoalba*) along the coasts of Italy and Greece in 1991 (Van Bressem et al. 1993). Porpoise morbillivirus was isolated from common porpoises (*Phocoena phocoena*) from the coast of England and Scotland (Kennedy et al. 1992a; Thompson et al. 1992). The dolphin and porpoise morbilliviruses are closely related to each other and more distantly related to PDV (Barrett et al. 1993), and the probable parent of these viruses is CDV. The presence of two closely related pathogenic morbilliviruses that may circulate together or separately complicates the epidemiology of cetacean morbilliviral diseases.

The viruses CDV, PDV, DMV, and PMV all infect multiple species, which may increase an agent's geographical distribution. An agent's distribution may also be enhanced if it causes high morbidity and low mortality in at least some of the affected hosts. The agent is especially effective if the ailing and contagious host continues a gregarious behavior.

As demonstrated by intensive serologic surveys on a wide number of pinniped species (Duignan et al. 1994, 1995a; Geraci and Duignan 1993), morbilliviruses were present in many species prior to the die-offs. The interactions of the newly infected species could result in the appearance of both mass mortality and new reservoir hosts in new geographic areas. Antibodies to PDV were found in harp seals (*Phoca groenlandica*), hooded seals (*Cystophora cristata*), and ringed seals (*Phoca hispida*) in Greenland collected prior to the 1988 harbor seal epizootic (Duignan et al. 1997a) and in harbor seals (Duignan et al. 1993) and gray seals

(*Halichoerus grypus*) along the Atlantic coast (Duignan et al. 1997b), illustrating the wide geographic distribution and wide host range of PDV. Infection by PDV results in low mortality in the gregarious harp seals, aiding its spread. As of 1992–93, PDV was confined to the Atlantic: antibodies to PDV were not detected in serum collected from 80 Pacific Ocean harbor seals along the northwest coast of the USA during 1992–93 (Duignan et al. 1995b).

Duignan et al. (1995c) hypothesized that two species of pilot whales (*Globicephala* sp.) enzootically infected with morbilliviruses could act as long-distance vectors between America and Europe. The distribution of DMV and PMV infection in the U.S. epizootics also provides evidence that the 1992–93 epizootic may have been initiated by rare contact between PMV-infected Atlantic dolphins and immunologically naive Gulf of Mexico dolphins. Reidarson and colleagues (1998) first reported DMV-infected animals along the northwest Pacific coast of the United States. Six of 18 stranded common dolphins (*Delphinus delphis*) in southern California tested positive for antibodies to DMV, and morbilliviral RNA was detected in three of these six by reverse transcriptase polymerase chain reaction. The possibility of dolphin-to-seal contact and transfer of infection cannot be dismissed as a potential pathway for transmission to other naive marine mammal populations.

Serological evidence of morbillivirus infection was recently found in small cetaceans from the southeastern Pacific off the coast of Peru (Van Bressem et al. 1998). Although one group proposed a morbillivirus as the cause of the mortality of Mediterranean monk seals (Hernandez et al. 1998; Osterhaus et al. 1992, 1997, 1998), another group suggested that those deaths were caused by toxins from algal blooms (Costas and Lopez-Rodas 1998).

The disease agents of marine mammals have a simple and clear advantage over the agents of terrestrial animals: there are no vaccination programs due to the impossible logistics of the marine environment. Vaccines have been proposed for captive marine mammals, but traditional efficacy and safety testing is not possible. Though two inactivated CDV vaccines tested in harbor seals provided protection after challenge (Visser et al. 1989), there are many reasons why vaccine application is limited in the wild. The proposed morbillivirus vaccine has many inherent problems, including stress of handling and an impractical multiple-dose regimen, apart from the basic ethical question of whether vaccination of wild animals should even be considered.

9.2 The Host

Both inbreeding and stress can alter susceptibility of the host species. As populations of certain marine mammals dwindle, the gene pool may become limited and species ability to resist disease may decrease. Environmental conditions can cause stress that leads to immunosuppression, making a free-ranging population more susceptible to disease. When a host is stressed by any number of factors (starvation, pollution, infection), the efficacy of the animal's immune system is decreased.

Herpesviruses are large, double-stranded DNA viruses that integrate their genetic material into the host genome and recrudesce during periods of stress to the host. Many species of pinnipeds are infected by herpesviruses. Phocine herpesvirus was first identified as a cause of morbidity and mortality during an epizootic of PDV in harbor seals in 1985 (Osterhaus et al. 1985). The high mortality observed (over 17,000 seals) has been correlated to coinfection of at least these two viruses. The organism from that epizootic has since been identified as phocine herpesvirus-1 (PHV-1). A second seal herpesvirus (PHV-2) has been isolated from a captive California sea lion (Kennedy-Stoskopf et al. 1986), and free-ranging harbor seals from the Atlantic Ocean coast of the United States (Harder et al. 1996). Collectively, the herpesviruses are usually associated with lesions in the oral and nasal mucosae, pneumonia, ocular disease, and abortion. Neurotrophic strains of herpesvirus are common and have been described in a harbor porpoise (*Phocoena phocoena*) (Kennedy et al. 1992b). More recently, two stranded immature bottlenose dolphins demonstrated disseminated infections of two novel herpesvirus strains of unknown origin (Blanchard et al. 2001).

Antibodies to the phocine herpesviruses appear to be widely distributed. Antibodies to PHV-1 have been detected in Weddell seals (*Leptonychotes weddelli*) from the Antarctic (Stenvers et al. 1992). In a recent study conducted in Alaska, antibody to PHV-1 was found in 29–77% and to PHV-2 in 16–50% of a sample of pinnipeds collected from 1978 to 1994 (Zarnke et al. 1997). Species represented included walrus (*Odobenus rosmarus*), northern fur seal (*Callorhinitis ursinus*), harbor seal, spotted seal (*Phoca largha*), ribbon seal (*Histriophoca fasciata*), Steller sea lion (*Eumetopias jubatus*), bearded seal (*Erignathus barbatus*), and ringed seal (*Phoca hispida*). During a stranding event of 700 harbor seal pups on the central and northern coast of California, 379 pups died. Approximately one-half of the seals examined had lesions suggestive of a herpesvirus infection, including adrenocortical and hepatic necrosis associated with intranuclear inclusions and electron microscopic evidence of herpesvirus-like particles. A herpesvirus-like virus was isolated from adrenal tissue on cell culture (Gulland et al. 1997).

A herpesvirus has been suspected as the causative agent of an ocular syndrome (Aguirre et al. 2002). Twelve juvenile Hawaiian monk seals (*Monachus schauinslandi*) were brought into captivity from the French Frigate Shoals to Oahu in 1995 for rehabilitation. Ten animals developed corneal opacity. At least two blood samples were obtained from each of 10 of these captive juvenile seals and examined for antibody to PHV-1 (A. D. M. E. Osterhaus, unpublished data 1996). Using a cutoff of 1:20 for seropositivity, 8 of 10 seals were seropositive on at least one occasion, with titers ranging between 1:20 and 1:180. Overall, the data show fluctuating, low levels of antibody to PHV-1 throughout the period of observation (A. Aguirre, unpublished data 1996).

9.3 The Environment

Contamination of the marine ecosystem by chemicals and toxins may increase mortality during a disease outbreak. It has been speculated that marine mammals inhabiting polluted coastal areas accumulate higher levels of environmental pollutants through the food chain, thus becoming more susceptible to diseases. In controlled experiments, it was shown that the immune system of harbor seals fed fish from the Baltic Sea was impaired when compared to that of harbor seals fed fish from the North Atlantic Ocean, using in vivo and in vitro tests (DeSwart et al. 1994; Ross et al. 1995). It is not known, however, if these impaired functions made seals more susceptible to a die-off.

Environmental conditions can facilitate the spread of a disease agent within and among species. Weather conditions and loss of habitat can cause species to share breeding grounds and new feeding areas. Duignan et al. (1997a) have noted the increased prevalence of antibody to PDV in ringed seals sympatric with harp seals experiencing 83% seropositivity to PDV.

Laws protecting marine mammals have caused some species to increase to the point of overpopulation in some areas. Together with human overharvesting, this has resulted in depletion of fish populations and famine. Starving animals may have weakened immune systems, resulting in higher mortality during infections. They may also stray into new territories seeking food, further spreading infectious diseases.

Microenvironments such as sea parks, military bases with trained marine mammals, and rehabilitation centers create unnatural contacts between individuals and species. Stranded animals that are rescued and rehabilitated in a stranding facility or sea park are returned to the wild after being both stressed and exposed to other animals, perhaps other species. The rise of antibody levels to DMV in a rehabilitated animal has been documented: Reidarson and colleagues (1998) reported that one common dolphin rescued during a stranding event on the California coast became seropositive during the early weeks of its stay at a large aquarium. Morbilliviral RNA was identified in animals that died during the same stranding using reverse transcriptase polymerase chain reaction testing, indicating that DMV or a closely related virus may be present in the Pacific Ocean. Because no cohort of the seropositive animal became seropositive during his 14-month stay at the facility, the animal was released.

In other cases, the release of rehabilitated animals into the wild may represent a risk to the free-ranging populations. One famous orca held in a Mexican facility for many years presented active lesions of papillomavirus possibly related to stress caused by captivity (G. Bossart, personal commun. 1996). He may be released largely due to public sentiment. Additionally, animals such as this learn behaviors in the centers that are not suited to natural environments, such as eating dead fish and seeking human and canine contact, which may lead to increased exposure to disease agents.

Recently, programs allowing people to swim with dolphins have become a popular tourist attraction at some resorts. Wild dolphins are suspected of harboring a genital herpes virus. Increased dolphin–human contact may allow new disease opportunities.

9.4 Improved Observation

The majority of reported emerging diseases of marine mammals are the result of better opportunities to observe the animals and to apply diagnostic techniques. Rehabilitation centers provide stranded animals with short-term care and generate health information on wild animals in distressed situations. Stranded animals tend to be young, so the serological results of disease surveys conducted on serum samples from these animals do not reflect the entire population status. This information, however, is useful to indicate trends when it is compiled and interpreted properly, especially in light of passively transferred maternal antibodies. Serological data on the prevalence of antibodies to morbilliviruses proved helpful in establishing the release criteria of seropositive rehabilitated animals (Reidarson et al. 1998).

The dead and dying animals in an unexplained marine mammal mortality event often contribute information about adult animal disease status. California sea lions that died or became ill from domoic acid poisoning provided valuable baseline data on the prevalence of antibodies to morbilliviruses in the Pacific Ocean as well as another new isolate of San Miguel sea lion virus (SMSV) (F. Gulland, unpublished data 1998).

Sea parks have the advantage of commercial levels of funding and the ability to take sequential samples from a large variety of species for comparison. Arboviruses are transmitted by the bites of arthropods, and recently they have been reported to cause mortality in cetaceans. St. Louis encephalitis (SLE) virus was recently incriminated as the cause of death in a captive orca (*Orcinus orca*). The virus that causes SLE is thought to occur naturally in wild birds. It is transmitted among birds and between birds and other animals by *Culex* mosquitoes. More recently, a stranded gray whale (*Eschrichtius robustus*) in California was diagnosed with encephalitis, and Eastern equine encephalitis virus was isolated (F. Gulland, unpublished data 1999). Eastern equine encephalitis is an arboviral disease that is spread to inadvertent hosts such as horses and humans by infected *Culiseta melanura* mosquitoes. It is among the most serious of a group of mosquito-borne virus diseases and can affect the central nervous system and cause severe complications, even death. As with SLE, wild birds are the source of infection for mosquitoes, which can sometimes transmit the infection to horses, other animals, and, in rare cases, people. How these traditionally terrestrial, vector-borne viruses were introduced into the marine environment is subject to speculation.

Brucellosis is an important infectious disease of many terrestrial mammals, including humans. Infection with brucellae in gravid females is typically followed by abortion or stillbirth and by epididymitis, orchitis, and infertility in males. The disease can be spread horizontally by contact with infective tissues and discharges from aborting females. Members of the genus *Brucella* have recently been identified in several species of cetaceans and pinnipeds. *Brucella* strains have been isolated from harbor seals, a gray seal, a hooded seal, and several species of cetaceans (Ross et al. 1996; Foster et al. 1996). A *Brucella* species was isolated from the aborted fetal tissue of bottlenose dolphins along the

California coast (Ewalt et al. 1994). The isolates do not appear to be members of known species of brucellae, and a new nomenclature has been proposed (Jahans et al. 1997). The dolphin isolate is now termed *Brucella delphini* (Miller et al. 1999). Serological evidence supports the widespread nature of *Brucella* infections. Titers against brucellae were detected in 18 of 102 Pacific harbor seals and 4 of 50 California sea lions from Puget Sound, Washington, indicating relatively widespread infection among Pacific coast pinnipeds (Lambourn et al. 1996).

An agent achieves horizontal transmission if an animal infects its cohort. The most effective means of horizontal transmission between social animals are aerosols and contact, both of which are increased when animals encounter closer breeding and feeding conditions. Additional means of horizontal transmission are fomites, injections of infectious material, or vaccines containing contaminating microbes, or sexual transmission. A unique means of horizontal transmission has been reported for brucellae in pinnipeds. A brucella organism was recently isolated from *Parafilaroides* lungworms in a Pacific harbor seal, suggesting a potential role for the parasite as a mechanical vector (Garner et al. 1997). The role of the lungworm as a vector would explain the prevalence of antibody to brucellae in young animals. The serology may indeed reflect traditional infection, as the number of brucellae isolates is increasing.

Host range restriction is evident among influenza viruses (Webster et al. 1981), but interspecies transmission of influenza viruses is well known. Influenza viruses are members of the family Orthoparamyxoviridae. Each virion contains eight single-stranded RNA segments. The surface of the influenza virion is covered with spikes of hemagglutinating (HA) antigen, neuraminadase (NA) antigen, and matrix protein. Each influenza A virus is characterized by the composition of its HA and NA. Antigenic drifts and antigenic shifts are well-established features of human influenza viruses, creating the need for slightly different vaccines every year and substantially different vaccines about every decade.

An influenza A virus, A/Seal/Massachusetts/1/80(H7N7), was responsible for killing approximately 20% of the population of harbor seals in 1980 off the New England coast (Geraci et al. 1982). The virus was shown to have derived from one or more avian influenza viruses (Webster et al. 1981). Additional New England harbor seal mortality was caused by H4N5 in 1983, H4N6 in 1991, and H3N2 in 1992 (Hinshaw et al. 1992). Two influenza A viruses, H13N2 and H13N9, were isolated from a pilot whale (*Globicephala melaena*) that stranded off the coast of Maine in 1984 (Hinshaw et al. 1986). The relationship of the infections to stranding is unknown. The H13N9 whale virus was genetically related to avian influenza viruses (Hinshaw et al. 1986). In the South Pacific, influenza A virus (H1N3) was isolated from whales (Balaenopteridae species) (Lvov et al. 1978). Extensive sea bird populations that inhabit the breeding and resting habitats of many species of pinnipeds and cetaceans raise concern that genetic reassortment of an avian influenza virus might lead to an epizootic in marine mammals.

Military operations utilize trained California sea lions, beluga whales (*Delphinapterus leucas*), and dolphins. These valuable animals are closely observed for disease, as they are taken for missions in both the Atlantic and Pacific Oceans. Recently, an outbreak of a new calicivirus serotype was documented in California sea lions in a group of trained animals (Van Bonn et al. 2000). Caliciviruses are small, single-stranded RNA viruses that cause vesicular lesions on the skin of infected animals; in pinnipeds, these typically occur on the dorsal surfaces of the foreflippers (Gage et al. 1990). All available evidence indicates that caliciviruses isolated from the marine environment (Smith et al. 1973) circulate in numerous marine reservoirs and are the cause of vesicular exanthema of swine (VES). These viruses and their role in swine disease outbreaks in California have been well described (Madin 1973; Barlough et al. 1986). The SMSVs examined produce vesicular disease equivalent to VES. Calicivirus infections of marine mammals of coastal California are endemic, as shown by the numerous isolations of SMSVs. The prevalence of infection of marine mammals may be related to infection in the opaleye perch, which overlaps the host range of the California sea lion. Bowhead whales (*Balaena mysticetus*), at the top of the food chain, were sampled during a native subsistence hunt and found to exhibit antibodies to viruses last known to exist in the 1950s (O'Hara et al. 1998), but no disease was observed, and the significance of the antibodies is unknown.

9.5 Areas of Future Study

Increased interest in ecosystem health will benefit the marine ecosystem in general. Management of the system requires knowledge of what infectious agents are present in a population and what role that agent may play in causing a disease. The noted examples of emerging diseases in the marine environment are a reflection of our ability to diagnose their etiology. Most laboratories attempting to isolate microbial agents from marine mammal specimens use techniques developed for terrestrial animal viruses, which may be too harsh for viruses that evolve in a kinder marine environment. One major challenge is the availability of appropriate diagnostic systems. Samples taken are often from dead animals or surface swabs, requiring the addition of antibiotics to prevent the growth of contaminating bacteria. High levels of antibiotics may affect metabolic pathways in growing cells, decreasing the cells' ability to replicate a sensitive virus. Also, the number of primary cell cultures or cell lines derived from marine mammal tissues is limited, so many laboratories depend on standard, high-passage cell lines of terrestrial animal tissue origin for their isolation substrate. Currently, specific cells of marine mammals are being cultured for virus isolation and identification purposes (Lu et al. 1998, 2000). Serologic techniques (Gardner et al. 1996) adapted to marine mammal samples will provide the baseline information to this relatively new and expanding area of marine mammal health.

References

Aguirre, A.A., M. Hanson, and R. Braun. 2002. An ocular disease of unknown etiology in Hawaiian monk seals. J. Zoo Wildl Med. in press.

Barlough, J.E., E.S. Berry, D.E. Skilling, and A.W. Smith. 1986. Sea lions, caliciviruses and the sea. Avian/Exotic Pract 3:8–19.

Barrett T., I.K.G. Visser, L.V. Mamaev, L. Goatley, M.F. Van Bressem, and A.D.M.E. Osterhaus. 1993. Dolphin and porpoise morbilliviruses are genetically distinct from phocine distemper virus. Virology 193:1010–1012.

Blanchard, T.W., N.T. Santiago, T.P. Lipscomb, R.L. Garber, W.E. McFee, and S. Knowles. 2001. Two novel alpha herpesviruses associated with fatal disseminated infections in Atlantic bottlenose dolphins. J Wildl Dis 37:297–305.

Costas, E., and V. Lopez-Rodas. 1998. Paralytic phycotoxins in monk seal mass mortality. Vet Rec 142:643–644.

DeSwart, R.L., P.S. Ross, L.J. Vedder, H.H. Timmerman, S.H. Heisterkamp, H. Van-Loveren, J.G. Vos, P.J.H. Reijnders, and A.D.M.E. Osterhaus. 1994. Impairment of immune function in harbor seals (*Phoca vitulina*) feeding on fish from polluted waters. Ambio 23:155–159.

Domingo M.L., L. Ferrer, M. Pumarola, A. Marco, J. Plana, S. Kennedy, M. McAliskey, and B.K. Rima. 1990. Morbillivirus in dolphins. Nature 336:21.

Domingo M., J. Visa, M. Pumarola, A.J. Marco, L. Ferrer, R. Rabanal, and S. Kennedy. 1992. Pathologic and immunocytochemical studies of morbillivirus infection in striped dolphins (*Stenella longirostris*). Vet Pathol 27:463–464.

Duignan P.J., S. Sadove, J.T. Saliki, and J.R. Geraci. 1993. Phocine distemper in harbor seals (*Phoca vitulina*) from Long Island, New York. J Wildl Dis 29:465–469.

Duignan P.J., J.T. Saliki, D.J. St. Aubin, J.A. House, and J.R. Geraci. 1994. Neutralizing antibodies to phocine distemper virus in Atlantic walruses (*Odobenus rosmarus rosmarus*) from Arctic Canada. J Wildl Dis 30:90–94.

Duignan P.J., C. House, M.T. Walsh, T. Campbell, G.D. Gossart, N. Duffy, P.J. Fernandes, B.K. Rima, S. Wright, and J.R. Geraci. 1995a. Morbillivirus infection in manatees. Marine Mamm Sci 11:441–451.

Duignan P.J., J.T. Saliki, D.J. St. Aubin, G. Early, S. Sadove, J.A. House, K. Kovacs, and J.R. Geraci. 1995b. Epizootiology of morbillivirus infection in North American harbor seals (*Phoca vitulina*) and gray seals (*Halichoerus grypus*). J Wildl Dis 31:491–501.

Duignan P.J., C. House, J.R. Geraci, G. Early, H.G. Copland, M.T. Walsh, G.D. Bossart, C. Cray, S. Sadove, D.J. St. Aubin, and M. Moore. 1995c. Morbillivirus infection in two species of pilot whale (*Globicephala* sp.) from the western Atlantic. Marine Mamm Sci 11:150–162.

Duignan, P.J., C. House, D.K. Odell, R.S. Wells, L.J. Hansen, M.T. Walsh, D.J. St. Aubin, B.K. Rima, and J.R. Geraci. 1996. Morbillivirus infection in bottlenose dolphins: evidence for recurrent epizootics in the western Atlantic and Gulf of Mexico. Marine Mamm Sci 12:499–515.

Duignan, P.J., O. Nielsen, C. House, K.M. Kovacs, N. Duffy, G. Early, S. Sadove, D.J. St. Aubin, B.K. Rima, and J.R. Geraci. 1997a. Epizootiology of morbillivirus infection in harp, hooded, and ringed seals from the Canadian Arctic and western Atlantic. J Wildl Dis 33:7–19.

Duignan, P.J., N. Duffy, B.K. Rima, and J.R. Geraci. 1997b. Comparative antibody response in harbour and grey seals naturally infected by a morbillivirus. Vet Immunol Immunopathol 55:341–349.

Ewalt, D.R., J.P. Payeur, M.B. Martin, D.R. Cummins, and W.G. Miller. 1994. Characteristics of a *Brucella* species from a bottlenose dolphin (*Tursiops truncatus*). J Vet Diagn Invest 6:448–452.

Foster G., K.L. Jahans, R.J. Reid, and H.M. Ross. 1996. Isolation of *Brucella* species from cetaceans, seals, and an otter. Vet Rec 138:583–586.

Gage, L.J., L. Amaya-Sherman, J. Roletto, and S. Bently. 1990. Clinical sings of San Miguel sea lion virus in debilitated California sea lions. J Zoo Wildl Med 21:79–83.

Gardner, I.A, S. Hietala, and W.M. Boyce. 1996. Validity of using serological tests for diagnosis of diseases in wild animals. Rev Sci Tech 15:323–335.

Garner, M.M., D.M. Lambourn, S.J. Jeffries, P.B. Hall, J.C. Rhyan, D.R. Ewalt, L.M. Polzin, and N.F. Cheville. 1997. Evidence of *Brucella* infection in *Parafilaroides* lungworms in a Pacific harbor seal (*Phoca vitulina richardsi*). J Vet Diagn Invest 9: 303–306.

Geraci, J.R. 1989. Clinical investigation of the 1987–1988 mass mortality of bottlenose dolphins along the US central and south Atlantic coast. Final report to the National Marine Fisheries Service. U.S. Navy Office of Naval Research and Marine Mammal Commission, Washington, D.C.

Geraci, J.R., and P.J. Duignan. 1993. Survey for morbillivirus in pinnipeds along the northeastern coast. Final Report to National Oceanic and Atmospheric administration and National Marine Fisheries Service, Washington, D.C.

Geraci, J.R., D.J. St. Aubin, I.K. Barker, K.G. Webster, V.S. Hinshaw, W.J. Bean, H.L. Ruhnke, J.H. Prescott, G. Early, A.S. Baker, S. Madoff, and R.T. Schooley. 1982. Mass mortality of harbor seals: pneumonia associated with influenza A virus. Science 215:1129.

Grachev, M.A., V.P. Kumarev, L.V. Mamaev, V.L. Zorin, L.V. Baranova, N.N. Denikina, S.I. Belikova, E.A. Petrov, V.S. Kolesnik, V.N. Dorofeev, A.M. Beim, V.N. Kudelin, F.G. Nagiera, and V.N. Siderov. 1989. Distemper virus in Baikal seals. Nature 338: 209.

Gulland, F.M.D., L.J. Lowenstine, J.M. Lapointe, T. Spraker, and D.P. King. 1997. Herpesvirus infection in stranded Pacific harbor seals of coastal California. J Wildl Dis 33:450–458.

Harder, T.C., C.M. Harder, and H. Vos. 1996. Characterization of phocid herpesvirus-1 and-2 as putative alpha and gamma-herpesviruses of North American and European pinnipeds. J Gen Virol 77:27–35.

Hernandez, M., I. Robinson, A. Aguilar, L.M. Gonzalez, L.F. Lopez-Jurado, M.I. Reyero, E. Cacho, J. Franco, V. Lopez-Rodas, and E. Costas. 1998. Did algal toxins cause monk seal mortality? Nature 393:28–29.

Hinshaw, V.S, W.J. Bean, J.R. Geraci, P. Fiorella, G. Early, and R.G. Webster. 1986. Characterization of two influenza A viruses from a pilot whale. J Virol 58:655.

Hinshaw, V.S., B.C. Easterday, M. McGregor, D. Wentworth, and G. Early. 1992. Interspecies transmission of influenza A viruses. In Proceedings of the Third InterNational Symposium on Avian Influenza, pp. 106–110. University of Wisconsin, Madison.

Jahans, K.L., G. Foster, and E.S. Broughton. 1997. The characterization of *Brucella* strains isolated from marine mammals. Vet Microbiol 57:373–382.

Kennedy, S., J.A. Smyth, S.J. McCullough, G.M. Allen, F. McNeilly, and S. McQuaid. 1988. Confirmation of cause of recent seal deaths. Nature 335:404.

Kennedy, S., J.A. Smyth, P.F. Cush, P.J. Duignan, M. Platter, S.J. McCullough, and G.M. Allan. 1989. Histopathologic and immunocytochemical studies of distemper in seals. Vet Pathol 26:97–103.

Kennedy, S., T. Kuiken, H.M. Ross, M. McAliskey, D. Moffett, C.M. McNiven, and M. Carole. 1992a. Morbillivirus infection in two common porpoises (*Phocoena phocoena*) from the coast of England and Scotland. Vet Rec 131:286–290.

Kennedy, S., I.J. Lindstedt, M.M. McAliskey, D. Moffett, C.M. McNiven, and M. Carole. 1992b. Herpesviral encephalitis in a harbor porpoise (*Phocoena phocoena*). J Zoo Wildl Med 23:374–379.

Kennedy-Stoskopf, S., M.K. Stoskopf, M.A. Echaus, and J.D. Strandberg. 1986. Isolation of a retrovirus and a herpesvirus from a captive sea lion. J Wildl Dis 22:156–164.

Lambourn, D.M., S.J. Jeffries, and P.B. Hall. 1996. Evidence of brucellosis in Pacific harbor seals (*Phoca vitulina richardsi*) and California sea lions (*Zalophus californianus*) from Puget Sound, Washington. Program for the annual conference of the Wildlife Diseases Association, abstract 45, 21–25 July 1996, Fairbanks, Alaska.

Lipscomb, T.P., F.Y. Schulman, D. Moffatt, and S. Kennedy. 1994a. Morbilliviral disease in Atlantic dolphins (*Tursiops truncatus*) from the 1987–1988 epizootic. J Wildl Dis 30:567–571.

Lipscomb, T.P., S. Kennedy, D. Moffatt, and S. Kennedy. 1994b. Morbilliviral disease in an Atlantic bottlenose dolphin (*Tursiops truncatus*) from the Gulf of Mexico. J Wildl Dis 30:572–576.

Lipscomb, T.P., S. Kennedy, D. Moffatt, A. Krafft, B.A. Klaunberg, J.H. Lichy, G.A. Regan, G.A.J. Worthy, and J.K. Taubenbergen. 1996. Morbilliviral epizootic in bottlenose dolphins of the Gulf of Mexico. J Vet Diagn Invest 8:283–290.

Lu, Y., A.A. Aguirre, R.C. Braun and P. Loh. 1998. Establishment of monk seal cell lines. Vitro Cell Dev Biol 34:367–369.

Lu Y., A.A. Aguirre, C. Hamm, Y. Wang, Q. Yu, P.C. Loh, and R. Yanagihara. 2000. Establishment, cryopreservation, and growth of 11 cell lines prepared from a juvenile Hawaiian monk seal, *Monachus schauinslandi*. Meth Cell Sci 22:115–124.

Lvov, D.K, V.M. Zhdanov, A.A. Sazanov, N.A. Braude, E.V. Vladimirtcera, L.V. Agafonova, E.L. Skijanskaga, N.V. Kaverin, V.I. Reznik, T.V. Pysina, A.M. Oserovic, A.A. Berzin, I.A. Mjasnikova, R.Y. Podcernjaeva, S.M. Klimento, V.P. Andrejev, and M.A. Vakhno. 1978. Comparison of influenza viruses isolated from man and from whales. Bull WHO 56:923–930.

Madin, S.H. 1973. Pigs, sea lions and vesicular exanthema. *In* Proceedings of the second International conference on foot and mouth disease, pp. 78–81. Gustav Stern Foundation, New York.

Mamaev, L.V., N.N. Denikina, S.I. Belikov, V.E. Volchkov, I.K. Visser, M. Fleming, C. Kai, T.C. Harder, B. Liess, and A.D.M.E. Osterhaus. 1995. Characterisation of, morbilliviruses isolated from Lake Baikal seals (*Phoca sibirica*). Vet Microbiol 44:251–259.

Miller, W.G., L.G. Adams, T.A. Fitch, N.F. Cheville, J.P. Payeur, D.R. Harley, C. House, and S.H. Ridgway. 1999. *Brucella*-induced abortions and infection in bottlenose dolphins (*Tursiops truncatus*). J Zoo Wildl Med 30:100–110.

Morse, S.S. 1995. Factors in the emergence of infectious diseases. Emerg Infect Dis 1:7–15.

O'Hara, T.M., C. House, J.A. House, R.S. Suydam, and J.C. George. 1998. Viral serological survey of bowhead whales in Alaska. J Wildl Dis 34:39–46.

Osterhaus, A.D.M.E., H. Yang, and H. Spijkers. 1985. The isolation and partial characterization of a highly pathogenic herpesvirus from the harbour seal (*Phoca vitulina*). Arch Virol 86:239–251.

Osterhaus, A.D.M.E., J. Groen, H.E.M. Spijkers, H.W.J. Broeders, F.G.C.M. UytdeHaag, P. de Vries, J.S. Teppema, I.K.G. Visser, M.W.G. van de Bildt, and E.J. Vedder. 1990. Mass mortality in seals caused by a newly discovered morbillivirus. Vet Microbiol 23:343–350.

Osterhaus, A.D.M.E., I.K.G. Visser, M.V. de Swart, M.W. VanBressem, B.H.G. Van de Bildt, C. Orvelle, T. Barrett, and J.A. Raga. 1992. Morbillivirus threat to Mediterranean monk seals. Vet Rec 130:141–142.

Osterhaus, A.D.M.E., J. Groen, H. Niesters, M. van de Bildt, B. Martina, L. Vedder, J. Vos, H. van Egmond, B.A. Sidi, and M.E.O. Barham. 1997. Morbillivirus in monk seal mass mortality. Nature 388:838–839.

Osterhaus, A.D.M.E., M. van de Bildt, L. Vedder, B. Martina, H. Niesters, J. Vos, H. van Egmond, D. Liem, R. Baumann, E. Androukaki, S. Kotomatas, A. Komnenou, B.A. Sidi, A.B. Jiddou, and M.E.O. Barham. 1998. Monk seal mortality: virus or toxin? Vaccine 16:979–981.

Reidarson, T.H., J. McBain, C. House, D.P. King, J.L. Stott, A. Krafft, J.K. Taubenberger, J. Heyning and T.P. Lipscomb. 1998. Morbillivirus infection in stranded common dolphins from the Pacific Ocean. J Wildl Dis 34:771–776.

Ross, H.M., K.L. Jahans, A.P. MacMillan, R.J. Reid, P.M. Thompson, and G. Foster. 1996. *Brucella* species infection in North Sea seal and cetacean populations. Vet Rec 138: 647–648.

Ross, P.S., P.L. DeSwart, P.J.H. Rejnders, H. VanLoveren, J.G. Vos, and A.D.M.E. Osterhaus. 1995. Contaminant-related suppression of delayed-type hypersensitivity and antibody responses in harbor seals fed herring from the Baltic Sea. Environ Health Perspect 103:162–167.

Smith, A.W., T.G. Akers, S.H. Madin, and N.A. Vedros. 1973. San Miguel sea lion virus isolation, preliminary characterization and relationship to vesicular exanthema of swine virus. Nature 244:108–110.

Stenvers, O., J. Plotz, and H. Ludwig. 1992. Antarctic seals carry antibodies against seal herpesvirus. Arch Virol 123:421–424.

Taubenberger, J.K., M. Tsai, A.E. Krafft, J.H. Lichy, A.H. Reid, F.Y. Schulman, and T.P. Lipscomb. 1996. Two morbilliviruses implicated in bottlenose dolphin epizootics. Emerg Infect Dis 2:213–216.

Thompson, P.M., H.J.C. Cornwell, H.M. Ross, and D. Miller. 1992. A serological study of the prevalence of phocine distemper virus in a population of harbour seals in the Moray Firth, N.E. Scotland. J Wildl Dis 28:21–27.

Van Bonn, W., E.D. Jensen, C. House, T. Burrage, and D.A. Gregg. 2000. Epizootic vesicular disease in California sea lions. J Wildl Dis 36:500–507.

Van Bressem, M.F., I.K.G. Visser, R.L. de Swart, C. Orvell, L. Stanzani, E. Androukaki, K. Siakavara, and A.D.M.E. Osterhaus. 1993. Dolphin morbillivirus infection in different parts of the Mediterranean Sea. Arch Virol 129:235–242.

Van Bressem, M.F., K. Van Waerebeek, M. Fleming, and T. Barrett. 1998. Serological evidence of morbillivirus infection in small cetaceans from the Southeast Pacific. Vet Microbiol 59:89–98.

Visser, I.K.G., M.W.G. Van de Bildt, H.N. Brugge, P.J.H. Reijnders, E.J. Vedder, J. Kuiper, P. de Vries, J. Groen, H.C. Walvoort, F.G.C.M. UytdeHaag, and A.D.M.E. Osterhaus. 1989. Vaccination of harbour seals (*Phoca vitulina*) against phocid distemper with two different inactivated canine distemper virus (CDV) vaccines. Vaccine 7:521–526.

Webster, R.G., V.S. Hinshaw, W.J. Bean Jr., D.J. St. Aubin, and G. Petursson. 1981. Characterization of an influenza A virus from seals. Virology 113:712–724.

Zarnke, R.L., T.C. Harder, H.W. Vos, J.M. Ver Hoef, and A.D.M.E. Osterhaus. 1997. Serologic sruvey for phocid herpesvirus-1 and- 2 in marine mammals from Alaska and Russia. J Wildl Dis 33:459–465.

10

Viruses as Evolutionary Tools to Monitor Population Dynamics

Mary Poss
Roman Biek
Allen Rodrigo

The field of virology is a discipline integral to the medical sciences. The affiliation of virology with population and conservation biology may not be as apparent. Viruses, and in particular virus evolution, may significantly contribute to the understanding of changes in host population structure. The impact of viruses is most notable when infection results in high mortality. Viruses can also exist as persistent or latent infections in a population. Although individual carriers in the population may experience few clinical effects, chronic carriers of a virus enhance the potential for epizootic or zoonotic disease in contact populations. In addition to their role in disease, viruses may also be sensitive genetic indicators of host population structure. In this chapter, we present topics at the interface of virology and population biology, beginning with an evident case where the virus is a pathogen to the host and concluding with a theoretical framework for the use of viral genes as markers of host populations. The focus of this discussion is on wildlife populations, with reference to humans where zoonotic potential exists.

10.1 Viruses as Pathogens

10.1.1 Virus Infection and the Individual

To initiate an infection, a virus must first gain entry to the host by penetrating protective epithelial surfaces such as skin or mucosa. Viral replication can be restricted to these surfaces, or the virus can spread via the blood to other target organs in the body or via neuronal processes to the brain. Following systemic invasion, secondary or tertiary rounds of viral replication occur in target organs. Tissue damage can result either from direct cytopathic effects of virus replication in the cell or from the host immune response towards the virus. Because many viruses replicate primarily or secondarily in the respiratory or gastrointestinal tract, clinical signs referable to these organs are common. Tissue damage is frequently self-limiting, and the infected host can recover if supportive therapy is available. In wildlife populations, therapeutic intervention is not routinely available or practical. Viruses that cause only limited infection in humans or domestic animals may, therefore, be associated with high mortality in wildlife populations.

Although the most notable outcome of viral infection is disease or death, in fact, many infections may be subclinical. Often times, a virus may be cleared from the host because of preexisting immunity or because of an effective primary immune response. Alternatively, an infected individual may mount an immune response that is not sufficient to clear the virus. In this case, the virus may be maintained at a low level and the result of infection can either be asymptomatic or produce chronic disease. A third possible outcome of virus infection is that the virus may establish a latent infection. Reactivation and shedding of the virus can occur sporadically in a latently infected individual. In both persistent and latent infections, the host may be able to transmit the virus to others but have no outward indication of infection.

10.1.2 Viruses and Populations

The study of viral disease tends to focus on an infected individual; however, viruses are maintained in populations, not in individuals. Population size and structure are critical factors that determine the susceptibility of resident individuals to disease outbreaks. For a population to sustain a virus, an adequate number of susceptible hosts must be available (Finkenstadt et al. 1998a, b). This condition is met in large populations and also in smaller populations if immunity to the virus is short-lived. Thus, populations that experience a sudden increase in number could theoretically reach a critical carrying capacity for virus infection (Allen and Carmier 1996). In contrast, a population in decline may reach a point where there are insufficient susceptible hosts to maintain a virus.

Population behavior or structure can also determine the nature and success of a virus in the population (Foley et al. 1999). Species that live in herds or packs may maintain viruses that require close contact for spread. Large populations are capable of harboring more virulent viruses (Anderson 1995). Migratory animals may have seasonal exposure to viral pathogens, thus maintaining viruses that do not elicit long-term immunity. Migrants may also be conduits to disseminate viruses to new geographic areas (Anderson 1995).

The relationship between virus infection and population size is bilateral. On the one hand, viruses require minimum host population sizes for persistence. On the other hand, viral infections themselves are likely to affect both birth and death rates, which in turn determine population size. In wildlife populations, high mortality in adults may be the most evident result of infections. Both historical and recent examples of viral disease that affect wildlife populations are plentiful. One of the most notable historical examples is the decimation of wild ungulates following introduction of rinderpest virus into Africa in 1889 (Anderson 1995; Barrett and Rossiter 1999). An outbreak of canine distemper in Serengeti National Park, Tanzania, resulted in high mortality in lions and other carnivores in that park (Roelke-Parker et al. 1996), and the population of wild chimpanzees in Tai National Park, Ivory Coast, was severely depleted following an outbreak of Ebola virus (Formenty et al. 1999). In contrast, neonatal mortality or reproductive failure resulting from infectious agents may be more difficult to discern. In fact, infections with many viruses common in wildlife, including members of the families Parvoviridae, Herpesviridae, Paramyxoviridae, and Orbiviridae, can result in abortion or neonatal death. It is feasible, therefore, that population size can be modulated by virus infection in the absence of measurable adult mortality.

Thus, infection may lead to disease and death in some animals with a direct effect on population structure. The significance of viral infections in wildlife populations, however, is not limited to epizootic disease outbreaks. Viral infections may also result in a chronic carrier state, or reservoir, of virus in the population. Although persistently infected individuals may experience clinical signs, the significance of reservoir populations may manifest when there is contact with naive populations. Indeed, when considering the impact of disease on natural population dynamics, contact between viral carriers and unexposed individuals may be most important. In order to understand the effect of viruses on populations, it is essential to consider those factors that influence population interactions.

10.2 Population Dynamics

10.2.1 Habitat Fragmentation

In species conservation, the loss and fragmentation of natural habitats remain two of the biggest challenges to conservationists and wildlife managers. Viruses become a primary concern if they lead to epidemics in endangered species. In fact, the same features that affect population dynamics in a fragmented habitat are also critical for the introduction and maintenance of viruses in these populations. Hence, the role of viral infections in a broader spatial context deserves more attention.

Habitat loss has a detrimental effect on living communities. In the long term, the disruption of formerly contiguous landscapes, or fragmentation, may have equally profound consequences for populations. Where intact habitat is fragmented, a human-altered matrix surrounds remnants. For example, many formerly forested landscapes now exist as patches of forest surrounded by farmland and human settlements. For many forest species, this matrix does not provide suitable habitat. Thus, populations become effectively isolated from each other because they are confined to forested fragments. As available habitat declines, so does population size, making these species vulnerable to local extinction (Gonzalez et al. 1998; Saccheri et al. 1998). Local extinction of single populations does not preclude overall persistence of a species on the landscape level. According to the metapopulation concept (Levins 1969), populations will persist if not all subpopulations become extinct simultaneously and recolonization of empty patches can occur. Thus, both the frequency of local extinctions and the degree of connectivity between patches are important determinants of species persistence. An important implication of these concepts is that maintenance of the status quo might not be sufficient to prevent the loss of species on a landscape scale. That is, even if all remaining subpopulations and their habitat are protected as nature reserves, local extinction might occur at a rate fast enough that it cannot be balanced by immigration (Hanski et al. 1996). Quite clearly, factors that tend to accelerate local extinction rates such as new pathogens will worsen the prospect of survival.

Of course, the image of a metapopulation consisting of numerous equal subunits is a crude oversimplification. Under natural conditions subpopulations are likely to differ from each other in many respects, including total number of individuals and the degree of individual exchange. One consequence of subpopulation heterogeneity is that populations on some patches will produce sufficient offspring to outweigh mortality, resulting in an excess of individuals leaving the patch as emigrants. Such "source" populations are likely to have a much greater influence on overall species persistence than "sink" populations, that is, those that are maintained only by continuous immigration (Pulliam 1988; Murphy et al. 1990). Obviously, such specific demographic information on metapopulation structure is critical to managing and preserving wildlife populations effectively. In addition, the absolute size of habitat fragments is not the only determinant of population viability. Depending on its shape, a given remnant might not contain undisturbed core areas, leading to the exclusion of sensitive species (Temple 1986). While edge effects can be due to a range of factors, including altered light or moisture regimes, many studies point at the influence of new species entering habitat fragments from the edge and affecting native communities by predation or competition. For example, nest success near forest edges appears to be reduced in many bird communities due to predators and nest parasites accessing forest fragments from the surrounding matrix (Rich et al. 1994; Robinson et al. 1995). It is plausible, therefore, that increased contact between native populations in a fragmented landscape with species inhabiting matrix could result in the introduction of new pathogens, including viruses (Holmes 1996). This may lead to metapopulations in a landscape that have different levels of immunity and, hence, different levels of susceptibility to future exposure with a virus. Demographic variability in source–sink relationships may be of critical importance for the prospects of species survival in fragmented landscapes.

10.2.2 Habitat Changes and Emergent Infections

Although some infections can be initiated by new viruses, evidence strongly supports the premise that emergent infections more commonly arise from contact of a susceptible population with a known infectious agent. In this manner, a virus endemic in one population may emerge as a virulent pathogen if it is transmitted to susceptible, naive hosts. If the virus is maintained in a reservoir population, then the endemic area for the disease is defined by the geographic distribution of the carrier species. It follows, therefore, that changes in reservoir population demographics may have significant impact on zoonotic and epizootic infections. Habitat fragmentation can affect population distribution by clustering susceptible hosts and increasing their contact with reservoir hosts inhabiting the surrounding matrix. The recent outbreaks in African mammals of both canine distemper (Harder et al. 1995; Roelke-Parker et al. 1996) and rabies (Cleaveland and Dye 1995) resulted from transmission of these viruses from domestic dog reservoirs in conjunction with increasing encroachment of human populations on wildlife reserves.

The importance of understanding reservoir host ecology has been underscored by recent examples of emergent infections in humans that are directly associated with change in habitat or in population structure of the carrier species. For example, in Argentina an epidemic of hemorrhagic fever was related to conversion of pampas to agricultural land. Hemorrhagic fever in humans may be caused by members of a variety of virus families, including Hantaviridae and Arenaviridae, which are maintained in rodent reservoirs. Within these virus families, each individual virus chronically infects a single rodent species (family Muridae) (Levis et al. 1998). Species of mice that thrived on new vegetation arising in cultivated fields and roadsides replaced native species associated with pampas vegetation (Mills et al. 1994). The invading rodents were hosts for the arenavirus Junin virus, responsible for Argentina hemorrhagic fever. The endemic disease area reflected the range of the rodent host (Mills and Childs 1998). The geographic distribution of the reservoir host, however, was larger than the endemic disease area, suggesting that subpopulation dynamics of the host may influence disease outbreaks.

How can we identify the source of the pathogen that causes an epidemic? Animals and insects that are reservoirs for the virus often exhibit no clinical signs to indicate that they are infected. Recently, the genetics of the virus itself has been used to trace the source of disease outbreaks (Holmes et al. 1995; Holmes 1998). Viruses can undergo several hundred generations annually and thus leave a measurable evolutionary record over a short time period. The genotype of a virus recovered from a recently infected host will be most closely related to the sequence of the virus present in the reservoir host. Based on genetic analysis of viral sequences, it was determined that the closest relative of the virus isolated from African lions that succumbed to distemper was derived from domestic dogs (Harder et al. 1995; Roelke-Parker et al. 1996; Carpenter et al. 1998). A comparison of rabies virus genomes recovered following an outbreak of rabies in African mammals was similarly most closely related to domestic dogs (Nel et al. 1997). This approach has also been used to discern the animal origin of important human pathogens such as human immunodeficiency virus (HIV) (Gao et al. 1999), influenza virus (Schafer et al. 1993; Shu et al. 1994; Scholtissek 1995; Brown et al. 1998; Kawaoka et al. 1998), and rotavirus (Shirane and Nakagomi 1994).

10.2.3 Viral Evolution and Host Population Dynamics

The same properties that lead to the success of a virus as a pathogen may make these organisms ideal markers of host population structure. Viruses are dependent on host cell processes for most aspects of the viral life cycle; therefore, these "parasitic" organisms must be well adapted to their host species. The short generation time and rapid rate of evolution allow a virus to respond quickly to a changing host environment. Indeed, viruses have been shown to reflect the phylogenic relationships of their hosts. For example, papillomavirus phylogeny mirrors the origin and emigration of human ancestors from Africa (Bernard 1994; Chan et al. 1997). Similarly, the phylogeny of African primates is mirrored by the phylogeny of the simian immunodeficiency virus (SIV) that infects each member of this group (Hirsch et al. 1993; Gao et al. 1999).

Flaviviruses are RNA viruses that are transmitted by arthropod vectors. Divergence of this group of viruses reflects the generation time of the vector with which a specific virus is associated. For example, tick-associated flaviviruses diverge more slowly than do flaviviruses that use mosquito vectors, indicating the longer generation time of ticks as compared to mosquitoes (Marin et al. 1995; Zanotto et al. 1995, 1996).

10.2.4 Retroviruses and Host Population Dynamics

Because of the rapid evolutionary potential of RNA viruses and evidence of host–virus coevolution, RNA viruses may indeed be good indicators of recent changes in a host population. Members of the family Retroviridae are particularly well suited to be genetic markers of their host because of their unique life style. Retroviruses evolve rapidly because of the error-prone replication process inherent in viral-encoded replicates (Coffin 1986) and rapid viral turnover. For the best studied retrovirus, HIV-1, there are typically 0.01 substitutions per site per year in the gene encoding the envelope protein (Ricchetti and Buc 1990; Coffin 1995), with the virus completing approximately 300 generations per year. Compare this to the average eukaryotic or prokaryotic gene, which accumulates substitutions at the rate of 0.01 per site per four million years, and one begins to appreciate the enormous potential for mutability that retroviruses have. Unlike other RNA viruses, retroviruses integrate into the host genome, leaving a permanent record of infection. Furthermore, most retrovirus infections are persistent and nonpathogenic and therefore do not substantively affect the longevity of their host. Finally, most vertebrate species are host to retroviruses (Tristem et al. 1996; Herniou et al. 1998; Martin et al. 1999). The combination of rapid evolution, a persistent, stable genetic record of viral infection, and wide distribution in vertebrates suggests that retrovirus genomes may be good candidates for study of recent changes in host population structure.

If one looks at well-studied retroviruses, for example, HIV-1, it becomes apparent that much can be learned about the hosts themselves. For instance, Leitner and colleagues (1996) showed that sequences encoding the HIV envelope gene obtained from different individuals in a transmission cluster very closely matched the known transmission history of the group. Court cases involving the transmission of HIV across individuals (Ou et al. 1992) also testify to the potential for tracking host-to-host contact by resolving HIV retrovirus evolutionary histories (phylogenies). On a broader scale, the existence of geographically related clusters of retroviral sequences, seen most readily with HIV (Kuiken and Korber 1994) and also with feline immunodeficiency virus (FIV) (Bachmann et al. 1997), points to the fact that barriers to movement in the host population can be inferred from the phylogenies of the retroviruses themselves. This happens because the probability of host contact across a barrier is lower than that of host contact within a region where no barrier exists. Consequently, viruses obtained from hosts that are isolated from other hosts will tend to cluster together on an evolutionary tree.

It is reasonable to suggest that retroviral genomes can be used as proxy genetic markers of hosts and provide as much, if not more, information about historical host movement and dynamics as do host genes. Consequently, there is a theoretical imperative to consider the possibility that integrated retroviral genomes may be treated as proxy genetic markers of their hosts. The fact that these genomes evolve so rapidly further implies that, unlike eukaryotic genes, the evolution of retrovirus proxies may be tracked over a reasonable period of time (i.e., historical time, rather than evolutionary time). Therefore, a model of host population dynamics that is based on inferences derived from retrovirus evolution may allow us to track the movement of hosts, changes in host population size, and the extent of host population subdivision. As pointed out above, such information is of critical value in the effort to predict and prevent the loss of species.

10.3 A Simple Model of Viral Evolution

Consider a population of mammals that is threatened or endangered. To define a comprehensive management strategy for such a population, it is necessary to know its size, the rate at which numbers are growing or (more likely) falling, its geographic home range or, if the population is fragmented, the home range of the constituent subpopulations, and the extent of contact between animals from one subpopulation and another. Of course, there are methods of sampling and tracking such populations that have been applied extensively by wildlife and conservation biologists. To these methods, the population geneticist can add tools that will allow the same processes to be inferred using genetic data, for example, microsatellite markers or mitochondrial DNA haplotype frequencies. What retroviruses offer are alternative genetic data that, by virtue of the rapidity of their evolution, can allow these same processes to be studied over a span of time that is consistent with field studies. Think of the retroviruses that infect each host as a unique marker for that host. When that host infects another, it transmits its retroviral marker, which then begins to change in the newly infected individual so that it now becomes a marker for that individual. The movement of these hosts can be tracked by the movement of these markers. Consequently, the types of dynamics that conservation biologists study—migration and growth—may be measurable over historical rather than evolutionary time.

In this section we describe a very simple model of viral evolution to demonstrate how it may be used to estimate host population size. Consider a constant-sized population of N individual potential hosts in which there is an equilibrium prevalence of infection, with N_h as the number of infected individuals. Let us select two individuals at random from this population. Under a stochastic Poisson model of infection where each infectious individual has an equal probability of transmission, the probability that both were infected from the same individual in the previous infection cycle (i.e., the expected length of time it takes one individual to infect another) is $1/N_h$. If we trace back their route of infection, the probability (P) that this route began with the same individual in the kth infection cycle (where k is counted backward in time from present to past) is given by the product of the probability that there was no common infecting source for $k - 1$ infection cycles, and a common source is found in the kth cycle:

$$P(t = k) = \frac{1}{N_h}\left(1 - \frac{1}{N_h}\right)^{k-1}$$

which is a simple geometric distribution with a mean time of N_h. This formulation is directly equivalent to that applied in standard population genetics to derive the expected time to common ancestry of two randomly sampled individuals in a constant-sized population with no selection.

If a retroviral sequence is obtained from each of the two sampled individuals, then when the lineages come together in the infecting host, there is an additional period within the host before these two sequences coalesce, since some degree of divergence has occurred within the host itself.

For the purposes of estimation, it is convenient to rescale retroviral generations to infectious cycles, so that each unit of time (measured in infectious cycles) is g viral generations long, where g is the number of generations per infectious cycle. As a first approximation, we assume that an infectious cycle corresponds to the reproductive interval. Assuming that no selection acts on the viral population in vivo (perhaps a reasonable assumption with a minimally pathogenic retroviral population) and a constant within-host viral population size, N_v, the expected time to coalescence of two retroviruses within a host is $1/N_v$. Therefore, the expected total time $(E[t])$ for a pair of sequences, each sampled from a different host, to coalesce is:

$$E[t] = N_h + \frac{N_v}{g}$$

The expected nucleotide divergence between a pair of viral sequences obtained in the manner described above is $2\mu g E[t]$, where μ is the mutation rate of the virus per site per viral generation.

What all this means is that if we are able to obtain viral sequences from a sample of hosts and calculate the average differences (expressed as evolutionary distances) between pairs of sequences, we would be in a position to obtain estimates for the above parameters. N_v could be estimated from the within-host viral variation, and we would perhaps make intelligent guesses about g (possibly working within a Bayesian estimation framework), leaving us with an estimate of N_h, the number of infected hosts. But the idea is not to stop here; rather, these same types of models can be extended to take account of host population subdivision and migration, as well as changes in host population size. Finally, with retroviruses, it is possible to sample over time and actually track the accumulation of mutations along the viral genome in a span of time that is consistent with field studies, unlike genetic studies of wildlife populations, where mutations occur over the course of thousands of years.

Of course, there are always difficulties with simplistic theoretical frameworks and natural biological systems. For example, if we are to make inferences about the effective size of the potential host population, N, we would have to assume that N_h, defined above as the number of infected hosts, is some constant proportion of N. This is equivalent to assuming that prevalence is at equilibrium and changes in size of the potential host population are reflected in changes in the numbers of infected hosts. Also, we have excluded the possibility that there is some kind of selection acting on either our host population or the viral population within the host. To the extent that significant selective forces act on either of these populations, our model may be more or less robust.

Here, we have described just one simple model to illustrate how one might go about using viral diversity to tell us something about the host population. However, there are many population genetic methods and models available. One class of these—coalescent methods—models the genealogies of viruses and hosts (Rodrigo and Felsenstein 1998) and is particularly useful when estimation is based on small samples of hosts/viruses drawn from much larger populations. It is sufficient to note here that there is an impressive mass of population genetics theory available to the conservation biologist. The difficulty has been finding the right genetic marker for populations of conservation interest. Retroviruses appear to be a theoretically defensible choice.

10.4 Conclusions

This chapter examines the role of viruses both as forces that can shape populations and as tools that can be used to monitor population dynamics. There is no doubt that viruses can cause high mortality and sometimes extinction of populations. For a virus to persist, however, it must have access to susceptible hosts. Hence, host demographics play an important role in sustaining the virus population. Because viruses are so dependent on the host, and because viruses are capable of extraordinary mutation rates by eukaryotic standards, it also follows that viruses will closely track changes in host populations. Thus, the phylogeny of viral genes may mirror the phylogeny of the infected host population. The theoretical use of retroviruses as a rapidly evolving host gene to discern recent changes in host demographics is introduced in this chapter. If viruses can be used to ascertain current rather than historical changes in population structure, then the relationship between virology and conservation biology may take on new significance.

References

Allen, L.J., and P.J. Cormier. 1996. Environmentally driven epizootics. Math Biosci 131: 51–80.
Anderson, E.C. 1995. Morbillivirus infections in wildlife (in relation to their population biology and disease control in domestic animals). Vet Microbiol 44:319–332.

Bachmann, M.H., C. Mathiason-Dubard, G.H. Learn, A.G. Rodrigo, D.L. Sodora, P. Mazzetti, E.A. Hoover, and J.I. Mullins. 1997. Genetic diversity of feline immunodeficiency virus: dual infection, recombination, and distinct evolutionary rates among envelope sequence clades. J Virol 71:4241–4253.

Barrett, T., and P.B. Rossiter. 1999. Rinderpest: the disease and its impact on humans and animals. Adv Virus Res 53:89–110.

Bernard, H.U. 1994. Coevolution of papillomaviruses with human populations. Trends Microbiol 2:140–143.

Brown, I.H., P.A. Harris, J.W. McCauley, and D.J. Alexander. 1998. Multiple genetic reassortment of avian and human influenza A viruses in European pigs, resulting in the emergence of an H1N2 virus of novel genotype. J Gen Virol 79:2947–2955.

Carpenter, M.A., M.J. Appel, M.E. Roelke-Parker, L. Munson, H. Hofer, M. East, and S.J. O'Brien. 1998. Genetic characterization of canine distemper virus in Serengeti carnivores. Vet Immunol Immunopathol 65:259–266.

Chan, S.Y., H.U. Bernard, M. Ratterree, T.A. Birkebak, A.J. Faras, and R.S. Ostrow. 1997. Genomic diversity and evolution of papillomaviruses in rhesus monkeys. J Virol 71: 4938–4943.

Cleaveland, S., and C. Dye. 1995. Maintenance of a microparasite infecting several host species: rabies in the Serengeti. Parasitology 111:S33–S47.

Coffin, J.M. 1986. Genetic variation in AIDS viruses. Cell 46:1–4.

Coffin, J.M. 1995. HIV population dynamics in vivo: implications for genetic variation, pathogenesis, and therapy. Science 267:483–489.

Finkenstadt, B., and B. Grenfell. 1998a. Empirical determinants of measles metapopulation dynamics in England and Wales. Proc R Soc Lond B 265:211–220.

Finkenstadt, B., M. Keeling, and B. Grenfell. 1998b. Patterns of density dependence in measles dynamics. Proc R Soc Lond B 265:753–762.

Foley, J.E., P. Foley, and N. Pedersen. 1999. The persistence of a SIS disease in a metapopulation. J Appl Ecol 36:544–554.

Formenty, P., C. Boesch, M. Wyers, C. Steiner, F. Donati, F. Dind, F. Walker, and B. Le Guenno. 1999. Ebola virus outbreak among wild chimpanzees living in a rain forest of Cote d'Ivoire. J Infect Dis 179:S120–S126.

Gao, F., E. Bailes, D.L. Robertson, Y. Chen, C.M. Rodenburg, S.F. Michael, L.B. Cummins, L.O. Arthur, M. Peeters, G.M. Shaw, P.M. Sharp, and B.H. Hahn. 1999. Origin of HIV-1 in the chimpanzee *Pan troglodytes troglodytes*. Nature 397:436–441.

Gonzalez, A., J.H. Lawton, F.S. Gilbert, T.M. Blackburn, and I. Evans-Freke. 1998. Metapopulation dynamics, abundance, and distribution in a microecosystem. Science 281: 2045–2047.

Hanski, I., A. Moilanen and M. Gyllenberg. 1996. Minimum viable metapopulation size. Am Nat 147:527–541.

Harder, T.C., M. Kenter, M.J. Appel, M.E. Roelke-Parker, T. Barrett, and A.D. Osterhaus. 1995. Phylogenetic evidence of canine distemper virus in Serengeti's lions. Vaccine 13:521–523.

Herniou, E., J. Martin, K. Miller, J. Cook, M. Wilkinson, and M. Tristem. 1998. Retroviral diversity and distribution in vertebrates. J Virol 72:5955–5966.

Hirsch, V.M., C. McGann, G. Dapolito, S. Goldstein, A. Ogen-Odoi, B. Biryawaho, T. Lakwo, and P.R. Johnson. 1993. Identification of a new subgroup of SIVagm in tantalus monkeys. Virology 197:426–430.

Holmes, E.C. 1998. Molecular epidemiology and evolution of emerging infectious diseases. Br Med Bull 54:533–543.

Holmes, E.C., S. Nee, A. Rambaut, G.P. Garnett, and P.H. Harvey. 1995. Revealing the history of infectious disease epidemics through phylogenetic trees. Philos Trans R Soc Lond B 349:33–40.

Holmes, J.C. 1996. Parasites as threats to biodiversity in shrinking ecosystems. Biodiv Conserv 5:975–983.

Kawaoka, Y., O.T. Gorman, T. Ito, K. Wells, R.O. Donis, M.R. Castrucci, I. Donatelli, and R.G. Webster. 1998. Influence of host species on the evolution of the nonstructural (NS) gene of influenza A viruses. Virus Res 55:143–156.

Kuiken, C.L., and B. Korber. 1994. Epidemiological significance of intra- and inter-person variation of HIV-1. AIDS 8:S73-S83.

Leitner, T., D. Escanilla, C. Franzen, M. Uhlen, and J. Albert. 1996. Accurate reconstruction of a known HIV-1 transmission history by phylogenetic tree analysis. Proc Natl Acad Sci USA 93:10864–10869.

Levins, R. 1969. Some demographic and genetic consequences of environmental heterogeneity for biological control. Bull Entomol Soc Am 15:237–240.

Levis, S., S.P. Morzunov, J.E. Rowe, D. Enria, N. Pini, G. Calderon, M. Sabattini, and S.C. St. Jeor. 1998. Genetic diversity and epidemiology of hantaviruses in Argentina. J Infect Dis 177:529–538.

Marin, M.S., P.M. Zanotto, T.S. Gritsun, and E.A. Gould. 1995. Phylogeny of TYU, SRE, and CFA virus: different evolutionary rates in the genus *Flavivirus*. Virology 206: 1133–1139.

Martin, J., E. Herniou, J. Cook, R.W. O'Neill, and M. Tristem. 1999. Interclass transmission and phyletic host tracking in murine leukemia virus–related retroviruses. J Virol 73:2442–2449.

Mills, J.N., and J.E. Childs. 1998. Ecologic studies of rodent reservoirs: their relevance for human health. Emerg Infect Dis 4:529–537.

Mills, J.N., B.A. Ellis, J.E. Childs, K.T. McKee Jr., J.I. Maiztegui, C.J. Peters, T.G. Ksiazek, and P.B. Jahrling. 1994. Prevalence of infection with Junin virus in rodent populations in the epidemic area of Argentine hemorrhagic fever. Am J Trop Med Hyg 51:554–562.

Murphy, D.D., K.E. Freas, and S.B. Weiss. 1990. An environment-metapopulation approach to population viability analysis for a threatened invertebrate. Conserv Biol 4: 41–51.

Nel, L., J. Jacobs, J. Jaftha, and C. Meredith. 1997. Natural spillover of a distinctly Canidae-associated biotype of rabies virus into an expanded wildlife host range in southern Africa. Virus Genes 15:79–82.

Ou, C.Y., C.A. Ciesielski, G. Myers, C.I. Bandea, C.C. Luo, B.T. Korber, J.I. Mullins, G. Schochetman, R.L. Berkelman, A.N. Economou, et al. 1992. Molecular epidemiology of HIV transmission in a dental practice. Science 256:1165–1171.

Pulliam, H.R. 1988. Sources, sinks, and population regulation. Am Nat 132:652–661.

Ricchetti, M., and H. Buc. 1990. Reverse transcriptases and genomic variability: the accuracy of DNA replication is enzyme specific and sequence dependent. EMBO J 9: 1583–1593.

Rich, A.C., D.S. Dobkin, and L.J. Niles. 1994. Defining forest fragmentation by corridor width: the influence of narrow forest-dividing corridors on forest-nesting birds in southern New Jersey. Conserv Biol 8:1109–1121.

Robinson, S.K., F.R. Thompson III, T.M. Donovan, D.R. Whitehead, and J. Faaborg. 1995. Regional forest fragmentation and the nesting success of migratory birds. Science 267:1987–1990.

Rodrigo, A.G., and J. Felsenstein. 1998. Coalescent approaches to HIV population genetics. *In* Crandall, K.A., ed. The evolution of HIV, pp. 233–272. Johns Hopkins University Press, Baltimore.

Roelke-Parker, M.E., L. Munson, C. Packer, R. Kock, S. Cleveland, M. Carpenter, S.J. O'Brien, A. Pospischil, R. Hofmann-Lehmann, H. Lutz, et al. 1996. A canine distemper virus epidemic in Serengeti lions (*Panthera leo*). Nature 379:441–445.

Saccheri, I., M. Kuussaari, M. Kankare, P. Vikman, W. Fortelius, and I. Hanski. 1998. Inbreeding and extinction in a butterfly metapopulation. Nature 392:491–494.

Schafer, J.R., Y. Kawaoka, W.J. Bean, J. Suss, D. Senne, and R.G. Webster. 1993. Origin of the pandemic 1957 H2 influenza A virus and the persistence of its possible progenitors in the avian reservoir. Virology 194:781–788.

Scholtissek, C. 1995. Molecular evolution of influenza viruses. Virus Genes 11:209–215.

Shirane, K., and O. Nakagomi. 1994. Interspecies transmission of animal rotaviruses to humans as evidenced by phylogenetic analysis of the hypervariable region of the VP4 protein. Microbiol Immunol 38:823–826.

Shu, L.L., Y.P. Lin, S.M. Wright, K.F. Shortridge, and R.G. Webster. 1994. Evidence for interspecies transmission and reassortment of influenza A viruses in pigs in southern China. Virology 202:825–833.

Temple, S.A. 1986. Predicting impacts of habitat fragmentation on forest birds: a comparison of two models. *In* Verner, J., M.L. Morrison, and C.J. Ralph, eds. Wildlife 2000: modelling habitat relationships of terrestrial vertebrates, pp. 301–304. University of Wisconsin Press, Madison.

Tristem, M., E. Herniou, K. Summers, and J. Cook. 1996. Three retroviral sequences in amphibians are distinct from those in mammals and birds. J Virol 70:4864–4870.

Zanotto, P.M., G.F. Gao, T. Gritsun, M.S. Marin, W.R. Jiang, K. Venugopal, H.W. Reid, and E.A. Gould. 1995. An arbovirus cline across the northern hemisphere. Virology 210:152–159.

Zanotto, P.M., E.A. Gould, G.F. Gao, P.H. Harvey, and E.C. Holmes. 1996. Population dynamics of flaviviruses revealed by molecular phylogenies. Proc Natl Acad Sci USA 93:548–553.

11

Assessing Stress and Population Genetics Through Noninvasive Means

Samuel K. Wasser
Kathleen E. Hunt
Christine M. Clarke

Monitoring changes in animal abundance, the presence of pathogens, and physiological well-being of wildlife is critical for effective management of wildlife diseases. Monitoring animal abundance enables one to recognize when a potential die-off is occurring. Pathogen monitoring enables one to examine potential causes of the mortality. Monitoring physiological stress alerts investigators to conditions that could increase susceptibility to disease. Monitoring reproductive function can do this as well, in addition to alerting investigators to other problems such as endocrine disruption resulting from exposure to pollutants. These latter measures may also rule out alternative diseases that may present clinical signs similar to those of emergent diseases.

Obtaining this information can be difficult even for the most observable species. Available methods have been invasive, severely limiting the sampling frequencies required to effectively carry out such monitoring. Noninvasive techniques for acquiring DNA and hormones from wildlife living in remote areas can greatly facilitate the above monitoring needs. These techniques enhance sample accessibility, which is invaluable for a wide variety of monitoring purposes. DNA acquired from hair or feces can be used to identify individuals and can be applied to mark-recapture models to estimate animal abundances (Kohn et al. 1999; Woods et al. 1999; Mowat and Strobeck, 2000). Fecal DNA may also prove useful for detecting pathogens shed in feces, providing an early warning system for emergent diseases. Adrenal and gonadal hormones in feces can be used to measure physiological stress in response to environmental disturbances (Wasser 1996); they can also be used as indices of other disease states such as endocrine disruption in response to pollutants.

Fecal-based techniques can be enhanced further by new methods using specially trained scat detection dogs to increase sampling efficiency and reduce collection biases. Each of these methods can be integrated with geographic information system (GIS)-based technology, providing a landscape approach to some of these problems. This chapter describes the use of fecal-based technologies, providing examples of how they apply to both small-scale and large-scale wildlife disease management programs in remote areas.

11.1 Measurement of Fecal Hormones

Fecal steroid hormones have proven remarkably useful for noninvasive monitoring of reproductive and stress physiology. Feces typically contain large amounts of steroid hormones such as androgens, estrogens, and progestins; glucocorticoid stress hormones such as corticosterone, cortisol, and dehydroepiandrosterone (DHEA); and mineralocorticoids. The large protein hormones, such as luteinizing hormone and growth hormone, are destroyed in the gut and typically shed in urine and therefore are not measurable in feces. There are some additional small-molecular-weight hormones that may be measurable in feces, such as thyroid hormones and prostaglandins, but these groups have been little explored.

Steroid hormones are generally quite stable, though in the gut they are metabolized to various degrees by gut flora and intestinal enzymes. The steroid hormones are not distributed evenly in feces, and therefore fecal samples should be thoroughly mixed in the field, followed with further mixing or sifting at every subsequent stage of preparation or storage (Wasser et al. 1996). Steroids are also excreted in urine, usually after conjugation to water-soluble side groups by the liver. The proportion of each hormone voided in feces versus urine varies with the type of hormone and species (Palme et al. 1996). In cases in which a high proportion of steroid is voided in urine, animals should be observed to see if they urinate on their feces, as this is a potential source of contamination (Wasser et al. 1988).

Differences in diet (e.g., fiber intake, water intake) may affect fecal hormone levels (Wasser et al. 1993). Steroids are excreted into the gut via the bile after being cleared from the blood by the liver, and differences in fiber intake can change biliary excretion and rate of hormone metabolism. Fiber intake also affects total bulk of excreta, diluting or concentrating the hormones accordingly. In baboons fed differing amounts of fiber, resultant variation in fecal steroid excretion rates were mostly eliminated by freeze-drying the feces and expressing results as nanograms per grams of dried feces (Wasser et al. 1993). This approach has since proven useful in many species. Most of this dietary-related variation is due to greater fiber. This increases water content and associated fecal bulk of the samples, both of which can be eliminated by freeze-drying. In species with highly variable or seasonally changing diets (e.g., bears), it may still be advisable to examine different diets separately.

Once excreted, levels of detectable (immunoreactive) hormones may vary if hormones are metabolized to forms that have lesser or greater affinities for the assay antibody. Longevity of excreted hormones may be affected by location in the field (e.g., arid vs. moist conditions) and efficacy of short-term storage while transporting samples (e.g., silica or ethanol) and long-term storage in the laboratory (e.g., stored frozen with or without freeze-drying or oven drying). Longevity may also vary with metabolites, parent hormone, diet, and species (Whitten et al. 1998). Changes in detectable hormone levels are not necessarily a serious problem if all samples experience the same percentage change. For long-term storage in the laboratory, good results have been obtained for many species by simply freezing samples ($-20°C$) within a few hours of collection (Brockman et al. 1995; Berger et al. 1999). In African elephant and grizzly bear, feces preserved with freeze-drying, oven drying, silica, or no preservative (control) and held either at $-20°C$ or at room temperature had fairly stable levels of immunoreactive glucocorticoids for up to one year (E. Hunt and S.K. Wasser, unpublished data, 1999). It is encouraging that air-dried feces at room temperature can show stable hormone concentrations for many months, since this may mimic natural field conditions in open environments. In the same study, samples preserved with ethanol showed marked increases in cortisol metabolite concentrations over the first few months (E. Hunt and S.K. Wasser, unpublished data, 1999). Ethanol storage has been used successfully for progestins and estrogens in other species (e.g., yellow baboon; Wasser 1996). Optimal storage methods may vary with species and hormone.

Steroids can be extracted from feces using a variety of solvents, including ethanol, methanol, ethyl ether, dichloromethane, and others (Schwarzenberger et al. 1991, 1997; Wasser et al. 1991, 1994; Shideler et al. 1993; Palme and Möstl 1997). Many investigators use some variation of vortexing, shaking, or boiling a weighed amount of feces with several milliliters of 90% ethanol or methanol, for 20 minutes or more. The sample is then centrifuged, after which the alcohol supernatant can be stored, or dried down and reconstituted in assay buffer at the time of analysis. If a high proportion of the fecal metabolites are conjugated and the assay antibody does not recognize the conjugated forms, a subsequent solvolysis or hydrolysis step may be required to remove the conjugates.

Antibody choice for the assay is particularly important. Since the parent hormones in blood may be highly metabolized in feces, an antibody that detects the parent hormone may not reliably detect the fecal metabolites of that hormone. Progesterone, for example, is commonly metabolized into up to 18 different metabolites in mammals, present in varying proportions depending on species, and with varying affinities for different progesterone antibodies (Wasser et al. 1994; Schwarzenberger et al. 1997). Similar results have been obtained for cortisol (Goymann et al. 1999; Bahr et al. 2000). In such cases, antibodies that recognize a broad class of metabolites are therefore preferable to those that are specific for only one hormone. For example, the ICN corticosterone antibody (#07-120102, ICN Biomedicals Inc., Costa Mesa, Calif.) often has high affinities for cortisol metabolites in a wide variety of species, providing more reliable measures of adrenal activation than do some cortisol-specific antibodies.

Because of the strong species-specific differences in fecal metabolites, validations are necessary for each new species studied. Parallelism and accuracy tests ensure that the assay is accurately quantifying hormone across the tested range of concentrations and that there are no interfering substances in the feces (Diamandis and Christopoulos 1996). A variety of other validations can demonstrate that the fecal assay results have explanatory value—that the results correlate with physiological events preceding fecal excretion. One approach is to assess correlations of plasma hormone levels with fecal hormones excreted the next day (there is often about a one-day lag from plasma peak to excretion in feces, although the duration of this delay varies from species to species). However, blood sampling is invasive and often impractical, and with wild animals it is not always possible to obtain fecal samples from the day after blood sampling. Blood sampling is a particular problem for the glucocorticoids (stress hormones), because handling causes rapid increases (within two minutes) in plasma levels of these hormones. It should also be appreciated that plasma levels and fecal levels of hormones may represent different physiologic information. Plasma samples represent a snapshot of hormone level at one moment in time, whereas fecal samples are an integrated sum of hormone secretory events over the previous several hours or days. Therefore, an elevation in fecal levels may represent an increase in number of hormone peaks, an increase in amplitude of a single peak, or a chronic slight elevation in baseline hormone secretion. In some cases fecal measures may be more sensitive than blood, as the accumulation of slight differences in hormone secretion accentuates hormone concentrations measured in feces versus plasma (Wasser et al. 1995; Velloso et al. 1998).

A particularly useful validation is injection of a releasing hormone to increase endogenous release of the hormone of interest, with subsequent collection and assay of fecal samples. For example, challenges with injection of adrenocorticotropic hormone (ACTH) reliably cause increases in plasma levels of corticosterone or cortisol, the stress hormones of the adrenal gland, enabling validation of fecal stress hormone assays without the difficulties of collecting blood.

Injections of radiolabeled hormones can provide definitive answers about excretion lag times and, when coupled with chromatography, how the parent hormone is metabolized in feces, or whether the assay is measuring the predominant metabolite(s). The feces are collected for several days after radio label injection, with each sample counted for radioactivity. The radioactive metabolites can then be separated using high-performance liquid chromatography, and then each fraction containing unique metabolites analyzed separately for antibody affinity.

In addition, researchers can take advantage of known stressful events or reproductive events to validate fecal assays. An animal being moved to a new location or introduced to a new social group will usually show increases in stress hormones, while pregnant females usually show increases in fecal progestins.

Using the above techniques, fecal hormone assays have been validated for dozens of mammalian and avian species to date (Brown et al. 1997; Schwarzenberger et al. 1997; Wasser et al. 1988, 1993, 1994, 1996, 1997a, 2000). Fecal progestin, estrogen, and androgen assays have become particularly widespread due to their usefulness in assessing reproductive status. They have been successfully applied to free-living animals as diverse as moose (Berger et al. 1999), African elephant (Foley et al. 2001), American bison (Kirkpatrick et al. 1996), baboon (Wasser 1996), cotton-top tamarin (Savage et al. 1997), northern spotted owl (Wasser et al. 1997a), and kakapo parrot (Cockrem and Rounce 1995). Fecal glucocorticoid assays are increasingly being used to assess physiologic stress, and have been validated with ACTH challenges for approximately two dozen mammalian and avian species (Palme et al. 1996; Graham and Brown 1997; Goymann et al. 1999). Glucocorticoid levels increase in response to a wide variety of stressors, including starvation, habitat disruption, disease, and social disruption, and prolonged elevations in glucocorticoids are thought to interfere with reproduction and reduce resistance to disease (Moberg 1988; Wingfield et al. 1998). In field studies, fecal assays have demonstrated increases in glucocorticoids that correlate well with environmental disturbance (Wasser et al. 1997a; Millspaugh 1999), drought and social stress (Foley et al. 2001), cold stress (Harper and Austad 2000), and social aggression (Kotrschal et al. 1998; Wallner et al. 1999). These glucocorticoid assays may potentially be extremely useful in identifying populations that are directly stressed by disease or parasites, and/or are suffering from chronic stress that may reduce resistance to future disease.

A final important point is that normal values of fecal hormones remain unknown for most wild species. Thus, when investigating potentially stressed populations or populations that may be suffering reproductive dysfunction, it is essential to have a control or comparison group of relatively unstressed, healthy animals, ideally with the same diet, age structure, and reproductive structure as the potentially stressed population. As fecal assays come into increasingly widespread use, it will be valuable to develop a database of normal fecal hormone values for healthy, unstressed populations across seasons and diets.

11.2 Measurement of Fecal DNA

Most DNA obtained from scat tends to be small in size and quantity, owing to sample degradation. Nevertheless, it is possible to isolate mitochondrial DNA (mtDNA), microsatellite DNA, and single-copy nuclear DNA (scnDNA) from scat, each of which can provide a great deal of information about the individual and species in question. Of these various forms of DNA, mtDNA tends to be the easiest to amplify because it is substantially more abundant than nuclear DNA. mtDNA can be used to confirm species identity, determine geographic origins of populations, and assess rates of evolution (Cronin et al. 1991a, b; Avise 1993; Paetkau et al. 1995; Hamdrick and Avise 1996; Wooding and Ward 1997). Because mtDNA is maternally inherited (uniparental) and therefore haploid, its applications also have their limitations. The discriminating ability of DNA analysis can be further enhanced by restriction fragment length polymorphism (RFLP), which relies on restriction enzymes to highlight species differences by cleaving a DNA fragment at specific sequences within and between species, further highlighting species differences (Cronin et al. 1991a, b; Shields and Kocher 1991; Foran et al. 1997; Paxinos et al. 1997; Pilgrim et al. 1998). scnDNA can be used to establish the gender of the individual that left the sample (Aasen and Medrano 1990; Taberlet et al. 1993; Woods et al. 1999). Microsatellite DNA can also be used to establish individual identities, genetic relatedness (Morin et al. 1993; Blouin et al. 1996; Prodohl et al. 1998; Goodnight and Queller 1999), and geographic distributions (Paetkau et al. 1995).

Kohn and Wayne (1997) summarized past progress in the application of molecular scatology to wildlife biology. They argued that the challenges of the future for molecular scatology would be to develop better methods for reducing problems associated with polymerase chain reaction (PCR) inhibition, inconsistent and nonspecific results, and contamination. To this end, recent advances in fecal DNA preservation, extraction, and amplification, combined with statistical methods for calculating the probability of PCR errors, have been integral to further developments in molecular scatology and its application to the study of wildlife populations.

As DNA from feces is often degraded and low in quantity, the most difficult problem facing researchers has been the preservation and extraction of sufficient DNA to address the questions of population biology. Wasser and colleagues (1997b) found that both ethanol and silica preserved mtDNA and nuclear DNA. We recommend silica as a better overall choice for the field because of its relative ease for handling and transporting samples. Murphy and colleagues (unpublished data (1999) have noted the advantage of oven-drying silica-stored samples immediately upon collection to speed up the drying process. Frantzen and colleagues (1998) observed that a DMSO/EDTA/Tris/salt (DETs) solution was the most effective for preserving nuclear DNA; storage in 70% ethanol, freezing at $-20°C$, or drying performed nearly as well for mtDNA and short (<200 base pair) nuclear DNA fragments.

The species being studied also appears to affect the preservation of DNA material. This is probably a function of the amount of intestinal cells being shed, the bacterial flora in the gut that degrade DNA, and the presence of diet-related substances that might inhibit DNA amplification. Sample exposure also appears to affect DNA amplification success; samples found in open arid conditions are often much better preserved than those found in moist shaded conditions (Murphy et al. unpublished data 1999).

DNA extraction of fecal samples can also present a challenge, as inhibitors such as bilirubin, bile salts, polysaccharides, and Maillard reaction products can be copurified with the DNA and have a detrimental effect on the action of Taq polymerase (Cheah and Bernstein 1990; Constable et al. 1995; Deuter et al. 1995; Poinar et al. 1998). Deuter and colleagues (1995) demonstrated that the use of potato flour as an adsorption matrix in fecal extracts produced the most successful amplification from human stool samples, and Constable and colleagues (1995) used the detergent CTAB to break down polysaccharides present in the feces of omnivorous primates to amplify microsatellite DNA. Using gas chromatography mass spectrophotometry, Poinar and colleagues (1998) determined that PCR inhibition resulted from condensation products of reducing sugars with primary amines (the Maillard reaction) present in ground sloth coprolites. These could be eliminated by treatment with N-phenacylthiazolium bromide (PTB), which cleaves glucose-derived cross-linked proteins and may be of use in the extraction of non-fossilized fecal material as well.

Other researchers (Kohn et al. 1999; L. Olsen, personal commun. 2000) have improved fecal DNA extractions by using the commercially available DNA extraction kit Isoquick (Orca Research Inc., Bothell, Wash.). Flagstad et al. (1999) improved amplification success of microsatellites from ruminants with the use of a magnetic bead protocol (Dynabeads, Dynal AS, Oslo, Norway). We recently obtained markedly improved microsatellite amplification results from bear scat with a kit available from Qiagen (Qiagen Stool Kit, Qiagen Inc., Valencia, Calif.). The kit is the same as originally described by Wasser and colleagues (1997b) but with the addition of a proprietary stool lysis buffer and Inhibex (Qiagen Inc.) tablets to remove PCR inhibitors unique to feces (Clarke and Wasser 2000).

Contamination of PCR samples and the prospect of amplifying nonhost DNA sequences from ingested material have also been issues of concern in the application of fecal DNA technology. While contamination of PCR samples does occur, several containment strategies are useful in the elimination of such errors. It is recommended that the extraction of DNA and its subsequent amplification be physically restricted to different locations. Researchers should also consider the use of aerosol barrier pipette tips, and dedicated pipettes for sampling template DNA and another for amplified products is recommended. Mock extraction samples (all the extraction reagents except feces), and samples containing only PCR reagents should always be included in analyses, as these controls can pinpoint sources of contamination errors occurring "at the bench." The amplification of nonhost DNA in the feces of the species of interest has been limited, primarily through the use of species-specific mtDNA or microsatellite primers. Although Wasser and colleagues (1997b) demonstrated that hair directly added to fecal samples could result in the amplification of different individual's mtDNA haplotype, a less aggressive extraction procedure eliminated this source of contamination. At this point, it is still unknown if hair from prey animals or from grooming conspecifics can survive the acidic environment of the digestive tract, and researchers should be aware of this potential for error if they are studying such wildlife species.

The problem of inconsistent results using fecal DNA is much the same as when typing low-quantity DNA from hair or forensic samples. It is important to quantitate the DNA samples, preferably with Picogreen (Molecular Probes, Inc., Eugene, Ore.) and a DNA fluorometer (e.g., Versafluor, BioRad Co., Hercules, Calif.) to establish cutoff values for samples that may be too low in template to achieve a reliable genotype (Gagneux et al. 1997). Artifacts from PCR such as allelic dropout and the generation of false alleles can be mostly avoided by using a sufficient amount of template. When this is not possible, repeated amplifications (Taberlet et al. 1996; Goossens et al. 1998) and extractions are advised. Because DNA is not evenly distributed in a fecal sample, we have found that extracting samples in duplicate and amplifying each extractant twice is preferable to simply amplifying the same extractant multiple times. For the same reason, samples should be well mixed prior to taking the subsample for analysis, although some species may have higher concentrations of DNA on the outer part of fresh samples (Flagstad et al. 1999). Researchers are also advised to ensure that genotyping results are consistent with Hardy-Weinberg equilibrium (i.e., that they produced the expected number of homozygotes and heterozygotes in the entire sampled population). Another cross check is to calculate the correspondence between alleles isolated from comparing matched fecal and blood extracts for multiple loci across several animals (Kohn et al. 1999) or to calculate the probability of a PCR artifact as suggested by Gagneux et al. (1997).

When faced with low-quality DNA samples, it is also possible to combine methods to acquire the desired information. For example, when limited by the number of microsatellite loci that can be amplified in low-quality samples, it is possible to use RFLP techniques on more readily amplifiable mtDNA along with gender analyses using scnDNA to divide the population into smaller subgroups based on sex and RFLP type. This reduces the total number of individuals that need to be discriminated using the more difficult-to-amplify microsatellite DNA, potentially allowing discriminations to occur using a smaller number of microsatellite loci.

11.3 Fecal Hormone and DNA Applications to Wildlife Disease Monitoring

Rapid assessment of changes in animal abundance can be one of the first warnings that a potential epidemic is imminent. This is not so difficult for large, observable species in open habitat, but most species do not fit that description. Mark-recapture models have been commonly employed to estimate more difficult-to-observe species in remote areas. Animals are live-trapped, marked, released, and then retrapped. The recapture rate is then used to estimate population size (Skalski and Robson 1992). This method works well for small species but is also time-consuming. Traps must be constantly monitored to minimize associated mortality. The marking is invasive as well as time-consuming and all individuals must have an equal chance of being recaptured, that is, equal catchability. These problems are amplified for larger species.

More recently, investigators began using noninvasive methods of "capturing and marking" animals using DNA (Koehler et al. 1997; Kohn et al. 1999; Woods et al. 1999; Mowat and Strobeck 2000). One common technique is to use microsatellite DNA extracted from hair follicles, collected at hair snag stations distributed throughout the study area. Providing there is sufficient variability across the multiple loci being examined, microsatellite DNA can be used as a permanent mark of individual identities and hence applied to the mark-recapture study design. This appears to work well for some species; however, it also has its associated problems. The critical assumption of equal catchability in mark-recapture models may still be violated, depending on the collection method, species, gender, food availability, time of year, and habitat. For example, hair snag studies use scented lures such as aged cow's blood to draw animals under the chest-height barbed wire that pulls off a swath of hair. Depending on the species, the attractiveness of these lures may vary by the sex of the animal, its dominance rank in social species, its reproductive state, or the quality and availability of natural food sources (Woods et al. 1999; J. Pierce, unpublished data 2000). For example, in social species (e.g., wolves), dominance hierarchies may prevent subordinate individuals from gaining access to hair snag sites. In a variety of species, males are often more attracted to lures than are the more cautious females, and especially females with dependent young. Prior handling of animals by scientists may cause animals to avoid human scent associated with hair snags. Increased availability of high-quality natural foods may also limit the attractiveness of scented lures (Woods et al. 1999; J. Pierce, unpublished data 2000).

An alternative mark-recapture method uses DNA extracted from feces (e.g., Hoss et al. 1992; Constable et al. 1995; Gerloff et al. 1995; Kohn et al. 1995, 1999; Wasser et al. 1997b) to identify individuals. The mode of sample collection, however, can also introduce species-specific biases in capture probabilities. Visually searching for fecal samples of a particular species along roads, trails, or transect routes can produce biases in some species because particular individuals (e.g., females and immigrant males) may conceal their feces more so others (e.g., resident males). For example, male cheetahs defecate in conspicuous areas such as on top of rock outcroppings used as road signs in national parks; female cheetahs never defecate in such conspicuous locations (G. Mills, personal commun. 1999). These kinds of sex biases can be particularly problematic when attempting to determine the health of infanticidal species, such as large carnivore populations (e.g., grizzly bears). Rates of infanticide in such species can become especially high when the adult sex ratio is heavily skewed in favor of males (Wielgus and Bunnell 2000). From a disease standpoint, it is important to distinguish infanticidal causes of cub mortality from those that are disease based, especially since young are often the first to suffer mortality from a variety of different diseases.

Sophisticated computer models exist to accommodate violations of the equal catchability assumptions (Skalski and Robson 1992). The degree to which these biases occur and their effects on population estimates remain to be determined; the degree to which these biases can be reduced by modifying sampling techniques is also largely unknown.

One promising alternative scat sampling method we are pursuing relies on detection dog techniques, from narcotics, arson, and search-and-rescue work, to train domestic dogs to detect scat from selected wildlife species over large remote areas. The phenomenal ability of these dogs to detect specific odors is well documented. Trained dogs are able to detect odors as faint as three parts per million, odors can be detected from distances of half a mile downwind, and up to 18 different substances (species) can be detected at once (Bryson 1995; Syrotuck 1972; Tolhurst 1991). Perhaps most important, these high-drive, play-driven dogs are motivated by their expectation of receiving a play reward upon sample detection. They maximize their chances of receiving this reward by locating samples regardless of the subject's sex, rank, or reproductive condition. Thus, fecal sample detection occurs independent of these individual specific characteristics that might otherwise bias sample capture probabilities. The high sensitivity of this scent-driven sample detection method also makes levels of sample concealment relatively unlikely to affect detection and hence catchability rates. Biases resulting from avoidance of human scent by previously immobilized bears is similarly irrelevant to sample detection by dogs. Lastly, the dog's probability of sample detection can be regularly cross-checked. Field assistants set down samples of known sex and/or species in the study area, blind to the dog and handler, and note the rates at which the various samples are detected. Samples of nontarget species are set down as well to confirm that dogs are not mistakenly alerting to samples from these other species.

The above fecal sampling techniques can have pathogen-monitoring applications as well, providing the pathogens are shed in feces. In this case, the DNA being measured is from the pathogen rather than the host. For example, fecal DNA may prove useful in monitoring tuberculosis (TB) in lions because the TB lesions tend to be localized in the gut rather than the respiratory tract. Thus, TB infection rates could be monitored in lions without having to wait to acquire tissue from dead animals or randomly kill individuals as part of the sampling protocol. Mark-recapture approaches could also be applied to estimate the proportion of individuals in the population that are actually infected with a particular pathogen.

11.4 Monitoring Physiological Effects of Environmental Disturbance

Physiological stress can result in immunosuppression and associated reduced resistance to diseases. It therefore becomes important to know how much physiological stress associated with environmental disturbance increases the probability of disease outbreaks. This, too, requires monitoring and relatively high sample accessibility. Presently, most disturbance studies using fecal hormones rely on measurement of glucocorticoid metabolites (Wasser et al. 1997a; Millspaugh 1999; Foley et al. 2001). Another potential fecal hormone index of stress is DHEA. Concentrations of DHEA appear to be reduced in response to stress in some species—an effect opposite that of the glucocorticoids. Reproductive hormones such as progesterone, estradiol, and testosterone may decline in response to stress and their metabolites are also found in scat (Wasser et al. 1997a; Foley et al. 2001). Our laboratory has also measured thyroid hormones in feces (Hunt and Wasser, unpublished data, 2000), which may provide indices of starvation; this, too, could affect susceptibility to various disease states.

Fecal samples may be collected by direct observation for such analyses and correlated with associated behavioral measures; they may be collected cross-sectionally, to compare levels in known disturbed versus undisturbed areas; and they may be coupled with DNA measures, enabling repeated sampling of hormone levels from a given individual identified using microsatellite DNA extracted from the same sample. Sex differences in baseline stress hormone levels are common, with males typically having higher concentrations than do females. Thus, at the least it is often important to discriminate individuals by sex in cross-sectional studies involving stress hormones. Such discriminations can be done with DNA (Wasser et al. 1997a) or gonadal hormones (Kubokawa et al. 1992; Yamauchi et al. 1998). Repeat sampling is also vital to examine reproductive function. Since reproductive suppression is associated with physiological stress (Johnson et al. 1991; Wasser et al. 1995), its occurrence can also be used as an indirect measure of physiological disturbance.

Endocrine abnormalities can reflect other problems as well, such as endocrine disruption in response to pollutants. Repeat sampling is similarly necessary for indices of endocrine disruption. Identifying subpopulations experiencing endocrine disruption versus those that are reproductively normal can provide comparison groups that help pinpoint sources of these pollutants. For example, one might look for dietary differences between these subpopulations as possible sources of the pollutants in the food chain.

Lastly, all of these techniques can be made even more effective by coupling them with GIS technology. Disturbances, animal densities, habitat differences, extent of human encroachment—all can be georeferenced as disparate layers on a GIS. This approach is ideal for identifying the kinds of complex interrelationships that are so much a part of nature and the study of emergent diseases.

11.5 Conclusions

Wildlife monitoring is rapidly evolving, with new and improved techniques becoming available every year. We are only just beginning to realize the potential applications of these complementary techniques for a wide variety of conservation and management problems. Creative applications of these techniques to some of the more difficult wildlife problems should produce new advances in the coming years. And, the study of wildlife disease can very much benefit from this emerging technology.

References

Aasen, E., and J.F. Medrano. 1990. Amplification of the ZFY and ZFX genes for sex identification in humans, cattle, sheep and goats. Biotechnology 8:1279–1281.

Avise, J.C. 1993. Molecular markers, natural history and evolution. Chapman and Hall, New York.

Bahr, N.I., R. Palme, U. Mohle, J.K. Hodges, and M. Heistermann. 2000. Comparative aspects of metabolism and excretion of cortisol in three individual nonhuman primates. Gen Comp Endocrinol 117:427–438.

Berger, J., J.W. Testa, T. Roffe, and S.L. Monfort. 1999. Conservation endocrinology: a noninvasive tool to understand relationships between carnivore colonization and ecological carrying capacity. Conserv Bio 13:980–989.

Blouin, M.S., M. Parsons, V. Lacaille, and S. Lotz. 1996. Use of microsatellite loci to classify individuals by relatedness. Mol Ecol 5:393–401.

Brockman, D.K., P.L. Whitten, E. Russell, A.F. Richard, and M.K. Izard. 1995. Application of fecal steroid techniques to the reproductive endocrinology of female Verreaux's sifaka (*Propithecus verreauxi*). Am J Primatol 36:313–325.

Brown, J.L., S.K. Wasser, D.E. Wildt, L.H. Graham, and S.L. Monfort. 1997. Faecal steroid analysis for monitoring ovarian and testicular function in diverse wild carnivore, primate and ungulate species. Int J Mamm Biol (suppl. 2):27–31.

Bryson, S. 1995. Police dog tactics. McGraw Hill, New York.

Cheah, P.Y., and H. Bernstein. 1990. Colon cancer and dietary fiber: cellulose inhibits the DNA-damaging ability of bile acids. Nutr Cancer 13:51–57.

Cockrem, J.F., and J.R. Rounce. 1995. Non-invasive assessment of the annual gonadal cycle in free-living kakapo (*Strigops habroptilus*) using fecal steroid measurements. Auk 112:253–257.

Constable, J.J., C. Packer, D.A. Collins, and A.E. Pusey. 1995. Nuclear DNA from primate dung. Nature 373:393.

Cronin, M.A., S.C. Armstrup, and G.W. Garner. 1991a. Interspecific and intraspecific mitochondrial DNA variation in North American bears (*Ursus* spp.). Can J Zool 69: 2985–2992.

Cronin, M.A., D. Palmisciano, E. Vyse, and D.G. Cameron. 1991b. Mitochondrial DNA in wildlife forensic science: species identification of tissues. Wildl Soc Bull 19:94–105.

Deuter, R., S. Pietsch, S. Hertel, and O. Muller. 1995. A method for preparation of fecal DNA suitable for PCR. Nucleic Acids Res 23:3800–3801.

Diamandis, E.P., and Christopoulos, T.K. 1996. Immunoassay. Academic Press, San Diego.

Flagstad, O., K. Roed, J.E. Stacy, and K.S. Jakobsen. 1999. Reliable non-invasive genotyping based on excremental PCR of nuclear DNA purified with a magnetic bead protocol. Mol Ecol 8:879–883.

Foley, C.A.H., S. Papageorge, and S.K. Wasser. 2001. Non-invasive stress and reproductive measures of social and ecological pressures in free-ranging African elephants (*Loxodonta africana*). Conserv Biol 15:1134–1142.

Foran, D.R., K.R. Crooks, and S.C. Minta. 1997. Species identification from scat: an unambiguous genetic method. Wildl Soc Bull 25:835–839.

Frantzen, M.A., J.B. Silk, J.W. Ferguson, R.K. Wayne, and M.H. Kohn. 1998. Empirical evaluation of preservation methods for faecal DNA. Mol Ecol 7:1423–1428.

Gagneux, P., C. Boesch, and D.S. Woodruff. 1997. Microsatellite scoring errors associated with noninvasive genotyping based on nuclear DNA amplified from shed hair. Mol Ecol 6:861–868.

Gerloff, U., C. Schlotterer, K. Rassmann, I. Rambold, G. Hohmann, B. Fruth, and D. Tautz. 1995. Amplification of hypervariable simple sequence repeats (microsatellites) from excremental DNA of wild living bonobos (*Pan paniscus*). Mol Ecol 4:515–518.

Goodnight, K.F., and D.C. Queller. 1999. Computer software for performing likelihood tests of pedigree relationship using genetic markers. Mol Ecol 8:1231–1234.

Goossens, B., L.P. Waits, and P. Taberlet. 1998. Plucked hair samples as a source of DNA: reliability of dinucleotide microsatellite genotyping. Mol Ecol 7:1237–1241.

Goymann, W., E. Möstl, T. van't Hof, M.L. East, and H. Hofer. 1999. Non-invasive fecal monitoring of glucocorticoids in spotted hyaenas, *Crocuta crocuta*. Gen Comp Endocrinol 114:340–348.

Graham, L.H., and J.L. Brown. 1997. Non-invasive assessment of gonadal and adrenocortical function in felid species via faecal steroid analysis. Int J Mamm Biol (suppl. 2):78–82.

Hamdrick, J.L., and J.C. Avise, eds. 1996. Conservation genetics: case histories from nature. Chapman and Hall, New York.

Harper, J.M., and S.N. Austad. 2000. Fecal glucocorticoids: a noninvasive method of measuring adrenal activity in wild and captive rodents. Physiol Biochem Zool 73:12–22.

Hoss, M., M. Kohn, S. Paabo, K. Knauer, and W. Schroder. 1992. Excrement analysis by PCR. Nature 359:199.

Johnson, E.O., T.C. Kamilaris, S. Carter, P.W. Gold, and G.P. Chrousos. 1991. Environmental stress and reproductive success in the common marmoset *Callithrix jacchus jacchus*. Am J Primatol 25:191–201.

Kirkpatrick, J.F., J.C. McCarthy, D.F. Gudermuth, S.E. Shideler, and B.L. Lasley. 1996. An assessment of the reproductive biology of Yellowstone bison (*Bison bison*) subpopulations using noncapture methods. Can J Zool 74:8–14.

Koehler, G.M., S.K. Wasser, C.S. Houston, D.J. Pierce, and J.P. Skalski. 1997. DNA mark-recapture estimator of black bear numbers: is the hair ball technique a hair brain estimator? *In* Abstracts of the sixth western black bear workshop, p. 7. Washington Department of Fish and Game, Ocean Shores.

Kohn, M.H., and R.K. Wayne. 1997. Facts from feces revisited. TREE 12:223–227.

Kohn, M., F. Knauer, A. Stoffella, W. Schroder, and S. Paabo. 1995. Conservation genetics of the European brown bear—a study using excremental PCR of nuclear and mitochondrial sequences. Mol Ecol 4:95–103.

Kohn, M.H., E.C. York, D.A. Kamradt, G. Haught, R.M. Sauvajot, and R.K. Wayne. 1999. Estimating population size by genotyping faeces. Proc R Soc Lond B 266:657–663.

Kotrschal, K., K. Hirschenhauser, and E. Möstl. 1998. The relationship between social stress and dominance is seasonal in greylag geese. Anim Behav 55:171–176.

Kubokawa, K., S. Ishii, H. Tajima, K. Saitou, and K. Tanabe. 1992. Analysis of sex steroids in feces of giant pandas. Zool Sci 9:1017–1023.

Millspaugh, J.J. 1999. Behavioral and physiological responses of elk to human disturbances in the southern Black Hills, South Dakota. Ph.D. thesis, University of Washington, Seattle.

Moberg, G.P., ed. 1988. Animal stress. Oxford University Press, New York.

Morin, P.A., J. Wallis, J.J. Moore, R. Chakaborty, and D.S. Woodruff. 1993. Non-invasive sampling and DNA amplification for paternity exclusion, community structure and phylogeography in wild chimpanzees. Primates 34:347–356.

Mowat, G., and C. Strobeck. 2000. Estimating population size of grizzly bears using hair capture, DNA profiling and mark-recapture analysis. J Wildl Manage 64:183–193.

Paetkau, D., W. Calvert, I. Stirling, and C. Strobeck. 1995. Microsatellite analysis of population structure in Canadian polar bears. Mol Ecol 4:347–354.

Palme, R., and E. Möstl. 1997. Measurement of cortisol metabolites in faeces of sheep as a parameter of cortisol concentration in blood. Int J Mamm Biol (suppl. 2):192–197.

Palme, R., P. Fischer, H. Schildorfer, and M.N. Ismail. 1996. Excretion of infused ^{14}C-steroid hormones via faeces and urine in domestic livestock. Anim Reprod Sci 43: 43–63.

Paxinos, E., D. McIntosh, K. Ralls, and R. Fleischer. 1997. A non-invasive method for distinguishing among canid species: amplification and enzyme restriction of DNA from dung. Mol Ecol 6:483–486.

Pilgrim, K.P., D.K. Boyd, and S.H. Forbes. 1998. Testing for wolf–coyote hybridization in the Rocky Mountains using mitochondrial DNA. J Wildl Manage 62:683–689.

Poinar, H.N., M. Hofreiter, W.G. Spaulding, P.S. Martin, B.A. Stankiewics, H. Bland, R.P. Evershed, G. Possner, and S. Paabo. 1998. Molecular coproscopy: dung and diet of the extinct ground sloth *Nothrotheriops shastensis*. Science 281:402–406.

Prodohl, P.A., W.J. Loughry, C.M. McDonough, W.S. Nelson, E.A. Thompson, and J.C. Avise. 1998. Genetic maternity and paternity in a local population of armadillos assessed by microsatellite DNA markers and field data. Am Nat 151:7–19.

Savage, A., S.E. Shideler, L.H. Soto, J. Causado, L.H. Giraldo, B.L. Lasley, and C.T. Snowdon. 1997. Reproductive events of wild cotton-top tamarins (*Saguinus oedipus*) in Colombia. Am J Primatol 43:329–337.

Schwarzenberger, F., E. Möstl, E. Bamberg, J. Pammer, and O. Schmehlik. 1991. Concentrations of progestagens and oestrogens in the faeces of pregnant Lipizzan, trotter and thoroughbred mares. J Reprod Fertil 44 (Suppl):489–499.

Schwarzenberger, F., R. Palme, E. Bamberg, and E. Möstl. 1997. A review of faecal progesterone metabolite analysis for non-invasive monitoring of reproductive function in mammals. Int J Mamm Biol 62 (suppl 2):214–221.

Shideler, S.E., A.M. Ortuno, F.M. Moran, E.A. Moorman, and B.L. Lasley. 1993. Simple extraction and enzyme immunoassays for estrogen and progesterone metabolites in the feces of Macaca fascicularis during non-conceptive and conceptive ovarian cycles. Biol Reprod 48:1290–1298.

Shields, G.F., and T.D. Kocher. 1991. Phylogenetic relationships of North American ursids based on analysis of mitochondrial DNA. Evolution 45:218–221.

Skalski, J.R., and D.S. Robson. 1992. Techniques for wildlife investigations, design and analysis of capture data. Academic Press, San Diego.

Syrotuck, W.G. 1972. Scent and the scenting dog, 3rd ed. Barkleigh Productions, Mechanicsburg, Pa.

Taberlet, P., H. Mattock, C. Dubois-Paganon, and J. Bouvet. 1993. Sexing free-ranging brown bears, Ursus arctos, using hairs found in the field. Mol Ecol 2:399–403.

Taberlet, P., S. Griffin, B. Goossens, S. Questiau, V. Manceau, N. Escaravage, L.P. Waits, and J. Bouvet. 1996. Reliable genotyping of samples with very low DNA quantities using PCR. Nucleic Acids Res 24:3189–3194.

Tolhurst, B. 1991. The police textbook for dog handlers. Sharp Printing, Rome, N.Y.

Velloso, A.L., S.K. Wasser, S.L. Monfort, and J.M. Dietz. 1998. Longitudinal fecal steroid secretion in maned wolves (Chrysocyon brachyurus). Gen Comp Endocrinol 112:96–107.

Wallner, B., E. Möstl, J. Dittami, and H. Prossinger. 1999. Fecal glucocorticoids document stress in female Barbary macaques (Macaca sylvanus). Gen Comp Endocrinol 113:80–86.

Wasser, S.K. 1996. Reproductive control in wild baboons measured by fecal steroids. Biol Reprod 55:393–399.

Wasser, S.K., L. Risler, and R.A. Steiner. 1988. Excreted steroids in primate feces over the menstrual cycle and pregnancy. Biol Reprod 39:862–872.

Wasser, S.K., S.L. Monfort, and D.E. Wildt. 1991. Rapid extraction of faecal steroids for measuring reproductive cyclicity and early pregnancy in free-ranging yellow baboons (Papio cynocephalus cynocephalus). J Reprod Fertil 92:415–423.

Wasser, S.K., R. Thomas, P.P. Nair, C. Guidry, J. Southers, J. Lucas, D.E. Wildt, and S.L. Monfort. 1993. Effects of dietary fibre on faecal steroid measurements in baboons (Papio cynocephalus cynocephalus). J Reprod Fertil 97:569–574.

Wasser, S.K., S.L. Monfort, J. Southers, and D.E. Wildt. 1994. Excretion rates and metabolites of oestradiol and progesterone in baboon (Papio cynocephalus cynocephalus) faeces. J Reprod Fertil 101:213–220.

Wasser, S.K., A. De Lemos Velloso, and M.D. Rodden. 1995. Using fecal steroids to evaluate reproductive function in female maned wolves. J Wildl Manage 59:889–894.

Wasser, S.K., S. Papageorge, C. Foley, and J.L. Brown. 1996. Excretory fate of estradiol and progesterone in the African elephant (Loxodonta africana) and patterns of fecal steroid concentrations throughout the estrous cycle. Gen Comp Endocrinol 102:255–262.

Wasser, S.K., K. Bevis, G. King, and E. Hanson. 1997a. Noninvasive physiological measures of disturbance in the northern spotted owl. Conserv Biol 11:1019–1022.

Wasser, S.K., C.S. Houston, G.M. Koehler, G.G. Cadd, and S.R. Fain. 1997b. Techniques for application of faecal DNA methods to field studies of ursids. Mol Ecol 6:1091–1097.

Wasser, S.K., K.E. Hunt, J.L. Brown, K. Cooper, C.M. Crockett, U. Bechert, J.J. Millspaugh, S. Larson, and S.L. Monfort. 2000. A generalized fecal glucocorticoid assay for use in a diverse array of non-domestic mammalian and avian species. Gen Comp Endocrinol 120:260–275.

Whitten, P.L., D.K. Brockman, and R.C. Stavisky. 1998. Recent advances in noninvasive techniques to monitor hormone-behavior interactions. Yrbk Phys Anthropol 41:1–23.

Wielgus, R.B., and F.L. Bunnell. 2000. Possible negative effects of adult male mortality on female grizzly bear reproduction. Biol Conserv 93:145–154.

Wingfield, J.C., D.L. Maney, C.W. Breuner, J.D. Jacobs, S. Lynn, M. Ramenofsky, and R.D. Richardson. 1998. Ecological bases of hormone-behavior interactions: the "emergency life history stage." Am Zool 38:191–206.

Wooding, S., and R. Ward. 1997. Phylogeography and Pleistocene evolution of the North American black bear. Mol Biol Evol 14:1096–1105.

Woods, J.G., D. Paetkau, D. Lewis, B.N. McLellan, M. Proctor, and C. Strobeck. 1999. Genetic tagging free-ranging black and brown bears. Wildl Soc Bull 27:616–627.

Yamauchi, K., S.-I. Hamasaki, K. Iwatsuki, T. Nakamura, K. Siota, Y. Takeuchi, and Y. Mori. 1998. Sex determination of sika deer (*Cervus nippon*) using fecal DNA analysis. Abstr Jpn Soc Zoo Wildl Med 4:59.

12

Animal Behavior as a Tool in Conservation Biology

J. Michael Reed

The behavior of animals is important to conservation biology because behavior affects species persistence through a wide variety of mechanisms. These mechanisms can include social disruption of breeding, dispersal and settlement decisions, learned and socially facilitated foraging, translocation success, and canalized behavior that is maladaptive (Reed 1999). In addition, understanding behavior can be critical to solving problems such as reserve design. As an example, one of the current controversies in conservation biology is how to create proper corridors to facilitate dispersal among protected areas (Beier and Noss 1998). Haddad (1999) showed how corridor use might be predicted from animal behavior at habitat boundaries. If we understood the behaviors of endangered species as well as we do those of some domestic species (e.g., domestic sheep do not like walking into their own shadows; Kilgour and Dalton 1984), we could solve one aspect of corridor design. Behavior also can be critical to determining species management goals. For example, a recent study of pilot whales (*Globicephala melas*) showed that its unusual group structure and mating system require management for many pods rather than management for large numbers of individuals within a single pod (Amos et al. 1993). Beyond setting management goals, behavior sometimes can be manipulated to achieve a particular goal. Just as predators take advantage of prey behavior (Jabłoński 1999), species behaviors can be utilized to achieve conservation goals.

Despite the importance of behavior to species conservation, animal behaviorists only recently have entered the field of conservation biology (Clemmons and Buchholz 1997; Caro 1998; Gosling and Sutherland 2000). Sutherland (1998) clearly demonstrated the lack of integration between the fields of animal behavior and conservation biology. He reviewed the subject matter of papers published in 1996

Table 12.1. Examples of how various species are censused and the behaviors exploited by the census method to increase effectiveness

Species (or taxon)	Census Method	Behavior(s) Exploited	Citation
Black bear (*Ursus americanus*)	Counting opened cans of sardines that were nailed to trees	Olfactory prey detection	Powell et al. (1996)
Red-cockaded woodpeckers (*Picoides borealis*)	Counting "active" cavity trees	Excavating resin wells and scraping bark from nest and roost trees	Reed et al. (1988)
Most frogs	Counting calling males	Attracting mates by calling	Zimmerman (1994)

in *Animal Behaviour* and in *Conservation Biology*. None of the 229 papers in the former journal dealt with conservation and only 9 of 97 in the latter focused on animal behavior. It is possible that this is a result of a lack of understanding of the myriad ways in which animal behavior can influence species conservation. Of all areas of conservation biology, censusing and monitoring are the only ones where the methods center on species behaviors. All animal censusing, monitoring, and inventory methods take advantage of some behavior (table 12.1). When census methods are refined for particular target species, refinements often take advantage of specific behaviors, and the effectiveness of inventory and monitoring depend on the effectiveness of these methods (Bibby et al. 1992; Reed 1996).

In this chapter I present examples of ways in which behavior can be used specifically as a tool to achieve conservation goals. Although most examples come primarily from birds and terrestrial mammals, behavior can be a tool in conservation of other taxa. Because recent symposia cover several of these topics, (Clemmons and Buchholz 1997; Caro 1998; Gosling and Sutherland 2000), I do not review all the ways behavior can relate to species conservation. Rather, I address several important issues in conservation biology: finding behavioral indicators of ecosystem health, increasing reproductive success and survival, facilitating movement, improving population viability analysis, facilitating species translocation or reintroduction, and reducing loss of genetic variability. For each of these issues, I provide examples of how behavior might be manipulated to achieve specific goals. It is my intent to stimulate interest in considering and applying behavior to solve specific conservation problems.

12.1 Behaviors as Indicators of Ecosystem Health

Indicators are relatively simple measures intended to reflect complex phenomena (Bibby 1999). Popular biological indicators of ecosystem health or integrity include the presence or abundance of particular pollutants in wildlife, specific habitat structures, and the presence or abundance of particular species (e.g., Furness and Greenwood 1993; Karr 1995). The disadvantages of many indicators include high economic cost of measurement, destruction of the individual being tested, necessity of long-term data, and a long time lag between the introduction of a problem and response of the indicator. Although not often used, particular behaviors or disruption of typical behaviors also can be indicators of ecosystem health, and they sometimes avoid the problems associated with other bioindicators (Warner et al. 1966). I briefly review the use of behaviors to indicate the presence of pollutants, and the potential of these behaviors to act as indicators of other types of environmental effects. The principles I present also can be applied to the presence of diseases that can alter behavior or produce specific atypical behaviors.

Although most research on ecotoxicity focuses on determining lethal doses of toxicants, sublethal effects can result in population decline by disrupting normal behaviors (Cohn and MacPhail 1996; table 12.2). A wide variety of effects of pollutants on normal behaviors have been reported, including tremors in limbs, altered time/energy budgets, disrupted motor and sensory functions, and decreased performance in learned tasks (Peakall 1985; Døving 1991; Kulig et al. 1996). For example, Galindo et al. (1985) found northern bobwhites (*Colinus virginianus*) exposed to methyl parathion were more susceptible to predation. Nocera and Taylor (1998) found that very young common loons (*Gavia immer*) exposed to mercury spent less time brooding, resulting in decreased fledging rates. Despite the lethal effects of toxicants, however, not all studies find behavioral responses to sublethal doses. Consequently, if one wants to determine specific responses or lack of response to particular chemicals, experimental work is needed. If one is interested in determining only if a toxicant in general is a problem in an ecosystem, rather than initially determining the specific toxicant, observing a suite of behaviors to determine deviation from typical behaviors can be a sensitive indicator. Determining what constitutes atypical behavior requires fairly extensive data on the behaviors of healthy individuals for use in comparison (Døving 1991; Cohn and MacPhail 1996; Kulig et al. 1996). Given the rapidly expanding number of chemicals released into the environment, often with unknown effects (Colborn et al. 1998), this type of research should receive a high priority.

Although they have not been addressed formally, behaviors could be used as indicators of other types of ecosystem health. For example, it is reasonable to

Table 12.2. Population decline can result from a variety of sublethal effects on behaviors

Type of Behavior	Effect on Population
Parental care	Fewer offspring produced
Mating and reproduction	Fewer matings, fewer offspring
Predator avoidance and alarm response	Increased mortality
Behavior during development	Increased mortality
Predatory	Increased starvation
Migration or dispersal	Increased mortality, decreased colonization

Source: Adapted from Døving (1991) and Cohn and MacPhail (1996).

assume that population density and habitat quality would be related for a given species. More detailed studies, however, have demonstrated this to be incorrect in many circumstances. For example, high population density might be associated with nonterritorial (not breeding) animals, rather than with territorial (breeding) individuals (Van Horne 1983). Higher densities also could be associated with social disruption and, ultimately, reduced reproductive success (Purcell and Verner 1998). In many species, reproduction can be difficult to assess, and behavioral indicators of reproductive status might be used (e.g., Vickery et al. 1992). Changes in time budgets, such as time spent foraging, also can reflect changes in environmental condition, such as sea bird foraging time reflecting fish prey stocks (Montevecchi 1993).

Potential downsides to using behaviors as indicators include the necessity of deciding on discrete, quantifiable behaviors to measure, and the difficulty of determining specific cause and effect, particularly for identifying particular toxicants (Kulig et al. 1996; Peakall 1996). If the goal to determine *if* there is a problem, rather than what the specific problem is, these concerns are less important. Developing behavioral indicators of ecosystem health will involve an interesting blending of experimental psychology, ethology, and field studies (Cohn and MacPhail 1996).

12.2 Increasing Reproductive Success

When the intrinsic rate of increase (r) of a closed population (one with no emigration or immigration) is less than zero, its numbers decline over time. Intrinsic rate of increase is a composite variable that combines death rate and birth rate. The birth rate in some species is dependent on behaviors that lend themselves to manipulation.

Many of the mechanisms by which behavior can affect reproductive success can be grouped under the rubric of social facilitation. Of conservation interest is the breakdown of these social factors at small population sizes, which can result in a sudden reduction in reproductive success or survival (Fowler and Baker 1991). Another behavioral problem related to small population size is increased frequency of hybridization, which results from inadequate behavioral barriers to mating (Rhymer and Simberloff 1996). The obvious solution to problems of inadequate population size is to increase its size. This option is not always available, such as with rare species or with some captive populations. Researchers working on captive populations of flamingos (*Phoenicopterus minor*) found a unique solution to the problem. Flamingos are social breeders, and small groups of birds do not breed (Stevens 1991). Pickering and Duverge (1992) were able to stimulate prereproductive displays in a captive flock by putting up mirrors. This suggests the possibility of other methods for social stimulation, particularly in captive populations, such as showing still or videotaped pictures of conspecifics. Some species of birds and lizards are known to respond to projected images (e.g., Clark et al. 1997), though to my knowledge, this method has not been used to facilitate breeding.

12.3 Increasing Survival

Survival rate is the other component of the intrinsic rate of increase in a closed population. Behavior contributes to individual survival in terrestrial vertebrates primarily through learned behaviors or responses. In this section I show examples of two behaviors that affect survival rate, namely, predator avoidance and learned foraging skills, and briefly discuss how they might be manipulated to increase survival rate.

Predator avoidance is key to survival for most species. For some vertebrates, effective predator avoidance is related to local habitat familiarity (Isbell et al. 1990; Clarke et al. 1993). For other species, predator defense can be a social behavior that depends on a minimum group size. For example, when groups of pronghorn (*Antilocapra americana*) drop below 12–15 individuals, their typical defensive behavior breaks down, making them more susceptible to predation (Leopold 1933). These behaviors are important to decisions about how to release individuals when reintroducing them to a new site, but are not readily manipulated in wild populations. In contrast, for some species, recognizing predators and developing escape reactions are cultural skills that must be learned when the animals are young. This can be a problem when translocating animals or when introducing naive (captively bred) animals to the wild because these skills might be lacking. Recently, researchers have used models of predators to condition wild and captive-reared animals to predators (e.g., McLean et al. 1995; Bunin and Jamieson 1996). Any increase in predator recognition escape proficiency could increase survival rate. In many species, foraging efficiency increases with practice and learning (Kamil and Yoerg 1981) and social learning plays an important role in diet learning by juveniles (Provenza and Balph 1987). Work on domestic species (ruminants, goats, sheep), in particular, has shown the importance of social learning of food selection. For example, experience early in life affects food selection by adults (Distal and Provenza 1991), food preference by mothers affects that of offspring (Mirza and Provenza 1990), and aversions or preferences to novel foods can be transmitted culturally to offspring (Thorhallsdottir et al. 1990). An extreme example of social learning affecting diet selection of young comes from research on domestic cats. Wyrwicka (1981) electrically stimulated pleasure areas in cat brains while feeding them unusual foods (e.g., bananas). This resulted in a food preference for the cats, and this preference was learned by their offspring.

Social facilitation also affects proximate foraging behaviors. For example, group size can affect prey selection in species that sometimes cooperatively forage. In some species there is conspecific attraction to other foraging animals (Pöysä 1991b; Beauchamp et al. 1997), and decreased predation risk associated with large foraging groups can increase the amount of time spent foraging (Pöysä 1991b). Foraging behaviors in many species have already been manipulated to increase survival probabilities, particularly in the use of feeding stations (Helender 1978). For example, although diurnal cycles and weather affect white-tailed deer (*Odocoileus virginanus*) foraging patterns, Henke (1997) showed that deer enclosed in natural areas could be conditioned to feed at any time of day at feeding stations. Feeding stations can be used for more than just increasing food availability. They also could be used to train individuals on a new food type, to train aversion to a food type that is associated with increased mortality (such as crops, or a food type that has become poisonous through pollution), or to train animals to avoid high-risk areas. For species that feed in groups, decoys can be used to attract species to new feeding areas (Kear 1990).

Castro and colleagues (1994) used supplemental feeding to enhance habitat quality for endangered hihi (*Notiomystis cincta*, a honeyeater) that had been translocated to an island for protection. They found that in the new habitat type, with the altered distribution of food (feeding stations), the birds adopted polygamous mating patterns not normally exhibited in this monogamous species. These accidental alterations of behavior could have negative effects on species and should be guarded against. For example, polygamous mating results in a lower effective population size than does monogamous mating of the same number of males and females, thus increasing the rate of loss of genetic variability (Falconer 1989).

12.4 Facilitating Movement

The primary cause of species extinction worldwide is habitat loss and fragmentation. These mechanisms put species at risk by reducing population size and through edge effects, which are biotic and abiotic intrusion into interior habitat from an ecotone (Harris 1984). The isolation associated with habitat fragmentation has two components, physical isolation and psychological isolation. Physical isolation is a function of distance (MacArthur and Wilson 1967). Psychological isolation is based on behavior and has received less attention. Despite the capacity to disperse a given distance, many species will not cross a habitat gap or an ecotone (Desrochers and Hannon 1997; St. Clair et al. 1998). Not knowing the behavioral response of individuals can lead to incorrect predictions of the effects of habitat isolation. For example, bighorn sheep (*Ovis canadensis*) tend to avoid entering timber when moving among feeding sites (Geist 1971). A resource manager unaware of this behavioral tendency would make incorrect predictions of species responses to some reserve designs and habitat management plans. Below I discuss three types of conservation problem associated with individuals moving among sites—dispersal, colonization, and corridor use—and methods that have been or might be used to manipulate behavior to help solve the problem.

12.4.1 Dispersal and Colonization

When a population becomes diminished and isolated, its chances of persistence are diminished because smaller populations are subject to stochastic factors affecting genetic variability, demography, chance environmental events (Shaffer 1981), and reduced opportunities for population rescue through immigration (Brown and Kodric-Brown 1977). Here I am concerned with the problems of reduced dispersal among populations because it is the driving force in metapopulation dynamics (Wu et al. 1993). Dispersal among populations typically is treated in population biology as a mechanical or diffusional process, where an animal leaves its natal or breeding territory, travels in a random direction, and settles in the first appropriate site it finds (e.g., Johnson and Gaines 1990). Dispersal in many species, particularly vertebrates, involves behavioral decisions regarding if or when to disperse, what path to travel, and where to settle (Stamps 1991; Reed et al. 1999).

In addition, the nature of the landscape that defines population distributions can affect the behavioral decisions involved in dispersal (Lima and Zollner 1996). For example, behavior can affect emigration decisions, either by increasing dispersal in response to local interactions or by inhibiting emigration. Inhibition can be due a reluctance to cross an ecotone (Stamps et al. 1987; Hansen and di Castri 1992) or a gap in habitat (Desrochers and Hannon 1997; St. Clair et al. 1998), or avoidance of a particular habitat type (Geist 1971). An example of a behavioral inhibition to emigration comes from hyraxes (*Heterohyrax brucei* and *Procavia johnstoni*) living in the Serengeti (Hoeck 1989). These species live on isolated rock outcrops, and individuals tend to not disperse to other outcrops if they do no see or hear conspecifics at that outcrop (other examples are given by Reed 1999). Similar behavioral mechanisms can inhibit immigration once an individual has dispersed, such as failure (or inhibition) to settle in suitable habitat because of the absence of some cue correlated with habitat quality (e.g., Stamps 1991; Weddell 1991). The most commonly studied behavioral mechanism affecting immigration is the effect of conspecific attraction on settlement decisions (e.g., Stamps 1987, 1991; Danchin et al. 1991). This can affect growth rates and persistence (Reed and Dobson 1993).

Behavior of emigration, immigration, and colonization (which I distinguish as a special case of immigration to empty habitats) can be exploited in conservation settings. Colonization of new or unused areas could be encouraged if the cues used by a particular species to assess habitat suitability were known and could be altered. Hunters have used cues to manipulate behavior for centuries, for example, in attracting waterfowl by means of calls and decoys (e.g., Kear 1990). Along the same lines, nest boxes have been employed to attract cavity-nesting ducks to areas from which they had been absent (Kadlec and Smith 1992). Decoys and playback also have been used to induce colonially nesting birds to establish new colonies (e.g., Kress 1983), and dispersing griffon vultures (*Gyps fulvus*) were attracted to former breeding sites by spreading white paint to simulate droppings (Sarrazin et al. 1996).

Immigration rate or success could be enhanced for some species by creating breeding sites (e.g., Blanco et al. 1997), but this represents a change in habitat suitability rather than being a behavioral manipulation. The best examples of manipulating behavior to encourage natural colonization of suitable sites come from colonially nesting birds. This work was pioneered by Kress (1983), who presented painted decoys and played sound recordings of least terns (*Sterna antillarum*) to attract potential breeders to sites where he wanted colonies to be formed. These methods have been used successfully with at least seven other species of colonial sea birds (Kress 1997; Veen 1977) and have potential for species with other breeding systems (e.g., Rodgers 1992). The goal of this work is to provide a conspecific cue of habitat suitability. This takes advantage of prospecting behavior, which is exhibited by species across the animal kingdom, whereby animals gather information about potential breeding sites and make decisions based on expected reproductive success (Reed et al. 1999). Cues might include as widely differing actions as playing conspecific songs (Verner 1992) and spreading white paint to simulate droppings (Sarrazin et al. 1996). Results from colonial waterbirds have been dramatic, and the approach used has potential for success for any territorial species.

Considering movement behavior and how it might be manipulated to solve conservation problems is poorly developed, for the most part, but has many possibilities. For example, threats to amphibian persistence include roads bisecting breeding and adult habitats, and breeding pond loss. Efforts to provide tunnels under roads have shown that species can be very particular in their requirements for use. Behavioral tendencies and restrictions should be considered in reserve designs (Schultz 1998) and in so-called sustainable development to ensure target species can take advantage of the landscape design.

12.4.2 Corridor Use

Another way to facilitate dispersal among patches of habitat is through corridors. Reserve designs and sustainable development plans often depend on corridors because of their theoretical importance to sustaining otherwise isolated populations (Saunders and Hobbs 1991). The efficacy of corridors in different situations, particularly what constitutes a corridor for different species, is controversial, but a recent review concluded that corridors can be effective (Beier and Noss 1998). Corridors can be as short as a highway underpass (Reed et al. 1975) or tens of kilometers long, such as buffer zones along railroads or highways (Walters et al. 1988). What constitutes a suitable corridor depends in a large part on behavioral decisions, which means corridor construction needs to take target species behaviors into account.

Understanding how an animal uses its environment can help predict what will constitute a corridor. For example, some species use roads for foraging or dispersal (Munguira and Thomas 1992; Seabrook and Dettmann 1996), so roads might make useful corridors for some species of bat. In other taxa, roads can be barriers to dispersal (Richardson et al. 1997). Studies of movement behaviors and their consequences are critical to understanding corridor use. Rosenberg et al. (1998) found that *Ensatina eschscholtzii*, a plethodontid salamander, moves across bare soil as well as across ground covered by vegetation, but the movements across the former habitat were much quicker. It is unknown, however, which cover type is better for the salamander. What actually constitutes a corridor for different species remains controversial (Beier and Noss 1998). The best approach for determining this is to develop replicated studies with different treatments for each species in question. In reality, time and money constraints often preclude this. Recently, Haddad (1999) provided a method that could shorten this process by showing that corridor use might be predicted from behaviors at ecotones. In looking at the behaviors of three species of butterfly, Haddad (1999) found that habitat specialists responded to ecotones by changing course and that they used corridors to move among patches, while habitat generalists moved regardless of ecotones. Using simulation modeling of movement behavior, he showed that some increases in corridor width increased movement among patches, but that after a certain width the added benefit was minimal. Using his methods, combining behavior observations with simulation models, one might predict the efficacy of corridors for habitat specialists.

Behavior might even be influenced to encourage corridor use. White-tailed deer become more nocturnal and avoid roads and certain habitat types during the hunting season (Kilgo et al. 1998). For this species, placing corridors away from roads could result in increased use during the hunting season. The same type of detailed knowledge of behavior for species targeted for conservation might provide similar management tools.

12.5 Improving Population Viability Analysis

One basic problem in conservation biology is predicting population persistence. This can be done for some species through population viability analysis (PVA), a modeling tool that uses demographic data (survival and reproductive rates) and their relationships with environmental factors to project population sizes over time (Beissinger and Westphal 1998). Incorporating behavior in these models can increase prediction accuracy for some species. Because Reed (1999) recently discussed incorporating behavior into PVA models, here I will only highlight some salient points.

Effects of behavior on survival and reproduction can be incorporated directly as a model parameter or indirectly by modifying the probability density function of a parameter. For example, density-dependent effects on survival or reproduction due to behavioral processes such as finding a mate, predator defense, or mating disruption can be incorporated as a parameter that modifies reproduction or survival. Models based on PVA also can be modified to incorporate the results of altered mating preferences or, in metapopulation models, the results of altered dispersal patterns (e.g., Boulinier and Danchin 1997). Spatially explicit population viability models can be developed for single populations and for metapopulations (Dunning et al. 1995), and if the model is individually based, it has great potential for incorporating behavior. For example, one could incorporate decision making during dispersal, mating preference, mate encounter rates, effects of habitat shape on encounter rate and selection, and so on.

Incorporating behavioral data into PVA, however, is simple only in theory. Incorporating behavior into demographic models requires translating behaviors into demographic or spatial consequences (Beissinger 1997), which can be difficult because the data required for this incorporation normally do not exist and are difficult to gather. As a consequence, one often must rely on best guesses of the effects of different behaviors on model parameters, and PVA predictions can be very sensitive to small changes in some parameter values (e.g., Stacey and Taper 1992). In addition, even if there is a demonstrated effect of behavior on any demographic parameter, it is not always necessary to incorporate it into a

PVA. Hanski (1994) developed a metapopulation model that can incorporate behavior in multiple ways, but the model predictions appear to be insensitive to its incorporation. Sensitivity analysis or, more specifically, elasticity analysis (de Kroon et al. 1986) can be used to anticipate the potential importance of incorporating behavioral effects in modeling population persistence. If predicted persistence is insensitive to the degree of change in demographic parameters anticipated from incorporating behavior, it should not be incorporated into the PVA.

12.6 Successful Translocation and Reintroduction

Translocation and reintroduction are used commonly for species conservation (e.g., Griffith et al. 1989). They are used to create or augment populations and increase genetic variability. They can be done artificially, by moving adults, juveniles, or eggs (including by implanting ova), or more naturally by manipulating behaviors. The importance of behaviors in these processes varies by species and sometimes by individual, and could manifest at many stages. For example, behavioral adaptations of translocated individuals might not be suitable for local conditions (Warren et al. 1996), introduced animals might not be incorporated into local populations (Clarke and Schedvin 1997), or hybridization with related species could occur (Parkin 1996). How animals are reared in captivity also can have important effects on the success of released animals. Rearing nene (*Branta sandvicensis*) in captivity with adults increases survival of released birds over that experienced by young reared in the absence of adults (Marshall and Black 1992). Black-footed ferrets (*Mustela nigripes*) reared in seminatural pens have higher postrelease survival than do those reared in cages without pen experience (Biggins et al. 1998). This pattern has been observed in other species, as well, and can be a particular problem for species with complex social interactions (Watts and Meder 1996).

12.7 Decreasing Loss of Genetic Variability

Loss of genetic variability threatens species persistence by increasing the frequency of homozygosity of deleterious alleles and by decreasing the variability available for selection (Frankel and Soulé 1981). The rate of loss of genetic variability from a population is predicted by its effective size (N_e), which is determined by population size, sex ratio, skew in reproductive success, mating system, and similar traits (Parker and Waite 1997). Of most immediate concern to conservation biologists are problems associated with increased homozygosity within a population, which can result in inbreeding depression, which increases extinction risks (Allendorf and Leary 1986). One way to slow the rate of loss of genetic variability from a population is through immigration. The methods described above to increase dispersal or colonization would increase genetic variability. It is important to be aware of the potential problems created by these actions, such as outbreeding depression or disruption of local adaptation (e.g., Frankel and Soulé 1981). Consequently, benefits of these actions need to be weighed against potential detriments.

Another method of decreasing the rate of loss of genetic variability is through selective breeding. By allowing more animals to mate and by reducing variance in reproductive success among individuals, a greater amount of variability is passed to the next generation (Ballou and Foose 1996). This possibility seems relatively simple for captive populations, but unlikely for wild populations. Observations of mating preferences based on characteristics that can be manipulated suggest avenues for altering mate selection in wild populations of some species. Research on zebra finches (*Poephila guttata*) showed that colored leg bands could affect mate selection and reproduction (Burley 1986a). Reproductive success was twice as high for "attractive" birds (i.e., those banded with "attractive" colors) (Burley 1986a) and the sex ratio of offspring favored the "attractive" mate (Burley 1986b, c). In addition, mortality rates were higher for "unattractive" birds (60% vs. 13% for "attractive" birds during the time of the experiment; Burley 1985). Patterns of mating preference or reproductive success associated with artificial marking have been observed in other species (Hagan and Reed 1988), but not in all (Weatherhead et al. 1991; Cristol et al. 1992). Whether band color affects reproduction and survival more generally is controversial, as are the proposed mechanisms driving these patterns (Ratcliffe and Boag 1987; Burley 1986b). Recent manipulative experiments where unique adornments were provided to breeders reinforce some of Burley's ideas. For example, Witte and Curio (1999) provided novel traits (a colored feather to the head, altered bill color, and stripes under the tail) to Javanese mannikins (*Lonchura leucogastroides*) and found they affected mate choice and that males and females responded differently to the traits.

From a conservation perspective, if individual "attractiveness" could be manipulated in wild populations of a target species, it is theoretically possible to increase effective population size with a minimum of intrusion. This would allow slowing the rate of loss of genetic variability through two mechanisms, altering the breeding sex ratio and decreasing variance in reproductive success among individuals, which would reduce inbreeding depression. Blumstein (1998) modeled simple scenarios of the effects of female mate selection preferences on effective population size. He focused strictly on the effect of mate selection on breeding sex ratio, and some of his results can be simply shown using the following expression for calculating effective population size (N_e):

$$N_e = \frac{4N_m N_f}{N_m + N_f}$$

where N_m and N_f are the effective number of males and females in a population, that is, approximately the number of males and females breeding (Falconer 1989). For a fixed number of breeders, as the sex ratio becomes more skewed, the effective size decreases (figure 12.1). Altering mating preference also would affect variance among individuals in reproductive success. This relationship can be shown simply as

$$N_e = \frac{4N - 2}{V_k + 2}$$

where N is the number of breeders and V_k is the variance in gamete production, or variance in family size (Falconer 1989). If all individuals have the same reproductive success ($V_k = 0$), N_e is maximized at $2N - 1$, and N_e declines with increased variance in reproductive success among individuals (figure 12.1).

Based on the above models, manipulating mate preference and relative reproductive success among individuals can increase effective population size. In fact,

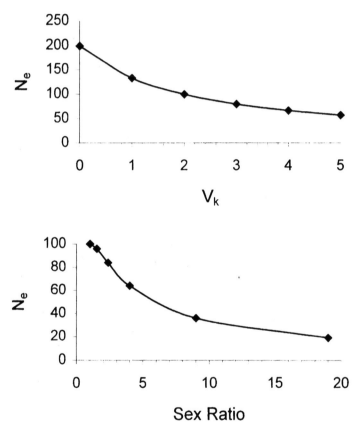

Figure 12.1 Effective population size (N_e) estimates for populations under different mating and reproductive scenarios. N_e is a measure of the rate of loss of genetic variability; the smaller N_e is the faster genetic variability is lost. Top: N_e for 100 breeders with different variances among individuals in reproductive success (V_k). Bottom: N_e for a population with 100 breeders with different sex ratios of breeders (M/F). In both graphs it is assumed that populations are ideal except for the variable being altered on the x-axis and that generations do not overlap; see text for equations.

even habitat manipulation can affect mating patterns in some species (Apollonio et al. 1998). Based on experimental manipulations of mate choice, these methods have the potential for working in both sexually dimorphic and monomorphic species. Each species should be tested for an effect, and it should be determined whether the effect, if found, is enough to merit manipulation.

12.8 Conclusions

In this chapter I have shown that there are important problems in conservation biology that can be solved or ameliorated for some species by manipulating behaviors. Some of the solutions discussed have been implemented for active conservation, while others are at the planning stage. My goal was to show the myriad and often quite creative uses to which behavior might be incorporated or manipulated to achieve specific conservation goals. More attention needs to be paid to the behaviors of wild animals, with particular reference to altering the environment to enhance reproductive performance, survival, and dispersal. I also encourage researchers to review methods used in domestic animal care and captive-breeding programs to look for other ways in which behavior might be manipulated. Researchers in these areas have long recognized the importance of animal behavior in their work and have tools or approaches that could be adopted for in situ and ex situ animal conservation (e.g., Kilgour and Dalton 1984). Information on animal behavior relevant to conservation, particularly for captive breeding of wild animals, exists for invertebrates (Demarest and Bradley 1995), fishes (Francis-Floyd and Williams 1995), reptiles and amphibians (Burghardt and Milostan 1995), birds (Hutchins et al. 1995), and mammals (terrestrial, Gittlemen and McMillan 1995; marine, Ellis 1995). Greater interaction among conservation biologists, animal behaviorists, captive breeders, and domestic breeders would advance the field quickly. Not coincidentally, it is this type of disciplinary crossover that is necessary to develop the field of conservation medicine. Veterinarians, physicians, and biologists have been working in related and often overlapping fields of research. Conservation medicine will draw on all the subdisciplines in these fields to develop the tools required to maintain ecosystem health.

References

Allendorf, F.W., and R.F. Leary. 1986. Heterozygosity and fitness in natural populations of animals. In Soulé, M., ed. Conservation biology: the science of scarcity and diversity, pp. 57–76. Sinauer, Sunderland, Mass.

Amos, B., C. Schlötterer, and D. Tautz. 1993. Social structure of pilot whales revealed by analytical DNA profiling. Science 260:670–672.

Apollonio, M., M. Festa-Bianchet, F. Mari, E. Bruno, and M. Locati. 1998. Habitat manipulation modifies lek use in fallow deer. Ethology 104:603–612.

Ballou, J.D., and T.J. Foose. 1996. Demographic and genetic management of captive populations. In Kleiman, D.G., M.E. Allen, K.V. Thompson, and S. Lumpkin, eds. Wild mammals in captivity: principles and techniques, pp. 263–283. University of Chicago Press.

Beauchamp, G., M. Bélisle, and L.-A. Giraldeau. 1997. Influence of conspecific attraction on the spatial distribution of learning foragers in a patchy habitat. J Anim Ecol 66: 671–682.

Bednarz, J.C. 1988. Cooperative hunting in Harris hawks (Parabuteo unicinctus). Science 239:1525–1527.

Beier, P., and R.F. Noss. 1998. Do habitat corridors provide connectivity? Conserv Biol 12:1241–1252.

Beissinger, S.R. 1997. Integrating behavior into conservation biology: potentials and limitations. In Clemmons, J.R., and R. Buchholz, eds. Behavioral approaches to conservation in the wild, pp. 3–22. Cambridge University Press, Cambridge.

Beissinger, S.R., and M.I. Westphal. 1998. On the use of demographic models of population viability analysis in endangered species management. J Wildl Manage 62:821–841.

Bibby, C.J. 1999. Making the most of birds as environmental indicators. Ostrich 70:81–88.

Bibby, C.J., N.D. Burgess, and D.A. Hill. 1992. Bird census techniques. Academic Press, London.

Biggins, D.E., J.L. Godbey, L.R. Hanebury, B. Luce, P.E. Marinari, M.R. Matchett, and A. Vargas. 1998. The effect of rearing methods on survival of reintroduced black-footed ferrets. J Wildl Manage 62:643–653.

Blanco, G., J.A. Fargallo, J.L. Tella, and J.A. Cuevas. 1997. Role of buildings as nest-sites in the range expansion and conservation of choughs Pyrrhocorax pyrrhocorax in Spain. Biol Conserv 79:117–122.

Blumstein, D.T. 1998. Female preferences and effective population size. Anim Conserv 1: 173–177.

Boulinier, T., and E. Danchin. 1997. The use of conspecific reproductive success for breeding patch selection in territorial migratory species. Evol Ecol 11:505–517.

Brown, J.H., and A. Kodric-Brown. 1977. Turnover rates in insular biogeography: effects of immigration on extinction. Ecology 58:445–449.

Bunin, J.S., and I.G. Jamieson. 1996. Responses to a model predator of New Zealand's endangered takahe and its closest relative, the pukeko. Conserv Biol 10:1463–1466.

Burghardt, G.M., and M.A. Milostan. 1995. Ethological studies on reptiles and amphibians: lessons for species survival plans. In Gibbons, E.F., Jr., B.S. Durrant, and J. Demarest, eds. Conservation of endangered species in captivity: an interdisciplinary approach, pp. 187–203. State University of New York Press, Albany.

Burley, N. 1985. Leg-band color and mortality patterns in captive breeding populations of zebra finches. Auk 102:647–651.

Burley, N. 1986a. Sexual selection for aesthetic traits in species with biparental care. Am Natu 127:415–445.

Burley, N. 1986b. Comparison of the band color preferences of two species of estrilid finches. Anim Behav 34:1732–1741.

Burley, N. 1986c. Sex-ratio manipulation in color banded populations of zebra finches. Evolution 40:1191–1206.

Caro, T.M. 1998. Behavioral ecology and conservation biology. Oxford University Press, New York.

Castro, I., E.O. Minot, and J.C. Alley. 1994. Translocation of hihi or stitchbirds *Notiomystis cincta* to Kapiti Island, New Zealand: transfer techniques and comparison of release strategies. *In* Serena, M., ed. Reintroduction biology of Australian and New Zealand fauna, pp. 113–120. Surrey Beatty, Chipping Norton, Australia.

Clark, D.L., J.M. Macedonia, and G.G. Rosenthal. 1997. Testing video playback to lizards in the field. Copeia 1997:421–423.

Clarke, M.F., and N. Schedvin. 1997. An experimental study of the translocation of noisy miners *Manorian melanocephala* and difficulties associated with dispersal. Biol Conserv 80:161–167.

Clarke, M.F., K.B. da Silva, H. Lair, R. Pocklington, D.L. Kramer, and R.L. McLaughlin. 1993. Site familiarity affects escape behaviour of the eastern chipmunk, *Tamias striatus*. Oikos 66:533–537.

Clemmons, J.R., and R. Buchholz. 1997. Behavioral approaches to conservation in the wild. Cambridge University Press, Cambridge.

Cohn, J., and R.C. MacPhail. 1996. Ethological and experimental approaches to behavior analysis: implications for ecotoxicology. Environ Health Perspect 104 (suppl 2):299–305.

Colborn, T., M.J. Smolen, and R. Rolland. 1998. Environmental neurotoxic effects: the search for new protocols in functional teratology. Toxicol Indust Health 14:9–23.

Cristol, D.A., C.S. Chiu, S.M. Peckham, and J.F. Stoll. 1992. Color bands to not affect dominance status in captive flocks of wintering dark-eyed juncos. Condor 94:537–539.

Danchin, E., B. Cadiou, J.-Y. Monnat, and R. Rodriguez Estrella. 1991. Recruitment in long-lived birds: conceptual framework and behavioural mechanisms. Proc Int Ornithol Congr 20:1641–1656.

de Kroon, H., A. Plaiser, J. van Groenendael, and H. Caswell. 1986. Elasticity: the relative contribution of demographic parameters to population growth rate. Ecology 67:1427–1431.

Demarest, J., and M. Bradley. 1995. Behavioral research with endangered invertebrate species. *In* Gibbons, E.F., Jr., B.S. Durrant, and J. Demarest, eds. Conservation of endangered species in captivity: an interdisciplinary approach, pp. 37–50. State University of New York Press, Albany.

Desrochers, A., and S.J. Hannon. 1997. Gap crossing decisions by forest songbirds during the post-fledging period. Conserv Biol 11:1204–1210.

Distal, R.A., and F.D. Provenza. 1991. Experience early in life affects voluntary intake of blackbrush by goats. J Chem Ecol 17:431–450.

Døving, K.B. 1991. Assessment of animal behaviour as a method to indicate environmental toxicity. Comp Biochem Physiol 100C:247–252.

Dunning, J.B., D.J. Stewart, B.J. Danielson, B.R. Noon, T.L. Root, R.H. Lamberson, and E.E. Stevens. 1995. Spatially explicit population models: current forms and future uses. Ecol Appl 5:3–12.

Ellis, S. 1995. Marine mammal behavior: conservation through research. In Gibbons, E.F., Jr., B.S. Durrant, and J. Demarest, eds. Conservation of endangered species in captivity: an interdisciplinary approach, pp. 441–463. State University of New York Press, Albany.

Falconer, D.S. 1989. Introduction to quantitative genetics, 3rd ed. Longman, New York.

Fowler, C.W., and J.D. Baker. 1991. A review of animal population dynamics at extremely reduced population levels. Rep Int Whaling Comm 41:545–554.

Francis-Floyd, R., and J.D. Williams. 1995. Behavioral research on captive endangered fishes of North America. *In* Gibbons, E.F., Jr., B.S. Durrant, and J. Demarest, eds.

Conservation of endangered species in captivity: an interdisciplinary approach, pp. 105–123. State University of New York Press, Albany.

Frankel, O.H., and M.E. Soulé. 1981. Conservation and evolution. Cambridge University Press, Cambridge.

Furness, R.W., and J.J.D. Greenwood. 1993. Birds as monitors of environmental change. Chapman and Hall, London.

Galindo, J., R.J. Kendall, C.J. Driver, and T. Lacher Jr. 1985. The effect of methyl parathion on the susceptibility of bobwhite quail (*Colinus virginianus*) to domestic cat predation. Behav Neural Biol 43:21–36.

Geist, V. 1971. Mountain sheep: a study in behavior and evolution. University of Chicago Press, Chicago.

Gittleman, J.L., and G.C. McMillan. 1995. Mammalian behavior: lessons from captive studies. In Gibbons, E.F., Jr., B.S. Durrant, and J. Demarest, eds. Conservation of endangered species in captivity: an interdisciplinary approach, pp. 355–376. State University of New York Press, Albany.

Gosling, L.M., and W.J. Sutherland. 2000. Behaviour and conservation. Cambridge University Press, Cambridge.

Griffith, B., J.M. Scott, J.W. Carpenter, and C. Reed. 1989. Translocation as a species conservation tool: status and strategy. Science 245:477–480.

Haddad, N.M. 1999. Corridor use predicted from behaviors at habitat boundaries. Am Nat 153:215–227.

Hagan, J.M., and J.M. Reed. 1988. Red color bands reduce fledging success in red-cockaded woodpeckers. Auk 105:498–503.

Hansen, A.J., and F. de Castri. 1992. Landscape boundaries: consequence for biotic diversity and ecological flows. Springer, New York.

Hanski, I. 1994. A practical model of metapopulation dynamics. J Anim Ecol 63:151–162.

Harris, L.D. 1984. The fragmented forest: island biogeography theory and the preservation of biotic diversity. University of Chicago Press, Chicago.

Helender, B. 1978. Feeding white-tailed sea eagles in Sweden. In Temple, S.A., ed. Endangered birds: management techniques for preserving threatened species, pp. 149–159. Croom Helm, Madison, Wisc.

Henke, S.E. 1997. Do white-tailed deer react to the dinner bell? An experiment in classical conditioning. Wildl Soc Bull 25:291–295.

Hoeck, H.N. 1989. Demography and competition in hyrax: a 17 year study. Oecologia 79:353–360.

Hutchins, M., C. Sheppard, A.M. Lyles, and G. Casadei. 1995. Behavioral considerations in the captive management, propagation, and reintroduction of endangered birds. In Gibbons, E.F., Jr., B.S. Durrant, and J. Demarest, eds. Conservation of endangered species in captivity: an interdisciplinary approach, pp. 263–289. State University of New York Press, Albany.

Isbell, L.A., D.L. Cheney, and R.M. Seyfarth. 1990. Costs and benefits of home range shifts among vervet monkeys (*Cercopithecus aethiops*) in Amboseli National Park, Kenya. Behav Ecol Sociobiol 27:351–358.

Jabłoński, P.G. 1999. A rare predator exploits prey escape behavior: the role of tail-fanning and plumage contrast in foraging of the painted redstart (*Myioborus pictus*). Behav Ecol 10:7–14.

Johnson, M.L., and M.S. Gaines. 1990. Evolution of dispersal: theoretical models and empirical tests using birds and mammals. Annu Rev Ecol Syst 21:449–480.

Kadlec, J.A., and L.M. Smith. 1992. Habitat management for breeding areas. *In* Batt,

B.D.J., A.D. Afton, M.G. Anderson, C.D. Ankney, D.H. Johnson, J.A. Kadlec, and G.L. Krapu, eds. Ecology and management of breeding waterfowl, pp. 590–610. University of Minnesota Press, Minneapolis.

Kamil, A.C., and S.I. Yoerg. 1981. Learning and foraging behavior. *In* Bateson, P.P., and P.H. Klopfer, eds. Perspectives in ethology, pp. 325–364. Plenum, New York.

Karr, J.R., 1995. Using biological criteria to protect ecological health. *In* Rapport, D.J., C. Gaudet, and P. Calow, eds. Evaluating and monitoring the health of large scale ecosystems, pp. 137–152. Springer, New York.

Kear, J. 1990. Man and wildfowl. Poyser, London.

Kilgo, J.C., R.F. Labisky, and D.E. Fritzen. 1998. Influences of hunting on the behavior of white-tailed deer: implications for conservation of the Florida panther. Conserv Biol 12:1359–1364.

Kilgour, R., and C. Dalton. 1984. Livestock behaviour: a practical guide. Westview Press, Boulder, Colorado.

Kress, S.W. 1983. The use of decoys, sound recordings, and gull control for re-establishing a tern colony in Maine. Colonial Waterbirds 6:185–196.

Kress, S.W. 1997. Using animal behavior for conservation: case studies in seabird restoration from the Maine coast. Yamashina Inst Ornithol 29:1–26.

Kulig, B., E. Alleva, G. Bignami, J. Cohn, D. Cory-Slechta, V. Landa, J. O'Donaghue, and D. Peakall. 1996. Animal behavioral methods in neurotoxicity assessment. Environ Health Perspect 104(suppl 2): 193–204.

Leopold, A. 1933. Game management. Scribner's, New York.

Lima, S.L., and P.A. Zollner. 1996. Towards a behavioral ecology of ecological landscapes. TREE 11:131–135.

MacArthur, R.H., and E.O. Wilson. 1967. The theory of island biogeography. Princeton University Press, Princeton, N.J.

Marshall, A.P., and J.M. Black. 1992. The effect of rearing experience on subsequent behavioural traits in Hawaiian Geese *Branta sandvicensis*: implications for the recovery programme. Bird Conserv Int 2:131–147.

McLean, K.G., G. Lundie-Jenings, and P.J. Jarman. 1995. Teaching an endangered mammal to recognise predators. Biol Conserv 75:51–62.

Mirza, S.N., and F.D. Provenza. 1990. Preference of the mother affects selection and avoidance of foods by lambs differeing in age. Appl Anim Behav Sci 28:255–263.

Montevecchi, W.A. 1993. Birds as indicators of change in marine prey stocks. In Furness, R.W., and J.J.D. Greenwood, eds. Birds as monitors of environmental change, pp. 217–266. Chapman and Hall, London.

Munguira, M.L., and J.A. Thomas. 1992. Use of road verges by butterfly and burnet populations, and the effect of roads on adult dispersal and mortality. J Appl Ecol 29: 316–329.

Nocera, J.J., and P.D. Taylor. 1998. In situ behavioral response of common loons associated with elevated mercury exposure. Conserv Ecol 2(2):art. 10.

Parker, P.G., and T.A. Waite. 1997. Mating systems, effective population size, and conservation of natural populations. *In* Clemmons, J.R., and R. Buchholz, eds. Behavioral approaches to conservation in the wild, pp. 243–261. Cambridge University Press, Cambridge.

Parkin, D. 1996. Colonisation and hybridisation in birds. *In* Holmes, J.S., and J.R. Simons, eds. The introduction and naturalisation of birds, pp. 25–35. Stationary Office, London.

Peakall, D.B. 1985. Behavioral responses of birds to pesticides and other contaminants. Residue Rev 96:45–77.

Peakall, D.B. 1996. Disrupted patterns of behavior in natural populations as an index of ecotoxicity. Environ Health Perspect 104(suppl 2):331–335.

Pickering, S.P.C., and L. Duverge. 1992. The influence of visual stimuli provided by mirrors on the marching displays of lesser flamingos (*Phoencopterus minor*). Anim Behav 43:1048–1050.

Powell, R.A., J.W. Zimmerman, D.E. Seaman, and J.F. Gilliam. 1996. Demographic analyses of a hunted black bear population with a refuge. Conserv Biol 10:1–11.

Pöysä, H. 1991a. Does the attractiveness of teal foraging groups depend on the posture of group members? Ornis Scand 22:167–169.

Pöysä, H. 1991b. Effects of predation risk and patch quality on the formation and attractiveness of foraging groups of teal, *Anas crecca*. Anim Behav 41:285–294.

Provenza, F.D., and D.F. Balph. 1987. Diet learning by domestic ruminants: theory, evidence and practical implications. Appl Anim Behav Sci 18:211–232.

Purcell, K.L., and J. Verner. 1998. Density and reproductive success of California towhees. Conserv Biol 12:442–450.

Ratcliffe, L.M., and P.T. Boag. 1987. Effects of colour bands on male competition and sexual attractiveness in zebra finches. Can J Zool 65:333–338.

Reed, D.F., T.N. Woodard, and T.M. Pojar. 1975. Behavioral response to mule deer to a highway underpass. J Wildl Manage 39:361–367.

Reed, J.M. 1996. Using statistical probability to increase confidence of inferring species extinction. Conserv Biol 10:1283–1285.

Reed, J.M. 1999. The role of behavior in recent avian extinctions and endangerments. Conserv Biol 13:232–241.

Reed, J.M., and A.P. Dobson. 1993. Behavioural constraints and conservation biology: conspecific attraction and recruitment. TREE 8:253–256.

Reed, J.M., J.H. Carter III, J.R. Walters, and P.D. Doerr. 1988. An evaluation of indices of red-cockaded woodpecker populations. Wildl Soc Bull 16:406–410.

Reed, J.M., T. Boulinier, E. Danchin, and L.W. Oring. 1999. Informed dispersal: prospecting by birds for breeding sites. Curr Ornithol 15: 189–259.

Richardson, J.H., R.F. Shore, and J.R. Treweek. 1997. Are major roads a barrier to small mammals? J Zool Lond 243:840–846.

Rhymer, J.M., and D. Simberloff. 1996. Extinction by hybridization and introgression. Annu Rev Ecol Syst 27:83–109.

Rodgers, R.D. 1992. A technique for establishing sharp-tailed grouse in unoccupied range. Wildl Soc Bull 20:101–106.

Rosenberg, D.K., B.R. Noon, J.W. Megahan, and E.C. Meslow. 1998. Compensatory behavior of *Ensatina eschscholtzii* in biological corridors: a field experiment. Can J Zool 76:117–133.

Sarrazin, F., C. Bagnolini, J.-L. Pinna, and E. Danchin. 1996. Breeding biology during establishment of a reintroduced griffon vulture (*Gyps fulvus*) population. Ibis 138: 315–325.

Saunders, D.A., and R.J. Hobbs. 1991. Nature conservation. Vol. 2: The role of corridors. Surrey Beatty, Sydney.

Schultz, C.B. 1998. Dispersal behavior and its implications for reserve design in a rare Oregon butterfly. Conserv Biol 12:284–292.

Seabrook, W.A., and E.B. Dettmann. 1996. Roads as activity corridors for cane toads in Australia. J Wildl Manage 60:363–368.

Shaffer, M.L. 1981. Minimum population sizes for species conservation. Bioscience 31: 131–134.

Stacey, P.B., and M. Taper. 1992. Environmental variation and the persistence of small populations. Ecol Appl 2:18–29.

Stamps, J.A. 1987. Conspecifics as cues to territory quality: a preference of juvenile lizards (*Anolis aeneus*) for previously used territories. Am Nat 129:629–642.

Stamps, J.A. 1991. The effect of conspecifics on habitat selection in territorial species. Behav Ecol Sociobiol 28:29–36.

Stamps, J.A., Buechner, M., and V.V. Krishnan. 1987. The effects of edge permeability and habitat geometry on emigration from patches of habitat. Am Nat 129:533–552.

St. Clair, C.C., M. Belisle, A. Desrochers, and S. Hannon. 1998. Winter responses of forest birds to habitat corridors and gaps. Conserv Ecol 2:13.

Stevens, E.F. 1991. Flamingo breeding: the role of group displays. Zoo Biol 10:53–64.

Sutherland, W.J. 1998. The importance of behavioural studies in conservation biology. Anim Behav 56:801–809.

Thorhallsdottir, A.G., F.D. Provenza, and D.F. Balph. 1990. Ability of lambs to learn about novel foods while observing or participating with social models. Appl Anim Behav Sci 25:25–33.

Van Horne, B. 1983. Density as a misleading indicator of habitat quality. J Wildl Manage 47:893–901.

Veen, J. 1977. Functional and causal aspects of nest distribution in colonies of the Sandwich tern (*Sterna sandvicensis* lath). Behaviour 20 (suppl):1–193.

Verner, J. 1992. Data needs for avian conservation biology: have we avoided critical research? Condor 94:301–303.

Vickery, P.D., M.L. Hunter, and J.V. Wells. 1992. Use of a new reproductive index to evaluate relationship between habitat quality and breeding success. Auk 109:697–705.

Walters, J.R., S.K. Hansen, J.H. Carter, P.D. Manor, and R.J. Blue. 1988. Long-distance dispersal of an adult red-cockaded woodpecker. Wilson Bull 100:494–496.

Warner, R.E., K.K. Peterson, and L. Borgman. 1966. Behavioural pathology in fish: a quantitative study of sublethal pesticide toxification. J Appl Ecol 3 (suppl):223–247.

Warren, C.D., J.M. Peek, G.L. Servheen, and P. Zagers. 1996. Habitat use and movements of two ecotypes of translocated caribou in Idaho and British Columbia. Conserv Biol 10:547–553.

Watts, E., and A. Meder. 1996. Introduction and socialization techniques for primates. *In* Kleiman, D.G. M.E. Allen, K.V. Thompson, and S. Lumpkin, eds. Wild mammals in captivity: principles and techniques, pp. 67–77. University of Chicago Press, Chicago.

Weatherhead, P.J., D.J. Hoysak, K.J. Metz, and C.G. Eckert. 1991. A retrospective analysis of red-band effects on red-winged blackbirds. Condor 93:1013–1016.

Weddell, B.J. 1991. Distribution and movements of Columbian ground squirrels (*Spermophilus columbianus* (Ord)): are habitat patches like islands? J Biogeogr 18:385–394.

Witte, K., and E. Curio. 1999. Sexes of a monomorphic species differ in preference for mates with a novel trait. Behav Ecol 10:15–21.

Wu, J., J.L. Vankat, and Y. Barlas. 1993. Effects of patch connectivity and arrangement on animal metapopulation dynamics: a simulation study. Ecol Mod 65:221–254.

Wyrwicka, W. 1981. The development of food preferences. Thomas, Springfield, Ill.

Zimmerman, B.L. 1994. Audio strip transects. *In* Heyer, W.R., M.A. Donnelly, R.W. McDiarmid, L.-A.C. Hayek, and M.S. Foster, eds. Measuring and monitoring biological diversity: standard menthods for amphibians, pp. 92–97. Smithsonian Institution Press, Washington, D.C.

Part III

Ecological Health and Humans

13

Global Ecological Change
and Human Health

Jonathan A. Patz
Nathan D. Wolfe

Environmental changes and ecological disturbances, due to both natural phenomena and human activities, have exerted and can be expected to continue to exert a strong influence on the incidence, proliferation, and emergence of disease. This chapter focuses on both near-term and sequential effects that environmental factors and their changes have on ecosystems and, consequently, on human health. The types of environmental conditions and changes that contribute to proliferation of disease are discussed, as well as the effects on health of changes in climate and land use, especially deforestation and its many sequelae: logging, road building, hunting, human settlement patterns, agricultural and commercial development, and formation of water bodies (including water control systems. An illustrative example of an illness (malaria) resulting from ecological change is also presented.

13.1 Environmental Conditions and Proliferation Of Disease

13.1.1 Climate

Climate consists of variations in interactions among components of the climate system, including the atmosphere, oceans, sea ice, and land features. Changes in any of the climate system components, whether natural or from external (human-induced) forcing, can cause climatic variation and change. Change in climate can result in increased temperature, rise in sea level, extremes in the hydrologic cycle (Kattenberg et al. 1996), and potentially, accelerated ozone depletion (Shindell et al. 1998; Kirk-Davidoff et al. 1999). Small changes in the mean climate can produce relatively large changes in the frequency and severity of extreme weather events (Karl and Knight 1997).

Until rather recently, climatic changes occurred solely through natural phenomena such as changes in the output of the sun or slowing of ocean circulation. Today, human activities also influence climate change: fossil-fuel combustion (coal, oil, and gas), which increases carbon dioxide (CO_2) concentration; deforestation, which reduces CO_2 absorption capacity of forested areas; industrial processes and biomass burning, which produce emissions of gases such as CO_2 and volatile organic compounds (Schimel et al. 1996). The current levels of atmospheric methane (CH_4) are the highest they have ever been (Chappellaz et al. 1990). Halogenated compounds, which destroy the stratospheric ozone layer and have greenhouse warming properties, are now found in the atmosphere although they do not exist naturally (WMO 1995). The latest assessment by the United Nations' Intergovernmental Panel on Climate Change (IPCC) now projects that average global temperatures will rise by between 1.5°C and 6.0°C by the year 2100. Epstein (chap. 4) gives a more detailed discussion of the implications of climate change on infectious disease.

13.1.2 Seasonality

Many diseases follow a seasonal pattern in their incidence. On the one hand, this suggests an association with seasonal weather variability; on the other hand, when studying the effect of weather per se, it is important to obtain observations across all seasons to factor in "seasonal effects" that are not necessarily weather related (e.g., school calendar). Weather seasonality, vector development, and abundance are affected by the amount of rain, length of rainy and dry seasons, and the interval between the seasons. The different *Shistosoma* blood fluke species, which transmit shistosomiasis, vary in their optimal seasonal conditions. Parasite abundance is directly proportional to abundance of snails, which serve as intermediate hosts. The snails lay eggs in proportion to the amount of vegetation present, which in turn depends, in part, on seasonal variation in rain, water temperature, and flow (Jobin 1999). In China, with construction of the Three Gorges high dam, the snail distribution and shistosomiasis prevalence vary with annual rainfall and evaporation, water levels in the Yangzte River, and ground altitude (Xu et al. 1999). In the Plateau State in Nigeria, onchocerciasis (river blindness), caused by the *Onchocerca volvulus* parasite and transmitted by the bite of the blackfly *Simulium*, peaks at the beginning and end of the rainy season, as biting density occurs at the height of the rainy season (Nwoke et al. 1992). In sub-Saharan Africa, epidemics of meningococcal meningitis consistently erupt during the hot dry season and subside soon after the onset of the rainy season (Moore 1992).

13.1.3 Temperature

Extreme hot and cold temperatures have been associated with increased morbidity and mortality of people (Kilbourne 1998), especially during periods of several consecutive days' persistent high overnight temperatures (Ramlow and Kuller

1990). The very young, the elderly, those with preexisting cardiovascular and respiratory disorders are at particular risk for heat exhaustion and heat stroke (Larsen 1990a, b; Kilbourne 1998).

Temperature and ultraviolet radiation are key variables in the formation of ground-level ozone, or photochemical smog. Higher temperatures accompanying global warming are posited to increase the frequency of ozone pollution episodes (Grey et al. 1987; Morris et al. 1989; EPA 1989). High levels of ozone usually last for a period of three to four days and extend over areas greater than 600,000 km². The presence of ozone increases the sensitivity of asthmatics to allergens and impairs lung function, particularly in children and the elderly (Beckett 1991; Schwartz 1994; Koren and Bromberg 1995). In addition, the combination of high temperatures, sulfur dioxide, and particulate air pollution may result in higher mortality than that caused by the pollutants alone (Bobak and Roberts 1997; Katsouyanni et al. 1997).

13.1.4 Hydrological Cycles, Rainfall, Floods, and Sea Level Rise

Changes in the hydrological cycle, including severe storms, floods, and droughts, can promote the emergence and proliferation of disease or occurrence of injuries and fatalities. Heavy rainfall, especially in areas of deforestation and/or erosion where the soil is not absorbent, can cause damaging floods. Direct injury and death are caused by people being swept away and by drowning (Noji 1997). Flooding triggers a series of sequential effects: disruption of water purification and sewage systems, resulting in water contamination and disease; release and dissemination of toxic chemicals from storage and waste disposal sites; contamination of food crops and disorganization of resource distribution, resulting in malnutrition, especially in children; formation of stagnant water bodies, resulting in increased incidence of vector-borne disease; and crowding within shelters, with increased respiratory disease (NCDC 1999). High-density populations in developing countries with settlements in unprotected flood plains, inadequate or inaccessible shelters, and inadequate early warning systems are especially vulnerable to the effects of severe storms. Sea surface temperatures must be higher than 26°C for hurricanes to form (Gray 1979); according to model simulations, a 2°C warming of the ocean's surface intensifies hurricane wind speeds by up to 12%, or 7 m/s (Beneson 1990). Also, sea surface warming will cause a rise in sea level estimated to be between 14 to 80 cm by the year 2100, sufficient to inundate many low-lying coastal areas.

13.1.5 Drought

Rises in temperatures can lead to increased precipitation through changes in the hydrological cycle, but, ironically, higher mean temperatures will likely lead to reduced soil moisture and drought due to enhanced evaporation (Kattenberg et al. 1996). Drought can cause a series of domino effects deleterious to health, with increased morbidity and mortality: reduction of food production, leading to malnutrition and famine and their disease sequelae; forced migration, with its accompanying poor sanitation conditions; and depletion of water resources, leading to breakdown of sanitation and fecal contamination of drinking, bathing, and dishwashing water supplies. These disruptions, in turn, lead to such diseases as marasmus (multiple nutrient deprivation), kwashiorkor (severe protein deficiency), opportune respiratory infections, and parasitic, bacterial, and viral diseases. Drought can contribute to the triggering of wildfires with their accompanying injury, death, and degradation of air quality. Dispersed fire smoke, which can travel far and wide, increases the incidence of cardiac and respiratory problems, especially asthma and chronic obstructive pulmonary disease.

13.1.6 El Niño

Extreme climate variability stemming from the El Niño events vary geographically and climatically. This present-day driver of climate extremes has been used to shed light on potential health effects of future climate change (Hales et al. 2000): torrential rains with subsequent epidemic malaria (southwestern Uganda highlands, 1997–98; Kilian et al. 1999); drought-exacerbated fires causing air pollution (Indonesia, 1997–98; Patz et al. 2000a); a greater than 5°C increase in ambient temperature, with a doubling in number of hospital admissions for pediatric diarrhea in Lima, Peru, during the 1997–98 El Niño; unseasonable rains during usually dry summer months, leading to increased population of deer mice and subsequent outbreak of hantavirus pulmonary syndrome in the Four Corners region of U.S. Southwest during 1993 (Engelthaler et al. 1999; Glass et al. 2000); flooding, affording submersion and development of mosquito eggs and subsequent 1998 outbreak of Rift Valley fever in Kenya (Linthicum et al. 1999); seasonal warming of sea surface temperatures enhancing plankton blooms of copepods, reservoirs for *Vibrio cholerae*, corresponding with cholera frequency correlated with El Niño events in Bangladesh over an 18-year period (Pascal et al. 2000).

13.2 Deforestation and Its Ecological Sequelae

Deforestation typically involves a series of ecologically disruptive changes—logging, road construction, hunting, agricultural and commercial development, changing patterns of human settlement and animal domestication, new close contacts between humans and animals, and formation of bodies of water—all of which, singly and in combination, affect morbidity and mortality in the indigenous, emigrating, and incoming populations.

13.2.1 Logging

Logging industry practices vary among geographic regions and have effects on the spread of infectious diseases. Clear-cutting (as in Southeast Asia) leads to a rapid decrease in biological diversity, desertification, and/or vast monocultures (as in Borneo's oil palm plantations). Resulting monocultures may be highly susceptible to the emergence of plant diseases, particularly when bordered by fragments of high botanical biodiversity with their variety of plant pathogens (Wolfe et al. 2000). Selective extraction of highly valued timber species (as in Central Africa), which is more profitable in the long-term than clear-cutting, often involves extensive interface between humans and regions of high vertebrate and microbial biodiversity, a situation with substantial potential for the emergence of novel pathogens (Wolfe et al. 2000).

13.2.2 Road Building

Construction of roads for transportation of trees from the forest facilitates the introduction of hunting, new settlements, crop farming, ranching, mining, commercial development, tourism, and building of dams and hydroelectric plants, with all their attendant disruption of ecological systems (Patz et al. 2000b). The building of roads often leads to compromises in biodiversity, habitat fragmentation, and hazards to wildlife, hunters, and vehicles. Access to heretofore pristine forests and newly deforested areas facilitates exposure of nonimmune, nonprotected populations—loggers, farmers, construction workers, miners, tourists, and conservationists—to indigenous and newly arrived vectors and pathogens. These new populations bring with them, and introduce to existing vectors and settlers along the forested/deforested interface, new vector species and infective agents from their far-flung points of origin. Association between humans and animals is increased through establishment of animal conservation and rehabilitation centers, national parks, and wildlife reserves (Fowler 1996; Hunter 1996; Graczyk et al. 1999; Nizeyi et al. 1999), zoos, aquaria, and seaworlds (Cambre and Buick 1996; Schultz et al. 1996; Williams and Thorne 1996; Hannah 1998; Michalak et al. 1998), industrial animal production, nontraditional agriculture (Hunter 1996; Tully and Shane 1996; Gavazzi et al. 1997), aquaculture (Pedersen et al. 1997), and free-ranging/farmed game species and hunting (Kapel 1997; Wilson et al. 1997; Wolfe et al. 2000). Therefore, although deforestation may lead to decreased microbial diversity, the frequency of microbial contact will, in many instances, be increased.

13.2.3 Hunting

Initial logging activities in pristine regions entail the building of roads to remote regions, increased transportation access provided by logging trucks, and an increase in logger-associated settlements. In response to new logger demand and access to previously inaccessible markets via logging trucks, the pattern of hunting activities changes from largely subsistence based to more market-based commercial hunting. Historically, hunting activities radiated from isolated villages in a circular pattern, decreasing at the periphery of the hunting range. With roads in place, hunting activities assume a linear pattern, with traps set along and equidistant from the road, providing hunting at more points and over a greater area. Therefore, there is potentially increased contact between humans and the microbial pathogens in the remains of the primary forests (Weise and Burke 2000).

Wild animal hunting and butchering necessitate extensive contact with diverse body tissues of the prey and the microbial organisms they carry. On occasion, disease can be transmitted to distantly related species; however, when the prey is phylogenetically similar to humans, such as nonhuman primates, hunters are particularly susceptible to transmission of infection by novel microbes. Indeed, chimpanzees share nearly identical infectious disease susceptibilities with humans and, as active hunters themselves, are near the top of the food chain (Wrangham et al. 2000). The infected hunter, in turn, serves as a reservoir, or carrier, for secondary transmission of the novel microbes to the immediately surrounding and, ultimately, more remote human populations involved in the transportation, sale, purchase, and consumption of meat. Recombination of previously present and new acquired (through hunting) pathogens in hunters holds the potential for producing completely novel pathogens; of particular concern is risk of retroviruses (Burke 1998). Today's expanding network of logging roads, along with other contemporary human activities (e.g., air travel), enhances the speed and extent of secondary transmission subsequent to a primary infection, increasing the probability that locally emerging microbes will become global.

The practice of bushmeat hunting in Africa has led to local emergence episodes (Wolfe et al. 2000): workers collecting and preparing chimpanzee meat have become infected with Ebola (WHO 1996); the initiation of a local epidemic of monkeypox (an orthopoxvirus similar to smallpox), which continued for four generations of human-to-human contact, has been attributed to the hunting of a red colobus monkey (Jezek et al. 1986). There are other specific human activities that pose risks similar to those of wildlife hunting and butchering. For example, in the Tai forest, a Swiss researcher performing a necropsy on a chimpanzee contracted Ebola (Le Guenno et al. 1995).

13.2.4 Cross-Species Infection

Contact between humans and other animals can provide the opportunity for cross-species transmission and the emergence of novel microbes into the human population. The taxonomic transmission rule (TTR) states that the probability of successful cross-species infection increases with decreased genetic distance between hosts (Wolfe et al. 2000). And indeed, nonhuman primates and humans do have similar susceptibilities to many infective microbes. Cross-species transmission by humans suggests that endangered nonhuman primate species are vulnerable to contact with human pathogens. Interestingly, the parasite *Giardia* was introduced to the Ugandan mountain *Gorilla gorilla beringei* by humans through their ecotourism and conservation activities (Nizeyi et al. 1999; Cranfield et al. chap. 22). Surveillance of nonhuman primates and other reservoirs of infection may help flag risk of emerging pathogens (Wolfe et al. 1998).

Although surprisingly little is known about the underlying patterns of microbial biodiversity, it is reasonable to assume that microbial diversity parallels the extent of substrate diversity: most diverse soil-associated microbes in regions with diverse soil types, most diverse plant-associated microbes in regions with great botanical diversity, high diversity of vertebrate-associated microbes with high diversity of vertebrates. Even as global biodiversity shrinks, with contemporary movement of humans and human communities, individual humans come into contact with a more diverse global sampling of microbes today than at any other time in history. And, simultaneously, a single emergence event will have increased potential for global spread.

Transmission of infectious disease between nonhumans and humans is also affected by the degree of vector competence, genetic diversity of vectors, and frequency in chromosomal inversion of both vector and pathogen, which vary by species, geographic region, and ecology of the habitat. As the environment changes, some vectors convert from being primarily zoophyllic (affinity to animals) to primarily anthrophyllic (affinity to humans) (Graczyk 1997). Parasites eliminated from human and domestic animal populations can survive in sylvatic habitats (e.g., national parks or wildlife refuges), later returning to human and domestic animal reservoirs. Lax public health policies, sanitation standards, and disease control activities facilitate the interchange (McCrindle et al 1996; Graczyk and Fried 1998).

13.3 Change in Human Settlement Patterns

The combination of changed ecology and influx of new human and/or reservoir populations creates an auspicious dynamic for increased proliferation, incidence, and emergence of disease. The introduction of logging, hunting, road construction, farming, and ranching attracts new populations, who then expand the demand for food supplies—and thereby, for further increased hunting, farming, and ranching—and commercial development. Existing pathogens and their hosts either adapt to the changed conditions or, unable to adapt, leave. Their vacated niches are then filled with other, new-to-the-region species for which the environment is favorable for breeding, feeding, and development.

The newly arriving populations are highly vulnerable to both the veteran and newly arriving pathogens. They have not yet developed immunity to either. Further, the new settlers can serve as reservoirs of infectious agents they have carried from both their former home areas and those they have picked up on their journeys to their new homes. As the settlements become more densely populated, the interactions among vectors and human reservoirs of pathogens increase substantially the exchange and transmission of infection. The new settlers' risk for morbidity may be further increased by a lack of knowledge about effective strategies for protecting themselves from becoming feed sources for the extant vectors (e.g., use of bed nets and remaining indoors during peak biting hours as protections against mosquito hosts of malaria-causing parasites). The burgeoning development of settlements in areas previously occupied by primary forests introduces new materials and their accompanying pathogens to the environment: pathogens of environmental reservoirs of soil, wood, and decaying vegetation are primarily fungal; the few of artificial media such as metal and plastics, bacterial; and those of water, arthropods, and vertebrates include parasites, viruses, bacteria, and fungi (WHO 1996). As the risk for microbial emergence is positively associated with the extent of regional microbial diversity, host susceptibility, and frequency and intimacy of microbial contact (Wolfe et al. 2000), there is considerable potential for microbial emergence under these conditions of highly dense settlements with newly arriving diverse populations in close proximity to veteran and newly arriving infective agents. Thus, the ecological changes imposed by deforestation with successive habitat fragmentation, extensive ongoing immigration, and increased population density create conditions favorable to the emergence, proliferation, and transmission of pathogens, with consequent increase in human morbidity and mortality.

13.4 Forest Replacement by Human-Introduced Plants and Animals

13.4.1 Plants

The changed ecology created when new crops that are grown in previously forested areas can serve directly or indirectly to either support or, conversely, to inhibit existing and/or new incoming infective agents. For example, in Malaysia, 50 years of cyclical planting of rubber plants were paralleled by cycles of mosquito (*Aedes maculatus*)-transmitted malaria. In Trinidad, when erythrina trees were planted to shade cocoa plants, water-collecting bromeliads provided breeding sites for mosquitoes (*Ae. bellator*) and a malaria epidemic followed (removal of the bromeliads was followed by a reduction in malaria prevalence; Downs and Pittendrigh 1946). In groves of large shade trees among coffee plants in Africa, the phlebotomine sandfly hosts the *Leishmania* protozoa and leishmaniasis proliferates (Warburg et al. 1990), and the tall oil palms and mango crops provide hospitable habitats for colonization of the tsetse fly (genus *Glossina*), which transmits African trypanosomiasis, or sleeping sickness (Jobin 1999).

13.4.2 Animals

Cattle ranching provides a number of opportunities for transmission of disease to humans. Livestock can serve as a blood meal resource to vectors, which may reduce their feeding on humans. On the other hand, with plentiful livestock as blood meals, the vectors can become so copious that they seek out additional, human, blood meals (Giglioli 1963). Both the cattle and humans then become reservoirs for infection and transmission of pathogens.

The eating of beef can lead to emergent diseases. When cattle feed includes infected animal tissue, prion disease in the cattle can result (Scott et al. 1999). There is then a risk of secondary transmission of BSE (bovine spongiform encephalopathy) to the human upon consumption of the beef. During the processing and distribution of beef, contamination with pathogens, such as *Escherichia coli* 0157 can occur (Elder et al. 2000).

Domestic livestock can also pose a threat to the safety of drinking water. Cattle may harbor *Cryptosporidium* (the water-borne parasite causing cryptosporidiosis), which it sheds as infectious oocysts in feces (Fayer 1997). Then, agricultural runoff, flooding from heavy precipitation, or leaking septic tanks contaminate the drinking water with oocysts, which are resistant to chlorine treatment.

As with cattle, smaller, domestic and peridomestic animals, such as dogs, cats, armadillos, opossums, and rodents, can accompany the conversion of land use upon deforestation. They can harbor vectors and their resident pathogens, which can then be transmitted to humans causing such diseases as Chagas disease, cat-scratch disease, and Lyme disease. Some vertebrate hosts can also reduce the risk of human disease transmission, as is the case for lizards, which can dilute the risk of Lyme disease (Apperson et al. 1993).

In parts of Latin America, where some areas of the forest had been replaced with farmland, foxes moved in, serving as a reservoir host for the sandfly, the vector for visceral leishmaniasis kala-azar. In the Amazon region, *Leishmania*-infected dogs, accompanying their immigrant owners, triggered an increase in the prevalence of leishmaniasis (Lainson 1989). And, in Brazil's Para State, following deforestation and planting of pines and gmelina for a paper industry, the spiny rat, adapting to the changed habitat, became the infected reservoir host of a disfiguring type of cutaneous leishmaniasis parasite, maintaining the risk of human disease in a secondary forest setting (Ready et al. 1983).

13.4.3 Formation of Water Bodies

Human disturbance of the landscape can further enhance habitats that favor disease transmission through the formation of unnatural water bodies. The shape of a body of water and its proximity to the forest interface and to the road affect the breeding, sustainability, and ultimately, the abundance and competence of vectors. Also, the characteristics of water bodies are affected by the type of soil and their elevation (Patz et al. 2000b).

Cleared lands afford more sunlit puddles, favoring development of larvae of particular species (e.g., *Anopheles darlingi*). Puddles with pH <5.5 form in sandy soils when the soil is saturated; puddles with pH >5.5 form in areas with red clay (Encarnación 1993). Low salinity and conversion from acidic to alkaline water favor growth of freshwater parasite vectors such as snails (Southgate 1997).

13.4.4 Water Control Projects

There is a wide range of conditions under which specific vectors and their pathogens flourish. Some species prefer sunlit pools with turbid water, with little or no emergent vegetation, and others, the edges of clean, clear, gently moving streams; some, irrigation and hydroelectric reservoirs with frequent changes in water level, vertical shorelines, and emergent vegetation without organic material or salinity, and others, coastal areas with high salinity; still others, swamps with extensive vegetation cover and relatively permanent water bodies with organic material (Patz et al. 2000b; Jobin 1999).

Construction of water control systems (i.e., reservoirs, dams, and irrigation canals) can lead to a shift in vector populations and their infective agents. Breeding sites are created by contained waters: for mosquitoes, by excavation pits created during construction of dams and canals, and for fluke-transmitting aquatic snails, by tropical reservoirs and basin irrigation resulting from poor drainage, impounded water, and seepage. Some mosquitoes find stagnant surface water exposed to sunlight hospitable for breeding, such as irrigation ditches and riverine pools formed by flow diverted from river beds (WHO 1982). In Africa, rapids and white water on spillways and water control structures provide breeding sites for the blackfly *Simulium* that causes river blindness (Jobin 1999).

The mosquito with the *Wuchereria bancrofti* parasite, which causes filariasis, has adapted to several different regions and surroundings: in the Orient, to rice cultivation and flooding (Jobin 1999); in the Nile delta, to increases in soil moisture (Thompson et al. 1996); and in India, to standing water polluted by feces of domestic animals (Rajagopalan et al. 1990). In the Nile Delta below Cairo and in the Sudan, the intensity of irrigation is paralleled by the prevalence of bilharzia, as the snail vector is pumped along with the water. Similarly, in Iran, an improved irrigation system led to an upsurge of bilharzia, especially among those working and living within the immediate area (Jobin 1999).

13.5 Malaria as an Example

Malaria is ubiquitous in Africa, much of South America, and other tropical regions. Many species of malaria vectors and their parasites react sharply to changes in the ecology of their habitat: climate, deforestation, vegetation, human population density, bodies of water, and their locations.

Warm, wet, and humid environments provide favorable conditions for breeding, development, and thriving of the malaria vector, the *Anopheles* mosquito. Even slight rises in the minimum temperatures at which the *Anopheles* is sustained have been associated with substantial increases in malaria incidence (Lindsay and Birley 1996). As the temperature of heretofore malaria-free regions rises with global warming, malaria outbreaks can occur. In such regions, morbidity and mortality are exacerbated by the lack of immunity in the previously unexposed population. For example, although previously epidemic malaria (1934–36) has been eradicated in the Caucasus and Volga river basin (Bruce-Chwatt 1988), the current lack of exposure-induced immunity and concomitant declining public health system render the population vulnerable to epidemic malaria if and when the predicted warming occurs.

The combination of a rise in temperature and heavy rainfall accumulation has occasioned malaria outbreaks, even in regions at altitudes as high as 2000 meters—as in Western Kenya, when mean monthly temperature rose above 18°C and rainfall was over 150 mm per month (Malakooti et al 1998). Malaria epidemics in the Indian subcontinent and Sri Lanka have been attributed to the El Niño southern oscillation (Bouma and van der Kaay 1996).

Different malaria-transmitting mosquito species flourish in different environments. In Africa, *An. gambiae* requires sunlit pools with turbid water and little or no emergent vegetation. *Anopheles funestus* larvae concentrate in clear water, thriving in irrigation and hydroelectric reservoirs with their frequent changes in water level, and around shorelines with vertical, emergent vegetation without organic material or salinity. *Anopheles pharaoensis* require extensive vegetation cover, inhabiting swamps and relatively permanent water bodies with organic material. *Anopheles melas* occupies the coastal areas with high salinity. In rural India, *A. culcifacies* occurs in irrigation systems, whereas in Sri Lanka, the same species occupies the riverine pools created by diversion of flow out of river beds, breeding primarily in shallow, stagnant surface water exposed to sunlight. In the hilly regions of Burma, Thailand, and Indo-China, *An. balabacensis* deposits its eggs in deeply shaded pools and seepages in the rain forests, in footprints, mining pits, and irrigation ditches, and in excavated depressions in the open sunlight, and *An. minimus* occurs at the edges of clean, clear, gently moving streams (WHO 1982).

13.6 Looking Forward

It can be expected that human activity will continue to lead to ecological changes that can either increase or decrease human morbidity and mortality. Surveillance and analysis of the processes by which such changes proceed should afford better identification of regions at increased risk. Policies and programs for assisting communities in such regions need to be developed and implemented so as to minimize the emergence and spread of disease. The examples provided in this chapter show the broad implications of ecosystem change and human health; it also points to large gaps in our current knowledge. Expanded research efforts focused on understanding the specific causal relationships between environmental degradation and human disease is therefore of high public health priority.

References

Apperson, C.S., J.F. Levine, T.L. Evans, A. Braswell, and J. Heller. 1993. Relative utilization of reptiles and rodents as hosts by immature *Ixodes scapularis* (Acari: Ixodidae) in the coastal plain of North Carolina, USA. Exp Appl Acarol 17:719–731.
Beckett, W.S. 1991. Ozone, air pollution, and respiratory health. Yale J Biol Med 64:167–175.
Beneson, A.S. 1990. Control of communicable diseases in man. American Public Health Association, Washington, D.C.
Bobak, M., and A. Roberts. 1997. Heterogeneity of air pollution effects is related to average temperature. Br Med J 315:1161–1162.
Bouma, M.J., and H.J. van der Kaay. 1996. The El Niño southern oscillation and the historic malaria epidemics on the Indian subcontinent and Sri Lanka: an early warning system for future epidemics? Trop Med Int Health 1:86–96.
Bruce-Chwatt, L.J. 1998. History of malaria from prehistory to eradication. *In* Wernsdorfer, W.H., and I. McGregor, eds. Malaria: principles and practise of malariology, pp. 1–59. Churchhill Livingstone, Edinburgh.

Burke, D.S. 1998. The evolvability of emerging viruses. *In* Nelson, A.M., and C.R. Horsburgh Jr., eds. Pathology of emerging infections, vol. 2, pp. 1–12. American Society for Microbiology, Washington, D.C.

Cambre, R.C., and W.W. Buick. 1996. Special challenges of maintaining wild animals in captivity in North America. Rev Sci Tech 15:251–266.

Chappellaz, J., J.M. Barnola, D. Raynaud, Y.S. Korotkevich, and C. Lorius. 1990. Ice-core record of atmospheric methane over the past 160,000 years. Nature 345:127–131.

Downs, W.G., and C.S. Pittendrigh. 1946. Bromeliade malaria in Trinidad, British West Indies. Am J Trop Med Hyg 26:47–66.

Elder, R.O., J.E. Keen, G.R. Siragusa, G.A. Barkocy-Gallagher, M. Koohmaraie, and W.W. Laegreid. 2000. Correlation of enterohemorrhagic *Escherichia coli* O157 prevalence in feces, hides, and carcasses of beef cattle during processing. Proc Natl Acad Sci USA 97:2999–3003.

Encarnación, F. 1993. El bosque y las formaciones vegetales en la llanura amazonica del Peru. Alma Mater 6:95–114.

Engelthaler, D.M., D.G. Mosley, J.E. Cheek, et al. 1999. Climatic and environmental patterns associated with hantavirus pulmonary syndrome, Four Corners region, United States. Emerg Infect Dis 5:87–94.

Fayer, R., C.A. Speer, and J.P. Dubey. 1997. The general biology of *Cryptosporidium*. *In* Fayer, R., ed. *Cryptosporidium* and cryptosporidiosis, pp. 1–42. CRC Press, Boca Raton, Fla.

Fowler, M.E. 1996. Husbandry and disease of camelids. Rev Sci Tech 15:155–169.

Gavazzi, G., D. Prigent, J.M. Baudet, S. Banoita, and W. Daoud. 1997. Epidemiologic aspects of 42 cases of human brucellosis in the Republic of Djibouti. Med Trop 57: 365–368.

Giglioli, G. 1963. Ecological change as a factor in renewed malaria transmission in an eradicated area: a localized outbreak of *A. aquasalis*–transmitted on the Demerara River estuary, British Guinea, in the fifteenth year of *A. darlingi* and malaria eradication. Bull WHO 29:131–145.

Glass, G.E., J. Cheek, J.A. Patz, et al. 2000. Using remotely sensed data to identify areas at risk for hantavirus pulmonary syndrome. Emerg Infect Dis 6:238–246.

Graczyk, T.K. 1997. Immunobiology of trematodes in vertebrate hosts. *In* Fried, B., and T.K. Graczyk, eds. Advances in trematode biology, pp. 383–404. CRC Press, Boca Raton, Fla.

Graczyk, T.K., and B. Fried. 1998. Development of *Fasciola hepatica* in the intermediate host. *In* Dalton, J.P., ed. Fasciolosis, pp. 31–46. CAB International, New York.

Graczyk, T.K., and L.J. Lowenstine, and M.R. Cranfield. 1999. *Capillaria hepatica* (Nematoda) infections in human-habituated mountain gorillas (*Gorilla gorilla beringei*) of the Parc National de Volcans, Rwanda. J Parasitol 85:1168–1170.

Gray, W. 1979. Hurricanes: their formation, structure and likely role in the tropical circulation. *In* Shaw, D.B., ed. Meteorology over the tropical oceans, pp. 155–218. Royal Meteorology Society, London.

Grey, M., R. Edmond, and G. Whitten. 1987. Tropospheric ultraviolet radiation: assessment of existing data and effect on ozone formation. U.S. Environmental Protection Agency, Research Triangle Park, N.C.

Hales, S., S. Kovats, and A. Woodward. 2000. What El Niño can tell us about human health and global climate change. Glob Change Hum Health 1:66–77.

Hannah, H.W. 1998. Zoos and veterinarians: some legal issues. JAVMA 213:1559–1560.

Hunter, D.L. 1996. Tuberculosis in free-ranging, semi-ranging and captive cervids. Rev Sci Tech 15:171–181.

.Jezek, Z., I. Arita, M. Mutombo, C. Dunn, J.H. Nakano, and M. Szczeniowski. 1986. Four generations of probable person-to-person transmission of human monkeypox. Am J Epidemiol 123:1004–1012.

Jobin, W. 1999. Ecological design and health impacts of large dams, canals, and irrigation systems. E & FN Spon, London.

Kapel, C.M. 1997. *Trichinella* in Arctic, subarctic and temperate regions: Greenland, the Scandinavian countries and the Baltic States. SE Asian J Trop Med Public Health 28 (Suppl 1):14–19.

Karl, T.R., and R.W. Knight. 1997. The 1995 Chicago heat wave: how likely is a recurrence? Bull Am Meteorol Soc 78:1107–1119.

Katsouyanni, K., G. Touloumi, C. Spix, J. Schwartz, F. Balducci, S. Medina, G. Rossi, B. Wojtyniak, J. Sunyer, L. Bacharova, et al. 1997. Short-term effects of ambient sulphur dioxide and particulate matter on mortality in 12 European cities: results from time series data from the APHEA project. Br Med J 314:1658–1663.

Kattenberg, A., F. Giorgi, H. Grassl, G.A. Mehl, J.F.B. Mitchell, R.J. Stouffer, et al. 1996. Climate models projections of future climate. *In* Houghton, J.T., L.G. Meira-Filho, B.A. Callander, N. Harris, A. Kattenberg, and K. Maskell, ed. Climate change 1995. The science of climate change, pp. 285–357. Cambridge University Press, Cambridge.

Kilbourne, E.M. 1998. Illness due to thermal extremes. *In* Wallace, R.B., ed. Maxcy-Rosenau-Last Public health and preventive medicine, 14th ed., pp. 607–617. Appleton and Lange, Stamford, Conn.

Kilian, A.H.D., P. Langi, A. Talisuna, and G. Kabagambe. 1999. Rainfall pattern, El Niño and malaria in Uganda. Trans R Soc Trop Med Hyg 93:22–23.

Kirk-Davidoff, D.B., E.J. Hintsa, N.G. Anderson, and D.W. Keith. 1999. The effect of climate change on ozone depletion through changes in stratospheric water vapour. Nature 402:399–401.

Koren, H.S., and P.A. Bromberg. 1995. Respiratory responses of asthmatics to ozone. Int Arch Allergy Immunol 107:236–238.

Lainson, R. 1989. Demographic changes and their influence on the epidemiology of the American leishmaniasis. *In* Service, M.W., ed. Demography and vector-borne diseases, CRC Press, Boca Raton, Fla.

Larsen, U. 1990a. The effects of monthly temperature fluctuations on mortality in the United States from 1921 to 1985. Int J Biometeorol 34:136–145.

Larsen, U. 1990b. Short-term fluctuations in death by cause, temperature, and income in the United States 1930 to 1985. Soc Biol 37:172–187.

Le Guenno, B., P. Formenty, M. Wyers, P. Gounon, F. Walker, and C. Boesch. 1995. Isolation and partial characterization of a new strain of Ebola virus. Lancet 345:1271–1274.

Lindsay, S.W., and M.H. Birley. 1996. Climate change and malaria transmission. Ann Trop Med Parasitol 90:573–588.

Linthicum, K.J., A. Anyamba, C.J. Tucker, et al. 1999. Climate and satellite indicators to forecast Rift Valley fever epidemics in Kenya. Science 285:397–400.

Malakooti, M.A., K. Biomndo, and D.G. Shanks. 1998. Re-emergence of epidemic malaria in the highlands of Western Kenya. Emerg Infect Dis 4:671–676.

McCrindle, C.M., I.T. Hay, J.S. Odendaal, and E.M. Calitz. 1996. An investigation of the relative morbidity of zoonoses in paediatric patients admitted to GA-Rankuwa Hospital. J S Afr Vet Assoc 67:151–154.

Michalak, K., C. Austin, S. Diesel, M.J. Bacon, P. Zimmerman, and J.N. Maslow. 1998. *Mycobacterium tuberculosis* infection as a zoonotic disease: transmission between humans and elephants. Emerg Infect Dis 4:283–287.

Moore, P.S. 1992. Meningococcal meningitis in sub-Saharan Africa: a model for the epidemic process. Clin Infect Dis 14:515–525.

Morris, R.E., M.W. Gery, M.K. Liu, G.E. Moore, C. Daly, and S.M. Greenfield. 1989. Sensitivity of a regional oxidant model to variations in climate parameters. U.S. Environmental Protection Agency, Washington, D.C.

National Climatic Data Center (NCDC). 1999. Mitch: the deadliest Atlantic hurricane since 1780. Web site: http//www.ncdc.noaa.gov/oa/reports/mitch/mitch.htmt.html, accessed p. 314 Oct. 6, 2001.

Nizeyi, J.B., R. Mwebe, A. Nanteza, M.R. Cranfield, G.R.N.N. Kalema, and T.K. Graczyk. 1999. *Cryptosporidium* sp. and *Giardia* sp. infections in mountain gorillas (*Gorilla gorilla beringei*) of the Bwindi Impenetrable National Park, Uganda. J Parasitol 85: 1084–1088.

Noji, E.K., 1997. The public health consequences of disasters. Oxford University Press, New York.

Nwoke, B.E., C.O. Onwuliri, and G.O. Ufomadu. 1992. Onchocerciasis in Plateau State, Nigeria: ecological background, local disease perception and treatment; and vector/parasite dynamics. J Hyg Epidemol Microbiol Immunol 36:153–160.

Pascal, M., X. Rodo, S.P. Ellner, R. Colwell, and M.J. Bouma. 2000. Cholera dynamics and El Niño-southern oscillation. Science 289:1766–1769.

Patz, J.A., D. Engelberg, and J. Last. 2000a. The effects of changing weather on public health. Annu Rev Public Health 21:271–307.

Patz, J.A., T.K. Graczyk, N. Geller, and A.Y. Vittor. 2000b. Effects of environmental change on emerging parasitic diseases. Internat J Parasitol 30:1395–1405.

Pedersen, K., I. Dalsgaard, and J.L. Larsen. 1997. *Vibrio damsela* associated with diseased fish in Denmark. Appl Environ Microbiol 63:3711–3715.

Rajagopalan, P.K., P.K. Das, K.N. Paniker, R. Reuben, D.R. Rao, L.S. Self, and J.D. Lines. 1990. Environmental and water management for mosquito control. In Curtis, C.F., ed. Appropriate technology and vector control. CRC Press, Boca Baton, Fla.

Ramlow, J.M., L.H. Kuller. 1990. Effects of the summer heat wave of 1988 on daily mortality in Allegheny County, PA. Public Health Rep 105:283–289.

Ready, P.D., R. Lainson, and J.J. Shaw. 1983. Leishmaniasis in Brazil: XX. Prevalence of "enzootic rodent leishmaniasis" (*Leishmania mexicana amazonensis*), and apparent absence of "pian bois" (*Le. braziliensis guyanensis*), in plantations of introduced tree species and in other non-climax forests in eastern Amazonia. Trans R Soc Trop Med Hyg 77:775–785.

Schimel, D., D. Alves, I. Enting, M. Heimann, F. Joos, D. Raynaud et al. 1996. Radiative forcing of climate change. In Houghton, J.T., L.G. Meira-Filho, B.A. Callander, N. Harris, A. Kattenberg, and K. Maskell, eds. Climate change 1995: The science of climate change, pp. 65–131. Cambridge University Press, Cambridge.

Schultz, D.J., I.J. Hough, and W. Boardman. 1996. Special challenge of maintaining wild animals in captivity in Australia and New Zealand. Rev Sci Tech 15:289–308.

Schwartz, J. 1994. Air pollution and daily mortality: a review and meta-analysis. Environ Res 64:36–52.

Scott, M.R., R. Will, J. Ironside, H.O.B. Nguyen, P. Tremblay, S.J. DeArmond, and S.B. Prusiner. 1999. Compelling transgenetic evidence for transmission of bovine spongiform encephalopathy prions to humans. Proc Natl Acad Sci USA 96:15137–15142.

Shindell, D.T., D. Rind and P. Lonergan. 1998. Increased polar stratospheric ozone losses and delayed eventual recovery owing to increasing greenhouse-gas concentrations. Nature 392:589–592.

Southgate, V.R. 1997. Shistosomiasis in the Senegal River basin: before and after the construction of dams at Diama, Senegal and Manantali, Mali and future prospects. J Helminthol 71:125–132.

Thompson, D.F., J.B. Malone, M. Harb, et al. 1996. Bancroftian filariasis distribution in the southern Nile delta: correlation with diurnal temperature differences from satellite imagery. Emerg Infect Dis 3:234–235

Tully, T.N., and S.M. Shane. 1996. Husbandry practices as related to infectious and parasitic diseases of farmed ratites. Rev Sci Tech 15:73–89.

U.S. Environmental Protection Agency (EPA). 1989. In Smith, J.B, and D.A. and Tirpak, eds. The potential effects of global climate change on the United States. Report No. EPA 230-05-89-057. Office of Policy, Planning and Evaluation, Washington, D.C.

Warburg, A., J. Montoya-Lerma, C. Jaramillo, A.L. Cruz-Ruiz, and K. Ostrovska. 1990. Leishmaniasis vector potential *Lutzomyia* spp., in Colombian coffee plantation. Med Vet Entomol 5:9–16.

Weise, S.F., and D. Burke. 2000. Deforestation, hunting and the ecology of microbial emergence. Global Change Hum Health 1:10–25.

Williams, E.S., and E.T. Thorne. 1996. Infectious and parasitic diseases of captive carnivores, with special emphasis on the black-footed ferret (*Mustella nigripes*). Rev Sci Tech 15:91–114.

Wilson, M.L., P.M. Bretsky, G.H. Cooper, S.H. Egberson, H.J. Van Kruiningen, and M.L. Cartter. 1997. Emergence of raccoon rabies in Connecticut, 1991–1994: spatial and temporal characteristics of animal infection and human contact. Am J Trop Med Hyg 57:457–463.

Wolfe, N.D., A.A. Escalate, W.B. Karesh, A.M. Kilbourne, A. Spielman, and A.A. Lal. 1998. Wild primate populations in emerging infectious disease: the missing link? Emerg Infect Dis 4:149–158.

Wolfe, N.D., M.N. Eitel, J. Gockowski, P.K. Muchaal, C. Nolte, A.T. Prosser, J.N. Torimiro, S.F. Weisy, and D.S. Burke. 2000. Deforestation, hunting and the ecology of microbial emergence. Global Change Hum Health 1:10–25.

World Health Organization (WHO). 1982. Manual on environmental management for mosquito control. Offset Publication No. 66. WHO, Geneva.

World Health Organization (WHO). 1996. Outbreak of Ebola haemorrhagic fever in Gabon officially declared over. Wkly Epidemiol Rec 71:125–126.

World Meteorological Organization (WMO) United Nations Environment Programme. 1995. Scientific assessment of ozone depletion: 1994. Global Ozone Research and Monitoring Project Report No. 25. WMO, Geneva.

Wrangham, R., M. Wilson, B. Hare, and N.D. Wolfe. 2000. Chimpanzee predation and the ecology of microbial exchange. Microb Ecol Health Dis 12:186–188.

Xu, X.J., X.X. Yang, Y.H. Dai, G.Y. Yu, L.Y. Chen, and Z.M. Su. 1999: Impact of environmental change and schistosomiasis transmission in the middle reaches of the Yangtze River following the Three Gorges construction project. SE Asian J Trop Med Public Health 30:549–555.

14

Biodiversity and Human Health

Eric Chivian
Sara Sullivan

The extinction of species may be the most destructive and permanent of all assaults on the global environment, for it is the only one that is truly irreversible (Wilson 1989). While species stocks may recover following mass extinction events over the course of millions of years of evolution (Myers et al. 2000), and may fill ecological niches that had once been occupied, it must be understood that these replacing species are new ones and that the unique information encoded by the DNA of the lost species is gone forever. This fact adds to the gravity of the current extinction crisis, the first that is the result of human activity.

Many have looked at the question of what will convince people to alter their behavior so as to preserve other species. Clearly, spiritual, ethical, and aesthetic values, as well as economic factors, are powerful motivating forces. But it may be argued that these considerations are not persuasive enough to result in any significant changes in personal behaviors or public policy, that they are too abstract and hard for most people to relate to, and that it will take an understanding of the direct, concrete, and personal risks to peoples' health and lives, and particularly to the health and lives of their children, for them to be motivated enough to protect the global environment. Such is the goal of this chapter—to present in summary form the potential human health consequences of species loss and the resultant disruption of ecosystems, to demonstrate that human health is inextricably linked to the health of the global environment, to make clear the central message of this new field of conservation medicine—that "man is embedded in Nature" (Thomas, 1974).

When *Homo sapiens* evolved some 120,000 years ago, the number of species on the earth was the largest ever (Wilson 1989), but human activity has resulted in species extinction rates that are currently 100 to 1000 times those of prehuman levels (Pimm et al. 1995). While the record demonstrates that humans hunted to extinction scores of large mammals and birds as early as tens of thousands of years ago (Miller et al. 1999; Flannery 1999), it is only in recent times that these extinctions have spread to virtually every part of the planet and to almost every phylum. Species numbers are now being reduced so rapidly that one quarter or more of all species now alive may become extinct during the next 50 years if these rates persist (Ehrlich and Wilson 1991; IUCN 1997). Such losses have prompted biologists to refer to the present period as "the sixth extinction" (Pimm and Brooks 2000) (the last great extinction event, the fifth, was 65 million years ago, at the end of the Cretaceous Period when dinosaurs became extinct).

Global climate change (Peters and Lovejoy 1992), stratospheric ozone depletion (Blaustein et al. 1994), chemical pollution (Colburn et al. 1993), acid rain (Schindler 1998), the introduction of alien species (Lovei 1997), and the over-hunting of species (Pauly et al. 1998) all threaten biodiversity (the total complement of different living organisms in an environment), but it is the degradation, reduction, and fragmentation of habitats that is the greatest threat, particularly in species-rich areas such as tropical rain forests and coral reefs (Pimm and Raven 2000).

14.1 Potential Medicines

With a loss of species, we are losing plants, animals, and microorganisms, perhaps most of which are still undiscovered, that may contain valuable new medicines (only about 1.5 million species have been identified [May 1988], but there may be 10 or even 100 times that number [Pimm et al. 1995]). Over the course of millions of years of evolution, species have developed chemicals that have allowed them to fight infections, tumors, and other diseases and to capture prey and avoid being eaten, chemicals that have become some of today's most important pharmaceuticals. Tropical rainforest organisms, for example, have given us D-tubocurarine (from the chondrodendron vine *Chondrodendron tomentosum*), synthetic analogues of which are indispensable for achieving deep muscle relaxation during general surgery; quinine and quinidine (from the cinchona tree *Cinchona officianalis*), used in treating malaria and cardiac arrhythmias, respectively; vinblastine and vincristine (from the rosy periwinkle plant *Vinca rosea*), which have revolutionized the treatment of leukemias and Hodgkin's lymphomas; and erythromycin, neomycin, and amphotericin (from soil microbes of the genus *Streptomyces*), essential broad-spectrum antibiotics. Temperate species have yielded some of our most useful drugs as well; aspirin was originally extracted from the willow tree (*Salix alba*) and digitalis (from which digoxin and digitoxin were derived, used in treating congestive heart failure and atrial arrhythmias) originally came from the foxglove plant (*Digitalis purpurea*). And medications have been developed from marine species, such as cytarabine from a Caribbean sponge (*Tethya crypta*), which is the single most effective agent for inducing remission in acute myelocytic leukemia (Chabner et al. 1996). A recent survey of the 150 most frequently prescribed drugs in the United States demonstrated that 57% of them contained compounds, or were patterned after compounds, derived from other species (Grifo et al. 1997).

14.1.1 Taxol

The story of taxol and the Pacific yew tree (*Taxus brevifolia*) illustrates how we may be losing new medicines before species have been analyzed for their chemical content. The commercially useless Pacific yew was routinely discarded as a trash tree during logging of old-growth forests in the Pacific Northwest of the United States until it was found to contain the compound taxol, a substance that kills cancer cells by a mechanism unlike that of other known chemotherapeutic agents—preventing cell division by inhibiting the disassembly of the mitotic spindle (Cowden and Paterson 1997). The discovery of the complex molecule taxol and its novel mechanism of action has led to the synthesis of several taxol-like compounds that are even more effective than the natural compound (McGuire et al. 1989), illustrating how a clue from nature has led to the discovery of a whole new class of medicines that would have been extremely difficult to discover in the lab. Early clinical trials have reported that taxol was able to induce remission in advanced ovarian cancer cases unresponsive to other treatments (Nicolaou et al. 1996); subsequent experience has shown that taxol may be one of the most promising new medicines available for breast and ovarian cancer (McGuire et al. 1989). Other species with biomedical potential like the Pacific yew are most likely being lost to deforestation and other forms of habitat destruction.

14.1.2 Cone Snails

The diversity of peptide compounds in the venoms of cone snails, a genus of predatory snails (numbering about 500 species) inhabiting tropical coral reef communities, is so great that it may rival that of alkaloids in higher plants and secondary metabolites in microorganisms (Olivera et al. 1990). Some of these compounds, which have been shown to block a wide variety of ion channels, receptors, and pumps in neuromuscular systems, have such selectivity that they have become important tools in neurophysiological research and may become invaluable to clinical medicine. One class of peptides, the ω-conotoxins, bind with such specificity to neuronal calcium channels that they have been instrumental in identifying various subtypes of calcium channels (Olivera et al. 1994) and are now commonly used by molecular biologists for labeling purposes (Zapp 1996).

Because voltage-sensitive calcium channels are integral to the pain-signaling mechanism of the nervous system (Atanassoff et al. 2000), conotoxins hold particular promise for the management of pain. In fact, one *Conus* peptide calcium channel antagonist (designated ω-conotoxin MVIIA, ziconotide, or SNX-111 in its synthetic form) has demonstrated unprecedented success in treating the chronic intractable pain of patients with AIDS, cancer, and peripheral neuropathies (Miljanich 1997) and has recently received approvable status by the FDA as an analgesic (G. Miljanich, personal commun. 1999). Ziconotide has 1,000 times the potency of morphine, but unlike morphine, it does not lead to the development of tolerance or addiction or to a clouding of consciousness (Bowersox et al. 1998). Other therapeutic applications being explored for this novel drug include its use as a neuroprotective agent following head injury, stroke, or coronary bypass surgery (Miljanich 1997) (as nerve cell death from insufficient blood flow seems to occur after a cascade of events initiated by the influx of calcium ions), as well as a treatment for spasticity as a result of spinal cord injury (Ridgeway et al. 2000). As coral reefs are increasingly threatened in many parts of the world (Hayes and Goreau 1998), the existence of reef-dwelling organisms such as cone snails is similarly threatened.

14.2 Research Models

Species loss also leads to the loss of invaluable research models that help us understand human physiology and disease. Biomedical research has long relied on mice, rats, and guinea pigs as experimental subjects, and on a host of other organisms possessing unique structures or physiologies, for example, fruit flies (Rubin and Lewis 2000) and *Escherichia coli* in genetics research, and horseshoe crabs and squid for the study of nerve cells. Three groups of animals that are endangered—bears, sharks, and chimpanzees—as well as a microscopic roundworm provide examples of the kinds of biological secrets that could be lost to medical science with the loss of species.

14.2.1 Bears

Some bear species are threatened because of destruction of their habitats, and because of overhunting, which is in part secondary to the high prices their organs, reputed to have medicinal value, bring in Asian black markets (Mainka and Mills 1995). Bear gallbladders, for example, are worth 18 times their weight in gold. Yet living bears are far more valuable than the sum of all their body parts. For example, in winter months black bears (*Ursus americanus*) enter a period of three to seven months of hibernation called "denning," in which they neither eat, drink, urinate, nor defecate, yet they can deliver young and nurse them, do not lose bone or lean body mass, and do not become ketotic or uremic (Hasan et al. 1997). During this period, bears derive energy and water through selective fat catabolism (Herminghuysen et al. 1995). Insight into the mechanisms that regulate fat metabolism in bears could offer a greater understanding of obesity (Herminghuysen et al. 1995), a growing public health problem that affects 55% of American adults and is associated with an increased risk of such illnesses as diabetes, heart disease, hypertension, and stroke. Despite a lack of weight bearing for months, black bears do not suffer from osteoporosis, a condition that occurs during periods of immobility, or a lack of weight bearing, in all other mammalian species that have been studied, including humans (Floyd et al. 1990). Understanding how bears prevent bone resorption could lead to new ways of preventing and treating osteoporosis, an overwhelming public health problem resulting in billions of dollars in direct health care costs and lost productivity and in tens of thousands of deaths each year in the United States alone (National Osteoporosis Foundation 2000). Similarly, learning how denning black bears avoid becoming uremic—because of their ability to convert urea into protein—could help in the development of effective treatments for renal failure (Nelson 1987), which can only be treated presently by renal dialysis and kidney transplantation. Although black bears are not threatened or endangered (in fact, their numbers are dramatically increasing in regions like New England), it seems likely that other species of bears that are imperiled, such as the polar bear (*Ursus maritimus*) and the Asiatic black bear (*Ursus thibetanus* (IUCN 2000) may similarly represent physiological models of human diseases.

14.2.2 Sharks

Why sharks, the first vertebrates to have developed antibodies and a full complement of immune proteins, seem to be less prone than many other marine species to infections or tumors has prompted intensive investigation into the nature of their immune systems (Litman 1996; Mestel 1996). The answer may be that sharks produce powerful infection- and cancer-fighting molecules in their tissues that boost their immune response, presumably the result of 450 million years of successful evolutionary experiments. One such substance, the aminosteroid squalamine, has demonstrated anti-angiogenic properties and has advanced to phase II clinical trials for the treatment of lung cancer (Rao et al. 2000). The potential of squalamine for preventing neovascularization has prompted research into other conditions characterized by abnormal blood vessel growth, such as retinopathy, where it has already proven effective in animal models (Higgins et al. 2000). Squalamine has also shown broad-spectrum antibiotic activity against Gram-negative and Gram-positive bacteria (with a potency comparable to ampicillin), as well as to some fungi and protozoa (Moore et al. 1993). As many shark species including the basking shark (*Cetorhinus maximus*), the leopard shark (*Triakis semifasciata*), and the great white shark (*Carcharodon carcharias*) are endangered (IUCN 2000) by overfishing, by the demand for shark fin soup and for shark cartilage, and by outright slaughter, our ability to understand their unique immune systems could be endangered as well (Malakoff 1997).

14.2.3 Chimpanzees

Like some bear and shark species, chimpanzees are threatened by the lucrative trade in their body parts—chimpanzees and other higher primates are being slaughtered and sold as bushmeat for premium prices in restaurants in Africa—in addition to the unceasing destruction of their habitats (Weiss 1999). As our closest primate relatives (with whom we share 98.5% of our DNA), chimpanzees have long been studied in an attempt to understand our early ancestors and to offer insights into the evolution of human behavior and social structure. They also play a critical role in understanding our biology. A subspecies of chimpanzee, *Pan troglodytes troglodytes*, for example, has been identified as the original host and reservoir of HIV-1, the most prevalent type of the AIDS viruses (Hahn et al. 2000). The HIV-1 virus has caused a global crisis, infecting more than 50 million people worldwide with an estimated rate of new infections at nearly six million per year (Hahn et al. 2000). While there have been other zoonotic, or cross-species, viral transmissions between simians and humans before (e.g., simian/human T-cell leukemia virus, simian foamy virus), none has been subsequently transmitted from the new host to other humans (Korber et al. 2000). Discovering what enabled the virus to be a successful trans-species infection, how it evolved, and why simians can carry the virus without becoming ill in contrast to humans are all vital questions that we need to answer if we are to find a cure for HIV/AIDS and avert another possible global epidemic in the future. Answers will need to come not only from research labs, but also from free-living chimpanzees in the wild, where the geographical and evolutionary history provided essential clues to the SIV$_{cpz}$ and HIV-1 discovery (Gao et al. 1999). Scientists have determined 24 other primate species infected with related viruses, and should these viruses be undetectable by current blood tests for HIV-1 and HIV-2 and be capable of infecting humans, we face the possibility of another pandemic (Hahn et al. 2000). Due to the globalization of trade, commerce, and transportation, in addition to an increasing and interconnected population, the world is a much "smaller" place than it used to be, and routes of viral transmission have expanded commensurately. As we continue to encroach upon natural habitats, we doubtless will encounter more zoonotic infections.

14.2.4 Caenorhabditis elegans

The mapping of the complete genome of the microscopic roundworm *Caenorhabditis elegans* presents scientists with both a wealth of information and a powerful analytic tool. Though nematodes and mammals diverged from a common ancestor some 700–800 million years ago (Roush 1997), the genetic codes regulating many fundamental biological processes have been conserved through evolution. As a result of this shared biology, *C. elegans* is an excellent model to investigate human genetics and disease. The roundworm exhibits a fair degree of complexity, such as a nervous system and developmental mechanisms similar to our own, yet it is simple enough—the worm has only 959 cells and 19,000 genes (Pennisi 1998)—to make it an ideal research organism. Scientists have designed experiments to learn how the worm's genes control its metabolism of glucose (Sze et al. 2000), the mechanisms of its programmed cell death (Chen et al. 2000), and its longevity (Kimura et al. 1997; Taub et al. 1999), the results of which may yield insights into human diabetes; diseases in which there are disturbances in programmed cell death, such as cancer, autoimmune disease, and neurodegenerative states; obesity; and aging. Moreover, the functional analyses of *C. elegans* orthologues of human disease genes, particularly those associated with congenital disorders such as muscular dystrophy and Alzheimer disease, could offer invaluable information about disease mechanisms and, consequently, possible therapeutic interventions (Culetto and Sattelle 2000). Although *C. elegans* is not an endangered species, it is included in this discussion to illustrate the vast amounts of information, enormously useful to biomedical science, that may be contained in other species, some of which are becoming extinct without having ever been identified or studied.

14.3 Ecosystem Services

An ecosystem is made up of the sum total of all the species and their actions and interactions with each other and with the nonliving matter within a particular environment. How ecosystems provide the services that sustain all life on this planet remains one of the most complex and poorly understood areas of biological science (Daily 1997; Cohen and Tilman 1996).

Ecosystem services include such vital functions as regulating the concentration of oxygen, carbon dioxide, and water vapor in the atmosphere; filtering pollutants from drinking water; regulating global temperature and precipitation; forming soil and keeping it fertile; pollinating plants; and providing food and fuel (Daily 1997; Costanza et al. 1997). We cannot live without these life support services, whose functioning can be disrupted by species loss (Vitousek et al. 1997).

The regulation of the gaseous composition of the atmosphere, which mediates all energy that enters and leaves the earth (Daily 1997), is arguably the most fundamental ecosystem service. Billions of years ago, chemical reactions and physical processes on the earth created an environment that made life possible. Living organisms, in turn, transformed the atmosphere and, through complex and intricate webs of interaction, continue to play an essential role in maintaining the atmosphere's gaseous constituents of carbon dioxide, oxygen, nitrogen, and water vapor. A simplified look at a plant, for example, illustrates the dynamic link between life and the atmosphere. During photosynthesis, the plant takes up carbon dioxide, fixes the carbon, and releases oxygen back into the atmosphere. Bacteria living on the roots of the plant use this oxygen to transform the gaseous nitrogen into forms that can be assimilated by plants and converted into proteins (Ricklefs and Miller 2000). Water vapor in the atmosphere that has condensed and fallen as precipitation is returned through the plant's leaves in the process of transpiration. Thus, the earth's atmosphere and biota are interdependent and interact to continually cycle essential elements and water through the environment.

One critically important service is the role that biodiversity within ecosystems may play in controlling the emergence and spread of infectious diseases in plants, animals, and humans. This protective function of biodiversity has only recently begun to be appreciated (Anderson and Morales 1994; Daszak et al. 2000; Ostfeld and Keesing 2000; Epstein 1997; Chivian 1997). Examples of human infectious diseases that can be affected by natural or human-induced distrubances to biodiversity include malaria and leishmaniasis, through deforestation (Walsh et al. 1993); Lyme disease, by changes in the number of acorns and in the populations of black-legged ticks, white-footed mice, and white-tailed deer (Ostfeld et al. 1998); Argentine hemorrhagic fever, by the replacement of natural grasslands with corn monoculture (Morse 1995); and cholera by algal blooms, secondary, in part, to warming seas and to fertilizer and sewage discharge (Colwell 1996).

14.3.1 Lyme Disease

The emergence of Lyme disease, which is now the most common vector-borne illness in the United States (CDC 2000), illustrates how the loss of biodiversity can increase the risk of infectious disease. Lyme disease is transmitted to humans via ticks carrying the spirochete bacterium *Borrelia burgdorferi*. Ticks acquire the disease-causing spirochete when they take a blood meal from an infected competent host (an organism that effectively transmits the infectious agent), which in the central and eastern United States is most frequently the white-footed mouse (*Peromyscus leucopus*). White-footed mice are particularly abundant in forested areas, and in species-poor communities they have a disproportionately high representation; thus, when a tick attaches to a host to take a blood meal, it is most likely an infected white-footed mouse (Ostfeld and Keesing 2000). In studying the relationship between species diversity and the incidence of Lyme disease, researchers have found a negative correlation between the species richness of terrestrial small mammals and reported cases of Lyme disease per capita, a consequence of what they are calling the "dilution effect": as more species of incompetent hosts are added to the community from which ticks can feed, the number of blood meals from infected white-footed mice declines (Ostfeld and Keesing 2000). With fewer ticks acquiring the spirochete, the risk of human exposure diminishes as well. While the function of biodiversity in this model does not necessarily apply to vector-borne illnesses involving numerous competent hosts, other infectious diseases with similar elements, such as cutaneous leishmaniasis, Chagas disease, babesiosis, and plague, (Ostfeld and Keesing 2000), can be investigated in light of this new evidence to explore the relationship between biodiversity and disease risk.

14.3.2 Hantavirus

The case of the hantavirus pulmonary syndrome outbreak in the American Southwest provides a valuable model of how altered numbers of a prey species carrying an infectious agent can result in the emergence of a lethal human infectious disease. Six years of drought in the Four Corners area ended in the late winter and spring of 1993 with unusually heavy snow and rainfall. The drought had reduced the populations of the natural predators of the native deer mouse (owls, snakes, coyotes, and foxes), while the increased precipitation led to a very heavy crop of pine nuts and grasshoppers, food for the mice (Wenzel 1994). As a result, there was a 10-fold increase over baseline levels in the deer mouse population. During this time, 17 previously healthy people, most of them Native Americans, rapidly developed a severe respiratory syndrome; 13 of them died (Duchin et al. 1994). It was subsequently learned that the deer mice carried a hantavirus that they shed in their saliva and excreta, and that with the greater numbers of mice, there was a greater chance for people to come in contact with their excreta and thus to become infected. In this case, a change in climate triggered the outbreak of a highly lethal infectious disease, but other alterations in species numbers, for example, through overhunting of a major predator of a human infectious disease carrier, might do so as well. It is not known how many viruses or other infectious agents in the environment, potentially harmful to man, are being held in check by the natural regulation afforded by biodiversity.

14.4 Conclusions

Despite an avowed reverence for life, human beings continue to destroy other species at an alarming rate, rivaling the great extinctions of the geologic past. In the process, we are foreclosing the possibility of discovering the secrets they contain for the development of life-saving new medicines and of invaluable models for medical research, and we are beginning to disrupt the vital functioning of ecosystems on which all life depends. We may also be losing some species so uniquely sensitive to environmental degradation that they may serve as our "canaries," warning us of future human health threats.

It may be that policy makers and the public will give protection of the global environment their highest priority only when they have begun to understand that human health and life ultimately depend on the health of other species and on the integrity of global ecosystems.

Physicians and other health professionals, veterinarians, conservation biologists, and other environmental scientists need to become knowledgeable about the human health dimensions of species loss and ecosystem disruption, for they may be the most powerful voices in helping to promote this understanding.

References

Anderson, P.K., and F.J. Morales. 1994. The emergence of new plant diseases: the case of insect-transmitted plant viruses. In Wilson, M.E., R. Levins, and A. Spielman, eds. Disease in evolution: global changes and emergence of infectious diseases. Ann NY Acad Sci 740:181–194.

Atanassoff, P.G., M.W.B. Hartmannsgruber. J. Thrasher, et al. 2000. Ziconotide, a new N-type calcium channel blocker, administered intrathecally for acute postoperative pain. Reg Anesth Pain Med 25:274–278.

Blaustein, A.R., P.D. Hoffman, D.G. Hokit, and J.M. Kiesecker. 1994. UV repair and resistance to solar UV-B in amphibian eggs: a link to population declines. Proc Natl Acad Sci USA 91:1791–1795.

Bowersox, S. et al. 1998. SNX-111: antinociceptive agent N-type voltage-sensitive calcium channel blocker. Drugs Fut 23:152–160.

Centers for Disease Control and Prevention (CDC). 2000. Surveillance for Lyme disease—United States, 1992–1998. Morbid Mortal Wkly Rep 49:1–11.

Chabner, B.A., C.J. Allegra, G.A. Curt, and P. Calabresi. 1996. Antineoplastic agents. In Hardman, J.G., L.E. Limbird, P.B. Molinoff, R.W. Ruddon, et al., eds. Goodman and Gilman's The pharmacological basis of therapeutics, 9th ed., p. 1251. McGraw-Hill, New York.

Chen, F., et al. 2000. Translocation of C. elegans CED-4 to nuclear membranes during programmed cell death. Science 287:1485–1489.

Chivian, E. 1997. Global environmental degradation and species loss: implications for human health. *In* Grifo, F., and J. Rosenthal, eds. Biodiversity and human health, pp. 7–38. Island Press, Washington, D.C.

Cohen, J.E., and D. Tilman. 1996. Biosphere 2 and biodiversity: the lessons so far. Science 274:1150–1151.

Colburn, T., R.S. vom Saal, and A.M. Soto. 1993. Developmental effects of endocrine-disrupting chemicals in wildlife and humans. Environ. Health Perspect 101:378–384.

Colwell, R.R. 1996. Global climate and infectious disease: the cholera paradigm. Science 274:2025–2031.

Costanza, R., R. d'Arge, R. de Groot, et al. 1997. The value of the world's ecosystem services and natural capital. Nature 387:253–260

Cowden C.J., and I. Paterson. 1997. Cancer drugs better than taxol? Nature 387:238–239.

Culetto, E., and D.B. Sattelle. 2000. A role for *Caenorhabditis elegans* in understanding the function and interactions of human disease genes. Hum Mol Genet 9:869–877.

Daily, G.C., (ed.). 1997. Nature's services: societal dependence on natural ecosystems. Island Press, Washington, D.C.

Daszak, P. A.A. Cunningham, and A.D. Hyatt. 2000. Emerging infectious diseases of wildlife—threats to biodiversity and human health. Science 287:443–449.

Duchin, J.S., F.T. Koster, C.J. Peters, et al. 1994. Hantavirus pulmonary syndrome: a clinical description of 17 patients with a newly recognized disease. N Eng J Med 330:949–955.

Ehrlich, P.R., and E.O. Wilson. 1991. Biodiversity studies: science and policy. Science 253:758–762.

Epstein, P.R. 1997. Climate, ecology, and human health. Consequences 3:2–19.

Flannery, T.F. 1999. Paleontology: debating extinction. Science 283:182–183.

Floyd, T., R.A. Nelson, and G.F. Wynne. 1990. Calcium and bone metabolic homeostasis in active and denning black bears. Clin Orthopaed Rel Res 255:301–309.

Gao, F., E. Bailes, D.L. Robertson, et al. 1999. Origin of HIV-1 in the chimpanzee *Pan troglodytes troglodytes*. Nature 397:436–441.

Grifo, R., D. Newman, A.S. Fairfield, B. Bhattacharya, and J.T. Grupenhoff. 1997. The origins of prescription drugs. *In* Grifo, F., and J. Rosenthal, eds. Biodiversity and human health. Island Press, Washington, D.C.

Hahn, B.H., G.M. Shaw, K.M. De Cock, and P.M. Sharp. 2000. AIDS as a zoonosis: scientific and public health implications. Science 287:607–614.

Hasan, Y., R.L. Morgan-Boyd, A. Dickson, C. Meyer, and R.A. Nelson. 1997. Plasma carnitine levels in black bears [abstract]. Fed Am Soc Exp Biol J 11:A371.

Hayes, R.L., and N.I. Goreau. 1998. The significance of emerging diseases in the tropical coral reef ecosystem. Rev Biol Trop 46 suppl 5:173–185.

Herminghuysen, D., M. Vaughn, R.M. Pace, and G. Bagby. 1995. Measurement and seasonal variations of black bear adipose lipoprotein lipase activity. Physiol Behav 57: 271–275.

Higgins, R.D.; R.J. Sanders, Y. Yan, M. Zasloff, and J.I. Williams. 2000. Squalamine decreases retinal neovascularization. Invest Ophthalmol Vis Sci 41:1507–1512.

International Union for the Conservation of Nature (IUCN). 1997. 1996 Red list of threatened animals. Baillie J, and T. Groombridge, eds. IUCN, Gland, Switzerland.

International Union for the Conservation of Nature (IUCN), 2000. 2000 red list of threatened animals. IUCN, Gland, Switzerland.

Kimura, K.D., et al. 1997. *Daf-s*, an insulin receptor-like gene that regulates longevity and diapause in *Caenorhabditis elegans*. Science 277:942–946.

Korber, B., M. Muldoon, J. Theiler, et al. 2000. Timing the ancestor of the HIV-1 pandemic strains. *Science* 2000 288:1789–1796.

Litman, G.W. 1996. Sharks and the origins of vertebrate immunity. Sci Am 275:67–71.

Lovei, G.L. 1997. Global change through invasion. Nature 388:627–628.

Mainka, S.A. and J.A. Mills. 1995. Wildlife and traditional Chinese medicine—supply and demand for wildlife species. J Zoo Wildl Med 26:193–200.

Malakoff, D. 1997. Extinction on the high seas. Science 277:486–488.

May, R.M. 1988. How many species are there? Science 241:1441–1449.

McGuire, W.P. et al. 1989. Taxol: a unique antineoplastic agent with significant activity in advanced ovarian epithelial neoplasms. Ann Int Med 111:273–279.

Mestel, R. 1996. Sharks' healing powers. Nat His 105:40–47.

Miljanich, G.P. 1997. Venom peptides as human pharmaceuticals. Sci Med 4:6–15.

Miller, G.H. et al. 1999. Pleistocene extinction of *Benyornis newtoni*: human impact on Australian megafauna. Science 283:205–208.

Montgomery, S. Grisly trade imperils world's bears. Boston Globe, 2 March, 1992. p. 345.

Moore, K.S., S. Wehrli, H. Roder, et al. 1993. Squalamine: an aminosterol antibiotic from the shark. Proc Natl Acad Sci USA 90:1354–1358.

Morse, S.S. 1995. Factors in the emergence of infectious diseases. Emerg Infect Dis 1:7–15.

Myers, N., et al. 2000. Biodiversity hotspots for conservation priorities. Nature 403:853–858.

National Osteoporosis Foundation (NOF). 2000. Fast facts on osteoporosis. NOF, Washington, D.C.

Nelson, R.A. 1987. Black bears and polar bears—still metabolic marvels. Mayo Clin Proc 62:850–853.

Nicolaou, K.C., R.K. Guy, and P. Potier. 1996. Taxoids: new weapons against cancer. Sci Am 274:94–98.

Olivera, B.M., J. Rivier, C. Clark, et al. 1990. Diversity of *Conus* neuropeptides. Science 249:257–263.

Olivera, B.M., G.P. Miljanich, J. Ramachandran, and M. Adams. 1994. Calcium channel diversity and neurotransmitter release: the ω-conotoxins and ω-agatoxins. Ann Rev Biochem 63:823–867.

Ostfeld, R.S., and F. Keesing. 2000. Biodiversity and disease risk: the case of Lyme disease. Conserv Biol 14:722–728.

Ostfeld, R.S., et al. 1998. Integrative ecology and the dynamics of species in oak forests. Integr Biol 1:178–186.

Pauly, D., V. Christensen, J. Dalsgaard, R. Froese, F. Torres. 1998. Fishing down marine food webs. Science 279:860–863.

Pennisi, E. 1998. Worming secrets from the *C. elegans* genome. Science 282:1972–1974.

Peters, R.L., and T.E. Lovejoy, (eds.). 1992. Global warming and biological diversity. Yale University Press, New Haven, Conn.

Pimm, S.L., and T. Brooks. 2000. The extinction: how large, where, and when? *In* Raven, P.H., ed. Nature and human society, pp. 46–62. National Academy of Sciences Press, Washington, D.C.

Pimm, S.L., and Raven. R. 2000. Extinction by numbers. Nature 403:843–844.

Pimm, S.L., G.J. Russell, J.L. Gittleman, and T. Brooks. 1995. The future of biodiversity. Science 1995. 269:347–350.

Rao, M.N., A.E. Shinner, L.A. Noecker, et al. 2000. Aminosterols from the dogfish shark, *Squalus acanthias*. J Nat Prod 63:631–635.

Ricklefs, R., and G. Miller. 2000. Ecology, 4th ed. Freeman, New York.

Ridgeway, B, M. Wallace, A. Gerayli. 2000. Ziconotide for the treatment of severe spasticity after spinal cord injury. Pain 85:287–289.

Roush W. 1997. Worm longevity gene cloned. Science 277:897–898.

Rubin, G.M., and E.B. Lewis. 2000. A brief history of *Drosophila*'s contributions to genome research. Science 287:2216–2218.

Schindler, D.W. 1998. A dim future for boreal waters and landscapes. Bioscience 48:157–164.

Sze, J.Y., et al. 2000. Food and metabolic signaling defects in a *Caenorhabditis elegans* serotonin-synthesis mutant. Nature 403:560–564.

Taub, J., et al. 1999. A cytosolic catalase is needed to extend adult lifespan in *C. elegans daf-C* and *clk-l* mutants. Nature 399:162–166.

Thomas, L. 1974 The lives of a cell: notes of a biology watcher. Viking Press, New York.

Vitousek, P.M., et al. 1997. Human domination of Earth's ecosystems. Science 277:494–499.

Walsh, J.F., D.H. Molyneux, and M.H. Birley. 1993. Deforestation: effects on vector-borne disease. Parasitology 106:S55–S75.

Weiss, R.A., and R.W. Wrangham. 1999. From *Pan* to pandemic. Nature 397:385–386.

Wenzel, R.P. 1994. A new hantavirus infection in North America. N Engl J Med 330:1004–1005.

Wilson, E.O. 1989. Threats to biodiversity. Sci Am 261:108–116.

Zapp, A. 1996. "Ghoulish" potion shows nervous system function. NIH Record, available at www.nih.gov/news/NIH-Record/10_08_6/main/htm (accessed 10/4/01).

15

Vector-Borne Infections and Health Related to Landscape Changes

David H. Molyneux

The major vector-borne infections and the diseases caused by them are focused in the tropics; there is a significant overlap between the distribution of the majority of important vectors of human and animal diseases and the richly biodiverse tropical rain forest ecosystems, woodland savannas, and the edges of these ecosystems between the tropics of Cancer and Capricorn. The major insect vector groups—*Anopheles, Aedes, Culex* and *Mansonia* mosquitoes; *Simulium* blackflies; new-world vectors of *Leishmania* (*Lutzomyia*); *Chrysops* vector of *Loa loa*; and *Glossina* species, which transmit trypanosomes—all have species that are dependent on forest, woodland savanna, or riverine forest ecosystems. It is the erosion of these ecosystems, the behavior of the vectors at the forest edge, the impact of reforestation on the interactions between humans with vectors, and reservoir hosts at the interface that determine the epidemiology of human infective agents (table 15.1). Additional factors are the degrees of immunity of local or migrant populations and their behavior, the interaction with and behavior of reservoir hosts, and the behavior and adaptability of insect vectors. Walsh and colleagues (1993) extensively reviewed the knowledge of vector-borne parasitic infections and the deforestation process. Molyneux (1997) provided an update on the patterns of change in vector-borne infections, identifying key components of changes in epidemiology in addition to deforestation. These include urbanization, antimicrobial and insecticide resistance, conflict-related events, and the many water-related issues—dam construction, irrigation, microdams, bore-hole constructions, inadequate waste-water disposal, and long-term water resource deficits. Over the last decade, remarkable advances have been made in the use and development of remote sensing techniques and data that, together with geographical information systems, provide exciting new tools for the bio-

Table 15.1. Major forest-associated vectors and disease

Mosquitoes	Distribution	Remarks
Malaria		
Anopheles dirus *An. minimus* *An balabacensis* *An. leucosphyrus*	Southeast Asia	Both *An. dirus* and *An. minimus* adapt to new plantations.
An. fluviatilis	India, Nepal	Replaced by *An. culicifacies* following deforestation.
An. darlingi	Amazon basin	Increase in abundance following forest disturbance.
An. gambiae forest form	West Africa	Deforestation provides temporary pools for larval habitats.
Lymphatic Filariasis		
Anopheles maculatus	Malaysia	*Wuchereria bancrofti* vector.
Aedes niveus group	Thailand	Forest vector breeds in tree holes and bamboo.
Mansonia bonneae *M. dives*	Thailand	Swamp forest vectors of *Brugia malayi*; adapted to rubber plantations.
Yellow Fever		
Aedes africanus—forest *A. simpsoni*—interface *A. aegypti*—urban	Africa (Uganda)	Contact between monkeys and humans initially via forest, then agriculture at forest edges and urban environment.
Onchocerciasis		
Simulium sanctipauli *S. squamosum* *S. yahense* *S. soubrense* *S. neavei*	West Africa Central Africa	Vectors of *Onchocerca volvulus*, "forest" forms.
S. woodi *S. exiguum*	East Africa	Controlled or eradicated as forest cover lost.
S. guianense	Ecuador	Increase in transmission of *O. volvulus* as forest clearance occurs.
S. ochraceum	Guatemala, Mexico	Associated with coffee fincas; larval habitat, small streams with thick cover.
Leishmaniasis		
Lutzomyia longipalpis	S. America	Adaptation to peridomestic habitats in deforested areas create new foci of visceral leishmaniasis caused by *Leishmania chagasi*.
Lut. whitmani	Brazil	Highly adaptable vector: formally strictly sylvatic, has become peridomestic and anthropophilic as *Le braziliensis* vector.
Lu. (Psychodopygus) wellcomei	Brazil	Vector of *Le. braziliensis* (Brazil).
Lu. flaviscutellata	Brazil	Highly anthropophilic forest vector.
Lu. umbratilis	Brazil, Guyana	Vector of *Le. amazonensis*. Adaptable wetter forest "varzea" species primarily zoophilic on rodents.

(continued)

Table 15.1. Continued

Mosquitoes	Distribution	Remarks
Lu. olmeca	Central America	Vector of *Le. guyanensis*. Association of humans with forest edge where edentates (sloths and anteaters) occur and where opposums can move into peridomestic habitats.
Phlebotomus orientalis	Sudan	Vector of *Le. mexicana*. Human cases of chicleros ulcer through exposure at night to sand fly. Vector of *L. donovani*. Increased vector abundance and transmission associated with maturity of *Acacia/Balanites* woodland.
Loaisis		
Chrysops dimidiata	Central/West	Vector of *Loa loa* restricted to dense forest and
C. silacea	Africa	forest edge. Loss of forest reduces prevalence as breeding sites in forest swamp destroyed. Vectors have restricted range outside canopy. Adapted to rubber plantations.
Trypanosomiasis		
Glossina palpalis	*West and Central*	*Vectors of Trypanosoma brucei gambiense*. Rap-
G. fuscipes	*Africa*	idly adapted to plantations and peridomestic
G. tachinoides		habitats from riverine or lacustine habitats. Clearing vegetation eliminates tsetse.
G. pallidipes	East/Southern	Destruction of savanna woodland removes habitat.
G. morsitans	Africa	*G. pallidipes* adapts to *Gurelina* plantations.
G. fuscia group	Central/West Africa, rainforest	Populations retreat in face of deforestation; limited evidence of repopulation of reafforested areas.

sciences (Rogers and Williams, 1993; Hay et al. 1996, Thomson et al. 1996; 1999; Thomson and Connor 2000) in relation to vector-borne infections, changing ecosystems, and medium- to long-term prediction capacity.

Climate modeling is being developed to predict ecological change with various scenarios of global warming due to increased carbon dioxide (CO_2) emissions (together with other greenhouse gases), sea-level rises, water resource stress, global crop productivity, and change in forest biomass over the next decades. These models emphasize the rate of change against various scenarios of greenhouse gas levels, from "unmitigated increase" to maintenance of 1990s levels at emissions of 750 and 550 parts per million (ppm) (Meteorological Office 1999).

There has been a widespread recognition of the importance of conserving the diversity represented in the tropical forests as a global resource. However, the rate of forest destruction progresses in the face of increased populations and the desire to exploit natural resources without regard for sustainability, a situation that prevails in Amazonia, equatorial Africa, and Southeast Asia. The impact of El Niño events in reducing rainfall in Southeast Asia, leading to extensive and uncontrolled forest fires, greatly increased the rate of reduction in forest cover. The consequences of that event on vector-borne infections and health in the previously forested areas have not yet been determined. The trend in increased frequency of violent weather patterns, maintained particularly in the Americas and attributed to El Niño/La Niña cycles, has persisted through 1999.

Over the next two decades it is anticipated that noncommunicable diseases will contribute proportionately to the increased global burden of disease (Murray and Lopez 1996). Many of the impacts on disease burden over the next two decades will result from increased urbanization, increased tobacco consumption, and life-style changes associated with urban living. As a consequence, incidence of cardiovascular diseases and diabetes will increase in older age groups, and the demographic transition will produce an increasingly older population with expectations of improved health care in Western developed countries. Against this scenario of an increase in noncommunicable disease burden, the burden associated with exploitation of the natural forest resources will be relatively small. Exploitation of forest resources will, however, result in increased trauma, violence-related injury, and snake bite, which will have potentially more serious consequences in the absence of appropriate and accessible health facilities or a significant public health care system.

Nonvector-borne communicable infectious diseases will pose a threat through contact with as yet unknown or ill-defined potential zoonotic agents present in animal reservoir hosts, of which the most significant potential threat are simian viruses or viruses in the diverse, abundant, and cyclical rodent populations, whose adaptive capabilities and high reproductive rate enable them to exploit changed environments. In addition, the activities associated with deforestation bring human viruses from "developed" societies into indigenous populations, with disastrous consequences to such nonimmune communities.

The European Commission TREES (Tropical Ecosystem Environment observations by Satellites) project provides via the Internet a global tropical forest map at 1-km resolution. This project is a collaboration between the Joint Research Centre and the European Space Agency and will provide an operational tropical forest monitoring system, thereby directing attention to active areas of deforestation to allow more detailed observations (further information can be obtained from www.mtr.sai.jre.it/progjhects/treeswww/treesopref.html. To date, limited if any of these valuable data have been used by the health fields; important opportunities for assessing health issues on the ground from highlighted "active areas" of deforestation clearly present themselves to government health agencies or nongovernmental organizations. These data have been used for predictive epidemiology purposes by Thomson et al. (2000) for analysis of *L. loa* epidemiology using forest and forest edge as the determinants of distribution of *Chrysops* vectors of *L. loa*.

The effects of deforestation and associated reafforestation can be stratified via the following principles (table 15.2).

1. The roles and behaviors of animal reservoir hosts, vectors, and humans (including their immune status) are key determinants in the transmission of in-

Table 15.2. Deforestation, reafforestation, and disease: interactive factors and components of epidemiology, health, and transmission

Migration		
Controlled	Transmigration, Indonesian planned migration, and colonization re-settlement of urban populations in Brazil; uncontrolled emigration to urban centers after failure of crops, loss of soil nutrients, and erosion	
Uncontrolled	Conflict-associated migration into forests	
	Rwanda → Congo (highlands)	
	Illegal mining in S. America and S.E. Asia	Immune condition of host S.E. Asia spread of multidrug-resistant *P. falciparum* due to forest/interface activities
Return migration following loss of agricultural productivity and dependence on staples	Reduced nutritional status	Epidemic malaria due to absence of immunity
Forest Fires in Southeast Asia	Acute increase in regional respiratory disease, pressure on urban hospitals	No prediction to date of impact on mosquito vectors
Company Projects	Controlled labor force but no social structure in community	Company health care
	Increased violence and drug/alcohol use, injuries, sexually transmitted diseases in absence of family structure, use of commercial sex workers, snake bite	
Forest Development Roads, mining (legal and illegal) Oil/gas extraction Hydropower Reafforestation projects Timber extraction	Peripheral or uncontrolled communities develop.	No government or private health care provision; control of vector-borne diseases in forests or deforested areas not feasible.
	Contact with animal reservoirs and exposure to new infectious agents; key source of emergent diseases	

fection to humans exposed within the forest and at the interface. Examples include capacity of animal reservoirs to exist in close proximity to humans (Brug's filariasis, Mak et al. 1982; visceral leishmaniasis, Lainson et al. 1983); the ability of insect vectors to adapt to human hosts as a source of blood meals, particularly if the animal reservoir hosts are destroyed or migrate (*Psychodopygus wellcomeii*, Ward et al. 1973; *Lutzomyia whitmani*, Ryan et al. 1990); the biting behavior of vectors in relation to human behavior, for example, human availability, flight range of vector, biting frequency, and biting times in relation to human habits (*Anoph-*

eles darlingi, Marques 1987, Charlwood and Alecrim 1989; *An. leucosphyrus*, Marwoto and Arbani 1991).

2. Forest-related activities increase and change exposure to human infective vector-borne infections. There are several activities, both planned and unplanned, that result in increased human disease in environments when there is limited, if any, structured health care. This is particularly well defined in migrant populations of nonimmunes in Amazonia and Indonesia (Prothero 1991). In Brazil, malaria transmission has been associated with deforested environments that provide many suitable breeding sites for *Anopheles darlingi* (borrow pits associated with road building and gold mining; Marques 1987), while expansion of *An. darlingi* populations in parallel with deforestation will maintain malaria as a significant health problem in Amazonia. In Southeast Asia, "border" malaria associated with migrants and gem mining has maintained and possibly enhanced the spread of multiple-drug-resistant malaria in border areas of Thailand, Myanmar, Cambodia and Laos with changing patterns of vector roles and abundance. *Anopheles minimus* dominated the more open scrub after forest habitats were destroyed, reducing *An. dirus* complex populations (Suvannadabba 1991). In Indonesia, transmigration policy facilitates malaria by allowing deforestation, which is associated with the displacement of *An. balabacensis* and *An. leucosphyrus*. However, many transmigrants from overpopulated Java are nonimmune and enter the forest at night to exploit the forest resources, hence become exposed to forest *Anopheles* (Marwoto and Arbani 1991).

Leishmaniasis epidemics are closely linked with increased exposure of humans to *Lutzomyia* species in the New World, where settlements are associated with road building, mining, oil and gas exploration, and other activities that bring humans into the forest (gathering natural forest products). In amazonian Brazil, *Lu. whitmani* and *Psychodopygus wellcomei* sand flies are associated with *Leishmania brasiliensis* infections, *Lu. umbratilis* with *Le. guyanensis*, and in Central America, *Lu. olmeca* with *Le. mexicana* (resulting in characteristic lesions of chichleros ulcer; see review by Lainson and Shaw 1998). All these associations depend on forest-related activities and anthropophilic sand flies. The degree of human exposure is therefore related to the degree of anthropophily and adaptability of the vector to human hosts; the same vector species may vary in its capacity to feed on humans (Tesh et al. 1972; Mayrink et al. 1979).

3. The destruction of forest habitat has resulted in the removal, displacement, or eradication of vector species. *Simulium neavei* blackflies, vectors of *Onchocerca volvulus*, were eradicated from Nyanza Province, Kenya, through clearing of undergrowth and forest (Raybould and White 1979). Tanzanian populations of closely related species, *S. woodi*, declined rapidly after deforestation in the Usambara mountains and was paralleled by the reduction in prevalence of *Onchocerca volvulus* (Muro and Msiray 1990; Muro and Raybould 1990).

Declines in population of *Glossina palpalis* and *G. tachinoides* through deliberate riverine forest destruction in West Africa has a rapid effect on the vector species. In eastern Africa forest/savanna, tree clearing using mechanical means removed the habitat of *G. pallidipes* and *G. morsitans* and was an accepted method of tsetse, and hence trypanosomiasis, control (Leak 1998).

Similarly, in parts of West Africa, the vectors of *Loa loa, Chrysops,* have significantly declined as forest degradation has progressed (Duke 1972; Sasa 1976). Recent studies using remotely sensed forest data and soil profiles are being used to map the distribution of *Chrysops* in areas of co-endemicity with *Onchocerca volvulus* (Thomson et al. 2000). This is required given that control of onchocerciasis by the drug Mectizan (ivermectin) is compromised by severe adverse reactions in populations where *Loa* and *Onchocerca* are co-endemic.

4. Deforestation is accompanied by common patterns of change in the distribution of vectors. The process of forest loss, particularly in West Africa, has accelerated over the last two decades, where extensive long-term studies of *Anopheles gambiae* species complex (Coluzzi et al. 1985; Touré et al. 1998) and the *Simulium damnosum* group of species have been undertaken. The *Anopheles* studies have been undertaken through extensive mapping of cytospecies of *An. gambiae* in longitudinal studies (Coluzzi et al. 1985; Touré et al. 1998) and *S. damnosum* cytospecies as part of the evaluation process of the Onchocerciasis Control Programme (OCP) in 11 countries of West Africa (WHO 1994). These studies have clearly demonstrated links between deforestation and increased savanna and scrub vegetation with the distribution of cytospecies more prevalent in savanna and arid environments spreading to southerly latitudes.

Similarly, riverine tsetse fly species *Glossina palpalis* and *G. tachinoides* are less common in northerly parts of their range, as deforestation of savanna woodland in Guinea has progressed over some four decades. These species have rapidly adapted to peridomestic and plantation environments in more southerly parts of their range (Laveissiere et al. 1984, 1985).

5. Reafforestation is associated with rapid capacity of vectors to adapt to new frequently non-indigenous climax vegetation. Destruction of primary forest with the introduction of commercial cash crops is common throughout the tropics. In Southeast Asia, natural forests have been replaced by tea, coffee, rubber, eucalyptus, and various types of fruit orchards (Singhasivanon et al. 1999). These new plantations have sustained primary vectors of malaria *An. dirus* and *An. minimus.* Suvannadabba (1991) suggested coffee and rubber plantations favored the resurgence of *An. minimus,* but Rosenberg and colleagues (1990) emphasized that *An. dirus,* usually associated with natural forest, has been focused in scrub vegetation and tea and rubber plantations. It is probable that variations in observed behavior of *An. dirus* are attributable to variation in dominant cytospecies, each of which has well-defined habitats in relation to the forest ecosystem of Southeast Asia (Baimai 1984).

In South America, reforestation with *Gmelina molinas* and *Pinus caribea* plantations in the Jari valley following removal of low-lying forest demonstrated that the reservoir of *Leishmania amazonensis,* the rodent *Proechimys,* and its vector *Lu. flaviscutellata* proved particularly well adapted to both types of exotic plantations (Ready et al. 1983). *Gmelina* plantations in Kenya on the edges of the Lambwe Valley provided habitats for *Glossina pallidipes,* a tsetse fly found in abundance in the dense thickets of the valley floor but that also rapidly adapted— presumably with its preferred natural host, the bushbuck *Tragelaphus scriptus,* to the *Gmelina*-forested hillsides (Turner 1981). *Glossina palpalis* group species in West Africa, particularly in the Ivory Coast, have colonized coffee and cocoa plantations, moving away from the classical riverine vegetation that provided the only areas of high humidity before commercial plantations were developed (Laveissiere et al. 1984, 1985). The *Glossina fusca* group populations of tsetse associated with dense West and Central Africa forests, although not significant as vectors of human trypanosomiasis, are retreating, although there is no evidence of *G. fusca* group flies adapting to commercial plantations to the same extent as has the *G. palpalis* group (Baldrey and Molyneux 1980). This may be due to the well-known indiscrimate feeding habits of *G. palpalis* group flies, which are highly adaptable compared with *G. fusca* group, the latter having clear host preferences (Leak 1998).

6. Forest ecosystems are recognized as rich sources of plant and animal biodiversity; this biodiversity is also observed in insect vectors and in parasites infecting humans. The richness of tropical forest ecosystems is undisputed. Studies on the animal fauna, particularly insects, have revealed high species richness often associated with single tree species. Studies on human parasites and their vectors have also revealed a diversity hitherto unrecognized; this has been achieved in particular through advances in characterization techniques enabling morphologically similar organisms to be identified and separated, thereby providing a better understanding of epidemiology, clinical disease, and distinct etiologies. This is exemplified by the in-depth studies on *Leishmania* epidemiology and sand fly biology in the Amazon, where several new species of *Leishmania* have been identified (Lainson and Shaw 1998), but also the heterogenecity of morphologically identical sand fly species has been revealed by biochemical, chemical, and molecular techniques (Ward et al. 1988). Similarly, in South and Central America, where the causative agent of Chagas disease, *Trypanosoma cruzi*, is transmitted by triatomine bugs, the biological diversity of the many forms of *T. cruzi* has been revealed. Only some of these forms may be infective to humans. The diversity in forms of *T. cruzi* is reflected in its presence in over 100 different mammalian hosts and several genera of Triatominae (Morel 1998). Human infectivity and distinctiveness of these ecological forms is as yet unknown but exemplifies the biodiversity of the parasites, reservoir hosts, and vector systems in tropical forests, particularly of "protozoan" parasites—trypanosomes, *Leishmania*, malaria—and their vectors.

Studies on *Anopheles* mosquitoes throughout the tropics have revealed the existence of cytospecies complexes through biochemical and molecular techniques (Coluzzi et al. 1985; Touré et al. 1998; Donnelly et al. 1999). Most of the major vector complexes of *Anopheles* have distinct behaviors potentially reflecting differences in vectoral importance. Cytogenetic studies have revealed forms of *An. gambiae* associated with forest, savanna, and semidesert habitat. In Southeast Asia, *An. dirus* is made up of several different cytoforms, each with distinct biological characteristics (Baimai 1984). It is therefore likely that both vectors and the organisms they transmit will on closer study be shown to be more genetically diverse than hitherto recognized. The consequences of this will be that true species loss may not be detected, that we will underestimate adaptive capacity to changing environments, and that the true numbers of vector species not only have been underestimated but may never be known.

More recent studies of a time series from 1975 to 1997 by Wilson and colleagues linking changes in riverine forest cover in Ghana to changes in cytospecies of *Simulium* over the same period confirm earlier predictions by Walsh and colleagues (1993). Wilson and colleagues compared proportions of forest, degraded forest savanna, and urban areas of the Tano River area. The trends of decreasing forest cover affect the riverine habitats of the blackflies, as the percentages of savanna flies (cytospecies *S. damnosum* and *S. sirbanum*) compared with nonsavanna vectors (*S. santipauli, S. yahense*, and *S. squamosum*) showed a statistically significant trend for all seasons (Jan.–Mar., April–June, July–Sept., Oct.–Dec.) over four time periods: 1975–80, 1981–86, 1987–92, 1993–97.

Similar studies have been undertaken over similar time periods on the *Anopheles gambiae* complex in the same region (Coluzzi et al. 1985) and more recently in Mali (Touré et al. 1998). These studies have identified three karyotypes (Bamako, Mopti, Savanna) of *An. gambiae* sensu strictu and identified an association between these forms (using polytene chromosome identification of the 2R chromosome) and particular ecological conditions. *Anopheles arabiensis* is sympatric with *An. gambiae* and prevailed in arid savanna (Sahel and northern Sudan savannas), being able to withstand the most arid conditions of Saharan localities possibly breeding throughout the dry season, but being absent from irrigated and flooded areas. *Anopheles gambiae* Bamako and Savanna forms are more characteristic of humid savannas, breeding only in the rainy season. The Mopti taxon of *An. gambiae* is highly adapted to arid environments but is present in all ecological zones, including the Sahel and predesert environments. The Mopti form was the dominant, if not exclusive, form in flooded and irrigated areas and is capable of dryseason breeding. There was a positive correlation between rainfall and the frequency of Savanna and a negative correlation between rainfall and the frequency of Mopti. This latter correlation was confirmed by comparing the frequency of Mopti forms with normal difference vegetation indices as a measure of aridity by Thomson and colleagues (1997).

This study has demonstrated that detailed cytogenetic analyses over a period of two decades, and will allow a better understanding of the complexity of the ecology and genetics of these disease systems. It also demonstrates correlations between different cytoforms and habitats as characterized by rainfall, desertification and loss of savanna woodland and forest habitats. Such studies suggest what could be done in other regions where cytospecies are likely to respond to changes in habitat over time. However, such changes will only be detected if cytogenetics, remote sensing, meteorology, and ecology are brought together through long-term projects.

A recent U.K. Department of Environment Transport and the Regions report (Meteorological Office 1999) undertook various analyses to predict the impact of climate change based on unmitigated CO_2 emissions and a scenario of stabilization of emissions at 550 and 750 parts per million, around twice and three times preindustrial levels. With unmitigated emissions and even with a stabilization at

750 ppm, a significant loss of tropical rain forest will occur in South America, and an increase in water resource stress will occur in parts of Europe and the Middle East. Such effects would be avoided at 550 ppm stabilization. Climate change predictions suggest that with unmitigated emissions, global average temperature will rise by 3°C by 2080 and significant changes in precipitation will occur in the tropics. The effects under an unmitigated emission scenario will be a substantial dieback of tropical forests, particularly in northern South America and Central Southern Africa by the 2080s, while forests in North America, northern Asia, and China are predicted to grow considerably. Stabilization of CO_2 at 750 ppm will ensure this loss is substantially reduced even up to 2230. The absorption of CO_2 will increase during the next century under all scenarios, but as tropical vegetation dies back, the CO_2 sink is lost in 2070 if emissions, are unmitigated and about 100 years later if emissions are stabilized; without the sink, CO_2 concentrations will be higher than those assumed in the model, hence change will be greater. The models also predicts water resource stress and the changes in human health using *P. falciparum* malaria as an example. The models predicting substantial changes in *P. falciparum* distribution under all CO_2 scenarios, with additional numbers of people at risk of *P. falciparum* malaria in the various emissions scenarios of 150–220 million by the 2020s, 220–250 million by 2050, and up to 320 million in the 2080s. However, Rogers and Randolph (2000) have challenged the earlier predictions of an increased range of *P. falciparum* transmission associated with global climate change. They used an alternative statistical methodology based on the present distribution of *P. falciparum* to establish the current climatic constraints. Applied to future predicted scenarios, even the most extreme, they showed very limited changes to distribution of *P. falciparum*.

13.8 Conclusions

The predicted effects on forests and human activities associated with forests are clearly on a macroscale, and most changes that have been discussed in this chapter are at the microscale of the forest interface, the result of relatively small-scale changes in forest use and alternative land use. However, over recent decades these small changes have had a highly significant cumulative effect; that there will be a loss of forest biomass due to climate change (dieback) and increased human activities seems inevitable; patterns of change in rainfall and hence also water resource stress are predicted on a macroscale, while interannual variation is of course not predictable. However, medium-term predictions of up to four to six months will become increasingly reliable and can be used for epidemic prediction. Vectors, in particular, respond rapidly to ecological and environmental change; while over recent decades interactions between disease and the forest environment have been retrospectively documented, it may be possible to use models to predict change in forest ecosystems and thus assist in health planning and disease management. Tropical forests are not amenable to interventions by control of vectors, application of preventative or curative agents, or ideal development of effective health care systems to serve the needs of either naive or indigenous populations or migrants. If health care systems exist, they will be less able to respond to the changes over the next decades as resources diminish. While predictions are always difficult, one prediction can certainly be made—changes associated with diseases of the forest and its interface and the consequences of loss in human health will be less predictable than hitherto.

References

Baimai, V., C.A. Green, R.G. Andre, B.A. Harrison, and E.L. Peyton. 1984/1945. Cytogenetic studies of some species complexes of *Anopheles* in Thailand and Southeast Asia. Southeast Asian J of Trop Med Public Health 15:536–546.

Baldry, D.A.T., and D.H. Molyneux. 1980. Observations on the ecology and trypanosomiasis infection of a relict population of *Glossina medicorum* in the Komoe Valley of Upper Volta. Ann Trop Med Parasitol 74:79–91.

Charlwood, J.D., and W.A. Alecrim. 1989. Capture-recapture studies with South American malaria vector *Anopheles darlingi* Root. Ann Trop Med Parasitol 83:569–576.

Coluzzi, M., V. Petrarca, and M.A. Di Deco. 1985. Chromosomal inversion intergradation and incipient speciation in *Anopheles gambiae*. Zool Bull 52:45–63.

Donnelly, M.J., N. Cuamba, J.D. Charlwood, F.H. Collins, and H. Townson. 1999. Population structure in the malaria vector, *Anopheles arabiensis* Patton, in East Africa. Heredity 83:408–417.

Duke, B.O.L. 1972. Behavioral aspects of the life cycle of *Loa*. Zool J Linnean Soc 51 (suppl 1): 97–107.

Hay, S.I., C.J. Tucker, D.J. Rogers, and M.J. Picker. 1996. Remotely sensed surrogates of meteorological data for the study of the distribution and abundance of arthropod vectors of disease. Ann Trop Med Parasitol 90:1–19.

Lainson, R., and J.J. Shaw. 1998. New world leishmaniasis—the neotropical *Leishmania* species. *In* Cox, F.E.G, J.P. Kreier, and D. Wakelin, eds. Topley and Wilson Microbiology and microbial infections, 9th ed. Vol. 5, Parasitology, pp. 241–266. Arnold, London.

Lainson, R., J.J. Shaw, F.T. Silveira, and H. Fraiha. 1983. Leishmaniasis in Brazil. XIX. Visceral leishmaniasis in the Amazon region, and the presence of *Lutzomyia longipalpis* on the Island of Marajo, Para State. Trans R Soc Trop Med Hyg 77:323–330.

Laveissiere, C., T. Traore, and J.-P. Kienon. 1984. Ecologie de *Glossina tachinoides* Westwood 1850, en savane humide d'Afrique de l'ouest. Cahiers ORSTOM Se Entomol Med Parasitol 22:231–243.

Laveissiere, C., D. Couret, C. Staak, and J.-P. Hervouet. 1985. *Glossina palpalis* et ses hotes en secteur forestiere de Cote d'Ivoire. Relations avec l'epidemiologie de la trypanosomiase humaine. Cahiers ORSTOM Ser Entomol Med Parasitol 23:297–303.

Leak, S.G. 1998. Tsetse biology and ecology: their role in the epidemiology and control of trypanosomiasis. CABI Publishing, New York.

Mak J.W., W.H. Cheong, P.K.F. Yen, P.K.C. Lim, and W.C. Chan. 1982. Studies on the epidemiology of subperiodic *Brugia malayi* in Malaysia: problems in its control. Acta Trop 39:237–245.

Marques, A.C. 1987. Human migration and the spread of malaria in Brazil. Parasitol Today 3:166–170.

Marwoto, H.A., and P.R. Arbani. 1991. Forest malaria in Indonesia. *In* Sharma, V.P. and

A.V. Kondrashin, eds. Forest malaria in Southeast Asia, pp. 115–131. WHO/MRC, New Delhi.

Mayrink, W., P. Williams, M.V. Coelho, M. Dias, and A.V. Martins. 1979. Epidemiology of dermal leishmaniasis in the Rio Doce Valley, State of Minas Gerais. Ann Trop Med Parasitol 73:123–137.

Meteorological Office. 1999. Climate change and its impacts: stabilisation of CO_2 in the atmosphere. Health Centre for Climate Prediction and Research, London.

Molyneux, D.H. 1997. Patterns of change in vector-borne diseases. Ann Trop Med Parasitol 91:827–839.

Morel, C.M. 1998. Chagas disease: from discovery to control and beyond—history, myths and lessons to take home. Advisory Committee on Health Research, WHO/RPS/ACHR35, Rome, Italy.

Muro, A.I.S., and N.R. Msiray. 1990. Decline in onchocerciasis in the eastern Usambara mountains, north east Tanzania, and its possible relationship to deforestation. Acta Leidensia 58:141–150.

Muro, A.I.S., and J.N. Raybould. 1990. Population decline of *Simulium woodi* and reduced onchocerciasis transmission of Amani, Tanzania, in relation to deforestation. Acta Leidensia 59:153–159.

Murray, C.J.L., and A.D. Lopez. 1996. The global burden of diseases. World Health Organization, Geneva.

Prothero, R.M. 1991. Resettlement and health: Amazonia in tropical perspective. *In* A desordem ecologica na amazonica, pp. 161–182. Universidade Federal do Para (UFPA), Belem, Para, Brazil.

Raybould, J.N., and G.B. White. 1979. The distribution, bionomics and control of onchocerciasis vectors (Diptera: Simuliidae) in Eastern Africa and the Yemen Tropenmed Parasitol 30:505–547.

Ready, P.D., R. Lainson, and J.J. Shaw. 1983. Leishmaniasis in Brazil: XX. Prevalence of "enzootic roden leishmaniasis" (*Leishmania mexicana amazonensis*) and apparent absence of "pian bois" (*Le braziliensis guyanensis*), in plantations of introduced tree species and in other non-climax forests in eastern Amazonia. Trans R Soc Trop Med Hyg 77:775–785.

Rogers, D.J., and S.E. Randolph. 2000. The global spread of malaria in a future, warmer world. Science 289:1763–1765.

Rogers, D.J., and B.G. Williams. 1993. Monitoring trypanosomiasis in space and time. Parasitology 106:577–592.

Rosenberg, R., R.G. Andre, and L. Somchit. 1990. Highly efficient dry season transmission of malaria in Thailand. Trans R Soc Med Hyg 84:22–28.

Ryan, L., A. Vexenat, P.D. Marsden, R. Lainson, and J.J. Shaw. 1990. The importance of rapid diagnosis of new cases of cutaneous leishmaniasis in pinpointing the sandfly vector. Trans R Soc Trop Med Hyg 84:786.

Sasa, M. 1976. Human filariasis. University Park Press, Baltimore.

Singhasivanon, P., et al. 1999. Malaria in tree crop plantations in South-Eastern and Western provinces in Thailand. SE Asian J Trop Med Public Health 30:399–404.

Suvannadabba, S. 1991. Deforestation for agriculture and its impact on malaria in southern Thailand. *In* Sharma, V.P., and A.V. Kondrashin, eds. Forest malaria in Southeast Asia, pp. 221–226. WHO/MRC, New Delhi.

Tesh, R.B., B.N. Chaniotis, B.R. Carrera, and K.M. Johnson. 1972. Further studies on the natural host preferences of Panamanian phlebotomine sandflies. Am J Epidemiol 95:88–93.

Thomson, M.C., S.J. Connor. 2000. Environmental information systems for the control of arthropod vectors of disease. Med Vet Entomol 14:227–244.

Thomson, M.C., S.J. Connor, and P.J.M. Milligan. 1996. The ecology of malaria—as seen from Earth-observation satellites. Ann Trop Med Parasitol 90:243–264.

Thomson, M.C., et al. 1997. Mapping malaria risk in Africa—what can satellite data contribute? Parasitol Today 13:313–318.

Thomson, M., et al. 1999. Towards a kala azar risk map for Sudan: mapping the potential distributon of *Phlebotomus orientalis* using digital data of environmental variables. Trop Med Int Health 4:105–113.

Thomson, M.C., et al. 2000. Satellite mapping of *Loa loa* prevalence in relation to ivermectin use in west and central Africa Lancet 356:1077–1078.

Touré Y.T., V. Petrarca, S.F. Traore, A. Coulibaly, H.M. Maiga, O. Sankare, M. Sow, M.A. Di Deco, and M. Coluzzi. 1998. The distribution and inversion polymorphism of chromosomally recognized taxa of the *Anopheles gambiae* complex in Mali, West Africa. Parasitologia 40:477–511.

Turner, D.A. 1981. The colonization by the tsetse *Glossina pallidepes* Austen of a unique habitat—exotic coniferous plantations with special reference to the Lambwe Valley, Kenya. Insect Sci Appl 1:243–248.

Walsh, J.F., D.H. Molyneux, and M.H. Birley. 1993. Deforestation effects on vector-borne disease. Parasitology 106(suppl): S55–S75.

Ward, R.D., J.J. Shaw, R. Lainson, and H. Fraiha. 1973. Leishmaniasis in Brazil: VIII. Observations on the phlebotomine fauna of an area highly endemic for cutaneous leishmaniasis in the Serra dos Carajas, para State. Trans R Soc Trop Med Hyg 67: 174–183.

Ward, R.D., et al. 1988. The *Lutzomyia longipalpis* complex: reproduction and distribution. Biosys Haematophagus Insects 37:257–269.

World Health Organization (WHO) 1994. 20 years of onchocerciasis control. Onchocerciasis Control Programme in West Africa, WHO, Rome, Italy

16

Ecological Context of Lyme Disease
Biodiversity, Habitat Fragmentation, and Risk of Infection

Richard S. Ostfeld

Felicia Keesing

Eric M. Schauber

Kenneth A. Schmidt

Rapid, human-induced erosion of biological diversity may have enormous consequences for the transmission of infectious diseases to humans. The consequences of biodiversity loss for human health are only beginning to be understood. The development of theory linking biodiversity to risk of infectious diseases has been slow, and empirical studies based on a solid theoretical foundation are few. How might biodiversity loss influence the probability of human exposure to infectious diseases? At one extreme, imagine that all the natural habitats surrounding a town are converted into parking lots, resulting in a massive loss of local biodiversity. Such a change might be expected to wipe out many diseases simply because the pathogens and their reservoirs and vectors would cease to exist. In such a case, biodiversity loss would reduce disease risk. However, another scenario might entail the conversion of diverse natural vegetation to much less diverse agricultural fields, optimizing the habitat for outbreaks of commensal mice and leading to epidemics of diseases for which the mice are reservoirs (e.g., Mills and Childs 1998; Keesing 2000). Loss of biodiversity in this landscape would increase disease risk. Whether the loss of biodiversity decreases or increases disease risk is thus likely to depend on which species are lost or reduced, the spatial patterns and magnitude of diversity loss, and the consequences of diversity loss for the dynamics of the remaining species, including the pathogens and parasites. Progress in understanding and predicting the consequences of eroding biodiversity for human health will require the development of a strong theoretical foundation combined with rigorous empirical studies.

In this chapter, we focus on linkages between community ecology and risk of human exposure to zoonotic diseases. Zoonotic diseases are those in which a pathogen normally resides in one or more species of animal hosts but may be transmitted to humans either directly (e.g., via a bite or scratch) or indirectly (e.g., via aerosols, contaminated water or food, or an arthropod vector). We focus on Lyme disease, a tick-borne spirochetal disease of humans, as a model system. As a vector-borne zoonosis, Lyme disease is an appropriate model of how complex interactions within a diverse community of organisms, and between the organisms and their habitats, influence human health.

Below, we describe the natural history of Lyme disease and introduce the key organisms involved in the epizootic. We then describe how biodiversity and the interactions among species in ecological communities may influence risk of human exposure to this disease, and the effects of habitat destruction and conversion on some key species of hosts for ticks, with some possible outcomes for disease incidence in humans. Finally, we discuss how the likelihood that animal reservoirs will transmit diseases directly or indirectly to humans (i.e., their reservoir competence) may be influenced by the composition of host communities. Our general objectives are to explore the issue of how interactions among species in ecological communities may influence disease risk and stimulate further multidisciplinary research into the intersection of community ecology and epidemiology.

16.1 Natural History of Lyme Disease

Lyme disease is caused by a spirochete bacterium, *Borrelia burgdorferi*, which is transmitted to humans during blood meals taken by some species of ixodid ticks. In eastern and central North America, the vector is the black-legged tick, *Ixodes scapularis*, whereas in western North America, Europe, and Asia, the most common vectors are *I. pacificus*, *I. ricinus*, and *I. persulcatus*, respectively (Lane et al. 1991; Piesman and Gray 1994; Randolph and Craine 1995). All these ixodid hosts have three active life stages: larva, nymph, and adult, each of which takes a single blood meal from a vertebrate host before dropping off the host and either molting into the next life stage (in the case of larvae and nymphs) or reproducing and dying (in the case of adults). These blood meals represent opportunities both for the tick to acquire an infection from its host and for it to transmit an infection to its host (Mather 1993; Ostfeld 1997).

The phenology of tick activity and infection is fundamental to the existence of Lyme disease. Due to highly inefficient transovarial passage of the spirochete, larval ticks typically hatch from eggs uninfected with *B. burgdorferi* (Piesman et al. 1986; Patrican 1997). Newly hatched larval *I. scapularis* seek a host in midsummer and will feed from a wide variety of mammalian, avian, and reptilian hosts (Lane et al. 1991; Fish 1993). After their blood meal, replete larvae drop off the host and molt into nymphs, which undergo winter diapause and begin to seek hosts the following spring or summer. Similar to larvae, nymphs are generalist parasites that will feed on dozens of species of host. Fed nymphs drop off their hosts and molt into adults, which are active three to four months later, in the autumn. Adult ticks are much more specialized than either larvae or nymphs, feeding predominantly on white-tailed deer (Fish 1993; Ostfeld et al. 1996, Wilson 1998).

A larva that feeds on an infected host may acquire an infection and molt into an infected nymph, which is likely to transmit the infection to a host during its nymphal meal. Between 0% and 40% of host-seeking nymphal black-legged ticks are infected with *B. burgdorferi*, depending on locality (Ginsberg 1993; Piesman and Gray 1994). This percentage is higher in host-seeking adult ticks, because adults have had two opportunities (one each during their larval and nymphal meals) to acquire an infection. Therefore, both adults and nymphs are capable of transmitting Lyme disease to humans. However, despite the generally higher infection prevalence of adults, the majority of Lyme disease cases are transmitted by nymphs, because nymphs are (1) active in summer, when human outdoor activity is highest; (2) considerably smaller and more difficult to detect before and during blood meals than are adults; and (3) less specialized in host selection and more likely to attempt a blood meal from a human host (Barbour and Fish 1993; Dister et al. 1997).

What determines whether a larval tick will acquire an infection during its blood meal and thus molt into an infected nymph? Different species of tick hosts tend to have different probabilities of transmitting an infection to a feeding tick. In eastern and central North America, the host most likely to transmit an infection to a feeding tick is the white-footed mouse (*Peromyscus leucopus*) (Levine et al. 1985; Mather et al. 1989; Schmidt and Ostfeld 2001). Recent studies have revealed that between 40% and 90% of larval ticks feeding on white-footed mice become infected (Mather et al. 1989; Schmidt and Ostfeld 2001). Another host with a high reservoir competence is the eastern chipmunk (*Tamias striatus*), which infects between 20% and 80% of feeding larvae (Mather et al. 1989; Schmidt and Ostfeld 2001). Most other mammalian, avian, and reptilian hosts have a considerably lower reservoir competence.

16.2 Metrics of Lyme Disease Risk

Two ecological metrics of disease risk are relevant to the epidemiology of Lyme disease. The clearest ecological risk factor is the density of infected nymphs (DIN) occurring in areas that people use recreationally or domestically (Maupin et al. 1991; Barbour and Fish 1993). This metric determines the probability that a person will encounter a tick capable of spreading infection, assuming the person enters tick habitat. However, DIN incorporates two factors, density and infection prevalence, which may be determined by different biological processes. Density of nymphs may be a complex function of (1) composition of the community of hosts for larvae, which determines average feeding success for larvae and therefore survival to the nymphal stage; (2) abundance of white-tailed deer, which is the predominant host for adult black-legged ticks (Wilson 1998), and which therefore is critical for maintaining a tick population (Telford, chap. 24 in this volume); and (3) abiotic (climatic) variables that influence tick survival when off hosts (Bertrand and Wilson 1997).

The other primary ecological risk factor is nymphal infection prevalence (NIP), which is largely independent of the population density of ticks (for an exception, see Reservoir Competence, below) and should be a simple function of the distribution of tick meals on the host community. Because variation in NIP will determine the probability that a given tick bite will result in a case of Lyme disease, NIP is an important risk factor. Both NIP and DIN are valuable in determining ecological causes of variation in risk of exposure to Lyme disease and to other vector-borne diseases.

16.3 Biodiversity, Community Ecology, and the Dilution Effect

In forested landscapes of the eastern and central United States, the white-footed mouse is typically one of the most abundant vertebrates. Imagine a larval tick that has recently hatched and is waiting on the forest floor, nearly immobile, for a potential vertebrate host to approach. In habitats or years of high *P. leucopus* density, the tick will have a high probability of encountering a mouse. Owing to the high reservoir competence of mice, this tick will likely become infected during its larval meal and molt into an infected nymph, dangerous to people the next spring or summer. Two situations should reduce the probability that the questing larval tick will encounter a white-footed mouse. One is a reduction in the population density of mice. In fact, multiyear monitoring of the population density of both ticks and white-footed mice in oak forests of New York State reveals a positive correlation between mouse density one year and the infection prevalence of nymphs the following year, indicating that fewer larval ticks feed on mice in years of low mouse abundance (R. Ostfeld et al., unpublished data 2001). The other situation that will reduce encounter probabilities with mice is an increase in the number (or relative abundance) of nonmouse hosts in the forest. Such an increase in host diversity would enhance the probability that the tick will feed on a nonmouse host. Larvae that feed on other mammalian, avian, or reptilian hosts are less likely to become infected with *B. burgdorferi* and therefore less likely to become dangerous when they feed as nymphs one year later.

Ostfeld and Keesing (2000) termed this second situation the "dilution effect," because the presence of a high diversity of hosts for ticks dilutes the impact of white-footed mice and reduces disease risk, as measured by nymphal infection prevalence. Some empirical support exists for the dilution effect. First, Schmidt et al. (1999) reported that, in years of high abundance of the eastern chipmunk (which is another common tick host), fewer larval ticks infested mice, which suggests that chipmunks are capable of diluting the effect of mice. More generally, this observation indicates that when nonmouse hosts are abundant, they may deflect tick meals away from mice. The dilution effect concept is related to, but distinct from, the concept of zooprophylaxis. Zooprophylaxis, or the use of animals (typically livestock) to deflect vector (typically mosquito) meals away from humans, differs from the dilution effect in the following ways. (1) Zooprophylaxis is concerned with reducing the probability that the vector will bite a human, whereas the dilution effect is concerned with the distribution of vector bites on more competent versus less competent reservoir hosts, and assumes no change in the probability that an individual vector will attack a human. (2) Zooprophylaxis does not apply to infection prevalence of vectors. (3) Zooprophylaxis does not apply to the diversity of hosts, only to substitutes for humans. (4) Zooprophylaxis applies to anthroponotic diseases (i.e., those in which humans are the prinicipal disease reservoir), whereas the dilution effect applies to zoonotic diseases.

Second, Schmidt and Ostfeld (2001) compared the observed infection prevalence of host-seeking nymphs (NIP = 38%) at their New York study sites to the infection prevalence expected if all larvae fed only on the most competent reservoir species at those sites (between 69% and 93%). This expected value of NIP was derived empirically from the reservoir competence of mice and chipmunks measured at the same sites. From the discrepancy between observed and expected NIP, they inferred that 61% of larval tick meals must be taken from poor reservoir species, that is, from hosts other than mice and chipmunks. This inference demonstrates that the impact of white-footed mice on NIP must be diluted by a group of nonmouse hosts.

Third, Ostfeld and Keesing (2000) analyzed the relationship between the number of species (species richness) of ground-dwelling birds, small mammals, and lizards in the states of the eastern United States and Lyme disease incidence in those states. They regressed the number of cases of Lyme disease per 100,000 population, as reported by state health departments to the Centers for Disease Control and Prevention, against species richness of the major vertebrate hosts for ticks in each state. They found that states with few species of small mammals and lizards had variable, but often high per capita incidence of Lyme disease. Those states with the highest species richness consistently had the lowest incidence of Lyme disease, suggesting that the dilution effect operates at broad geographical scales.

To summarize, several lines of evidence strongly suggest that the dilution effect operates to reduce NIP and therefore the probability that Lyme disease will result from any particular tick bite. These include the ability of increasing numbers of chipmunks to reduce average tick burdens on white-footed mice, the observation that NIP is dramatically lower than would be expected if larvae fed exclusively on the most competent reservoirs, and a strong reduction in Lyme disease incidence in regions with high small-mammal diversity.

Nevertheless, the possibility remains that high diversity in the community of vertebrate hosts will increase the total number of feeding opportunities for larval ticks, and thus increase tick survival probability from the larval to the nymphal stage. If so, the density of nymphs may be higher where host diversity is higher, even though their infection prevalence will be reduced, and the net effect could potentially be an increase in DIN. Schmidt and Ostfeld (2001) analyzed this possibility in a series of mathematical and simulation models. A simple simulation model assembled vertebrate communities consisting of 6–12 species of hosts (plus mice and chipmunks), assuming that the species do not interact with one another and that numerical dominance in the community is unrelated to reservoir competence. The simulated species were drawn from a pool of hosts having a uniform distribution of reservoir competence (between 0.0 and 0.20) and an equal probability of providing a blood meal to larval ticks. Under these simplifying assumptions, indeed, communities with greater species richness provided the tick population with more feeding opportunities and contributed to higher DIN.

However, under a set of more realistic assumptions, the effect of increasing species richness on DIN reversed. The species most likely to be added to communities of forest floor vertebrates as richness increases are carnivorous and omnivorous mammals, granivorous and insectivorous birds, and insectivorous lizards. These species are likely to have a net antagonistic effect on white-footed mice via either predation on mice (e.g., carnivorous mammals) or competition with mice (omnivorous and granivorous species). Therefore, in these more realistic simulations, each species added reduced the relative dominance of white-footed mice by 10% on average. The result was that DIN decreased dramatically with increasing species richness. Interestingly, the relationship between host richness and DIN was curvilinear, with a steeply increasing slope when species were added to species-poor communities, but a more shallow slope when species were added to species-rich communities. If this simulation accurately represents natural dynamics, then loss of species from already species-poor communities will have a disproportionately heavy impact on disease risk.

The dilution effect appears to be a general phenomenon, not restricted to the Lyme disease system (R.S. Ostfeld and F. Keesing, unpublished data 2001). For the dilution effect to apply to a vector-borne zoonosis, the following conditions must hold: (1) the vector must be a generalist that parasitizes at least several species of host, including humans; (2) hosts parasitized by the vector must vary strongly in their reservoir competence, such that some are highly infective and others are dilution hosts; (3) vectors must acquire the pathogen via blood meals rather than relying predominantly on transovarial transmission; and (4) the most competent reservoir host(s) must be dominant members of the host community, feeding a high proportion of the tick population. A correlary of condition is that (4) rarer host species that will tend to occur only in more diverse communities will tend to be poor reservoirs for the pathogen. The extent to which these conditions are met is the subject of ongoing assessments.

16.4 Habitat Destruction and Alteration

Many forested landscapes in eastern and central North America are fragmented as a result of conversion of forest into suburban developments and agricultural fields. The result is a series of landscapes in which a gradient of forest patches exists, from small wooded lots under a hectare to expanses of continuous forest. Recent field studies in Indiana and Illinois indicate that population densities of white-footed mice are considerably higher in forest patches than in continuous forest (Nupp and Swihart 1996, 1998; Krohne and Hoch 1999), and that mouse density tends to be inversely correlated with patch size (Krohne and Hoch 1999). The ecological mechanisms behind this pattern are not entirely clear, but two main possibilities exist. The first is that the proliferation of edge habitat that occurs when landscapes become fragmented increases overall habitat quality for white-footed mice. This may be particularly pronounced in agricultural landscapes, in which mice take refuge in wooded lots but are able to exploit both the forest and adjoining agricultural fields for foraging (Van Buskirk and Ostfeld 1998). The second possible mechanism is that natural enemies of mice, such as carnivores and raptors, decline or disappear from forest habitat in highly fragmented landscapes (Rosenblatt et al. 1999).

Studies by Nupp and Swihart (1996, 1998) and Rosenblatt and colleagues (1999) reveal that some rodents, such as chipmunks, gray squirrels (*Sciurus carolinensis*), and fox squirrels (*S. niger*), which may compete strongly with mice for food, decline or disappear in small forest patches. Similarly, mammalian predators on mice, such as long-tailed weasels (*Mustela frenata*), red foxes (*Vulpes vulpes*), gray foxes (*Urocyon cinereoargenteus*), and coyotes (*Canis latrans*), do not occupy small wooded lots, requiring larger expanses of forest (Rosenblatt et al. 1999). Other studies suggest that avian predators on mice, such as barred owls (*Strix varia*), are less abundant in highly fragmented landscapes than in more continuous old-growth forest (Haney 1997; Mazur et al. 1998).

Such a reduced diversity of vertebrates in highly fragmented landscapes may affect Lyme disease dynamics through two different pathways. First, the loss of vertebrate diversity may result in a high proportion of tick meals being taken from mice, thus increasing NIP. As a result, habitat alteration reduces host diversity and weakens the dilution effect. Second, reductions in predators on and competitors with mice may be responsible for increased absolute mouse density, which should raise both NIP and DIN and thus disease risk. Further studies of the interactions among habitat fragmentation, vertebrate communities, and disease risk are warranted.

16.5 Reservoir Competence as a Community Phenomenon

In the case of a vector-borne zoonosis, reservoir competence can be defined operationally as the probability that an uninfected vector feeding on an infected host will acquire an infection. Each species of host is often assumed to be characterized by a specific reservoir competence value; however, for many hosts, considerable variation exists in estimates of species-specific competence values. The causes of variation among species in reservoir competence are thought to be related to the ability of the pathogen to survive and proliferate in the tissues of hosts, which in turn is at least partly determined by the host's immune system. Variation in reservoir competence among populations of the same species is less well studied and poorly understood. Intraspecific variation in estimated reservoir competence of rodents and ground-dwelling songbirds for Lyme disease throughout the northeastern United States was recently described by Giardina and colleagues (pers. comm. 2000).

Reservoir competence for any given species of host can be described by two parameters. One is the maximal probability that a feeding vector will acquire an infection from an infected host. For pathogens that are vulnerable to attack by the host's immune system and rapidly cleared from the bloodstream, this maximal value typically will be achieved during or immediately after inoculation by an infected vector. In the weeks and months after inoculation, reservoir competence typically declines, as a result of immune response by the host and dispersal of pathogens from the bloodstream to other tissues. Therefore, the second key parameter describing reservoir competence is the rate of decay with time since inoculation.

Geographic areas with the highest human incidence of Lyme disease typically are those in which populations of ticks show a specific seasonality in activity patterns, such that larvae from one generation reach their activity peak within a month or two after the nymphs from the prior generation reached their peak (Fish 1993). An example of this pattern is the northeastern United States, where vertebrate hosts become infected by one-year-old nymphs in May–July, and then newly hatched larval ticks feed from those infected hosts in August and September. Altering the timing of larval meals with respect to either nymphal activity or reservoir decay rates would probably change the risk of Lyme disease. For instance, in Europe, seasonality in feeding of both larval and nymphal *I. ricinus* is less pronounced than for *I. scapularis* in North America, possibly contributing to the lower infection prevalence in nymphal *I. ricinus* (Sonenshine 1993; Randolph and Craine 1995; Tälleklint and Jaenson 1994).

If the reservoir competence of a host decays rapidly with time since inoculation (Nakao and Miyamoto 1993; Lindsay et al. 1997; Markowski et al. 1998), it seems likely that the average competence of a host over a season, or its *realized reservoir competence*, will be influenced strongly by how frequently it is inoculated by infected vectors. The more often it is inoculated, the more likely it is to have a reservoir competence near the maximum value, because repeated exposures may reestablish infections that would otherwise disappear. For some hosts, repeated inoculations may result in protective immunity that will prevent the return of the realized competence to the maximum value. The rate of new exposures to Lyme disease infections for any given host should be positively correlated with the abundance of infected nymphs that are seeking hosts. The DIN, in turn, is related to the probability that larval ticks fed the prior year on highly competent hosts, such as white-footed mice, versus less competent hosts. Thus, numerical dominance of a highly competent reservoir, like the white-footed mouse, within the host community may strongly affect the frequency of repeated inoculations and influence the realized reservoir competence of other host species.

Schauber and Ostfeld (2002) explored the nature of these interactions and their possible effects on both reservoir competence and NIP using mathematical models that simulate the Lyme disease system. The model assumed initially that (1) two types of hosts (A and B) exist; (2) tick and host densities are constant from year to year; (3) one cohort of each host species exists per year with no reproduction or mortality; (4) ticks are distributed among hosts at random; (5) activity of questing ticks is normally distributed across weeks (SD = 3 weeks), with a larval peak (1 larva/m^2) in late August and a nymphal peak (0.1 nymphs/m^2) in mid June; and (6) the weekly encounter rate between ticks and hosts, 0.005, is the same for both types of host and for both life stages. Host A was modeled after the white-footed mouse, with a high maximum reservoir competence (0.9) and a slow rate of competence decay (0.1 week). Host B represented what may be a more typical host, with lower maximum competence (0.5) and a faster rate of competence decay (0.5 week).

The model indicated that, at a constant abundance of host B, increasing abundance of white-footed mice increases the realized reservoir competence of host B. The shape of the curve relating host B competence to proportional abundance of host A is nonlinear, with a rapid rate of increase in host B competence as abundance of mice increases from 0 to 0.2 of the host community, but a saturating increase as mice become relatively more abundant (figure 16.1). Intuitively, the presence of abundant mice causes more frequent inoculation of host B individuals, returning them frequently toward their maximum values for reservoir competence. Including reservoir competence decay in the model also made realized reservoir competence and NIP highly sensitive to the abundance of nymphs. A high abundance of nymphs is likely to increase the frequency at which hosts are reinfected and thus keep hosts near their maximum competence.

The results of this modeling exercise suggest that field measurements of reservoir competence may vary among years and sites depending on the abundance

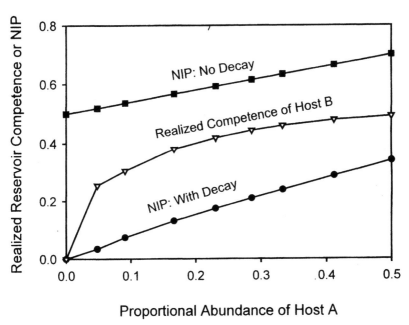

Figure 16.1 Output of a simulation model incorporating the effects of reservoir competence decay on a Lyme disease enzootic. Competence decay makes the realized competence (triangles) of a poor reservoir host, host B, sensitive to the abundance of a highly competent reservoir host, host A. In addition, competence decay reduces nymphal infection prevalence (NIP) at equilibrium (circles) far below what the model predicts without accounting for reservoir decay (squares). To produce this graph, we held the abundance of host B constant at 10 per hectare while varying the abundance of host A. See text for model details.

of ticks and other hosts. In addition, the model suggests that management actions that reduce the abundance of nymphal black-legged ticks may reap the additional benefit of reducing NIP. We highlight the need for further studies of the nature and temporal dynamics of reservoir competence. In particular, the generality of reservoir decay functions, the effect of repeated inoculations on reservoir competence, and the effects of one host species on the reservoir competence of others are in need of empirical exploration. The model demonstrates the plausibility that, under certain assumptions, reservoir competence of each species of host may be influenced by factors extrinsic to the host, such as the relative abundance of other reservoir hosts. Interestingly, demographic turnover in host populations may have similar effects on Lyme disease epidemiology as decaying reservoir competence (Schauber and Ostfeld 2002).

16.6 Conclusions

Conservation medicine is a new, multidisciplinary arena for analyzing and applying linkages between conservation biology and diseases of humans and other animals. We have described what we consider a case study of conservation medicine consisting of the interactions between diversity of forest floor vertebrates and risk of exposure to Lyme disease. We have attempted to draw connections between the causes and consequences of variation in biodiversity, which is the purview of conservation biology, and the risk of exposure to disease, which falls under the rubric of epidemiology.

We have highlighted three situations in which issues of conservation concern apply to human health. First, we discussed the dilution effect, whereby the addition of species to a community of tick hosts dilutes the impact of a dominant community host species, the white-footed mouse, and thus reduces infection prevalence of ticks. We expect the dilution effect to be a general phenomenon in vector-borne zoonoses whenever the following conditions hold: (1) the vector is a generalist parasite; (2) hosts vary substantially in their reservoir competence; (3) vectors acquire the pathogen largely through horizontal, rather than vertical, transmission; and (4) the most competent reservoir host is a dominant member of the host community. Research that assesses the generality of these four conditions is warranted.

Second, we outlined some mechanisms by which habitat destruction and fragmentation alter the composition of vertebrate communities and described the possible consequences for Lyme disease risk. Because some vertebrates that are predators on or competitors with white-footed mice are negatively affected by forest destruction, and because disturbed landscapes that contain much edge habitat may constitute high habitat quality for mice, habitat destruction may increase the number or proportion of tick meals that are taken from the most competent reservoir. Loss of some vertebrates from disturbed landscapes may have synergistic effects on disease risk by both reducing the power of the dilution effect and increasing the absolute population size of white-footed mice.

Third, we discussed a potential mechanism for a positive feedback loop between the loss of vertebrate diversity and the reservoir competence of hosts. According to this scenario, as either proportional or absolute density of white-footed mice increases, so too should the density and infection prevalence of ticks. As the density of infected ticks increases, the rate of exposure of nonmouse hosts to Lyme disease infection should increase. If poor reservoir competence is a result of a rapidly decreasing spirochetemia in hosts following inoculation, then repeated inoculations should result in a higher average, or realized, reservoir competence by maintaining elevated spirochetemias. Higher realized reservoir competence of nonmouse hosts will increase the infection prevalence of the tick population, completing the feedback loop. This argument is necessarily speculative, because few data exist to evaluate most of the assumptions.

We hope that our conceptual models of the potential impacts of community diversity and habitat destruction on disease risk will stimulate direct testing of the assumptions and predictions. Multidisciplinary interactions under the aegis of conservation medicine should result in a deeper understanding of both the ecological determinants of disease risk and practical mechanisms of preventing disease.

References

Barbour, A.G., and D. Fish. 1993. The biological and social phenomenon of Lyme disease. Science 260:1610–1616.

Bertrand, M.R., and M.L. Wilson. 1997. Microhabitat-independent regional differences in survival of unfed *Ixodes scapularis* nymphs (Acari: Ixodidae) in Connecticut. J Med Entomol 34:167–172.

Dister, S.W., D. Fish, S.M. Bros, D.H. Frank, and B.L. Wood. 1997. Landscape characterization of peridomestic risk for Lyme disease using satellite imagery. Am J Trop Med Hyg 57:687–692.

Fish, D. 1993. Population ecology of *Ixodes dammini*. *In* Ginsberg, H.S. ed. Ecology and environmental management of Lyme disease, pp. 25–42. Rutgers University Press, New Brunswick, N.J.

Ginsberg, H.S. 1993. Geographical spread of *Ixodes dammini* and *Borrelia burgdorferi*. *In* Ginsberg, H.S. ed. Ecology and environmental management of Lyme disease, pp. 63–82. Rutgers University Press, New Brunswick, N.J.

Haney, J.C. 1997. Spatial incidence of barred owl (*Strix varia*) reproduction in old-growth forest of the Appalachian Plateau. J Raptor Res 31:241–252.

Keesing, F. 2000. Cryptic consumers and the ecology of an African savanna. Bioscience 50:205–215.

Krohne, D.T., and G.A. Hoch. 1999. Demography of *Peromyscus leucopus* populations on habitat patches: the role of dispersal. Can J Zool 77:1247–1253.

Lane, R.S., J. Piesman, and W. Burgdorfer. 1991. Lyme borreliosis: relation of its causative agent to its vectors and hosts in North America and Europe. Annu Rev Entomol 36:587–609.

Levine, J.F., M.L. Wilson, and A. Spielman. 1985. Mice as reservoirs of the Lyme disease spirochete. Am J Trop Med Hyg 34:355–360.

Lindsay, L.R., I.K. Barker, G.A. Surgeoner, S.A. McEwen, and G.D. Campbell. 1997. Duration of *Borrelia burgdorferi* infectivity in white-footed mice for the tick vector *Ixodes scapularis* under laboratory and field conditions in Ontario. J Wildl Dis 33: 766–775.

Markowski, D., H.S. Ginsberg, K.E. Hyland, and R. Hu. 1998. Reservoir competence of the meadow vole (Rodentia: Cricetidae) for the Lyme disease spirochete *Borrelia burgdorferi*. J Med Entomol 35:804–808.

Mather, T.N. 1993. The dynamics of spirochete transmission between ticks and vertebrates. *In* Ginsberg, H.S., ed. Ecology and environmental management of Lyme disease, pp. 43–60. Rutgers University Press, New Brunswick, N.J.

Mather, T.N., M.L. Wilson, S.I. Moore, J.M.C. Ribeiro, and A. Spielman. 1989. Comparing the relative potential of rodents as reservoirs of the Lyme disease spirochete (*Borrelia burgdorferi*). Am J Epidemiol 130:143–150.

Maupin, G.O., D. Fish, J. Zultowsky, E.G. Campos, and J. Piesman. 1991. Landscape ecology of Lyme disease in a residential area of Westchester County, New York. Am J Epidemiol 133:1105–1113.

Mazur, K.M., S.D. Frith, and P.C. James. 1998. Barred owl home range and habitat selection in the boreal forest of central Saskatchewan. Auk 115:746–754.

Mills, J.N., and J.E. Childs. 1998. Ecological studies of rodent reservoirs: their relevance for human health. Emerg Infect Dis 4:529–537.

Nakao, M., and K. Miyamoto. 1993. Reservoir competence of the wood mouse, *Apodemus speciosus ainu*, for the Lyme disease spirochete, *Borrelia burgdorferi*, in Hokkaido, Japan. Jpn J Sanit Zool 44:69–84.

Nupp, T.E., and R.K. Swihart. 1996. Effect of forest patch area on population attributes of white-footed mice (*Peromyscus leucopus*) in fragmented landscapes. Can J Zool 74:467–472.

Nupp, T.E., and R.K. Swihart. 1998. Effects of forest fragmentation on population attributes of white-footed mice and eastern chipmunks. J Mammal 79:1234–1243.

Ostfeld, R.S. 1997. The ecology of Lyme-disease risk. Am Sci 85:338–346.

Ostfeld, R.S., and F. Keesing. 2000. Biodiversity and disease risk: the case of Lyme disease. Conserv Biol 14:1–7.

Ostfeld, R.S., C.G. Jones, and J.O. Wolff. 1996. Of mice and mast: ecological connections in eastern deciduous forests. Bioscience 46:323–330.

Patrican, L.A. 1997. Absence of Lyme disease spirochetes in larval progeny of naturally infected *Ixodes scapularis* (Acari: Ixodidae) fed on dogs. J Med Entomol 34:52–55.

Piesman, J., and J.S. Gray. 1994. Lyme disease/Lyme borreliosis. *In* Sonenshine, D.E., and T.N. Mather, ed. Ecological dynamics of tick-borne zoonoses, pp. 327–350. Oxford University Press, Oxford.

Piesman, J., J.G. Donahue, T.N. Mather, and A. Spielman. 1986. Transovarially acquired Lyme disease spirochetes (*Borrelia burgdorferi*) in field-collected larval *Ixodes dammini* (Acari: Ixodidae). J Med Entomol 23:219.

Randolph, S.E., and N.G. Craine. 1995. General framework for comparative quantitative studies on transmission of tick-borne diseases using Lyme borreliosis in Europe as an example. J Med Entomol 32:765–777.

Rosenblatt, D.L., E.J. Heske, S.L. Nelson, D.H. Barber, M.A. Miller, and B. MacAllister. 1999. Forest fragments in east-central Illinois: islands or habitat patches for mammals? Am Midland Nat 141:115–123.

Schauber, E. M. and R. S. Osterfeld. 2002. Modeling the effects of reservoir competence decay and demographic turnover in Lyme-disease ecology. Ecological Applications (in press).

Schmidt, K.A., and R.S. Ostfeld. In 2001. Biodiversity and the dilution effect in disease ecology. Ecology. 82:609–619.

Schmidt, K.A., R.S. Ostfeld, and E.M. Schauber. 1999. Infestation of *Peromyscus leucopus* and *Tamias striatus* by *Ixodes scapularis* (Acari: Ixodidae) in relation to the abundance of hosts and parasites. J Med Entomol 36:749–757.

Sonenshine, D.E. 1993. Biology of ticks, Vol. 2. Oxford University Press, New York.

Tälleklint, L., and T.G.T. Jaenson. 1994. Transmission of *Borrelia burgdorferi* s.l. from mammal reservoirs to the primary vector of Lyme borreliosis, *Ixodes ricinus* (Acari: Ixodidae), in Sweden. J Med Entomol 31:880–886.

Van Buskirk, J., and R.S. Ostfeld. 1998. Habitat heterogeneity, dispersal, and local risk of exposure to Lyme disease. Ecol Appl 8:365–378.

Wilson, M.L. 1998. Distribution and abundance of *Ixodes scapularis* (Acari: Ixodidae) in North America: ecological processes and spatial analysis. J Med Entomol 35:446–457.

17

Zoonotic Infections and Conservation

Thaddeus K. Graczyk

I n 1967, the Joint FAO/WHO Experts Committee defined zoonoses as "diseases and infections naturally transmitted between vertebrate animals and humans" (Soulsby 1974). Since then it has been recognized that zoonoses are problems in every country; they are no longer associated with rural and sylvatic environments, and represent the most frequent public health risk (Soulsby 1974). The term "anthropozoonoses" defines infections transmitted from man to animals (Soulsby 1974), for example, streptococcal meningitis (Graczyk et al. 1995), and "zooanthroponoses," infections transmitted from animals to man (Soulsby 1974). Both terms have been used interchangeably in the literature, which has generated considerable confusion; and therefore, the term "zoonoses" is commonly preferred. Zoonoses are classified into parasitic zoonoses, that is, protozooses and metazooses (trematodiases, cestodiases, nematodiases, pentastomidiases, and arthropodiases); microbial zoonoses: bacterioses, chlamydioses, rickettsioses, and viroses; and fungal zoonoses: mycoses. Obligate zoonoses include zoonoses transmitted from animals to humans, and facultative zoonoses include zoonotic infections transmitted predominantly among people or among animals (Soulsby 1974). Epidemiologically, zoonoses can be classified as (1) directly transmitted: food or water borne (Graczyk et al. 1997), contamination derived, transplacental transmitted, and blood borne (hepatitis E virus, human granulocytic ehrlichliosis [HGE]; Adachi et al. 1997; Balayan 1997); and (2) indirectly transmitted: vector borne (Strelkova 1996). Parasitic, microbial, and fungal zoonoses are common and increasingly more recognized as a significant threat to public health on a local and worldwide scale. There are several factors responsible for intensified transmission, and better detection, of life-threatening emerging and reemerging zoonotic parasites and pathogens at the end of the millennium:

- Advances in technology for human and animal transportation, which practically removed geographical barriers, have allowed humans to contact various species of animals in their natural habitats (Breitschwerdt and Dow 1999; Morris 1999).
- Changes, modifications, and thoughtless ecological interventions into stable ecosystems cause ecosystem disruption, loss of ecosystem stability and function, habitat fragmentation and degradation, and ecological disintegrity, which induce emergence of zoonotic diseases, such as Ebola virus infections (Galat and Galat-Luong 1997), or the reappearance of zoonoses believed to be eradicated (Strelkova 1996; Galat and Galat-Luong 1997; Kapel 1997; Meslin 1997; Weiss and Wrangham 1999).
- Contacts among animal species that did not interact spatially before have intensified, along with intensified association of humans with these animals through (1) animal conservation and rehabilitation centers, national parks and wildlife reserves (Fowler 1996; Hunter 1996; Graczyk et al. 1999; Nizeyi et al. 1999), zoos, aquaria, and seaworlds (Cambre and Buick 1996; Schultz et al. 1996; Williams and Thorne 1996; Hannah 1998; Michalak et al. 1998); (2) industrial animal production, nontraditional agriculture (Hunter 1996; Tully and Shane 1996; Gavazzi et al. 1997), and aquaculture (Pedersen et al. 1997); (3) increased urbanization, landscape fragmentation, and free-ranging/farmed game species and hunting (Bengis and Veary 1997; Fletcher 1997; Kapel 1997; Wilson et al. 1997; Weiss and Wrangham 1999); and (4) companion animals and pets (Graczyk et al. 1998; Premier and Jakob 1998).
- Locally and globally increased susceptibility of humans to various infectious agents as a complex epidemiology has resulted from (1) HIV pandemic and AIDS consequences (Breitschwerdt and Dow 1999; Weiss and Wrangham 1999), (2) advances in cancer diagnosis and intensification of immunosuppressive therapy for cancer and organ transplant patients (Morris 1999), (3) certain medical conditions (e.g., diabetes) that increase susceptibility to bacterioses (Lynch et al. 1998), and (4) effect of some agricultural and industrial pollutants, impairing human immune systems (Morris 1999).
- Significant shortage of donated human organs and medical advances in (1) genetically engineered organs produced in animals and (2) clinical xenotransplantation, which significantly increases potential for xenoses (= zoonoses acquired via xenotransplantation, e.g., transmissible bovine spongiform encephalopathies [TBSE], or HGE) in recipient posttransplant populations (Adachi et al. 1997; Borie et al. 1998; Platt 1998), can lead to generation of new pathogens (Auchincloss and Sachs 1998).
- There is a significant deficiency in management of zoonotic diseases (Franklin et al. 1997) in general (1) in the area of preventive medicine practices, which results from poor understanding and recognition of the reservoir(s) of zoonotic parasites and pathogens (i.e., natural host range; Franklin et al.

1997; Morris 1999), and (2) due to lack of public knowledge of transmission of some infectious agents, for example, TBSE, for economic and political reasons, hampering public access to scientific data (Kunkel et al. 1998).

- There is also a paucity of scientific data, which makes zoonoses poorly understood as a public health problem (Magras et al. 1997; Ferrari and Torres 1998), and lack of integration between public health and veterinary services (Eckert 1997a; Ralovich 1997; Breitschwerdt and Dow 1999). It is commonly assumed that veterinarians are responsible for animal and public health management regarding zoonoses (Agrimi and DiGuardo 1997; Eckert 1997a; Hannah 1998). Consequently, symptomatic rather than etiologic diagnoses are made by clinicians, particularly in pediatric and geriatric populations, and thus a considerable number of zoonotic disease cases go undiagnosed or misdiagnosed (McCrindle et al. 1996).

- Advanced molecular epidemiology techniques, for example polymerase chain reaction genotyping or DNA fingerprinting have been developed for diagnosis and identification of infectious agents, surveillance, and epidemiologic/epizootiologic investigations (Mercer et al. 1997; Heller et al. 1998; Michalak et al. 1998; Breitschwerd and Dow 1999; Sharp 1999).

- Funding of research focused on infectious diseases has been enhanced as a direct effect of recognition of the importance of the HIV pandemic (Breitschwerd and Dow 1999; Weiss and Wrangham 1999), as well as advances in dissemination of information on emerging and reemerging diseases and a common electronic access to information and databases (table 17.1).

The amount of available information and the dynamics of new information on emerging and reemerging zoonotic diseases make it difficult to present a comprehensive list, review, or update (Graczyk 1997a). It is suggested for fish or crustacean-borne zoonoses to consult Eckert (1997b) and Pedersen and colleagues (1997); reptile/amphibian-derived zoonoses, Ippen and Zwart (1996) and Johnson-Delaney (1996); avian zoonoses, Tully and Shane (1996) and Morris (1999).

Table 17.1. Web sites on infectious diseases and agents and emerging and reemerging pathogens

Web site	Address
Centers for Disease Control and Prevention (CDC)	http://www.cdc.gov
Morbidity and Mortality Weekly Reports	http://www.cdc.gov/mmwr/mmwr.html
Emerging Infectious Diseases	http://www.cdc.gov/ncidod/EID/eid.html
Emerging/Reemerging Infections Network	http://info.med.yale.edu/eph/eip/index.html
The Veterinary Medicine Virtual Library	http://netvet.wustl.edu/Vetmed.html
National Foundation for Infectious Diseases	http://www.medscape.com/Affiliates/NFID
Zoonosis references	http://med-med1.buedu/dshapiro/zooref.html
U.S. Department of Labor, Occupational Health, and Safety Administration (OSHA)	http://www.osha-scl.gov
Federation of American Scientists	http://www.healthnet.org/programs/pro-med.thml

Zoonoses represent significant public health problems in animal conservation and rehabilitation, animal agriculture, and the meat industry (Corry and Hinton 1997; Eckert 1997a, b; Patronek 1998). The competing conceptions of risk, a new approach proposed for animal agriculture, considers coping with risk and uncertainties related to transmission of zoonoses (mainly TBSE), microbial contamination of raw food products, and inactivation of zoonotic pathogens in waste management (Kunkel et al. 1998). The advantages of this approach rely on the recognition and characterization of these risks and subsequent development of proactive preventive practices (Kunkel et al. 1998). In terms of managing zoonotic infections, animal conservation and rehabilitation are challenged with similar issues of magnitude as the agricultural systems and the meat industry. However, vigilance against zoonotic disease in animal conservation and rehabilitation is maintained through industry and government-mandated sanitation standards, which are fortified by reporting regulations of local, regional, and federal health agencies (Cambre and Buick 1996; Hannah 1998). Therefore, animal conservation and rehabilitation will continue to be confronted, similarly as animal agriculture (Kunkel et al. 1998), with issues of policy in the face of uncertainties related to zoonotic diseases and their prevention (Cambre and Buick 1996; Schultz et al. 1996; Agrimi and DiGuardo 1997; Hannah 1998).

Conservation of wild animals in their natural habitats represents a unique challenge for management of zoonoses (Schultz et al. 1996; Graczyk and Cranfield 2000; Graczyk et al. 1999; Nizeyi et al. 1999). Most zoonotic parasites and pathogens (e.g., helminths) display three distinctive life cycles: sylvatic, zoonotic, and anthroponotic (Kapel 1997; Graczyk 1997b; Graczyk and Fried 1998a,b). Thus, the pathogens eradicated from human and domestic animal populations (e.g., through anthelmintic programs) can survive in sylvatic habitats of national parks or wildlife refuges and then subsequently reinvade human and domestic animal reservoirs (Kapel 1997; Graczyk and Fried 1998a,b). Most of the animal conservation operations are localized in developing countries, where public health policies and sanitation standards are not strongly established and disease control activities are often inadequate or unavailable (McCrindle et al. 1996; Cosivi et al. 1998). Therefore, many epidemiologic and public health aspects of zoonotic diseases in the areas of intensive and extensive animal conservation operations remain largely unknown (McCrindle et al. 1996; Cosivi et al. 1998; Graczyk and Cranfield 2000). There are several modes of transmission to humans of zoonotic parasites and pathogens specifically recognized in conservation and rehabilitation operations: (1) direct exposure and contact with animals through occupational and recreational contacts (Nizeyi et al. 1999); (2) exposure to vectors, (e.g., insects, ticks) transmitting zoonotic infectious agents (Ferrari and Torres 1998); (3) consumption or exposure to animal products, which includes illegal access to meat and improper utilization of wildlife products (Bengis and Veary 1997); (4) environmental contamination and insufficient or poor sanitation (Tully and Shane 1996; Graczyk et al. 1999; Nizeyi et al. 1999); and (5) unconventional transmission (Artois et al. 1996; Patronek 1998). It needs to be emphasized that many zoonotic opportunistic pathogens with well-recognized life cycles (e.g., leptospirosis, rabies, salmonellosis, toxoplasmosis, and tuberculosis) can infect humans via other means that are not well defined, for example scratching, inhalation, aerosolized urine, feces, or saliva, and blood transfusion (Artois et al. 1996).

Emergence and reemergence of zoonoses can be enhanced by climatic factors. Global warming is anticipated to have an effect on incidence and distribution of vector-borne diseases, (e.g., malaria, dengue, and arboviral encephalitis) through changing trends in average temperature and precipitation (Patz et al. 1996; Patz and Wolfe, chap. 13 in this volume; Epstein, chap. 4 in this volume). Warming trends can shift disease vectors and, consequently, distribution of vector-borne diseases to higher latitudes and altitudes (Patz et al. 1996, Molyneux, chap. 15 in this volume). Malaria is the most prevalent vector-borne disease worldwide. Local climatic changes observed in Rwanda, such as record high temperature and rainfalls in 1987, increased incidence of this disease and extended malaria occurrence to higher altitudes (Patz et al. 1996). During 1988 dengue reached an altitude of 1700 m during an unseasonably warm summer in Mexico (Patz et al. 1996). The most important predictor of dengue in Mexico is the median temperature during the rainy season; a fourfold increase is usually observed in the prevalence of this disease when the temperature rises from 17°C to 30°C (Patz et al. 1996). Temperature increase also affects the dynamics of dengue transmission by mosquitos; a direct relationship has been found between mosquito biting rates and rise of temperature (Patz et al. 1996). Epidemics of another mosquito-borne disease, arboviral encephalitis, generally occur south of the 20°C June isotherm; however, northerly outbreaks have been observed to occur during unseasonably warm years (Patz et al. 1996).

Waterborne diseases affected by global climate change include cryptosporidiosis (Graczyk et al. 2000), and diseases associated with temperature-sensitive marine ecosystem, for example, cholera and toxic algae (Patz et al. 1996). Waterborne epidemics of cryptosporidiosis, a zoonotic disease associated with livestock, coincide with unusually heavy spring rains, floods, and river swellings (Graczyk et al., 1997, 2000; Patz 1998). *Cryptosporidium parvum* is a protozoan parasite normally transmitted by the fecal–oral route that infects cattle and produces life-threatening zoonosis in people with impaired immune systems. The median infective dose (ID_{50}) for immunocompetent people is 132 oocysts, but the infection can be generated by as few as 30 oocysts (see Graczyk 1997a for review). The ID_{50} for immunosuppressed or immunocompromised individuals has not been established; however, it is thought that infection can be caused by a single oocyst (Rose 1997). The mortality rates due to *C. parvum* among these individuals varies from 52% to 68% (Rose 1997). Waterborne oocysts are long-lived and resistant to standard water disinfection (Graczyk et al. 1997). As a result, *C. parvum* has caused several massive waterborne epidemics and has become recognized as the most important biological water contaminants in the United States (Graczyk et al. 1997).

In marine environments, warm water and nitrogen favor blooms of dinoflagellates that can cause various types of shellfish poisoning diseases (Burkholder, chap. 18 in this volume; Patz 1998). Zooplankton, which feed on algae, can serve as reservoirs for *Vibrio cholerae* (Patz 1998). In Bangladesh, cholera follows seasonal warming of sea surface temperature and planktonic bloom (Colwell 1996).

Identification of the potential risk of zoonotic infections in animal conservation operations needs to be proactive in order for development of successful and self-sustainable prevention programs (Cambre and Buick 1996). Proactive identification of the zoonotic risk and the development of the best preventive management practices (BPMP) should consider (1) zoogeography of different climatic conditions (Gavazzi et al. 1997; Kapel 1997), (2) socioeconomy (Kapel 1997), (3) spatial and temporal characteristics of infectious diseases and their vectors present at the particular area (Kapel 1997), and (4) feasibility and environmental safety of preventive measures (Gavazzi et al. 1997; Kapel 1997; Stohr and Meslin 1997). At the level of national park management, there are specific issues related to wildlife conservation and ecosystem health, and zoonotic potential of many diseases. Fortunately for the animal conservation operations, some of the zoonotic disease preventive practices developed for livestock and wildlife in animal agriculture can be adopted with or without modification by the BPMP:

1. permanent surveillance of the disease and infectious agents (Meslin 1997; Ralovich 1997; Cosivi et al. 1998)
2. continuous education and special teaching focused on risk groups (Franklin et al. 1997; Cosivi et al. 1998; Graczyk et al. 1998)
3. large-scale baited vaccines and oral immunization of wildlife and feral animals against zoonotic pathogens (Stohr and Meslin 1997; Mackowiak et al. 1999; Olsen 1999)
4. mass treatment (e.g., immunization and anthelmintic actions) of livestock in areas adjacent to national parks and wildlife refuges (Eckert 1997a)
5. appropriate husbandry practices, prophylaxis, therapy, vaccines, and quarantine that prevent rather than promote zoonotic diseases (Hunter 1996; Agrimi and Guardo 1997; Olsen 1999; Williams and Thorne 1999)
6. antifertility or contraceptive vaccines for wildlife that reduce their abundance and risk for zoonotic transmission (Stohr and Meslin 1997)
7. prompt execution of public health policies, disease control activities, and sanitation standards (Franklin et al. 1997; Cosivi et al. 1998).

The future of conservation medicine in national parks will include the development of protocols for wildlife health monitoring and surveillance and evaluation of disease ecology issues related to zoonotic infections.

References

Adachi, J.A., E.M. Grimm, P. Johnson, M. Uthman, B. Kaplan, and R.M. Rakita. 1997. Human granulocytis ehrlichiosis in a renal transplant patient: a case report and review of the literature. Transplantation 64:1139–1142.

Agrimi, U., and G. DiGuardo. 1997. Study and management of zoonoses transmitted by wild animals. Ann Inst Sup Sanit 33:251–257.

Artois, M., F. Claro, M. Remond, and J. Blancou. 1996. Infectious pathology of Canidae and Felidae in zoological parks. Rev Sci Tech 15:115–140.

Auchincloss, H., and D.H. Sachs. 1998. Xenogeneic transplantation. Annu Rev Immunol 16:433–470.

Balayan, M.S. 1997. Epidemiology of hepatitis E infection. J Viral Hepat 4:155–165.

Bengis, R.G., and C.M. Veary. 1997. Public health risks with the utilization of wildlife products in certain regions of Africa. Rev Sci Tech 16:586–593.

Borie, D.C., D.V. Cramer, L. Phan-Thanh, J.L. Bequet, L. Makowka, and L. Hannoun. 1998. Microbiological hazards related to xenotransplantation of porcine organs into man. Infect Contr Hosp Epidemiol 19:355–365.

Breitschwerdt, E.G., and S.W. Dow. 1999. Why are infectious diseases emerging? *In* Bonagura, J.D., ed. Kirk's Current veterinary therapy, pp. 244–246. Saunders, Philadelphia.

Cambre, R.C., and W.W. Buick. 1996. Special challenges of maintaining wild animals in captivity in North America. Rev Sci Tech 15:251–266.

Colwell, R.R. 1996. Global climate and infectious diseases: the cholera paradigm. Science 274:2025–2031.

Corry, J.E., and M.H. Hinton. 1997. Zoonoses in the meat industry: a review. Acta Vet Hung 45:457–479.

Cosivi, O., J.M. Grange, C.J. Daborn, M.C. Raviglione, T. Fujikura, D. Cousins, R.A. Robinson, H.F. Huchzermeyer, I. de Kantor, and F.X. Meslin. 1998. Zoonotic tuberculosis due to *Mycobacterium bovis* in developing countries. Emerg Infect Dis 4:59–70.

Eckert, J. 1997a. Veterinary parasitology and human health. J Suisse Med 127:1598–1608.

Eckert, J. 1997b. Workshop summary: food safety: meat-and-fish-borne zoonoses. Vet Parasitol 64:143–147.

Ferrari, C.K., and E.A. Torres. 1998. Viral contamination of food products: a poorly understood public health problem. Pan Am J Publ Health 3:359–366.

Fletcher, T.J. 1997. European perspectives on the public health risk posed by farmed game mammals. Rev Sci Tech 16:571–578.

Fowler, M.E. 1996. Husbandry and disease of camelids. Rev Sci Tech 15:155–169.

Franklin, S., R. Pruitt, D., Willoughby, and K. Kemper. 1997. Animal exposure risk. Implications for occupational health nurses. AAOHN J 45:386–392.

Galat, G., and A. Galat-Luong. 1997. Virus transmission in the tropical environment, the socio-ecology of primates and the balance of ecosystems. Sante 7:81–87.

Gavazzi, G., D. Prigent, J.M. Baudet, S. Banoita, and W. Daoud. 1997. Epidemiologic aspects of 42 cases of human brucellosis in the Republic of Djibouti. Med Trop 57: 365–368.

Graczyk, T.K. 1997a. Epidemiology and epizootiology of *Cryptosporidium* infections. Rec Res Dev Microbiol 1:13–23.

Graczyk, T.K. 1997b. Immunobiology of trematodes in vertebrate hosts. *In* Fried, B., and T.K. Graczyk, eds. Advances in trematode biology, pp. 383–404. CRC Press, Boca Raton, Fla.

Graczyk, T.K., and M.R. Cranfield. 2001. Coprophagy and intestinal parasites: implications to mountain gorillas (*Gorilla gorilla beringei*). Primate Conserv 5:283–293.

Graczyk, T.K., and B. Fried. 1998a. Echinostomiasis: a common but forgotten food-borne disease. Am J Trop Med Hyg 58:501–504.

Graczyk, T.K., and B. Fried. 1998b. Development of *Fasciola hepatica* in the intermediate host. *In* Dalton, J.P., ed. Fasciolosis, pp. 31–46. CAB International, New York.

Graczyk, T.K., M.R. Cranfield, S.E. Kempske, and M.A. Eckhaus. 1995. Fulminant *Strep-*

tococcus pneumonia in a lion-tailed macaque (*Macaca silenus*) infant without meningeal signs. J Wildl Dis 31:75–78.

Graczyk, T.K., R. Fayer, and M.R. Cranfield. 1997. Zoonotic potential of cross-transmission of *Cryptosporidium parvum*: implications for waterborne cryptosporidiosis. Parasitol Today 13:348–351.

Graczyk, T.K., M.R. Cranfield, C. Dunning, and J.D. Strandberg. 1998. Fatal cryptosporidiosis in a juvenile captive African hedgehog (*Ateletrix albiventris*). J Parasitol 84: 178–180.

Graczyk, T.K., L.J. Lowenstine, and M.R. Cranfield. 1999. *Capillaria hepatica* (Nematoda) infections in human-habituated mountain gorillas (*Gorilla gorilla beringei*) of the Parc National de Volcans, Rwanda. J Parasitol 85:1168–1170.

Graczyk, T.K., B.M. Evans, C.J. Shiff, H.J., Karreman and J.A. Patz. 2000. Environmental and geographical factors contributing to watershed contamination with *Cryptosporidium parvum* oocysts. Environ Res 82:263–271.

Hannah, H.W. 1998. Zoos and veterinarians—some legal issues. JAVMA 213:1559–1560.

Heller, R., P. Riegel, Y. Hansmann, G. Delacour, D. Bermond, C., Dehio, F., Lamarque, H., Monteil, B. Chomel, and Y. Piemony. 1998. *Bartonella tribocorum* sp. nov., a new *Bartonella* species isolated from the blood of wild rats. Int J Syst Bacteriol 48: 1333–1339.

Hunter, D.L. 1996. Tuberculosis in free-ranging, semi-ranging and captive cervids. Rev Sci Tech 15:171–181.

Ippen, R., and P. Zwart. 1996. Infectious and parasitic diseases of captive reptiles and amphibians, with special emphasis on husbandry practices which prevent or promote diseases. Rev Sci Tech 15:43–54.

Johnson-Delaney, C.A. 1996. Reptile zoonoses and threat to public health. *In* Mader, D.R., ed. Reptile medicine and surgery, pp. 20–33. Saunders, Philadelphia.

Kapel, C.M. 1997. *Trichinella* in arctic, subarctic and temperate regions: Greenland, the Scandinavian countries and the Baltic States. SE Asian J Trop Med Publ Health 28: S14–S19.

Kunkel, H.O., P.B. Thompson, B.A. Miller, and C.L. Skaggs. 1998. Use of competing conceptions of risk in animal agriculture. J Anim Sci 76:706–713.

Lynch, M., J. O'Leary, D. Murnaghan, and B. Cryan. 1998. *Actinomyces pyogenes septis* arthritis in a diabetic farmer. J Infect 37:71–73.

Mackowiak, M., J. Maki, L. Motes-Kreimeyer, T. Harbin, and K. VanKampen. 1999. Vaccination of wildlife against rabies: successful use of a vectored vaccine obtained by recombination technology. Adv Vet Med 41:571–583.

Magras, C., M. Federighi, and C. Soulé. 1997. The danger for public health connected to the consumption of horse meat. Rev Sci Tech 16:554–563.

McCrindle, C.M., I.T. Hay, J.S. Odendaal, and E.M. Calitz. 1996. An investigation of the relative morbidity of zoonoses in paediatric patients admitted to GA-Rankuwa Hospital. J S Afr Vet Assoc 67:151–154.

Mercer, A., S. Fleming, A. Robinson, P. Nettleton, and H. Reid. 1997. Molecular genetic analyses of paramyxoviruses pathogenic for humans. Arch Virol 13:25–34.

Meslin, F.X. 1997. Emerging and re-emerging zoonoses, local and worldwide threat. Med Trop 57:7–9.

Michalak, K., C. Austin, S. Diesel, M.J. Bacon, P. Zimmerman, and J.N. Maslow. 1998. Mycobacterium tuberculosis infection as a zoonotic disease: transmission between humans and elephants. Emerg Infect Dis 4:283–287.

Morris, P.J. 1999. Zoonotic diseases of pet bird. *In* Bonagura, J.D., ed. Kirk's Current veterinary therapy, pp. 1113–1115. Saunders, Philadelphia.

Nizeyi, J.B., R. Mwebe, A. Nanteza, M.R. Cranfield, G.R.N.N. Kalema, and T.K. Graczyk. 1999. *Cryptosporidium* sp. and *Giardia* sp. infections in mountain gorillas (*Gorilla gorilla beringei*) of the Bwindi Impenetrable National Park, Uganda. J Parasitol 85: 1084–1088.

Olsen, C.W. 1999. Vaccination of cats against emerging and reemerging zoonotic pathogens. Adv Vet Med 41:333–346.

Patronek, G.J. 1998. Free-roaming feral cats—their impact on wildlife and human beings. JAVMA 212:218–226.

Patz, J.A. 1998. Climate change and Health: new research challenges. Health Environ Dig 12:49–53.

Patz, J.A., P.R. Epstein, T.A. Burke, and J.M. Balbus. 1996. Global climate change and emerging infectious diseases. JAMA 275:217–223.

Pedersen, K., I. Dalsgaard, and J.L. Larsen. 1997. *Vibrio damsela* associated with diseased fish in Denmark. Appl Environ Microbiol 63:3711–3715.

Platt, J.L. 1998. New directions for organ transplantation. Nature 392:S11–S17.

Premier, J., and W. Jakob. 1998. A case of *Echinococcus granulosis* hydatid cyst in a zoo born dromedary in the Berlin-Fridrichsfelde zoo. Berl Munch Tierarztl Wochensch 111:100–103.

Ralovich, B. 1997. Problems of microbial zoonoses in Hungary. Acta Microbiol Immunol Hung 44:197–221.

Rose, J.B. 1997. Environmental ecology of *Cryptosporidium* and public health implications. Annu Rev Public Health 18:135–161.

Schultz, D.J., I.J. Hough, and W. Boardman. 1996. Special challenge of maintaining wild animals in captivity in Australia and New Zealand. Rev Sci Tech 15:289–308.

Sharp, N.J.H. 1999. Molecular biology of infectious diseases. *In* Bonagura, J.D., ed. Kirk's Current veterinary therapy, pp. 246–250. Saunders, Philadelphia.

Soulsby, E.J.J. 1974. Parasitic zoonoses, clinical and experimental studies. Academic Press, New York.

Stohr, K., and F.X. Meslin. 1997. Zoonoses and fertility control in wildlife—requirements for vaccines. Reprod Fertil Dev 9:149–155.

Strelkova, M.V. 1996. Progress in studies on Central Asian foci of zoonotic cutaneous leishmaniasis: a review. Folia Parasitol 43:1–6.

Tully, T.N., and S.M. Shane. 1996. Husbandry practices as related to infectious and parasitic diseases of farmed ratites. Rev Sci Tech 15:73–89.

Weiss, R.A., and R.W. Wrangham. 1999. From *Pan* to pandemic. Nature 397:385–386.

Williams, E.S., and E.T. Thorne. 1996. Infectious and parasitic diseases of captive carnivores, with special emphasis on the black-footed ferret (*Mustellla nigripes*). Rev Sci Tech 15:91–114.

Wilson, M.L., P.M. Bretsky, G.H. Cooper, S.H. Egberson, H.J. VanKruiningen, and M.L. Cartter. 1997. Emergence of raccoon rabies on Connecticut, 1991–1994: spatial and temporal characteristics of animal infection and human contact. Am J Trop Med Hyg 57:457–463.

18

Chronic Effects of Toxic Microalgae on Finfish, Shellfish, and Human Health

JoAnn M. Burkholder

Near centers of increasing human population growth in estuarine and coastal areas throughout many regions of the world, outbreaks of toxic microscopic algae, or microaglae, increasingly have been noted in the past few decades (Hallegraeff 1993). Other toxic microalgae, the blue-green algae or cyanobacteria, historically have affected estuaries in certain regions, and freshwaters worldwide (Carmichael 1995). Scientists debate whether the incidence of toxic microalgal blooms is actually increasing or is perceived to be increasing because of improved reporting (Culotta 1992; Hallegraeff 1993). For example, the paralytic shellfish-poisoning dinoflagellates, which many cultures worldwide historically have known to be highly toxic, seem to have significantly increased in their geographic range within the past two decades (Hallegraeff 1993). Some scientists believe that this apparent increase reflects increased monitoring for these species, especially in economically depressed subtropical regions. Although such areas previously may well have been poorly assessed, it is difficult to believe the same of countries such as the United States, which has experienced a notable increase in the geographic range and occurrence of various toxic microalgae at least since the 1970s (figure 18.1).

The "signs of the times" might best be captured by the following statistics. In 1984 scientists recognized 22 species worldwide of the most notorious of the harmful microalgae, the dinoflagellates (Steidinger and Baden 1984). Within the past 15 years, that number has increased to more than 70 (Burkholder 1998; Hallegraeff 2000). Although some of these newly recognized toxic species are cryptic members of the "hidden flora" (Smayda 1989), many have made their presence known through major discoloration of the water, or massive fish kills, or serious health effects for people who consumed toxin-contaminated seafood—

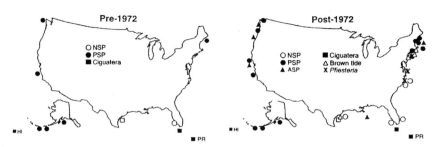

Figure 18.1

events that likely would have been unnoticed had they occurred previously in some of these geographic regions (Falconer 1993a; Burkholder 1998). Toxic microalgae other than dinoflagellates such as estuarine and marine diatoms of the toxic *Pseudonitzschia* complex—always previously considered beneficial, benign members of food webs—also have been detected because of serious, acute human health effects within the past 15 years, that have caused death of both humans and marine birds and mammals in the United States and Canada (Burkholder 1998).

18.1 Causative Factors and Management/Government Response

Among the 22 toxic dinoflagellate species that were recognized by the early 1980s, and in some cases much earlier, is *Gymnodinium breve*, the infamous red tide of Gulf Coast waters, which initiates bloom development offshore, where it appears to be little influenced by human activities (Steidinger 1993). Outbreaks of certain species in this group, and of various species among the newly recognized toxic dinoflagellates, are strongly correlated with nutrient pollution in quiet, poorly flushed coastal embayments and estuaries (Smayda 1989; Hallegraeff 1993). The affected areas generally have coincided with areas of increasing human population growth (Bryant et al. 1995) and associated nutrient pollution from poorly treated sewage and other sources (Burkholder 1998). Some species, such as the toxic *Pfiesteria* complex and many toxic cyanobacteria, are known to thrive in degraded waters overenriched with nutrients (Burkholder and Glasgow 1999). Other species, such as the toxic *Pseudonitzschia* complex, are most abundant in waters that have been enriched in nutrients from upwelling processes or from sewage and other anthropogenic nutrient additions (Burkholder 1998).

Aside from nutrient pollution—which clearly has positively influenced the growth of some toxic microalgae but has had little apparent effect on others—few explanations exist for the appearance of these and other newly known toxic algal species. Global climate change and the associated high variance of weather patterns (e.g., strong el Niño events) have been invoked as factors contributing to changes in aquatic ecosystem communities that could, perhaps, have enabled these organisms to become more easily established. Warm-optimal toxic microalgae, such as many cyanobacteria and certain dinoflagellates, may be stimulated by recent warming trends, leading to increased growth and range expansion.

In general, however, the environmental factors that are optimal for growth are poorly known for most toxic microalgae, other than for cyanobacteria, which have been well studied from the perspective of nutrient stimulation. Moreover, despite the fact that certain toxic microalgal species have been intensively researched for the past 50 years, the molecular and environmental signaling processes that interact to initiate and control toxin production remain unknown for all toxic microalgae (Burkholder and Springer 1999). Indeed, information as fundamental as the life cycles of many of these toxic organisms has remained beyond reach (Steidinger 1993; Burkholder 1998). Tracking tools amenable to field application, such as certain types of species-specific molecular probes, remain to be developed for many toxic algal species (Anderson 1995). Without this basic understanding, it is not yet possible to recognize various forms of the same species in field samples (Popovský and Pfiester 1990), and it is difficult or impossible to track those forms between the water column and the benthos or along major current trajectories.

Thus, the inevitable demands by local, regional, and federal governments to predict and control toxic outbreaks remain, at best, premature for most toxic microalgal species (Boesch et al. 1997). These outbreaks are often sporadic in time and space (Culotta 1992). Water resource protection is grossly underfunded worldwide, with low prioritization for such funding even in industrialized nations (World Resources Institute 1994; Miller 2000). Governments rapidly lose interest if blooms do not return year after year, and the funding is cut for research to determine the long-term presence of toxic microalgae, the chemical identity of their toxins, and the environmental factors conducive to their outbreaks (Culotta 1992; Burkholder 1998).

Compounding the inadequate funding issue is the fact that many countries have initially resisted or denied the unwelcome news of a toxic microalgal outbreak in their waters (e.g., Burkholder 1998). A common underlying reason is that the "economic halo" effect of such outbreaks can do serious damage to seafood, tourism, and other industries. For example, the first toxic *Alexandrium* outbreaks in New England during 1972 actually contaminated only certain shellfish species in localized coastal areas (Shumway 1990). However, consumers rapidly reached the false conclusion that their health would be protected only if they avoided all seafood from the United States and even from other nations. Grocery stores were unable to sell most seafood products, so the impact halo effect reached well beyond New England to other geographic regions. While agencies' reluctance to recognize and proactively address toxic microalgal outbreaks is perhaps understandable from an economic perspective, such posture can result in failure to protect public health, in this and other environmental issues (Burkholder 1998).

The sporadic and often haphazard attention given by regulatory agencies and legislatures during and shortly after the immediate crisis of a toxic outbreak characterizes the management strategy that historically has been practiced by many nations in response to certain groups of toxic microalgae (Anderson 1995; Burkholder 1998). Well-known exceptions involve monitoring programs that have been established in some countries (e.g., Sweden, France, Spain, some U.S. states, Canada, Japan, Australia) for certain marine toxic species that are known to cause human death or acute illness and death (Hungerford and Wekell 1993; Cembella and Todd 1993). The most extensive proactive monitoring programs for marine biotoxins in the United States target paralytic shellfish poisoning (PSP), caused by saxitoxin-producing dinoflagellates (Hungerford and Wekell 1993; table 18.1). These programs are designed to be expandable because of the cost of maintaining high effort in years without outbreaks, and because the outbreaks can be highly variable in timing and location. For example, in response to PSP outbreaks that occurred later in the year than usual, and in new areas, the West Coast states of the United States made rapid adjustments during the early 1990s in efforts to monitor for PSP by increasing the number of sampling sites and extending sampling periods (Hungerford and Wekell 1993). Outbreaks of amnesic shellfish poisoning (ASP), caused by toxic diatoms, were first known to occur in Canada during the late 1980s. The toxin involved, domoic acid, was identified within less than a week by Canadian scientists, and assay techniques were quickly developed and initiated to aid in monitoring or early warning systems (Cembella and Todd

Table 18.1. Some known intoxications of humans from microalgal toxins via seafood exposure route

Type	Acronym	Distribution	Causative Agent
Amnesic shellfish poisoning (domoic acid, a potent neurotransmitter)	ASP	Cosmopolitan (North America, Europe, Asia)	Diatoms (toxic *Pseudonitzschia* complex, including *Nitzschia* sp.; *Amphora coffeaeformis*)
Ciguatera finfish poisoning (polyether toxins and derivatives, mostly poorly characterized)	CFP	Tropical/subtropical cosmopolitan	Dinoflagellates (*Gambierdiscus toxicus*; cohorts *Prorocentrum lima*, *P. concavum*, *P. hoffmanianum*, *Ostreopsis lenticularis*, *O. siamensis*; blue-green *Trichodesmium erythraea*
Diarrhetic shellfish poisoning (polyether okadaic acid; dinophysistoxins, pectenotoxins, yessotoxins, and derivatives)	DSP	Cold- and warm-temperate Atlantic, Pacific, Indo-Pacific	Dinoflagellates (some *Dinophysis* spp., *Prorocentrum lima*)
Neurotoxic shellfish poisoning (ca. nine polyether brevetoxins and derivatives)	NSP	Subtropical/warm-temperate; recent NSP-like incidence in New Zealand	Dinoflagellates (*Gymnodinium breve*, at least one other *Gymnodinium* species)
Paralytic shellfish poisoning (at least 18 carbamate, decarmoyl, and sulfocarbamyl toxins and derivatives)	PSP	Cosmopolitan (cold- and warm-temperate, tropical/subtropical)	Dinoflagellates (toxic *Alexandrium* complex; *Gymnodinium catenatum*, *Lingulodinium polyedrum*, *Pyrodinium bahamense* var. *compressum*

Source: Modified from Burkholder (1998) and Hallegraeff (2000).

1993). Similar systems are now in place within affected areas of the United States and certain European nations.

However, as agency personnel in Alaska, California, and other states, and in various coastal areas of Canada have noted, even in relatively wealthy industrialized nations, the monitoring programs suffer from attrition when a toxic outbreak has not occurred in the previous year, and they are usually severely reduced or eliminated when two or more years elapse without a toxic outbreak (Anderson 1995; Hallegraeff et al. 1995). Furthermore, the ultimate destination of federal funding to address toxic microalgal outbreaks can be strongly politically directed away from where the toxic outbreaks actually occur.

Few monitoring programs exist for toxic algal outbreaks in developing nations. For example, ciguatera, a collection of symptoms caused by the toxins of certain dinoflagellates, is the primary known source of finfish seafood poisoning worldwide (Bagnis 1993). The lack of a generally accepted, sufficiently ciguatoxin-specific method for detection (Cembella et al. 1995a,b) and the expense involved in attempts to monitor ciguatoxin in economically depressed subtropical countries have prevented establishment of proactive monitoring programs in most affected regions. Instead, pamphlets advise tourists visiting such countries not to eat large predatory fish such as barracuda or grouper during seasons when ciguatoxin-related illnesses are known to occur (Bagnis 1993).

18.2 Recent Case Study: The *Pfiesteria* Issue

As an overall illustration of the above points, the *Pfiesteria* case has exemplified both flawed management and proactive management in one affected area. For the first seven years after toxic *Pfiesteria* was implicated in the death of well over a billion fish in the largest fish nursery ground on the U.S. Atlantic Seaboard, state health and environmental officials in North Carolina engaged in concerted attempts to undermine scientifically established effects of *Pfiesteria* (Burkholder 1998; Burkholder and Glasgow 1999; Burkholder et al. 1999b). In 1997, 1.2 million fish died in association with actively toxic *Pfiesteria* within the Albemarle-Pamlico Estuarine System in North Carolina, which had by that time sustained many toxic outbreaks involving the death of over a billion fish (Burkholder and Glasgow 1999). In that year, as well, 30,000 fish died in Chesapeake Bay, in close proximity to Washington, D.C. (Maryland Department of Natural Resources 1998). The state of Maryland acted responsibly and proactively to protect the health of its citizenry in the three small affected areas of the bay (Maryland Department of Natural Resources 1998). However, despite a massive public education campaign by the governor and his staff, people throughout the Chesapeake region erroneously concluded that seafood caught anywhere in the bay was unsafe to eat. The economic halo effect cost seafood, tourism, and other industries about US$60 million (Epstein 1998).

Of the approximately US$80 million that ostensibly were directed for the *Pfiesteria* issue by Congress and federal agencies during 1998–99, about 10% reached North Carolina the state that had sustained 99% of the toxic outbreaks (Burkholder and Glasgow 1999; records of NOAA-ECOHAB, Silver Spring, Md., 1998–99; the U.S. Environmental Protection Agency, [EPA], Washington D.C., 1998–99; the National Institutes for Environment Health Sciences [NIEHS], Research Triangle Park, N.C., 1998–99; the U.S. Centers for Disease Control and Prevention [CDC], Atlanta, Ga., 1998–99). Several laboratories received around $3 million over two years to develop molecular probes that cannot discern between toxic and benign *Pfiesteria*, while less than US$1 million was directed over five years to the one laboratory that was equipped for mass-producing *Pfiesteria* toxins and to an associated laboratory proficient in the toxin analysis. Without knowledge of the toxin identity, improved early warning systems with reliable assays for toxin detection cannot be developed, nor can improved diagnostics for toxin exposure or treatment methodologies be designed (Anderson 1995; Fairey et al. 1999). Although the waters repeatedly affected in North Carolina encompass an area more than 100 km^2, that state received about as much federal monitoring support as states where toxic *Pfiesteria* outbreaks have not been documented (NOAA Coastal Ocean Program and U.S. EPA records, 1998). The only toxic *Pfiesteria* outbreak verified in 1998 occurred in North Carolina waters during July and affected 500,000 fish (Burkholder et al. 1999a). The modest federal monitoring support that had been designated for that state early in the year did not arrive until autumn, after the "fish kill season" was over (N.C. Department of Environment and Natural Resources records, Raleigh, N.C., Sept. 1998).

18.3 Obvious, Acute Effects from Toxic Microalgae

From noxious blooms accompanied by oxygen depletion and fish kills in freshwater, estuarine, and marine aquatic food webs (Steidinger 1993; Carmichael 1995) to extreme illness and death of people who consume contaminated seafood from estuaries and marine waters (Falconer 1993a), acute effects on natural resources and human health from toxic microalgae have been documented in many geographic regions. Most historical emphasis on environmental effects from toxic microalgae has focused on finfish and shellfish kills from (1) direct toxin effects (e.g., *Gymnodinium breve* in the Gulf Coast and eastern Florida shores of the United States; the toxic *Pfiesteria* complex in Mid-Atlantic estuaries and aquaculture facilities; *Heterocapsa circularisquama* in coastal embayments used for intensive shellfish culture in Japan; *Chrysochromulina polylepis* in Scandinavian waters; Skjoldal and Dundas 1991; Steidinger 1993; Burkholder and Glasgow 1999; Matsuyama et al. 1998) or (2) indirect effects from nutrient stimulation and overgrowth of the microalgae (e.g., many species of toxic cyanobacteria in fresh waters; the cyanobacterium *Nodularia spumigena* in fresh waters and estuaries in Scandinavia and Australia; Carmichael and Falconer 1993; Skulberg et al. 1993; Carpenter and Carmichael 1995).

The most widespread form of acute human health effects from shellfish contamination by toxic microalgae is PSP, caused by certain dinoflagellates (table 18.1). Other human diseases linked to toxic marine microalgae include diarrhetic (DSP) and neurotoxic (NSP) shellfish poisoning as well as ASP (table 18.1).

These are also caused by dinoflagellates, with exception of ASP, caused by diaforms as mentioned above. A fourth group of poorly characterized dinoflagellate toxins, ciguatoxins, with exposure sustained through consumption of finfish, cause an array of physiological impacts in humans ranging from mild diarrhea to severe neurological dysfunction and death (ciguatera finfish poisoning or CFP; Bangis 1993; table 18.1).

The previously mentioned monitoring programs for PSP and ASP are in place because of repeated documentation of the potential for severe human illness or death from the toxic algae involved, which produce among the most potent biotoxins known (Falconer 1993a; Hallegraeff et al. 1995). For example, the progression of symptoms of acute, lethal exposure to PSP toxins involves numbness and tingling around lips and extremities, followed by vomiting, muscular weakening, and acute respiratory distress from paralysis of involuntary muscles (Kao 1993). If the person can stay alive for about 12 hours, there is a favorable prognosis for survival. Amnesic shellfish poisoning is named for one of the most striking symptoms caused by domoic acid, permanent memory loss similar to that found in Alzheimer disease. Exposure to near-lethal levels of this toxin has promoted copious saliva production that must frequently be cleared to prevent drowning, severe muscle cramping, and other extreme symptoms (Perl et al. 1990).

Less effort has been directed toward safeguarding human health from toxins of freshwater cyanobacteria, which under certain extreme conditions have proven lethal to humans. A recent incident in Brazil involved water that was taken from the midst of a toxic *Microcystis aeruginosa* bloom, poorly treated by a water treatment system that had not been rigorously maintained, and then given to people on dialysis for kidney failure (Jochimsen et al. 1998). The dialysis procedure effectively concentrated the toxins, leading to liver failure and death of almost 60 people. Most water treatment systems in industrialized as well as developing nations do not routinely test for the presence of cyanobacterial toxins in potable water supplies (exception, Australia; Choris and Bartram 1999). Furthermore, most water treatment systems do not use procedures that would remove these toxins from potable supplies (Carmichael 1995).

Other characterized algal toxins, such as the brevetoxins and DSP toxins, as well as partially characterized toxins such as those produced by *Gyrodinium aureolum*, some ciguatera dinoflagellates, and the toxic *Pfiesteria* complex in estuarine and marine waters, have not been known to cause human death (Falconer 1993a). However, as for PSP, ASP, and more potent CFP toxins, there is no known, reliable antidote for these toxins, so it is best to err on the side of caution in setting effectively "safe" levels of the algae that produce them. For example, closure of shellfish beds to harvest for human consumption occurs when cell densities of *G. breve* are at five or more cells per milliliter (Steidinger 1993). This approach is taken because rates of shellfish intoxification and detoxification (depuration) by toxic algae are, in most cases, directly related to the number of cells available to the animals (Shumway et al. 1995). It is also used because many algal toxins are extremely potent, and because the toxicity of individual shellfish in a given area is highly variable depending on the season, age of the animal, physiological status, and other factors (see the excellent review by Shumway et al. 1995 for additional information about acceptable toxin levels and monitoring programs worldwide). However, as Shumway and colleagues (1995) wrote regarding detection of algal toxins in seafood, "A big problem . . . is that reliable, validated chemical methods of analysis are not yet available, or cannot be easily operated because of the lack of standards and reference materials."

Shumway and colleagues (1995) also described attempts to conduct risk assessment on acute human health effects from exposure to algal toxins in seafood as largely based on little or no toxicological information. Without such data, accurate hazard assessment is not possible, even considering obvious, acute health effects. Despite many reported cases of human intoxication from consumption of toxin-laden shellfish, reliable human toxicity data are difficult to obtain because of variable sensitivity of the subjects—due to age, physiological condition, and so forth—and also because of high variability in the composition of individual toxins in the samples. Moreover, shellfish species handle toxins differently, so the exposure depends on the species of shellfish consumed and the environmental conditions in the area where they were harvested (Krogh 1988). In addition, many toxic algae produce mixtures of toxins that vary with season, age of the cell, nutritional status, and other factors (Hallegraeff et al. 1995).

18.4 Accumulating Evidence for Serious Chronic/Sublethal Effects on Fish and Humans

Chronic/sublethal effects of microalgal toxins on fish—on feeding activity, predator avoidance, osmoregulatory functioning, reproduction, recruitment (survival of the young), and disease resistance—would potentially be more serious at the population level than a localized kill from an acute exposure, which generally affects a small proportion of the population (table 18.2). The following examples illustrate some of these effects.

18.4.1 Shellfish

Removal (mortality) of parental stocks by a toxic red tide (*Gymnodinium breve*) was identified as the mechanism for recruitment failure of bay scallops (*Argopecten irradians*; Summerson and Peterson 1990). Blooms of the brown tide species *Aureococcus anophagefferens* were correlated with poor recruitment of bay scallops and slow recovery of scallop stocks, possibly through stimulation of gamete resorption in reproductive adults, failure or delay of larval settlement and metamorphosis, mortality of early life-history stages, or reduced growth of juveniles (Tettelbach and Wenczel 1993). *Aureococcus anophagefferens* also produces a bioactive dopamine-mimetic substance that can adversely affect shellfish.

Two major types of cancer in bivalve mollusks—neoplasias and germinomas (gonadal tumors)—have been related to dinoflagellate toxin exposure (Landsberg 1996). Disseminated neoplasia affects about 15 species worldwide and is common in softshell clams (*Mya arenaria*), cockles (*Cerastoderma edule*), and blue mussels (*Mytilus edulis*). This disease can contribute significantly to softshell clam

Table 18.2. Known chronic/sublethal effects of toxic microalgae and heterotrophic dinoflagellates on finfish and shellfish

Group	Effects
Finfish	
Cyanobacteria (blue-green algae: *Microcystis aeruginosa, Anabaena flosaquae, Nodularia spumigena*, many others)	Hepatic disease; clogging of gills, respiratory impairment
Diatoms (toxic *Pseudonitzschia* complex)	Hemocyte abnormalities; reduced fecundity of some zooplankton food species
Dinoflagellates (toxic *Alexandrium* complex and other saxitoxin producers; *Gymnodinium breve* and other brevetoxin producers; *Gambierdiscus toxicus* and other ciguateratoxin producers; toxic *Dinophysis* complex and other DSP spp.; toxic *Pfiesteria* complex)	Narcosis (prolonged lethargia); impaired swimming; loss of appetite; reduced fright response, enhanced vulnerability to predators; pathological changes in gill, liver, intestines; blanching of skin; disease (including gonadal tumors and neoplasias); enhanced mucus production; loss of desirable food organisms (food web effects)
Prymnesiophytes (*Chrysochromulina polylepis, Prymnesium parvum*)	Loss of selective cell permeability in gill tissue; increased mucus production; loss of desirable food organisms (food web effects)
Raphidophytes (toxic *Chattonella* complex, *Heterosigma akashiwo*)	Pathological changes in gill tissue; depressed heart rate and reduced blood circulation; edema and degenerative change of the epithelium
Shellfish	
Chrysophytes (brown-tide spp. *Aureococcus anophagefferens, Aureodoris lagunensis*)	Decreased activity of lateral cilia (apparently via a dopamine-mimetic substance); impaired recruitment timing; reduced reproduction and early development; loss of desirable food organisms (food web impacts)
Cyanobacteria (blue-green algae)	Clogging of gills; respiratory impairment
Diatoms (toxic *Pseudonitzschia* complex)	Hemocyte abnormalities
Dinoflagellates	Narcosis (prolonged lethargia); reduced fright response, reduced burrowing, enhanced vulnerability to predators; recruitment failure; direct predation on larvae; mortality of some phytoplankton species used as food (food web effects)
Prymnesiophytes	Loss of selective cell permeability in gill tissue

Source: Modified from Burkholder (1998).

population declines (e.g., in New England and Chesapeake Bays). Germinomas affect at least 10 shellfish species worldwide (Landsberg 1996). In some affected shellfish species, incidence of disseminated neoplasia has been positively correlated with outbreaks of PSP from the toxic *Alexandrium* complex (Landsberg 1996), and high levels of PSP toxins have been found in the tissues of the affected shellfish (Martin et al. 1990). Similarly, incidence of germinomas in softshell clams has been positively correlated with outbreaks of the toxic *Alexandrium* complex and of another toxic dinoflagellate, *Prorocentrum minimum* (venerupin shellfish poisoning), that is strongly stimulated by anthropogenic nutrient enrichment (e.g., Mallin 1994). Pathological effects (poorly developed digestive diverticula; attenuation of the epithelium with abnormal vacuolation and necrosis; large thrombi in the heart and in the open vascular system of the mantle, digestive diverticula, gill, and kidney tissues) and feeding inhibition have been experimentally demonstrated in bay scallops exposed to *P. minimum* (Wikfors and Smolowitz 1993, 1995).

18.4.2 Finfish

Bioaccumulation of saxitoxins and ciguatoxins, with depressed survival and recruitment, has been demonstrated experimentally for larval and adult finfish (White 1981a,b; Steidinger and Baden 1984; Robineau et al. 1991). This mechanism has been implicated in reducing finfish viability at the population level, leading to overall impairment of food web structure and function (Robineau et al. 1991; Landsberg 1995). Recent research suggests a chronic role of ciguatoxins in disease and reduced long-term survival of reef fishes (Landsberg 1995). Ingestion of the ciguatoxin producer *Gambierdiscus toxicus*, or ciguatera-contaminated tissues, has experimentally induced pathology in fish gills, liver, and intestine (Capra et al. 1988; Gonzalez et al. 1994) and behavioral changes (disorientation, loss of equilibrium, depressed feeding; Kelley et al. 1992; Magnelia et al. 1992). Landsberg (1995) invoked immune system suppression and other general physiological impairment (analogous to that observed in humans from ciguatoxin exposure; see below) in hypothesizing that ciguatoxins enable opportunistic bacteria and fungi to cause epizootics in reef fishes that consume macroalgae colonized by *G. toxicus*.

Effects on liver tissues of finfish from the estuarine blue-green alga *Nodularia spumigena* have been documented from chronic exposure to the hepatotoxin nodularin (table 18.2). This toxin inhibits phosphatases 1 and 2A, which can promote cytoskeletal disintegration, tissue injury, and tumor development at chronic levels (Runnegar et al. 1988; Falconer 1993b). In fresh waters, potent hepatotoxins from *Microcystis aeruginosa* have also been linked to fish disease.

Chronic/sublethal exposure to the toxic *Pfiesteria* complex in laboratory experiments causes a suite of symptoms in finfish. The toxins strip the epidermis and create lesions, also facilitating invasion by opportunistic microbial pathogens (bacteria, fungi, other protozoans). The osmoregulatory system, which is commonly located in the epidermis, is destroyed as well (Noga et al. 1996). Chronic exposure also significantly depresses the white blood cell count of test finfish and renders them more susceptible to infection by opportunistic microbial pathogens. A narcotizing effect of the toxins, along with depressed feeding and fright or escape response, has been experimentally observed in finfish as well (Burkholder et al. 1995)—not only from chronic exposure to *Pfiesteria* toxins, but also from chronic exposure to the toxins from PSP, NSP, and CFP dinoflagellates (Burkholder 1998; table 18.2).

18.4.3 Mammals

Most algal toxins are neurotoxins, but they also include an array of other substances with apparent multiple-organ targets (see review in Burkholder 1998; table 18.3). Known chronic effects of these toxins on mammalian health include immune system suppression, respiratory illnesses, central nervous system dysfunction with memory impairment and learning disabilities, lingering peripheral nervous system dysfunction, and promotion of malignant tumors. Experimental research with small mammals, and with both their tissues and human tissues, has yielded most of what is known about the modes of action of some of these toxins. However, the complete chemical structures and modes of action have yet to be determined for many algal toxins (e.g., Falconer 1993a; Fairey et al. 1999).

Saxitoxins and deriviatives (which cause PSP), brevetoxins (which cause NSP), and ciguatoxins (which cause CFP) all affect ion channel functioning, but each general group of toxins achieves that effect somewhat differently. Saxitoxins bind specifically to the voltage-sensitive sodium channel and block sodium ion influx in neurons. Brevetoxins cause substantial and persistent depolarization of nerve membranes by inducing a channel-mediated sodium ion influx. They depolarize muscle fiber membranes and depress acetylcholine-induced depolarization, and they induce differential release of neurotransmitters from cortical synaptosomes (Baden and Trainer 1993). In contrast, ciguatoxins open sodium channels at resting potential and prevent the open channels from being activated during subsequent depolarization. They also interfere with calcium functioning by replacing calcium ions at sites on neuroreceptors that control sodium permeability. Ciguatoxins can cause calcium ion–dependent contraction in smooth and skeletal muscle tissues; they can also promote release of norepinephrine and dopamine from pheochromocytoma cells, and they are immunogenic after coupling to serum protein (Hokama and Miyahara 1986; Baden and Trainer 1993; Bagnis 1993; Yasumoto and Murata 1993). Of these three toxin groups, ciguatoxins are considered the most widespread and least understood (Hokama and Miyahara 1986). Ciguatoxins can be neurotoxic, hemolytic, and/or hemoagglutinating (see review in Burkholder 1998). They induce an array of adverse chronic health effects in humans (Russell and Egen 1991; table 18.3). Long-term effects of CFP on humans may represent an immune sensitization phenomenon. Polycyclic ether ciguatoxins react with specific T-lymphocytes in producing abnormal immunoglobulin E, which in turn fixes onto mast cells to inhibit granule release of serotonin (Hokama and Miyahara 1986). Immunosuppression as well as tumor-promoting effects have also been documented from a chronic/sublethal exposure to a potent substance of the so-called DSP group of dinoflagellate toxins, okadaic acid, and to the blue-green algal toxin nodularin. Both of these toxins are potent protein phosphatase inhibitors (see review in Burkholder 1998). In mammalian (including human) tissues, okadaic acid suppresses interleukin-1 synthesis in peripheral blood monocytes, causes necrosis of the hippocampus region of the brain, acts as a non-phorbal ester-type tumor promotor, blocks protein synthesis in reticulocyte lysates, inhibits T-antigen– stimulated DNA replication; mimics stimulation of glucose transport

Table 18.3. Known chronic/sublethal effects of toxic microalgae and heterotrophic dinoflagellates on mammals

Group	Acronym	Effects	Linkage to Nutrient Pollution
Cyanobacteria (blue-green algae)	Cyanobacterial toxin poisoning	Malignant tumors (abdominal, uterine, thoracic); leukemia; hepatic disease; nausea, vomiting, asthma-like symptoms; swimmer's itch	+ (sewage, fertilizers, animal wastes, atmospheric fallout)
Diatoms (toxic *Pseudonitzschia* complex)	Amnesic shellfish poisoning	Central nervous system dysfunction, permanent Alzheimer-like memory loss, severe headaches, muscle spasms, mild to severe convulsions; nausea, vomiting, diarrhea	+ (sewage; N, P enrichment and silica depletion)
Dinoflagellates		Considered collectively:	
Gymnodinium breve, certain other *Gymnodinium* spp.	Neurotoxic shellfish poisoning*	Central, peripheral, and autonomic nervous system dysfunction; reversible short-term memory loss (learning disabilities and other cognitive dysfunction); malignant tumors; compromised immune system; hormonal dysfunction; asthmalike symptoms; severe headaches; joint pain; muscle spasms; altered blood pressure; nausea, vomiting, diarrhea	− (*G. breve*); n.a. for others
Certain *Dinophysis* spp., others	Diarrhetic shellfish poisoning		− (?)
Certain *Alexandrium* spp., *Pyrodinium bahamense, Gymnodinium catenatum*, others	Paralytic shellfish poisoning		− (?)
Gambierdiscus toxicus, Prorocentrum micans, others	Ciguatera finfish poisoning*		− (?)
Toxic *Pfiesteria* complex (*Pfiesteria piscicida, P. shumwayae*	*Pfiesteria* toxin poisoning		+ (sewage, swine wastes, fertilizers, N and P pollution)
Prymnesiophytes (*Chrysochromulina polylepis, Prymnesium parvum*)	—	Central nervous system dysfunction; blood disease; liver damage	? (conflicting evidence)

Source: From Falconer (1993a) and Burkholder (1998).

*Note that some cyanobacteria have saxitoxins (PSP-causing toxins) and ciguatoxins, but these cyanobacterial toxins have not been linked to mammalian disease thus far.

into adipocytes by insulin, increases production of prostaglandin E2 in peritoneal macrophages, and induces ornithine decarboxylase activity (Haystead et al. 1989; Aune and Yndestad 1993; Yasumoto and Murata 1993). The cyanobacterial toxin nodularin acts somewhat similarly and causes accumulation of tumor-promoting substances (liver, bronchogenic, abdominal carcinomas; uterine adenocarcinomas; thoracic lymphosarcomas). Nodularin, from estuarine cyanobacteria, is considered analogous to microcystin from freshwater species in chronic effects (Falconer 1993b; Ohta et al. 1994).

18.5 Pervasive Lack of Assessment of Chronic/Sublethal Effects

Chronic/sublethal effects of toxic microalgae on fish and other organisms in aquatic food webs are poorly understood, with few attempts to synthesize the fragmented data that exist on this broad subject (Bruslé 1995; Shumway 1995; Landsberg 1996; Burkholder 1998). Some fish disease monitoring programs have been initiated but usually have been maintained on a short-term basis in response to major catastrophic epizootics (e.g., Sindermann 1996). A notable exception is the fish disease monitoring network that was established in the mid 1990s at the Florida Department of Environmental Protection–Florida Marine Research Institute, working with other institutions and agencies throughout the Gulf of Mexico. More recently, toxic *Pfiesteria* outbreaks have increased public awareness that water quality, fish health, and human health can be strongly linked. Accordingly, fish disease monitoring programs have been initiated in various states of the Mid-Atlantic and Southeast United States, with similar programs planned for other Atlantic Coast states (U.S. EPA records, Washington, D.C.).

Consumption of contaminated seafood historically was regarded as the major route of human environmental exposure to neurotoxins from marine microalgae (Falconer 1993a). Toxic *Pfiesteria* outbreaks in the late 1990s broadened the focus to neurotoxic symptoms from contact with water or overlying air (Falconer 1993a; Glasgow et al. 1995; Grattan et al. 1998). Knowledge of chronic/sublethal impacts of toxic microalgae on human health will likely be strengthened from this broadened perspective about potential exposure routes. However, the information presently available is, surprisingly, almost as sparse as that for chronic effects on fish health. Biomarkers are surrogate, physiological effects known to be related to certain levels of exposure to a given toxin (Hulka et al. 1990). To determine the full range of chronic/sublethal effects on humans, they should be assessed over the long term (years) as well as the short term. Biomarkers are a critical component of any research to understand toxin effects on mammalian systems. However, in overviews of the status of knowledge about human diseases from algal toxin–contaminated seafood, Fleming et al. (1995, 1998) summarized the status of the issue:

> [T]he epidemiology of the human diseases caused by harmful marine phytoplankton is still in its infancy. In general, the epidemiology of these diseases has consisted of case reports of acute illness, sometimes as epidemic outbreaks, associated with the ingestion

of suspicious seafood. . . . Even these outbreaks are highly under-reported. . . . True incidence data are not available due to the lack of disease and exposure biomarkers in
humans, as well as the global lack of routine exposure and disease surveillance. . . .
Because diagnosis cannot be made accurately for either the clinical diseases or the
asymptomatic cases associated with these toxin exposures, it has not been possible to
investigate their true incidence in human populations. Nor has it been possible, without
human biomarkers, to . . . evaluate the true clinical course, treatment and prognosis. . . .
(p. 245)

To date, there are no programs to assess the potential for long-term (5–10 year)
effects from chronic/sublethal exposure to even the best known algal toxins for
which purified standards are available (e.g., Falconer 1993a; Fleming et al. 1995).
Some PSP cases have anecdotally reported malarialike symptoms for nearly a
decade following sublethal exposure, but the medical history of these people has
been poorly tracked. Similarly, symptoms of central nervous system dysfunction
from ciguatoxin poisoning have lingered or relapsed in affected subjects for many
years (e.g., Halstead 1988). Moreover, there is compelling evidence that ciguatera
cases, like other illnesses from microalgal toxins, are significantly underreported
because of failure to link the health symptoms to algal toxin–laden seafood or
failure to assign generalized symptoms to the correct cause (e.g., diarrhea, flulike
symptoms, severe stomach cramping, severe headaches). An additional reason is
poor communication linkages between local physicians and information centers.
For example, the contrast between the roughly 18,000 reports from physicians in
the Orient, the Caribbean, the South Pacific, and Australia and the cases actually
reported to "official" information clearinghouses indicates that the number of
"officially recognized" human cases of CFP is underestimated by half on an annual basis (Hokama and Miyahara 1986). Even surviving ASP cases—well over
100 people, some of whom were seriously affected in the toxic *Pseudonitzschia*
outbreaks in Canada during 1987—were not tracked to follow the degree of recovery from central nervous system dysfunction and other serious symptoms (Perl
et al. 1990; T. Perl, personal commun. 1998). Similarly, information is lacking
about the extent of chronic/sublethal effects from DSP toxins. Yet, as mentioned,
low-level, chronic exposure to some of the toxins in this general grouping, such
as okadaic acid, has been demonstrated to induce malignant tumors and immune
system suppression in both laboratory mammals (rats) and human tissues (Haystead et al. 1989; Sueoka and Fujiki 1998). Moreover, induction of malignancies
from exposure to other toxins (e.g., PCBs, DDT) takes time, often on the order
of 10 years or more (Miller 2000).

A few programs are currently assessing, to some degree, the potential for
chronic/sublethal effects from environmental exposure to toxic microalgae. One
of these was recently funded by CDC and NIEHS to quantify brevetoxin exposure from inhalation of the aerosolized toxins. The subjects are volunteers
who have been spending time on Florida beaches during *G. breve* red tides. Yet
brevetoxin exposure has been implicated in causing serious chronic effects such
as immune system compromise in lung tissue. This mechanism may have enabled attack by opportunistic bacteria that led to pneumonia and, eventually, to
the death of approximately 150 manatees (Bossart et al. 1998). Although breve-

toxins represent a well-known class of microalgal toxins, to date there has been no long-term effort to rigorously assess the potential for serious effects on human health from chronic, low-level environmental exposure (e.g., Fleming et al. 1995).

Another effort involves *Pfiesteria* and is more directly focused on assessment of chronic/sublethal effects. It was developed in late 1997 because the previously described toxic *Pfiesteria* outbreaks in the Chesapeake Bay attracted congressional attention. A multimillion dollar, three-year cohort study funded by the U.S. Congress through the CDC was initiated in several states to attempt to determine the extent of human health effects from environmental exposure to toxic *Pfiesteria*. However, as mentioned above, diagnostics needed to track and quantify toxin exposure and the range of effects are not yet available. *Pfiesteria* was relatively quiet during 1998–99 in comparison to its toxic activity in North Carolina waters during previous years (Burkholder and Glasgow 1999; Burkholder et al. 1995, Burkholder 1998), so no cases were known to have been exposed to verified toxic outbreaks in the first two years of the ongoing cohort study. Although most field exposure cases have not been studied (Burkholder 1998), several laboratory workers who were exposed to the putative neurotoxic aerosols from fish-killing *Pfiesteria* cultures have sustained lingering symptoms suggestive of nervous system and immune system compromise, ongoing for five to eight years following exposure (e.g., records at Duke Medical, Durham, N.C.). Thus, there is a critical need to characterize these toxins and to develop biomarkers for exposure, so that the toxin modes of action can be determined and treatment strategies can be developed.

A form of "outcome surveillance" has been established in some areas affected by harmful marine algal blooms, as central clearinghouses where people can call to report illnesses that they suspect to be related to seafood consumption. For example, from a seafood toxin hotline in Dade County, Florida, all cases fitting a clinical diagnosis of seafood toxin disease are reported to the Florida Health Department and, ultimately, to the CDC (Fleming et al. 1998). Hotlines are also available in various Mid-Atlantic and Southeast U.S. states to report incidence of fish disease or death and associated human sickness that might be related to toxic *Pfiesteria* outbreaks (U.S. EPA and NOAA-Coastal Ocean Program records, 1998–99).

18.6 Conclusions

Efforts to develop proactive, long-term, consistently funded measures to protect both people and aquatic resources from serious chronic/sublethal impacts of toxic microalgae have been lacking in industrialized as well as developing nations (Burkholder 1998). The lack of a concerted, consistent directive to assess the full range of health and resource effects from toxic microalgae is especially unfortunate, considering that there is exponential human population growth and increasing finfish and shellfish aquaculture along coastal estuarine and marine shores where many toxic algae occur (Hallegraeff 1993; DeFur and Rader 1995)—

growth that has been projected to continue for at least the next 20 years (Safina 1995). Similar population growth has been projected along major inland fresh waters used as potable supplies, where other toxic microalgae that can cause serious human health and natural resource effects (Adler et al. 1993; Carmichael and Falconer 1993; Chorus and Bartram 1999).

Beyond the lack of such programs to assess chronic/sublethal effects of toxic microalgae, and beyond the quagmire of politically rather than scientifically directed funds, is another serious impediment to progress in this issue. Unfortunately, governments frequently err on the side of economic gain in environmental health issues, despite potential or realized detriment to the health of the citizenry (e.g., O'Neill 1994; Burkholder 1998; Wilkinson 1998; Miller 2000). This problem especially affects chronic and insidious diseases, and it can translate into decades-long delays in progress on treatment (e.g., Shilts 1987; Garrett 1994). As a consequence, the management of both natural resources and human health is largely crisis driven (Miller 2000). Toxic microalgae increasingly have been at the forefront of citizens' concerns in coastal and freshwater resource management issues. The full range of their chronic/sublethal effects on fish and human health is an issue that merits concerted assessment from a proactive approach, rather than from crisis-driven actions.

References

Adler, R.W., J.C. Landman, and D.M. Cameron. 1993. The Clean Water Act 20 years later. National Resources Defense Council and Island Press, Washington, D.C.

Anderson, D.M. (ed.). 1995. ECOHAB—the Ecology and Oceanography of Harmful Algal Blooms, a national research agenda. National Oceanic and Atmospheric Administration and the National Science Foundation, Washington, D.C.

Aune, T., and M. Yndestad. 1993. Diarrhetic shellfish poisoning. *In* Falconer, I.R., ed. Algal toxins in seafood and drinking water, pp. 87–104. Academic Press, New York.

Baden, D.G., and V.L. Trainer. 1993. Mode of action of toxins in seafood poisoning. *In* Falconer, I.R. ed. Algal toxins in seafood and drinking water, pp. 49–74. Academic Press, New York.

Bagnis, R. 1993. Ciguatera fish poisoning. *In* Falconer, I.R., ed. Algal toxins in seafood and drinking water, pp. 105–115. Academic Press, New York.

Boesch, D.F., D.M. Anderson, R.A. Horner, S.E. Shumway, P.A. Tester, and T.E. Whitledge. 1997. Harmful algal blooms in coastal waters: options for prevention, control and mitigation. U.S. Department of Commerce and U.S. Department of the Interior, Washington, D.C.

Bossart, G.D., D.G. Baden, R.Y. Ewing, B. Roberts, and S.D. Wright. 1998. Brevetoxins in manatees (*Trichechus manatus latirostris*) from the 1996 epizootic: gross, histologic, and immunohistochemical features. Toxicol Pathol 26:276–282.

Bruslé, J., 1995. The impact of harmful algal blooms on finfish: occurrence of fish kills, pathology, toxicological mechanisms, ecological and economic impacts—a review. Institut Français de Recherche pour l'Exploitation de la Mer, Plouzané, France.

Bryant, D., E. Rodenburg, T. Cox, and D. Nielsen. 1995. Coastlines at risk: an index of potential development-related threats to coastal ecosystems. World Resources Institute, Washington, D.C.

Burkholder, J.M. 1998. Implications of harmful marine microalgae and heterotrophic dinoflagellates in management of sustainable marine fisheries. Ecol Appl 8 (suppl): S37–S62.

Burkholder, J.M., and H.B. Glasgow Jr. 1999. Science ethics and its role in early suppression of the *Pfiesteria* issue. Hum Org 58:443–455.

Burkholder, J.M. and J.J. Springer. 1999. Signalling in dinoflagellates, in microbial signaling and communication. *In* England, R.R., G. Hobbs, N.J. Bainton, and D.M. Roberts, eds. Fifty-seventh symposium of the Society for General Microbiology, pp. 220–238. Cambridge University Press, New York.

Burkholder, J.M., H.B. Glasgow Jr., and C.W. Hobbs. 1995. Distribution and environmental conditions for fish kills linked to a toxic ambush-predator dinoflagellate. Mar Ecol Prog Ser 124:43–61.

Burkholder, J.M., H.B. Glasgow Jr., N.J. Deamer-Melia, and E.K. Hannon. 1999a. The 1998 fish kill season: tracking a toxic *Pfiesteria* outbreak in the Neuse Estuary, North Carolina [abstract]. *In* Proceedings of the winter meeting of the American Society of Limnology and Oceanography (Santa Fe, N.M.). Available online at http://www.aslo.org//santafe99/abstracts/SSIOTH1045S.html

Burkholder, J.M., M.A. Mallin, and H.B. Glasgow Jr., 1999b. Fish kills, bottom-water hypoxia, and the toxic *Pfiesteria* complex in the Neuse River and Estuary. Mar Ecol Prog Ser 179:301–310.

Capra, M.F., J. Cameron, A.E. Flowers, I.F. Coombe, C.G. Blanton, and S.T. Hahn. 1988. The effects of ciguatoxin on teleosts. *In* Proceedings of the sixth International coral reef symposium, vol. 3, Executive Committee, Townsville Australia. pp. 7–41.

Carmichael, W.W. 1995. Cyanobacterial toxins. *In* Hallegraeff, G.M., D.M. Anderson, and A.D. Cembella, eds. Manual on harmful marine microalgae, pp. 163–175. Intergovernmental Oceanographic Commission, United Nations Educational, Scientific and Cultural Organization, Paris.

Carmichael, W.W., and I.R. Falconer. 1993. Disease related to freshwater algal blooms. *In* Falconer, I.R., ed. Algal toxins in seafood and drinking water, pp. 187–209. Academic Press, New York.

Carpenter, E.J., and W.W. Carmichael. 1995. Taxonomy of cyanobacteria. *In* Hallegraeff, G.M., D.M. Anderson, and A.D. Cembella, eds. Manual on harmful marine microalage, pp. 373–380. Intergovernmental Oceanographic Commission, United Nations Educational, Scientific and Cultural Organization, Paris.

Cembella, A.D., and E. Todd. 1993. Seafood toxins of algal origin and their control in Canada. *In* Falconer, I.R., ed. Algal toxins in seafood and drinking water, pp. 129–144. Academic Press, New York.

Cembella, A.D., L. Milenkovic, and G. Doucette. 1995a. Toxin analysis, Chap. 10, pt. A. *In vitro* biochemical and cellular assays. *In* Hallegraeff, G.M., D.M. Anderson, and A.D. Cembella, eds. Manual on harmful marine microalgae, pp. 181–211. Intergovernmental Oceanographic Commission, United Nations Educational, Scientific and Cultural Organization, Paris.

Cembella, A.D., L. Milenkovic, G. Doucette, and M. Fernandez. 1995b. Toxin analysis, chap. 10, *In vitro* biochemical methods and mammalian bioassays for phytotoxins. *In* Hallegraeff, G.M., D.M. Anderson, and A.D. Cembella, eds. manual on harmful marine microalgae, pp. 177–179. Intergovernmental Oceanographic Commission, United Nations Educational, Scientific and Cultural Organization, Paris.

Chorus, I., and J. Bartram, eds. 1999. Toxic cyanobacteria in water: a guide to their public health consequences, monitoring and management. World Health Organization and E & FN Spon, New York.

Culotta, E. 1992. Red menace in the world's oceans. Science 257:146–147.

DeFur, P.L., and D.N. Rader. 1995. Aquaculture in estuaries: feast or famine? Estuaries 18:2–9.

Epstein, P.R. 1998. Marine ecosystems: emerging diseases as indicators of change. Year of the Ocean special report. Center for Health and the Global Environment, Harvard Medical School, Boston.

Fairey, E.R., J.S.G. Edmunds, N.J. Deamer-Melia, H.B. Glasgow Jr., F.M. Johnson, P.R. Moeller, J.M. Burkholder, and J.S. Ramsdell. 1999. Reporter gene assay for fish killing activity produced by *Pfiesteria piscicida*. Environ Health Perspect 107:711–714.

Falconer, I.R. (ed.). 1993a. Algal toxins in seafood and drinking water. Academic Press, New York.

Falconer, I.R. 1993b. Mechanism of toxicity of cyclic peptide toxins from blue-green algae. *In* Falconer, I.R., ed. Algal toxins in seafood and drinking water, pp. 177–186. Academic Press, New York.

Fleming, L.E., J.A. Bean, and D.G. Baden. 1995. Epidemiology and public health. *In* Hallegraeff, G.M., D.M. Anderson, and A.D. Cembella, eds. Manual on harmful marine microalgae, pp. 475–486. Intergovernmental Oceanographic Commission, United Nations Educational, Scientific and Cultural Organization (UNESCO), Paris.

Fleming, L.E., D.G. Baden, J.A. Bean, R. Weisman, and D.G. Blythe. 1998. Seafood toxin diseases: issues in epidemiology and community outreach. *In* Reguera, B., J. Blanco, M.L. Fernandez, and T. Wyatt eds. Harmful microalgae, Proceedings of the 8th International Conference on Harmful Algal Blooms, pp. 245–248. Xunta de Galicia and Intergovernmental Oceanographic Commission, United Nations Educational, Scientific and Cultural Organization Paris.

Garrett, L. 1994. The coming plague—newly emerging diseases in a world out of balance. Penguin Books, New York.

Glasgow, H.B., Jr., J.M. Burkholder, D.E. Schmechel, P.A. Tester, and P.A. Rublee. 1995. Insidious effects of a toxic dinoflagellate on fish survival and human health. J Toxicol Environ Health 46:501–522.

Gonzalez, G., J. Brusle, and S. Crespo. 1994. Ultrastructural alterations of the cabrilla sea bass (*Serranus cabrilla*) liver related to experimental *Gambierdiscus toxicus* (dinoflagellate) ingestion. Dis Aquat Org 18:187–193.

Grattan, L.M., D. Oldach, T.M. Perl, M.H. Lowitt, D.L. Matuszak, C. Dickson, C. Parrott, R.C. Shoemacher, M.P. Wasserman, J.R. Hebel, P. Charache, and J.G. Morris Jr. 1998. Problems in learning and memory occur in persons with environmental exposure to waterways containing toxin-producing *Pfiesteria* or *Pfiesteria*-like dinoflagellates. Lancet 352:532–539.

Hallegraeff, G.M. 1993. A review of harmful algal blooms and their apparent global increase. Phycologia 32:79–99.

Hallegraeff, G.M. (ed.). 2000. Proceedings of the ninth international conference on harmful algal blooms. University of Tasmania, Hobart.

Hallegraeff, G.M., D.M. Anderson, and A.D. Cembella, eds. 1995. Manual of harmful marine microalgae. Intergovernmental Oceanographic Commission, United Nations Educational, Scientific and Cultural Organization (UNESCO), Paris.

Halstead, B.W. 1988. Poisonous and venomous marine animals of the world. Darwin Press, Princeton, N.J.

Haystead, T.J., A.T.R. Sim, D. Carling, R.C. Honnor, Y. Tsukitani, P. Cohen, and D.G. Hardie. 1989. Effects of the tumor promotor okadaic acid on intracellular protein phosphorylation and metabolism. Nature 337:78–81.

Hokama, Y., and J.T. Miyahara. 1986. Ciguatera poisoning: clinical and immunological aspects. J Toxicol Toxin Rev 5:25–62.

Hulka, B.S., J.D. Griffith, and T.C. Wilcosky, eds. 1990. Biological markers in epidemiology. Oxford University Press, New York.

Hungerford, J., and M. Wekell. 1993. Control measures in shellfish and finfish industries in the USA. In Falconer, I.R., ed. Algal toxins in seafood and drinking water, pp. 117–128. Academic Press, New York.

Jochimsen, E.M., W.W. Carmichael, A. Jisi, D.M. Cardo, S.T. Cookson, C.E.M. Holmes, M.B.D. Antunes, F.D.A. deMelo, T.M. Lyra, V.S.P. Barreto, S.F.M.O. Azevedo, and W.R. Jarvis, 1998. Liver failure and death after exposure to microcystins at a hemodialysis center in Brazil. N Engl J Med 338:873–878.

Kao, C.Y. 1993. Paralytic shellfish poisoning. In Falconer, I.R., ed. Algal toxins in seafood and drinking water, pp. 75–86. Academic Press, New York.

Kelley, A.M., C.C. Kohler, and D.R. Tindall. 1992. Are crustaceans linked to the ciguatera food chain? Environ Biol Fishes 33:275–286.

Krogh, P. 1988. Report of the Scientific Veterinary Committee (Section Public Health) on Paralytic Shellfish Poisons. Document VI/6492/88-EN-Rev.1, Commission of the European Communities, Directorate General for Aquaculture VI/B/II.2, Brussels.

Landsberg, J. 1995. Tropical reef-fish disease outbreaks and mass mortalities in Florida, U.S.A.: what is the role of dietary biological toxins? Dis Aquat Org 22:83–100.

Landsberg, J. 1996. Neoplasia and biotoxins in bivalves: is there a connection? J Shellfish Res 15:203–230.

Magnelia, S.J., C.C. Kohler, and D.R. Tindall. 1992. Acanthurids do not avoid consuming cultured toxic dinoflagellates yet do not become ciguatoxic. Trans Am Fish Soc 121: 737–745.

Mallin, M.A. 1994. Phytoplankton ecology of North Carolina estuaries. Estuaries 17:561–574.

Martin, J.L., A.W. White, and J.J. Sullivan. 1990. Anatomical distribution of paralytic shellfish toxins in softshell clams. In Granéli, E., B. Sundstrom, L. Edler, and D.M. Anderson eds. Toxic marine phytoplankton. pp. 379–384. Elsevier, New York.

Maryland Department of Natural Resources (MDNR). 1998. Water quality, habitat and biological conditions of river systems affected by Pfiesteria or Pfiesteria-like organisms on the lower eastern shore of Maryland: 1997 summary. Tidewater Ecosystem Assessment Division, Resource Assessment Service, MDNR, Annapolis, Md.

Masuyama, Y., T. Uchida, and T. Honjo. 1998. The effects of Heterocapsa circularisquama and Gymnodinium mikimotoi on clearance rate and survival of blue mussels, Mytilus galloprovincialis, in harmful microalgae. In Reguera, B., J. Blanco, M.L. Fernandez, and T. Wyatt, eds. Proceedings of the 8th international conference on harmful algal blooms, pp. 422–424. Xunta de Galicia and Intergovernmental Oceanographic Commission, United Nations Educational, Scientific and Cultural Organization, Paris.

Miller, G.T., Jr. 2000, Living in the environment, 11th ed. Brooks/Cole, New York.

Noga, E.J., L. Khoo, J.B. Stevens, Z. Fan, and J.M. Burkholder. 1996. Novel toxic dinoflagellate causes epidemic disease in estuarine fish. Mar Pollut Bull 22:219–224.

Ohta, T., E. Sueoka, N. Iida, A. Komori, M. Suganuma, R. Nishiwaki, M. Tatematsu, S.-J. Kim, W.W. Carmichael, and H. Fujiki. 1994. Nodularin, a potent inhibitor of protein phosphatases 1 and 2A, is a new environmental carcinogen in male F344 rat liver. Cancer Res 54:6402–6406.

O'Neill, D. 1994. The firecracker boys. St. Martin's Press, New York.

Perl, T.M., L. Bedard, R. Kosatsky, J.C. Hocking, E.C.D. Todd, and R.S. Remis. 1990. An outbreak of toxic encephalopathy caused by eating mussels contaminated with domoic acid N Eng J Med 322:1775–1780.

Popovský, J. and L.A. Pfiester. 1990. Dinophyceae (Dinoflagellida). Fischer, Stuttgart.

Robineau, B., J.A. Gagne, L. Fortier, and A.D. Cembella. 1991. Potential impact of a toxic dinoflagellate (*Alexandrium excavatum*) bloom on survival of fish and crustacean larvae. Mar Biol 108:293–301.

Runnegar, M.T.C., A.R.B. Jackson, and I.R. Falconer. 1988, Toxicity of the cyanobacterium *Nodularia spumigena* Mertens. Toxicon 26:143–151.

Russell, F.E. and N.E. Egen. 1991. Ciguateric fishes, ciguatoxin (CTX) and ciguatera poisoning. J Toxicol Toxin Rev 10:37–62.

Safina, C. 1995. The world's imperiled fish. Sci Am 273:46–53.

Shilts, R. 1987. And the band played on. St. Martin's Press, New York.

Shumway, S.E., 1990. A review of the effects of algal blooms on shellfish and aquaculture. J World Aquacult Soc 21:65–104.

Shumway, S.E. 1995. Phycotoxin-related shellfish poisoning: bivalve molluscs are not the only vectors. Rev Fish Sci 3:1–31.

Shumway, S.E., H.P. van Egmond, J.W. Hurst, and L.L. Bean. 1995. Management of shellfish resources. *In* Hallegraeff, G.M., D.M. Anderson, and A.D. Cembella, eds. Manual on harmful marine microalgae, pp. 433–459. Intergovernmental Oceanographic Commission, United Nations Educational, Scientific and Cultural Organization, Paris.

Sindermann, C.J. 1996. Ocean pollution—effects on living resources and humans. CRC Press, New York.

Skjoldal, H.R. and I. Dundas, eds. 1991. The *Chrysochromulina polylepis* bloom in the Skagerrak and the Kattegat in May–June 1988: environmental conditions, possible causes, and effects. Cooperative Research Report No. 175. International Council for the Exploration of the Sea, Paris.

Skulberg, O.M., W.W. Carmichael, G.A. Codd, and R. Skulberg. 1993. Taxonomy of toxic Cyanophyceae (cyanobacteria). *In* Falconer, I.R. ed. Algal toxins in seafood and drinking water, pp. 145–164. Academic Press, New York.

Smayda, T.J. 1989. Primary production and the global epidemic of phytoplankton blooms in the sea: a linkage? *In* Cosper, E.M., V.M. Bricelj, and E.J. Carpenter, eds. Coastal and Estuarine Studies No. 35, Novel phytoplankton blooms, pp. 449–483. Springer, New York.

Steidinger, K.A. 1993. Some taxonomic and biologic aspects of dinoflagellates. *In* Falconer, I.R., ed. Algal toxins in seafood and drinking water, pp. 1–28. Academic Press, New York.

Steidinger, K.A., and D.G. Baden. 1984. Toxic marine dinoflagellates. *In* Spector, D.L., ed. Dinoflagellates, pp. 201–261. Academic Press, New York.

Sueoka, E., and H. Fujiki. 1998. Carcinogenesis of okadaic acid class tumor promoters derived from marine natural products. *In* Reguera, B., J. Blanco, M.L. Fernandez, and T. Wyatt, eds. Harmful microalgae, Proceedings of the VIIIth international conference on harmful algal blooms, pp. 573–576. Xunta de Galicia and Intergovernmental Oceanographic Commission, United Nations Educational, Scientific and Cultural Organization, Paris.

Summerson, H.C. and C.H. Peterson. 1990. Recruitment failure of the bay scallop, *Argopecten irradians concentricus*, during the first red tide, *Ptychodiscus brevis*, outbreak recorded in North Carolina. Estuaries 13:322–331.

Tettelbach, S.T. and P. Wenczel. 1993. Reseeding efforts and the status of bay scallop *Argopecten irradians* (Lamarck, 1819) populations in New York following the occurrence of "brown tide" algal blooms. J Shellfish Res 12:423–431.

White, A.W. 1981a. Marine zooplankton can accumulate and retain dinoflagellate toxins and cause fish kills. Limnol Oceanogr 26:103–109.

White, A.W. 1981b. Sensitivity of marine fishes to toxins from the red-tide dinoflagellate *Gonyaulax excavata* and implications for fish kills. Mar Biol 65:255–260.

Wikfors, G.H., and R.M. Smolowitz. 1993. Detrimental effects of a *Prorocentrum* isolate upon hard clams and bay scallops in laboratory feeding studies. *In* Smayda, T.J. and Y. Shimizu, eds. Toxic phytoplankton blooms in the sea, pp. 447–452. Elsevier Science, New York.

Wikfors, G.H., and R.M. Smolowitz. 1995. Experimental and histological studies of four life-history stages of the eastern oyster, *Crassostrea virginica*, exposed to a cultured strain of the dinoflagellate *Prorocentrum minimum*. Biol Bull 188:313–328.

Wilkinson, T. 1998. Science under siege: the politicians' war on nature and truth. Johnson Book, Boulder, Colo.

World Resources Institute. 1994. Environmental almanac. Houghton Mifflin, Boston.

Yasumoto, T., and M. Murata. 1993. Marine toxins. Chem Rev 93:1897–1909.

Part IV

Implementing
Conservation Medicine

19

Ecological Health and Wildlife Disease Management in National Parks

Colin M. Gillin

Gary M. Tabor

A. Alonso Aguirre

Once considered to be shining examples of pristine nature, national parks and other conservation protected areas are facing increasing threats to their ecological integrity. Habitat fragmentation and degradation, habitat loss, species extinctions, alien species introductions, pollution, and recreational overusage represent some of the cumulative effects that are putting these areas of natural heritage under increasing stress. Long considered a footnote to protected area management, health concerns are gaining greater attention in protected areas as cumulative stresses are enhancing, and sometimes amplifying, conditions for disease and other health effects. In this chapter we discuss the influence of pollution and disease and other health effects on the integrity of parks and protected areas.

Protected areas are relatively undisturbed natural ecosystems set aside from human development and provided of legislative protection with the National Park Service Organic Act of 1916 (MacKintosh 1991). Despite protection from human influence to habitat and animal population numbers, the health of wildlife is intertwined with the influences of the encompassing ecosystem, both inside and outside park boundaries. Those influences will continue to increase as park visitor numbers increase, leading to behavioral stress on wildlife. Other environmental stressors including pollution from automobiles, snowmobiles, and upwind airborne contaminants lead to changes in the environment and degradation of the stability of these ecosystems.

The health and environmental stability of areas protected and conserved for natural resources can be measured by evaluating the pathologic changes threatening wildlife, humans, and plants within an ecosystem. In 1941, Aldo Leopold wrote, "A science of land health needs, first of all, a base-datum of normality, a picture of how healthy land maintains itself as an organism." Wagner and Colleagues (1995) and Krausman (1998) suggested that national parks could serve as a model to determine healthy, relatively pristine areas. Historical records may also provide the foundation to begin a database to determine the health of a protected ecosystem from which we can build upon through time and space. Collectively, a series of "snapshots" can provide a picture to document health issues throughout selected national parks in North America. In addition, long-term monitoring may be indicated in these protected areas to assess the effects of anthropogenic impact.

19.1 Historical Perspective

The National Park service administers approximately 80.7 million acres (32.6 million hectares) in the United States, of which more than 2.8 million acres remain in private ownership. The largest area is Wrangell–St. Elias National Park and Preserve, Alaska, at 13.2 million acres, 16.3% of the entire system. The smallest unit in the system is Thaddeus Kosciuszko National Memorial, Pennsylvania, at 0.02 of an acre.

In protecting and governing these federal lands, the Yellowstone National Park Act of 1872 was established and then strengthened with the National Park Service Organic Act of 1916. The National Park Act designated land set aside for the preservation of natural resources at the exclusion of human development and resource exploitation such as hunting, mining, and consumptive activities. In effect, these lands were set aside and protected from the influences of human development and change. Some influences and changes, however, do not fall within the scope of these acts.

19.2 National Parks and Conservation Medicine

A recent U.S. Environmental Protection Agency and National Park Service study found that snowmobile use in Yellowstone National Park produced unacceptable and damaging levels of air pollutants (U.S. Environmental Protection Agency 1999a,b,c; U.S. Department of the Interior 2000). This study led to a ban on snowmobile use as a means of winter travel in the parks.

An increasing number of diseases and their vectors in wildlife of national parks have most likely originated from contact with humans, domestic pets, and livestock within parks and in surrounding gateway communities (Aguirre et al. 1995a). Diseases, whether endemic or exotic, are capable of migrating with wildlife, particularly those that migrate long distances as with arboviruses in birds and bats (McLean 1991; Ubico and McLean 1995).

The threats and issues that affect protected areas within and beyond park boundaries are prime examples of bringing together the essence of defining the field of conservation medicine. The health of Yellowstone National Park, the first national park in North America, is the subject of a current issue that has caught the attention of local and national environmental activists, gateway communities, and government officials (Broscious 1998; Janofsky 1999). In 1995, the U.S. Department of Energy contracted with a nuclear waste company to build a radioactive and hazardous waste facility 90 miles northwest of the park. One of the primary issues was that prevailing winds could carry accidental or intentional emissions of radioactive particles into not only Yellowstone National Park but also Grand Teton National Park. The primary public concern was fear of radioactive exposure, threats to public health from increased cancer rates, and environmental contamination in one of the most unspoiled areas of North America.

Pollution effects can ultimately cause pathologic changes that are detrimental to the individual animal, the population, and eventually the ecosystem. A decrease in animal health and human health, a loss of the "unspoiled" nature of national parks through environmental degradation, and negative economic effects on the gateway communities all pose the potential to decrease ecological health. Although the project has been derailed, the health of people and park resources were sufficiently important to evaluate the interrelationships between ecological and environmental health. The issues evaluated in this scenario demonstrated that parks are not isolated ecologically and as such they experience effects in all the interconnected areas, perceived as natural or disturbed. Conservation medicine refers to the practice of ecological health and its connections with animals, humans, and the environment. It requires integrating health parameters across species, across landscapes, and across disciplines.

19.3 Health Issues in National Parks

National parks do not encapsulate pristine environs despite their legal protection. The health of wildlife communities in a national park is not guaranteed. Habitat fragmentation, changes in vegetative composition due to the consequences of "natural regulation" and behavior, and the increasing human pressures in and around parks have created a barrage of ecological health effects. This can manifest through the introduction of alien species, the presence of wildlife diseases, and the threat of waterborne and airborne pollutants. These effects are not new and have taken a toll on park resources and management since their inception (Clark 2000). There is evidence that most diseases diagnosed in wildlife, except for some native parasitic infections, first originated in human or domestic animal populations. Following eradication or control in domestic animals, many of those diseases have found a wildlife host or a sylvatic cycle (Aguirre et al. 1994). Diseases transmitted from wildlife to humans or to domestic animals may create crises in public health or animal welfare.

For example, brucellosis, a bacterial zoonosis caused by *Brucella abortus*, is characterized by abortions, arthritis, and reproductive failure in domestic livestock. The actual source of the disease in the Greater Yellowstone Area (GYA) is unknown; however, *B. abortus* was introduced in North America with cattle imported from Europe. *Brucella* may have been introduced in Yellowstone National Park as early as 1902 with bison (*Bison bison*) from Montana and Texas during efforts to improve the population numbers of an almost extinct park bison herd. The disease was first identified serologically in bison in the park in 1917 (Mohler 1917; Rush 1932). Meagher and Meyer (1994) concluded that *Brucella* was likely introduced to wildlife in Yellowstone National Park through cattle maintained for park employees. Both wapiti (*Cervus elaphus*) and bison are infected in the GYA (Nettles 1992). Wapiti on state-run winter feeding grounds maintain a seroprevalence rate of approximately 24%, and 50% of the bison herd are infected in the GYA (Thorne et al. 1978, 1991a,b; Williams et al. 1993). Approximately 46% of these bison will yield pure culture for *Brucella* (Roffe et al. 1999). Since 1925, bison at Wood Buffalo National Park, Northwest Territories, Canada, have been infected not only with *Brucella* but also with bacteria causing bovine tuberculosis, complicating the scenario (Joly et al. 1998; Tessaro et al. 1990). These diseases were transported from an introduced subspecies of bison relocated from Alberta, Canada. Johne's disease (*Mycobacterium paratuberculosis*), another chronic disease principally seen in livestock, was also diagnosed in this population (B. Elkin, Government of the Northwest Territories, personal commun. 2000).

There are several diseases of zoonotic importance that have been identified in North American national parks, including tick-borne relapsing fever in Grand Canyon National Park (Boyer et al. 1977), Colorado tick fever in Rocky Mountain National Park (McLean et al. 1989), and sylvatic plague (*Yersinia pestis*) in rodents in Lava Beds National Monument (Stark and Kinney 1969; Nelson and Smith 1976). Diseases affecting ungulates have been recorded in other national parks, including leptospirosis (*Leptospira* spp.) in white-tailed deer (*Odocoileus virginianus*) in Great Smokey Mountain National Park, Tennessee (New et al. 1993), and hemorrhagic septicemia in bighorn sheep (*Ovis canadensis*) in Rocky Mountain National Park, Colorado (Potts 1937). Parasitic nematode eye worms have been recorded in mule deer in Sequoia National Park, California, and studies of the intestinal parasites of small mammals were conducted in Grand Teton National Park (Oberhansley 1940; Pinter et al. 1988).

In a study conducted by Aguirre et al. (1993, 1995b), the most common wildlife diseases found in national parks were those affecting ungulates. These included lungworm-pneumonia complex in bighorn sheep and epizootic hemorrhagic dieases in white-tailed deer. Other ungulate diseases reported were meningeal worm (*Parelaphostrongylus*) infections in wapiti and deer, psoroptic scabies (*Psoroptes ovis*) in bighorn sheep, leptospirosis in deer, and pseudorabies in feral pigs (*Sus scrofa*). Carnivores in national parks were reported with rabies, sylvatic plague, canine distemper, borreliosis (*Borrelia burgdorferi*), and filarial (*Dirofilaria* spp.) infections. Other diseases have been documented as to be increasing in zoonotic importance in national parks, including tularemia (*Fransicella tularensis*) in rabbits (*Sylvilagus* spp.) and American beavers (*Castor canadensis*), and leptospirosis, giardiasis (*Giardia* spp.), and Rocky Mountain spotted fever in rodents (Aguirre et al. 1993, Aguirre and Starkey 1994).

Domestic livestock pathogens are commonly transmitted from cattle to wild cervids. In a survey of several western U.S. national parks, populations of mule deer (*Odocoileus hemionus*) and wapiti presented high titers to parainfluenza-3 virus, bovine herpesvirus-1, bovine viral diarrhea virus, bovine respiratory syncytial virus, blue tongue virus, and epizootic hemorrhagic disease virus (Aguirre et al. 1995b). During a similar study in Banff National Park, Canada, wapiti and moose (*Alces alces*) tested positive for antibodies to bovine viral diarrhea virus, bovine herpesvirus-1, and *Leptospira icterohemorrhagiae* (T. Shury, Parks Canada, personal commun. 1999).

In national parks of the western United States, gray wolves (*Canis lupus*) have been shown to carry antibodies to canine distemper, canine parvovirus, and infectious canine hepatitis (Johnson et al. 1994). Other canid species, including coyotes (*C. latrans*) and red fox (*Vulpes fulva*), occupying the same range are most likely exposed to similar diseases. Hantavirus and sylvatic plague have been documented in deer mouse (*Peromyscus maniculatus*) and vole (*Microtus* spp.) populations within Grand Teton and Yellowstone National Parks (Jannett 1996; M. Johnson, unpublished data 2000).

Diseases in national parks are not limited to terrestrial ecosystems. Whirling disease (*Myxobolus cerebralis*), a parasitic disease that kills salmonid fish species, including rainbow (*Oncorhynchus mykiss*) and cutthrout trout (*O. clarki*), was discovered in the Madison River of Yellowstone National Park in 1994 (Baldwin et al. 1997; Egan 1999). The disease is spread through introduction of fish reared in hatcheries.

Biologists have witnessed increased stress on wildlife populations due to lice infestations in Denali National Park, Alaska (P. Owens, Denali National Park Research and Resources Division, personal commun. 2000). Due to the large domestic dog population (including sled dogs) in the state and outside the park, canine parvovirus, canine distemper, and infectious canine hepatitis may be transmitted to wild carnivore populations (Elton 1931; Stephenson et al. 1982). Canine parvovirus was suspected to reduce gray wolf (*Canis lupus*) pup survival from 80% to 60% in the reintroduced populations of Central Idaho and the Yellowstone National Park during 1999 (E. Bangs, Wolf Recovery Coordinator, personal commun. 2000). Wolves of Glacier National Park, Montana, have also been affected by canine parvovirus and canine distemper (Johnson et al. 1994).

In many national parks of the Eastern Seaboard, tick-borne diseases such as Lyme disease (*Borrelia burgdorferi*) continue to present challenges to park management as human exposure increases with increased numbers of visitors (K. Anderson, Acadia National Park, personal commun. 2000). There is also an increased threat of rabies from infected raccoon populations as the disease epidemic has continued to spread northward from the southeastern United States since the 1970s (Rupprecht and Smith 1994). Hantavirus infection and sylvatic plague are at the forefront of zoonotic disease concerns in parks where the diseases likely exist and human contact with small mammals is evidently high (S. Cain, Grand Teton National Park, personal commun. 2000).

19.4 Management Implications

As a consequence to the presence of pathogens, park managers attempt to institute procedures to reduce risk of exposure to disease and zoonotic potential. For example, when researchers enter caves to study colonies of bats in Carlsbad Caverns National Park, New Mexico, they are required to wear respirators to reduce inhalation exposure from airborne pathogens. They also must have been previously immunized with a rabies vaccine (D. Pate, Carlsbad Caverns National Park, personal commun. 2000).

The National Park Service uses veterinary epidemiologists from the National Wildlife Health Center, the Biological Resources Division of the U.S. Department of Interior, and the Animal Plant Health Inspection Service of the U.S. Department of Agriculture (USDA) with diseases of economic and national importance such as brucellosis. Park managers surveyed in several national parks across the country, however, were not aware of contingency plans in the event of a major disease outbreak. In general, these managers plan to work in coordination with state and federal wildlife agencies, but only if they believe intervention is warranted (Aguirre et al. 1993; Aguirre and Starkey 1994). The current park policy on disease has evolved since the adoption of the National Park Service Management Policy (U.S. National Park Service 1988). In this policy, "intervention on the part of the Centers for Disease Control (CDC) will be used in cases of exotic species management when threatening situations pose a public health hazard as determined or a hazard to public safety." Now, however, disease intervention may also be used in pest management, including "many fungi, insects, rodents, diseases, and other species that may be perceived as pests and are native plants and animals existing under natural conditions as natural elements of the ecosystem." Management strategies may be implemented to control disease if a human health hazard as defined by the CDC exists or to protect against a significant threat to public safety. Diseases in native animals may also be controlled if the disease is an

> exotic organism; likely to be transmitted to livestock; a threat to human health; a natural disease, but conditions which would contribute to natural limitation of the spread of the disease are no longer present; if the disease threatens to extirpate a relict population from the park or threatens a threatened or endangered species; or upon completion of a plan approved by the Regional Director" (U.S. National Park Service 1994)

Natural regulation is the principal National Park Service policy concerning animal handling and disease issues. Handling of animals for disease treatment or specimen collection is generally not permitted (Zaugg et al. 1993). As part of the hands-off natural regulation policy, parks will not control diseases when they are endemic, are part of natural regulation of wildlife populations, and are not a threat to humans, livestock, or agriculture operations adjacent to parks (D. Barna, Chief of Public Affairs, National Park Service, Washington, D.C., personal commun. 2000). Under the NPS-77 guidelines (U.S. Park Service 1994), disease will not be controlled when the effort would prove impractical or would threaten other wildlife populations in the process. But if treatment is indicated, parks are required to institute a monitoring program to evaluate the effects of the treatment.

There is no documentation of eradication of any major disease in North American national parks. This is most likely due to a low emphasis on health and disease issues, a lack of personnel assigned or trained to the task, and inadequate record keeping/documentation. Historically, researchers have demonstrated diseases present in national parks that cause economic and human health risks. Many diseases are studied because they occur in epidemic proportions or a disease is identified incidentally. When mapped on a continental scale, disease distribution correlates to animal distributions, which are often fragmented in islands of habitat. Monitoring yields useful data whether positive or negative. For example, in 1999, veterinary researchers in Banff National Park tested sera from over 100 elk and approximately 40 other individual moose, bear (*Ursus americanus*), cougar (*Felis concolor*), coyote (*Canis latrans*), and lynx (*Lynx canadensis*). They found no animals seropositive for *Brucella abortus*, indicating that none of the animals were previously exposed (T. Shury, Parks Canada, personal commun. 1999). The uniformly negative results of this survey are valuable in defining the geographical extent of *B. abortus* infection.

In the Rocky Mountain region of the United States and Canada, wildlife veterinarians are mapping important wildlife diseases using geographical information systems (GIS) technology to identify known distribution of diseases, including brucellosis, chronic wasting disease, anthrax, bovine tuberculosis, sylvatic plague, and hantavirus (C. Gillin et al., unpublished data, 2000). This information will also identify information gaps in national parks throughout the Rockies and provide agency managers with a common database that spans jurisdictional boundaries. More important, this technology provides health professionals a means of evaluating ecological health involving human and animal well-being, the biotic environment (habitat quality and food supply), and physical environment (weather, climate, and edaphic conditions) and other data including socioeconomic or cultural information.

Health researchers and agency personnel in Canada recently participated in a workshop in Pacific Rim National Park on Vancouver Island, where the goal was to develop a set of measures to track the ecological integrity of the park (C. Stephen, Center for Coastal Health, personal commun. 2000). This workshop was established because Canada's national parks now have the mandate to protect and promote ecological integrity rather than tourism and recreation. During the proceedings participants identified the need for surveillance and monitoring systems to track disease and mortality as important indicators of ecological integrity. Other "indicators" included surveys of keystone or significant species, habitat mapping, and monitoring human impacts. Marine protected areas of Canada are also being looked at as important areas in ecological health. Socioeconomic and other values have been provided to create tools for the establishment of marine parks, including a proposed park in the southern Georgia Strait in British Columbia. Other provincial parks have elicited outside veterinary expertise in the evaluation of mass mortalities of wildlife. These included a large die-off at a gull colony and evidence of mass mortality of northern fulmars (*Fulmaris glacialis*) occurring along the Pacific coast and potentially affecting Pacific Rim National Park (C. Stephen, Center for Coastal Health, personal commun. 2000). Initial concerns involving these die-offs were related to toxin exposure with implications to animal and human risks.

19.5 Achieving Ecological Health in National Parks

In 1995, Aguirre et al. (1995a) argued for a national surveillance program for wildlife diseases in national parks. Due to dramatic increases in ungulate and other wildlife populations in many national parks, a disease surveillance program could help park managers approach decisions prior to a population crisis or epidemic (Porter et al. 1994). Wildlife populations serve as sentinels of the health and quality of the environment and are capable of transmitting and receiving diseases from domestic animals or humans. As park visitor days increase and human demographic patterns change, shifts in patterns of human exposure to diseases carried by wildlife, including rabies, plague, and tick-borne infections, increase (Boyce et al. 1992).

Basic research, monitoring, and surveillance programs should be implemented to document disease through serologic evaluation and identify parasitic infections of both native and exotic diseases and organisms in national parks of North America (Aguirre and Starkey 1994). Other needed health parameters include documenting introduced diseases and identifying disease vectors and related reservoirs, epizootics and emerging diseases, and the dynamics of disease within populations (Potts 1937; Oberhansley 1940; Nelson and Smith 1976; Jessup et al. 1981; Rabinowitz and Potgieter 1984; Pinter et al. 1988; Forrester 1990; Roelke 1990; New et al. 1993; Dunbar 1994; Aguirre et al. 1995a,b).

Despite the identified need for health surveillance and monitoring, the national parks of North America have been reluctant to implement comprehensive wildlife health programs. The American Association of Zoo Veterinarians estimates that 225–250 full-time or part-time contract veterinarians are working in the 185 accredited zoos in the United States (W. Amand, personal commun. 2000). The USDA Animal Welfare Act of 1994 requires a that staff veterinarian monitor the health of zoological collections. The National Park Service manages 379 properties, including national lakeshores, preserves, recreation areas, seashores, wild and scenic rivers, and 55 parks (D. Barna, National Park Service, personal commun. 1999). Currently, there is only *one* recently hired (July 2000) full-time veterinarian to monitor wildlife health of national parks in the United States. Yellowstone National Park has contracted veterinary support for specific projects, and Parks Canada hires one veterinarian over a six-month period each year on a contract basis. Similarly, Parks Canada has hired veterinarians for specific research projects, including Wood Buffalo National Park, where veterinarians were hired as meat inspectors during the large-scale culls of wood bison conducted during the 1950s and 1960s. In recent times, several national parks in Canada have begun using the Canadian Cooperative Wildlife Health Centres at the four veterinary colleges in Canada (T. Shury, Parks Canada, personal commun. 2000).

Health issues span taxa and disciplines. The park services must increase the capacity of their health-monitoring efforts. As the field of veterinary medicine expands into the environmental arena, it shares knowledge with an increasing number of conservation governmental and non-governmental agencies. Health professionals with clinical, epidemiology, and pathology skills will join biologists with backgrounds in ecology, conservation biology, biodiversity, and environmental ethics. This is where national parks, as laboratories of the natural world, will benefit most by integrating these professionals into wildlife programs using transdisciplinary approaches to problem solving.

Veterinary schools are preparing wildlife health professionals for environmental challenges involving global issues, ecological health, and problems facing complex management systems such as those found in national parks. Several university programs are incorporating a variety of wildlife and environmental health opportunities beyond traditional medicine. One of the first programs was the Southeastern Cooperative Wildlife Disease Study formed by the Southeastern Association of Fish and Wildlife Agencies in 1957 based at University of Georgia School of Veterinary Medicine. This remains a leading organization on wildlife disease research. In 1997, Tufts University School of Veterinary Medicine, Wildlife Trust (formerly Wildlife Preservation Trust International), and Harvard University Center for Health and Global Environment partnered to establish the Center for Conservation Medicine. Loyola University Chicago Stritch School of Medicine and the Chicago Zoological Society/Brookfield Zoo, in association with the University of Illinois College of Veterinary Medicine, have established the Conservation Medicine Center of Chicago. Other institutions, including the University of California Davis School of Veterinary Medicine's, Wildlife Health Program and the North Carolina State University College of Veterinary Medicine's, Environmental Medicine Consortium, have joined these efforts.

Wildlife health is an important part of environmental and public policy, as health of the environment is desired by both the ecologically aware public and natural resource management agencies. As we attempt to understand the role of interconnections in health, the continuum of life, and all processes within and outside a system, we begin to manage ecosystems, including those encompassing national parks, with an integrated approach involving wildlife and ecosystem health monitoring during the strategic planning of natural resources.

References

Aguirre, A.A., and E.E. Starkey. 1994. Wildlife disease in U.S. National Parks: historical and coevolutionary perspectives. Conserv Biol 8:654–661.

Aguirre, A.A., D.E. Hansen, and E.E. Starkey. 1993. Special initiative project: animal disease issues in the national park system. Cooperative Park Studies Unit Technical Report NPS/PNROSU/NRTR-93/16. U.S. Department of the Interior, National Park Service, Pacific Northwest Region, Denver, Colo.

Aguirre, A.A., E.E. Starkey, and D.E. Hansen. 1995a. Wildlife diseases in national park ecosystems. Wildl Soc Bull 23:415–419.

Aguirre A.A., D.E. Hansen, E.E. Starkey, and R.G. McLean. 1995b. Serologic survey of wild cervids for potential disease agents in selected national parks in the United States. Prev Vet Med 21:313–322.

Baldwin, T.J., J.E. Peterson, G.C. McGhee, K.G. Staigmiller, E. Motteram, and C. Downs. 1997. Distribution of whirling disease, caused by *Myxobolus cerebralis*, in salmonid fishes in Montana. *In* Research progress reports, Whirling disease symposium: expanding the database (6–8 March 1997 Logan, Utah), 45 pp. 19–26. The Whirling Disease Foundation, Bozeman, Mont.

Boyce, W., T. Yuill, J. Homan, and D. Jessup. 1992. A role for veterinarians in wildlife health and conservation biology. JAVMA 200:435–437.

Boyer, K.M., R.S. Munford, G.O. Maupin, C.P. Pattison, M.D. Fox, A.M. Barnes, W.L. Jones, and J.E. Maryland. 1977. Tick-borne relapsing fever: an interstate outbreak originating at Grand Canyon National Park. Am J Epidemiol 105:469–479.

Broscious, C., ed. 1998. Citizens' guide to the Idaho National Engineering Environmental Laboratory, 10th ed. Environmental Defense Institute, Troy, Idaho.

Clark, T.W. 2000. Wildlife resources: the elk of Jackson Hole, Wyoming. *In* Clark, T.W., D. Casey, and A. Halverson, eds. Developing sustainable management policy for the National Elk Refuge, Wyoming. Yale Sch Forest Environ Stud Bull Ser 104:171–187.

Dunbar, M.R. 1994. Florida panther biomedical investigations. In Jordan, D.B., ed. Florida panther conference proceedings (1–3 November 1994, Ft. Myers, Fla.), pp. 342–392.

Egan, D. 1999. Cutthroat face double danger in Yellowstone. Salt Lake Tribune, 7 Sept., p. A1.

Elton, C. 1931. Epidemics among sled dogs in the Canadian arctic and their relation to disease in the arctic fox. Can J Res 5:673–692.

Forrester, D.J. 1990. Studies on the health and diseases of white-tailed deer in Everglades National Park and the Stairsteps Unit of Big Cypress National Preserve. Coop Agreement CA-5000-7-8007. Florida Game and Fresh Water Fish Commission, Gainesville.

Jannett, F.J. 1996. Metapopulations of montane and long-tailed voles (*Microtus montanus* and *M. longicaudus*), and hantavirus and plague in the Jackson Hole small mammal community. *In* Annual Report, vol 20, pp. 52–56. University of Wyoming-National Park Service Research Center, Grand Teton National Park, Wyoming.

Janofsky, M. 1999. 2 groups fight nuclear incinerator project near Yellowstone. New York Times, 17 Sept., p. 12.

Jessup, D.A., B. Abbas, D. Behymer, and P. Gogan. 1981. Paratuberculosis in tule elk in California. JAVMA 179:252–254.

Johnson, M.R., D.K. Boyd, and D.H. Pletscher. 1994. Serologic investigations of canine parvovirus and canine distemper in relation to wolf (*Canis lupus*) pup mortalities. J Wildl Dis 30:270–273.

Joly, D.O., F.A. Leighton, and F. Messier. 1998. Tuberculosis and brucellosis infection of bison in Wood Buffalo National Park, Canada: preliminary results. *In* Irby, L., and J. Knight, eds. International symposium on bison ecology and management in North America, pp. 23–31. Montana State University, Bozeman.

Krausman, P.R. 1998. Conflicting views of ungulate management in North America's western national parks. Wildl Soc Bull 26:369–371.

Leopold, A. 1941. Wilderness as a land laboratory. Living Wilderness 6:3.

MacKintosh, B. 1991. The national parks: shaping the system. Division of Publications, National Park Service, Washington, D.C.

McLean, R.G. 1991. Arboviruses of wild birds and mammals. Bull Soc Vector Ecol 16:3–16.

McLean, R.G., R.B. Shriner, K.S. Pokorny, and G.S. Bowen. 1989. The ecology of Colorado tick fever in Rocky Mountain National Park in 1974. III. Habitats supporting the virus. Am J Trop Med Hyg 40:86–93.

Meagher, M., and M.E. Meyer. 1994. On the origin of brucellosis in bison in Yellowstone National Park: a review. Conserv Biol 8:645–653.

Mohler, J.R. 1917. Abortion disease. *In* Annual report of the Bureau of Animal Industry, U.S. Department of Agriculture, pp. 105–106. U.S. Government Printing Office, Washington, D.C.

Nelson, B.C., and C.R. Smith. 1976. Ecological effects of a plague epizootic on the activities of rodents inhabiting caves at Lava Beds National Monument, California. J Med Entomol 13:51–61.

Nettles, V.F. 1992. Wildlife diseases and population medicine. 1992. JAVMA 200:648–652.

New, J.C. Jr., W.G. Wathen, and S. Dlutkowski. 1993. Prevalence of *Leptospira* antibodies in white-tailed deer, Cades Cove, Great Smoky Mountains National Park, Tennessee, USA. J Wildl Dis 29:561–567.

Oberhansley, F.R. 1940. California mule deer: a host for nematode eye worms in Sequoia National Park. JAVMA 96:542.

Pinter, A.J., W.D. O'Dell, and R.A. Watkins. 1988. Intestinal parasites of small mammals from Grand Teton National Park. J Parasitol 74:187–188.

Porter, W.F., M.A. Coffey, and J. Hadidian. 1994. In search of a litmus test: wildlife management in U.S. national parks. Wildl Soc Bull 22:301–306.

Potts, M.K. 1937. Hemorrhagic septicemia in the bighorn of Rocky Mountain National Park. J Mammal 18:105–106.

Rabinowitz, A.R., and L.N.D. Potgieter. 1984. Serologic survey for selected viruses in a population of raccoons, *Procyon lotor* (L.), in the Great Smoky Mountains. J Wildl Dis 20:146–148.

Roelke, M.E. 1990. Florida panther recovery: Florida panther biomedical investigation final report, 1 July 1986–30 June 1990. Wildlife Research Laboratory Florida Game and Fresh Water Fish Commission, Gainesville, Fla.

Roffe, T.J., J.C. Rhyan, K. Aune, L.M. Philo, D.R. Ewalt, T. Gidlewski, and S.G. Hennager, 1999. Brucellosis in Yellowstone National Park bison: quantitative serology and infection. J Wildl Manage 63:1132–1137.

Rupprecht, C.E., and J.S. Smith. 1994. Raccoon rabies—the re-emergence of an epizootic in a densely populated area. Semin Virol 5:155–164.

Rush, W.M. 1932. Bang's disease in Yellowstone National Park buffalo and elk herds. J Mammal 13:371–372.

Stark, H.E., and A.R. Kinney. 1969. Abundance of rodents and fleas as related to plague in Lava Beds National Monument, California. J Med Entomol 6:287–294.

Stephenson, R.O., D.G. Ritter, and C.A. Nielsen. 1982. Serologic survey for canine distemper and infectious canine hepatitis in wolves in Alaska. J Wildl Dis 18:419–424.

Tessaro, S.V., L.B. Forbes, and C. Turcotte. 1990. A survey of brucellosis and tuberculosis in bison in and around Wood Buffalo National Park, Canada. Can Vet J 31:174–180.

Thorne E.T., J.K. Morton, and G.M. Thomas. 1978. Brucellosis in elk. 1. Serologic and bacteriologic survey in Wyoming. J Wildl Dis 14:74–81.

Thorne, E.T., J.D. Herriges Jr., and A.D. Reese. 1991a. Bovine brucellosis in elk: conflicts in the Greater Yellowstone Area. *In* Christiansen, A.G., L.J. Lyon, and T.N. Lonner, comps. Elk vulnerability symposium proceedings, pp. 296–303. Montana State University, Bozeman.

Thorne, E.T., M. Meagher, and R. Hillman. 1991b. Brucellosis in free-ranging bison: three perspectives. *In* Keiter, R.B., and M.S. Boyce, eds. The Greater Yellowstone Ecosystem—redefining America's wilderness heritage, pp. 275–287. Yale University Press, New Haven, Conn.

Ubico, S.R., and R.G. McLean. 1995. Serologic survey of neotropical bats in Guatemala for virus antibodies. J Wildl Dis 31:1–9.

U.S. Department of the Interior. 2000. Air quality concerns related to snowmobile usage in national parks. National Park Service, Air Resources Division, Denver, Colo.

U.S. Environmental Protection Agency. 1999a. Regulatory announcement: proposed finding on emission standards for new large spark-ignition nonroad engines. EPA420-F-99–004. EPA Office of Mobile Sources, Ann Arbor, Mich.

U.S. Environmental Protection Agency. 1999b. Regulatory Update. EPA's nonroad engine emissions control programs. EPA420-F-99–001. EPA Office of Mobile Sources, Ann Arbor, Mich.

U.S. Environmental Protection Agency. 1999c. Exhaust emission factors for nonroad engine modeling—spark ignition. Report No. NR-010b, EPA420-R-99–009. EPA Office of Mobile Sources, Ann Arbor, Mich.

U.S. National Park Service. 1988. Management policies. U.S. Department of the Interior, Washington, D.C.

U.S. National Park Service. 1994. Natural resources management guidelines. NPS-77. U.S. Department of the Interior, Washington, D.C.

Wagner, F.H., R. Foresta, R.B. Gill, D.R. McCullough, M.R. Pelton, W.F. Porter, and H. Salwasser. 1995. Wildlife policies in U.S. National Parks. Island Press, Washington, D.C.

Williams, E.S., E.T. Thorne, S.L. Anderson, and J.D. Herriges, Jr. 1993. Brucellosis in free-ranging bison (*Bison bison*) from Teton County, Wyoming. J Wildl Dis 29:118–122.

Zaugg J.L., S.K. Taylor, B.C. Anderson, D.L. Hunter, J. Ryder, and M. Divine. 1993. Hematologic, serologic values, histopathologic and fecal evaluations of bison from Yellowstone Park. J Wildl Dis 29:453–457.

20

Wildlife Health, Ecosystems, and Rural Livelihoods in Botswana

Michael D. Kock

Gary R. Mullins

Jeremy S. Perkins

Botswana has a wealth of biogeographical habitats, ranging from swamps to sand dunes, allowing for great diversity in its native animal population. The economic importance of livestock, both to rural livelihoods and to foreign exchange earnings, has precipitated the implementation of strict veterinary disease control measures. Vaccination, movement control, test and slaughter, blanket slaughter, and aerial spraying are methods of animal disease control that have been utilized over the past 40 years. In this chapter we examine the measures used to control foot-and-mouth disease, contagious bovine pleuropneumonia, and trypanosomiasis and the impact these measures have had on disease incidence, wildlife, ecosystems, and human health. We also suggest alternative strategies for future development of the wildlife and livestock sectors.

20.1 Background

Botswana's semi-arid climate, periodic droughts, poor soils, and lack of reliable water supplies severely limit its agricultural and other economic opportunities. In the 1960s, due to the discovery of large diamond reserves in the Kalahari (Morna 1979) and preferential trade policies with the European Union, the economic outlook of the country improved drastically. The human population is expected to reach 1.7 million by year 2001, with 80% of all Botswanans living in the eastern third of the country (Roodt 1998). However, the current population growth rate of over 3% may be reduced in the future due to one of the highest HIV prevalence rates in the world. Biogeographically, Botswana may be divided into two broad regions corresponding to the Zambezian and Kalahari-Highveld zones.

A wide range of habitats, from the permanent swamps of the Okavango Delta to the sand dunes of the southwest, gives rise to a great diversity of wildlife (WCMC 1991).

The northern Zambezian system contains the only year-round water in the country, the Okavango Delta and the Chobe/Linyanti Rivers, characterized by the occurrence of mopane (*Colophospermum mopane*) woodland, which can form dense stands in the highest rainfall areas (i.e., 650 mm per year), and in the southern reaches, forest reserves of teak (*Baikiaca plurijuga*). Large herbivores such as elephants (*Loxodonta africana*), buffalo (*Syncerus caffer*) and zebra (*Equus burchelli*) are essentially confined to the areas of permanent water in the late dry season but disperse following good rains that collect in pools throughout the mopane areas.

By contrast, the Kalahari system spans the more arid portion of the rainfall gradient and has no permanent surface water. Only species that have physiological or selective feeding adaptations to this lack of drinking water are found there (Knight 1991). The water-dependent wildlife community, composed of the blue wildebeest (*Connochaetes taurinus*) and the red hartebeest (*Alcephalus buselaphus*), migrate to survive, especially in dry years.

Botswana economic development and growth reflect the growing conflict in Africa between the demands of people through their crop and livestock production, and the conservation of natural resources (Perkins and Ringrose 1996). Threats to the maintenance of biodiversity (WCMC 1991) include the following.

The spread of pastoralism and an increase in cattle numbers

Botswana cattle population doubled from 1966 to 1985, when it reached 2.75 million head. Their range was expanded into the Kalahari by tapping fossil aquifers several hundred meters below the sand with borehole technology, thus breaking through the "age-old protection" (Cooke 1985) afforded by a lack of surface water.

The erection of veterinary cordon fences

Many of the fenced areas designated for livestock production were established in the late 1950s and early 1960s, before habitat requirements and migration patterns of wildlife were understood; ecozones and seasonal and cyclical movements of wildlife were not considered. Further fencing was erected throughout the 1980s and 1990s, in response to disease outbreaks and to meet the increasingly stringent health requirements for beef exports to the European Union.

Habitat degradation and illegal hunting

For example, overpopulation of elephants in the northeast over the last 20 years has placed enormous pressure on the riparian woodland, resulting in habitat destruction. The potential for poaching and other illegal hunting activities has been increased by the geographical expansion of people and livestock. Indeed, the remoteness of the areas typically opened up for new agricultural production creates pressure to misuse wildlife and other natural resources in the absence of commercially available food or subsistence means.

Environmental changes, including drought

Drought cycles form an integral part of the natural climatic pattern in Botswana (Bhalotra 1987). Additionally, there has been a drying trend during the 1990s in southern Africa that is predicted to continue. Many desert species can live without surface water, but water is ultimately essential for the food that many species need to survive and reproduce.

Failure to recognize wildlife as a viable alternative or parallel land use

The concept of wildlife management areas (WMAs) was formulated in the early 1970s to serve local communities primarily through sustainable wildlife utilization and to act as a buffer zone between the protected areas, ranches, and cattle posts. Instead of permitting *both* livestock and wildlife utilization, WMAs have hosted the rapid expansion of cattle posts within them.

Improved access to the Kalahari

With the advent of four-wheel-drive vehicles and deep borehole drilling, the Kalahari is longer accessible not only to the Bushmen, but also to the wider population.

20.2 Diseases and Disease Control Strategies

We focus in this chapter on three diseases that have been of primary concern to officials in Botswana—foot and mouth disease (FMD), contagious bovine pleuropneumonia (CBP), and trypanosomiasis (TRY)—and the control measures that have been employed to limit or eradicate them in livestock, wildlife, and in the case of TRY, humans.

Foot-and-mouth disease is a highly infectious viral disease of Arthyodactyla characterized by high morbidity and low mortality (Coetzer et al. 1994). In Africa, it is primarily spread by direct contact (Thomson 1996), and infected animals can become carriers for periods of three or more years (C. Foggin and E. Anderson, personal commun. 1999). Transmission between cattle is well known, and recent studies have provided evidence of transmission of FMD between carrier African buffalo (*Syncerus caffer*) and cattle under field conditions, although it appears to be a rare event (Thomson 1996; Foggin and Taylor 1996). The E.U. veterinary authorities require strict disease control measures in countries where FMD is endemic in order to avoid bringing the FMD virus into Europe through animals or animal products.

Contagious bovine pleuropneumonia is a mycoplasmal disease of cattle caused by infection with *Mycoplasma mycoides bovis.* The disease causes significant mortality, and infected animals can become carriers (Coetzer et al. 1994). Agricultural economies based primarily on cattle can be devastated by an outbreak of CBP. The disease had not been observed in Botswana since 1939, until it was reintroduced in 1995 through illegal movement of cattle from Namibia.

Trypanosomiasis is a disease of humans, livestock, and wildlife caused by hemoparasite infection with the protozoan *Trypanosoma* spp? *brucei gambiense* or *T. brucei rhodensiense* are the relevant subspecies occurring in sub-Saharan Africa, although only the latter is known in Botswana. The hemoparasite is transmitted by the bite of the tsetse fly (*Glossina* spp.). The common name for the disease in humans is "sleeping sickness," and in livestock, "nagana." Trypanosomiasis is fatal without treatment. According to district and the Centers for Disease Control health records, there have been no reported cases of sleeping sickness in Botswana since 1985. No bovine TRY had been recorded since 1987, but in 2000 numerous cases resulting in animal deaths erased earlier speculation about spread of the tsetse fly and confirmed the reemergence of TRY in Botswana. There has also been one confirmed report of an infection in a horse at a safari camp in the Delta (D. Jackson, personal commun. 1997).

The livestock disease control methods used in Botswana include vaccination and movement of controls (such as fencing), test and slaughter, and blanket slaughter. Although discontinued in 1992, aerial spraying was conducted for 20 years to control the spread of tsetse fly and TRY. A disease control option that the authors argue has been inadequately used in Botswana is alternative land use apart from livestock production (Child 1995; Kock 1996).

The choice of a disease control method for livestock is complicated by the fact that wildlife is both part of the problem and possibly part of the solution. Because of selective pressures during the process of evolution, wild species have adapted to their environments and to many infectious agents, including parasites, providing them with natural disease resistance. Adaptation may also result in wildlife being able to maintain parasite and infectious agent populations and serve as potential carriers of disease (Coetzer et al. 1994). Livestock introduced to southern Africa from Europe had not been exposed to many of the indigenous wildlife-associated diseases and therefore lacked disease resistance. Many of the breeds of livestock that have a long history in Africa now show evidence of disease resistance, for example, the *Trypanosoma*-tolerant Ndama cattle of central Africa. Conversely, domestic animals can also carry and transmit diseases to wildlife with fatal consequences (Alexander 1999). Monitoring and surveillance are therefore warranted for both livestock and wildlife.

Vaccination of livestock is often the first preventive method employed. There are few economically important infectious diseases for which vaccines are not available today (Coetzer et al. 1994). In Botswana in certain disease control zones, cattle are vaccinated two or even three times a year against FMD. But vaccination may fail to protect against a disease if the vaccine is not based on the appropriate serotype, is poorly handled, or is improperly administered or if too small of a percentage of the total population is vaccinated.

Movement controls have been used in Botswana both as a preventive and as emergency disease control measures. Preventively, movement of animals, including wildlife and some animal products, is strictly controlled in many instances through the requirement of veterinary permits through the Department of Animal Health and Production (DAHP).

Botswana is divided into 20 disease control zones. Official disease control strategy entails FMD vaccination zones, buffer or surveillance zones, FMD-free zones, zones that are designated stock free, and one area intended to contain the buffalo population in the country. Since 1954, over 2,500 km of fencing has been erected within or along the borders of these zones to control movement of animals. The fences were not evaluated in terms of environmental impact prior to their erection, creating significant controversy (Keene-Young 1999; Albertson 1998). Recently, the government of Botswana has agreed to a retrospective environmental impact assessment of the major disease control fences in the northern part of the country (D. Raborokgwe, personal commun. 1999). Notably, however, the installation of an emergency fencing system failed to contain the spread of CBP in Ngamiland in 1995, and blanket slaughter was implemented in 1996–97.

The test-and-slaughter method requires adequate manpower both in the field and in the laboratory, as well as efficient transport and communication. It presumes sensitive and specific laboratory testing methods and is costly, but it may conserve genetically valuable stock. Blanket slaughter is a final stage when all other methods have failed, and it can be extremely costly, logistically difficult, and fraught with potential socioeconomic ramifications.

20.3 Effects of Disease Control Methods

The use of control measures for promoting livestock health has benefited human health indirectly, by providing economic and nutritional benefits, as well as directly by reducing possible exposure to zoonotic diseases such as TRY. A single, cable-reinforced fence surrounds the Okavango Delta to keep the resident buffalo out of the FMD-vaccinated zones. Conversely, the fence has also served to prevent human encroachment into the Delta, preserving the ecosystem. Active vector control programs are conducted within the Delta to reduce the risk of human exposure to tsetse flies and thereby the trypanosomes they are known to carry. As well as for local residents, the risk of disease is an important consideration for overseas tourists visiting many of the wildlife lodges in the Delta.

Wildlife populations in Botswana have declined significantly over the last 20 years, with drought, fencing, hunting, and habitat loss due to the expansion of livestock production cited as contributing factors (Perkins 1996; Perkins and Ringrose 1996; Crowe 1995). The effects of the fences on wildlife, particularly those Kalahari species that migrate extensively seeking water and forage, have been difficult to measure due to the lack of baseline data (Keene-Young 1999; Albertson 1998; Williamson and Williamson 1984). Some reports have been anecdotal (Campbell 1984; Albertson 1998), while others have been based on predictive techniques (Williamson and Williamson 1984). Few objective, long-term scientific studies have ever been conducted on the effects of the fences on wildlife, most notably the blue wildebeest (Owens 1980, 1981; Williamson and Williamson 1984), despite claims to the contrary (Albertson 1998). As Williamson and Williamson (1984) observed, the difficulty of quantifying accurately the damage done by these cordon fences is complicated by the lack of information on wildlife populations before the fences were erected, and a failure to monitor their effects afterward. Rarely have they been evaluated comprehensively in terms of their negative or positive effects on wildlife health, ecosystems, and rural livelihoods (Perkins and Ringrose 1996). Although some have placed considerable emphasis on the effect of fences, they have tended to discuss other important factors such as land degradation, increase in cattle numbers, and deep borehole drilling (Albertson 1998; Perkins and Ringrose 1996). The Ngamiland Environmental Impact Assessment (1999–2000) is the first effort allowed by government to closely examine the effects of the veterinary cordon fences in over 45 years.

During the 1995 CBP outbreak, veterinary authorities in Botswana erected three cordon fences in an attempt to stop the spread of the disease southward and potentially into the E.U. export areas. As the fences were constructed under "emergency powers," they were exempt from the usual legal requirement mandating an ex-ante environmental impact assessment. The Setata CBP fence was constructed as a double cordon and was aligned through two major wildlife management areas, NG 4 and NG 5, in which an active community-based natural resource management program was being implemented. The communities close to the western end of the fence still practice a simple hunter-gatherer livelihood strategy (R. Hartley, pers. commun. 1999). The Setata fence had a direct and adverse impact on wildlife and the livelihoods of these hunter-gatherers. The effect of this fence on the distribution of key wildlife species is clear. Eland (*Taurotragus oryx*) and gemsbok (*Oryx gazella*) have a north–south movement in this area in response to needs for food and water, and the likely effect of this fence will be to disrupt normal migration patterns. The fence will further fragment the wildlife populations present, particularly an important subpopulation of giraffe (*Giraffa camelopardalis*) that exists in the area.

A double cordon, electrified fence with cable was constructed along the Botswana northern border with the Caprivi Strip in Namibia. Along the western border, the same double fence was then extended southward to link up with an existing double fence. The Okavango buffalo fence also was extended northward to link up with the aforementioned Namibia-Botswana border fence. These fences will have a significant negative impact on any future community-based programs, particularly among remote area dwellers both in Namibia and Botswana (Weaver and Newby 1996; Weaver 1997; K. Ross, Conservation International, personal commun. 1998).

After the fencing strategy failed to prevent the spread of CBP into the Haina Veld farms southeast of Maun, a blanket slaughter program for Ngamiland was approved and between April 1996 and January 1997, approximately 320,000 head of cattle were destroyed. The control of CBP cost the Botswana Government approximately US$100 million, and the ramifications of this epidemic for human health and its effects on ecosystem stability and wildlife health have been significant. In retrospect, the CBP epidemic should be evaluated critically in terms of whether livestock production was the appropriate form of land use in the first place, how fence alignment failed to consider neighboring land uses, and how to provide adequate risk protection for subsistence communities through economic diversification (Townsend and Sigwele 1998).

Tsetse fly control was officially introduced in Botswana in 1944 to protect the growing human population of the northwestern district of Ngamiland and its burgeoning livestock industry (Colonial Development Office circa 1930, in Davies 1980a,b). Over the next quarter century, more than 63,000 wild animals were destroyed, thousands of hectares of wilderness clear cut, trees ring-barked or burned, hundreds of kilometers of fence erected, and more than 30,000 liters of insecticide ground-sprayed in attempts to eliminate tsetse. Those efforts failed, although they were sufficiently successful in reducing tsetse populations to enable the livestock industry to survive. Indirectly, tsetse control is also undoubtedly responsible for making the Okavango Delta the flourishing tourist destination it is today; indeed, Mullins et al. (1997) concluded that tourists are today the main beneficiaries of tsetse control.

The tourist market has unquestionably affected methods of tsetse control. Large-scale aerial spraying using endosulfan began in 1973 and continued until 1992, when public concern about possible environmental effects led the government of Botswana to switch to the more environmentally friendly odor-baited, insecticide-impregnated targets (Willemse 1991; Vale et al. 1988).

20.4 Challenges for the Future

Present patterns of human economic activity in Botswana need to be reexamined with a view toward seeking those that are compatible in the long term with sustaining the health and integrity of the Kalahari ecosystem. Agricultural production systems such as extensive livestock production on semi-arid or arid rangelands requires major inputs in terms of disease control, artificial water points, and feed supplementation, among others, and it is clear that these systems are not viable without government subsidization.

The Kalahari is an ecological jewel that has the potential to contribute through wildlife utilization to the sustainable development of future generations of Botswana. It therefore behooves the authorities to develop and implement appropriate strategies through an integrated approach to the maintenance of healthy ecosystems, wildlife, and people. Tourism in Botswana already generates more foreign exchange than agriculture and is expected to grow by an average of 5% per annum through 2020, according to the Department of Tourism. With appropriate policies, more of this income and its benefits can be directed to currently disadvantaged rural populations, as was envisioned in 1980: "Enhanced game use is seen as the best way to raise the standard of living of the greatest number of people in the Kalahari, particularly those who are the poorest" (DHV Consulting 1980, p. 45).

An integrated approach toward economic development that recognizes the needs and limits of the environment as well as societal values is necessary, with the ultimate goal of "meeting the needs of the present without compromising the welfare of future generations." There are several key strategies that could contribute to maintaining biodiversity within the Kalahari ecosystem. These include

- linkage of protected areas through creation of movement corridors to ameliorate the effects of the insularization of wildlife populations that is occurring in the Kalahari as livestock numbers encroach on its periphery
- promotion of land uses other than livestock production, particulary those that are conservation related
- development of bioregional management plans: the Kgalagadi Transfrontier Park incorporating South Africa's Kalahari Gemsbok and Botswana's Gembok national parks is a recent example of a farsighted initiative between the two governments.

20.5 A More Enlightened and Sensitive Approach to Disease Control

The application of science should allow rational development of policy and decision making through the provision of knowledge in a form that the public and society understands (Rapport et al. 1998). New knowledge and experience are available that can guide and serve as the basis for environmentally friendly approaches to disease control. This is particularly important given the increasing value of wildlife through sustainable utilization and ecotourism.

New technologies such as geographic information systems (GIS) allow the effects of livestock and fires on rangeland to be visualized and documented more thoroughly (R. Stuart-Hill, BRIMP, personal commun. 1998). The epidemiology and transmission of economically important diseases in livestock and wildlife are being increasingly elucidated. We can help buffalo overcome the FMD virus by separating and hand-rearing calves (Foggin and Taylor 1996) or isolating animals into small herds under certain circumstances (G. Thomson, personal commun. 1999). Calves can be captured from their mothers before they are six months old, hand reared, and tested periodically. With several negative tests, including for Corridor disease (*Theileria* spp.), these animals could be declared disease free. With disease-free buffalo bringing as much as US$24,900 a head in 1998 it behoves agricultural authorities to explore the potential for commercial utilization of buffalo (centre for Wildlife Economics 1999).

A realistic evaluation of the future commercial livestock industry in Botswana must be made. Livestock production has failed both as a sustainable economic activity as well as a means of alleviating rural poverty. Simultaneously, livestock production and the policies supporting it have had severe direct and indirect effects on the environment. In the new millennium, Botswana's livestock industry will have to be far more competitive to survive in the global market (Mannathoko 1999; BMC 1998) and far more efficient in terms of monitoring and identification by source of each animal, a prerequisite in the aftermath of bovine spongiform encephalopathy in Europe (Raborokgwe 1999).

As shown in Zimbabwe, South Africa, and Namibia (Kock 1996; Child 1995), an ecologically sustainable alternative to livestock production in some parts of Botswana is wildlife ranching (Conybeare and Rozemeijer 1991; Conybeare 1998). Economic diversification is highlighted as an essential component of the country's future national development plan, encouraging exploration of alternatives for increasing income generation in rural areas. A wildlife ranching industry could diversify the rural economy by providing opportunities for ecotourism, live game sales, and sport hunting. The Department of Wildlife and National Parks (DWNP) has initiated an experimental game ranch program in the western part of Botswana that will form the basis for education and training (J. Broekhuis, DWNP, personal commun. 1999; Grossman Associates 1999). Development of wildlife quarantine facilities would enhance the ability of communities and ranchers to market wildlife within Botswana, and to export live game animals to lucrative markets elsewhere in southern Africa and even overseas, if changes in the present animal disease control legislation can be implemented (Conybeare and Rozemeijer 1991; Coneybeare 1998). The production of disease-free animals, particularly buffalo, would expedite these goals. If managed correctly, these opportunities should result in benefits to rural communities where they are needed most and where livestock production has failed.

20.6 Conclusions

Botswana faces a crisis in biodiversity in this century with issues of conservation biology and ecosystem health coming to the fore. Solutions to the issues of land use, including livestock production, wildlife, and natural resource utilization, require both a biological understanding and political support. Botswana must adopt an integrated approach to conservation that includes consideration not only of societal values and human well-being, but also of wildlife health and ecosystem stability. Past and present disease control strategies implemented without adequate regard for ecosystem health must be replaced with more sensitive and enlightened approaches.

References

Albertson, A. 1998. Northern Botswana veterinary fences: critical ecological impacts. Okavango Peoples Wildlife Trust, Maun, Botswana.

Alexander, K. 1999. Domestic animals: a historical problem for wildlife populations. In Proceedings of a national conference on conservation and management wildlife in Botswana: strategies for the twenty-first century, p. 153–156. Botswana Wildlife Service, Maun.

Bhalotra, Y.P.R. 1987. Climate of Botswana—Part II: elements of climate. Department of Meteorological Services, Ministry of Works and Communications, Gaborone, Botswana.

Botswana Meat Commission (BMC). 1998. Annual report. Lobatse, Botswana.

Campbell, A. 1984. A comment on Kalahari wildlife and the Khukhe fence. Botswana Notes Rec 13:111–118.

Centre for Wildlife Economics. 1999. Potchefstroom University, Potchefstroom, Republic of South Africa.

Child, G. 1995. Wildlife and People: the Zimbabwean success. WISDOM Foundation, Harare, Zimbabwe.

Coetzer, J., G. Thomson, and R. Tustin (eds.). 1994. Infectious diseases of livestock with special reference to southern Africa, vols. 1 and 2. Oxford University Press, New York.

Conybeare, A. 1998. Botswana multi-species range utilisation project. Department of Animal Health and Production, Gaborone, Botswana.

Conybeare, A., and N. Rozemeijer. 1991. Game ranching in Botswana: an assessment of the game ranching potential of eight controlled hunting areas. Ministry of Local Government and Lands, and Department of Wildlife and National Parks, Gaborone, Botswana.

Cooke, H.J. 1985. The Kalahari today: a case of conflict over resource use. Geogr J 151: 75–85.

Crowe, D. 1995. Status of selected wildlife resources in Botswana and recommendations for conservation actions. In The present status of wildlife and its Future in botswana, proceedings of a symposium, p. 11–31. Kalahari Conservation Society and Chobe Wildlife Trust, Maun.

Davies, J. 1980a. The history of tsetse fly control in Botswana. Ministry of Agriculture, Gaborone, Botswana.

Davies, J. 1980b. Sleeping sickness and the factors affecting it in Botswana. Ministry of Agriculture, Tsetse Control Division, Maun, Botswana.

DHV Consulting. 1980. Countrywide animal and range assessment project, 7 vols. European Development Fund and Ministry of Commerce and Industry, Gaborone, Botswana.

Foggin, C.M., and R.D. Taylor. 1996. Management and utilisation of the African buffalo in Zimbabwe. In Penzhorn, B.L., ed. Proceedings of the African buffalo as a game ranch animal, p. 144–158. South African Veterinary Association Wildlife Group, Pretoria.

Grossman Associates. 1999. Management plan for an experimental game ranch. Ministry of Commerce and Industry, Gaborone, Botswana.

Keene-Young, A.D. 1999. A thin line: Botswana's cattle fences. Africa Environ Wildl March/April, p. 72–79.

Knight, M.H. 1991. Ecology of the gemsbok (Oryx gazella gazella, Linnaeus) and blue wildebeest (Connochaetes taurinus, Burchell) in the southern Kalahari. Ph.D. thesis, University of Pretoria.

Kock, M.D. 1996. Zimbabwe: a model for the sustainable use of wildlife and development of innovative wildlife management practices. In Taylor, V.J., and N. Dunstone, eds. The exploitation of mammal populations, p. 229–249. Chapman and Hall, London.

Mannathoko, M. 1999. Agriculture, particularly the livestock industry as a form of sustainable land use. In Proceedings of a national conference on conservation and management of wildlife in Botswana: strategies for the twenty-first century, pp. 121–126. Botswana Wildlife Service, Maun.

Morna, C.L. 1979. Beyond the drought. Afr Rep 34(6):30–33.

Mullins, G., R. Allsopp, P. Nkhori, M. Kolanyane, and T. Phillemon-Motsu. 1997. The effects of tsetse fly and tsetse control on tourism in the Okavango Delta of Botswana. *In* Proceedings of the 24th meeting of the International Scientific Council for Trypanosomiasis Research and Control, 29 Sept.– 3 Oct. 1997, Maputo, Mozambique, ISCTRC, Maputo.

Owens, M.D. 1980. The fences of death. Afr Wildl 34(6):25–27.

Owens, M.D. 1981. Preliminary final report on central Kalahari predator research. Mimeographed report to the Department of Wildlife and National Parks, DWNP, Gaborone, Botswana.

Perkins, J.S. 1996. Botswana: fencing out the equity issue. Cattleposts and cattle ranches in the Kalahari desert. J. Arid Environ 33:503–517.

Perkins, J.S., and Ringrose, S.M. 1996. Development cooperation objectives and the beef protocol, the case of Botswana: a study of livestock, wildlife, tourism, degradation linkages. Metroeconomica Ltd., Gabarone, Botswana.

Raborokgwe, M.V. 1999. Cordon fences and wildlife issues. *In* Proceedings of a national conference on conservation and management of wildlife in botswana: strategies for the twenty-first century (Oct. 1999), p. 114–120. Botswana Wildlife Service, Maun.

Rapport, D., R. Costanza, P.R. Epstein, C. Gaudet, and R. Levins (eds.). 1998. Ecosystem health. Blackwell, London.

Roodt, V. 1998. The Shell tourist guide to Botswana. Shell Botswana, Gaborone, Botswana.

Thomson, G.R. 1996. Foot and mouth disease in the African buffalo. *In* Penzhorn, B.L., ed. Proceedings of The African buffalo as a game ranch animal, p. 113–120. South African Veterinary Association Wildlife Group, Pretoria.

Townsend, R.F., and H.K. Sigwele. 1998. Socio-economic cost-benefit analysis of action and alternatives for the control of contagious bovine pleuropneumonia in Ngamiland, Botswana. Department of Agricultural Economics, University of Pretoria.

Vale G., D. Lovemore, S. Flint, and G. Cockbill. 1988. Odour-baited targets to control tsetse flies, *Glossina* spp. (Diptera: Glossinidae), in Zimbabwe. Bull Entomol Res 78: 31–49.

Weaver, L.C. 1997. Background and potential impacts resulting from construction of a game and livestock proof fence by the government of Botswana south of the West Caprivi Game Reserve. World Wildlife Fund/Life programme, Windhoek, Namibia.

Weaver, L.C., and J. Newby, 1996. Potential ecological and social impacts of a veterinary control barrier between Botswana and Western Caprivi, Namibia. World Wildlife Fund, Gland, Switzerland.

Willemse, L. (1991) A trial of odour baited targets to control the tsetse fly, *Glossina morsitans centralis* (Diptera: Glossinidae) in west Zambia. Bull Entomol Res 81:351–357.

Williamson, D.T. and J.E. Williamson. 1984. An assessment of the impact of fences on large herbivore biomass in the Kalahari. Botswana Notes Rec 13:107–110.

World Conservation Monitoring Centre (WCMC). 1991. Biodiversity guide to Botswana. A report financed by the Commission of European Communities. WCMC, Cambridge.

21

Zoological Parks in Endangered Species Recovery and Conservation

Anthony Allchurch

The emergence of the scientific discipline of conservation medicine represents a timely response to increasing concerns about new or rediscovered diseases and their relationship to climate change, habitat degradation, and loss of biodiversity. Although primarily addressing issues affecting wildlife and human health, this new multidisciplinary science will have beneficial applications in the environment of captive animal management and may also profit from the wealth of experience in that domain.

Zoo medicine traditionally has considered disease a random process to be addressed through clinical response on an individual basis. This institutional focus on sustaining individual lives may be contrasted with the emphasis on population health favored by wildlife veterinarians (Hutchins et al. 1991). The concept of disease control in zoos through promoting a holistic system of health management is still gaining acceptance and may present unusual challenges that are peculiar to the dynamics of managing small populations in captivity (Miller 1992).

By comparison, proponents of conservation medicine are convinced that patterns of morbidity and mortality in wild animal and human populations may be predictable and consequential to changes in ecosystem health. The naturalistic habitats developed in modern zoos provide managed environments where systematic evaluation and the mapping and modeling of disease processes and host–parasite relationships may be studied in depth and, over time, support the basic tenets of this emerging field (Seal 1998).

Zoological institutions on the cutting edge of wildlife conservation maintain their captive populations of endangered species in support of in situ programs. The zoo veterinary community is presented with excellent opportunities to contribute its skills and services to successful and sustainable reintroduction programs. The health screening of captive wild animals for translocation between zoos is comprehensively prescribed through regulations and codes of practice. Zoo animals selected for reintroduction are subjected to even more stringent protocols for health evaluation and prophylactic conditioning for the release process.

Achieving zero tolerance for disease risks, however, may actually obstruct the progress of reintroduction programs (Ballou 1993) through compromising the competence of the individual to resist naturally occurring diseases. Indeed, it has been suggested (Lyles and Dobson 1993) that excessive protection of captive animals may increase their vulnerability to diseases in the wild. If this dilemma is the consequence of compliance with legislation and conformity with institutional standards and professional responsibilities, we must try to establish an intrinsic balance between the health status of free-living and captive communities of wild animals.

The cycle of transition from wild to captive living and ultimately to reintroduction into native habitats presents significant and complex health challenges. We need to understand the dynamic relationships between disease organisms and their hosts and the prevalent influences in the captive and wild environments that convert infection of the host into clinical disease. If the movement of captive animals is recognized as a stressor to initiate the disease process, then similar hazards may be experienced by free-living populations that become displaced through fragmentation or degradation of habitats.

Comparative analysis of the disease profiles of captive and free-living forms of a species may help in this endeavor. We have recently conducted a detailed study of the parasitic flora of the Mallorcan midwife toad (*Alytes muletensis*) using fecal material sampled from captive and free-living populations. Since 1985 this critically endangered species has been managed in captivity in Jersey Zoo, the headquarters of the Durrell Wildlife Conservation Trust, culminating in a series of multiple releases of captive bred specimens within their known historical range in Mallorca, the largest of the Balearic Islands located off the eastern seaboard of Spain. The success of this program has established it as a model for amphibian reintroduction, and no apparent disease consequences have been recorded. This could have been predicted by the results from fecal screening of the species, which established that there was no significant difference in the resident parasite profiles of the two populations and that the levels of pathogenic organisms were very low (C. Gonzales, unpublished data 2000).

The Medical Animal Record Keeping System (MedARKS) provides an excellent management tool for improving and refining healthcare (Teare 1991) and for maintaining individual specimen lifetime health records. New diagnostic technologies facilitate the comprehensive monitoring of captive animal populations for evidence of exposure to potential pathogens such as *Chlamydia psittaci* and *Yersinia pseudotuberculosis* and enable community health to be assessed on a quantitative basis. Spatial and temporal influences on disease patterns within the population may be considered within the framework of captive environmental health risks, which may emulate events occurring in free-living populations.

Over 400 birds in the collection of endangered species at Jersey Zoo have been closely monitored for evidence of exposure to *C. psittaci* during a five-year period. The initiative for this elaborate program originated from the loss of a number of specimens of pink pigeon, *Columba mayeri*, exported from Jersey to various institutions in the United States. Chlamydiosis was diagnosed as cause of death in each case, but vigorous testing of the captive population in Jersey has yielded only three transient positive results in a series of 136 tests on 74 pigeons. In total, 30 species of pigeons, parrots, pheasants, and waterfowl have been screened for *Chlamydia* using a polymerase chain reaction (PCR) detection system (Hewinson et al. 1991) and their individual histories and enclosures logged in MedARKS to provide a comprehensive model of the distribution of this organism in the bird population. Although clinical chlamydiosis has never been diagnosed in the collection in Jersey, the evidence of its existence has been established and appropriate strategies are employed to reduce the risk of clinical events occurring.

Similar efforts are now in progress to identify the distribution of the pathogenic organism *Yersinia pseudotuberculosis* in the unique collection of Callitrichidae at the Jersey Zoo. For many years yersiniosis has been the most persistent cause of acute or chronic fatal disease in marmosets and tamarins in captivity in Jersey (Bielli et al. 1999). A variety of measures have been implemented to control its incidence, including vaccination, prophylactic therapy, and improved husbandry. Now a PCR screen for *Y. pseudotuberculosis* is being tested for its sensitivity and specificity for identifying the organism in fecal material. Parallel screening of soil taken from enclosures housing the test animals is also being conducted in order to investigate the provenance of the disease in infected animals. Whenever possible, free-living forms of native wildlife such as small birds and rodents are also screened to determine their role in the epidemiology of yersiniosis in the Jersey collection. Once a reliable PCR method has been established, it will be extremely valuable for collaborative studies with those who are studying callitrichids in the wild.

More difficult to determine is the distribution of *Mycobacterium avium* within the bird collection. The malignant nature of avian tuberculosis caused by this organism and its relative obscurity within the host present exceptional challenges to its accurate and early identification. Nevertheless, once it is established as a causal agent of disease, risk factors can be assessed for bird populations exposed to the organism by direct contact or through contamination of habitats within the zoo.

Progressive husbandry practices in zoos increasingly favor free-living life styles for appropriate avian species. As a consequence, exposure to wild bird populations becomes more frequent and disease risks assume greater significance. Common enteric pathogens such as *Campylobacter, Salmonella*, and *Escherichia coli* are found in very high prevalences in gulls, crows, pigeons, and smaller bird species, and they have been shown to amplify environmental contamination with these organisms and increase the dissemination of infection to captive populations, which act as long-term reservoirs of infection (Pennycott 1999).

Similar consideration must be given to the relationship of parasitic diseases among free-living native species and the captive communities. Monitoring of zoo animals through fecal screening for parasites is the most frequently employed noninvasive procedure for health assessment. The concept of zero tolerance for endoparasites appears to enjoy universal acceptance in zoos even though wild animals invariably carry parasitic burdens without apparent clinical consequence. Records also reveal that whereas some species constantly provide positive results for parasites in spite of appropriate treatment, others remain negative for all parasites in the absence of any therapy. Unfortunately, there is also ample evidence of significant or even overwhelming infections in newborn or newly introduced animals that are exposed to parasitic challenges for the first time. Currently, a study of the incidence and distribution of gapeworms (*Syngamus* sp.) is in progress within the zoo grounds with particular focus on the vulnerability of infant crane chicks to infections with either *S. trachea* or *S. bronchialis* (J. Smith, unpublished data 2000). Rigorous sampling programs of bird fecal output, of soil in open aviaries, and of potential intermediate hosts such as the common earthworm are providing clues to the host–parasite relationship of these challenging parasites.

Open-plan zoos with natural features such as water systems, grassland, and abundant flora provide complex ecosystems that are essentially manageable and can be comprehensively monitored. Water quality can be measured quantitatively in free-flowing natural watercourses, and the effects of pollution and the influences of constant dilution can be identified by sequential sampling. Contained artificial water systems such as ponds and moats present particular problems, and in the absence of purification and/or filtration facilities, rapid deterioration in quality can occur, presenting significant health hazards. Local meteorological data can be recorded to create a survey of disease pathogens that may be temperature or weather dependent. Seasonal variations in disease incidence may be identified through such analysis. Weather data for Jersey recorded on a daily basis have been accumulated over at least a decade and can be analyzed for any potential influence in patterns of disease in the animal collection. Studies of *Y. pseudotuberculosis* confirm increased incidence of the disease during the winter period with consistent peaks in March each year (Brice 1995).

It is reasonable to postulate that a securely managed zoological park is effectively a closed community with controlled animal movements. Such a configuration provides a model for studying and evaluating disease processes, and when that zoo is located within an island such as Jersey, unique opportunities are presented for such endeavors. Surrounded by water, favored with a mild, temperate climate, and protected by controls that are stringent and historic, Jersey has been free of any communicable diseases such as tuberculosis or Newcastle disease for almost two decades. Ancient laws protecting the Jersey breed of cow have effectively prevented the importation of diseases common to domestic livestock elsewhere. Rabies regulations provide effective security against many infectious diseases of mammals, and the relative paucity of native wildlife, including arthropod vectors of disease, contributes to the unusually benign health climate in the human and animal populations.

The animal collection at Jersey Zoo is exceptionally dynamic, with constant movement of animals to other zoos and for reintroduction purposes and occasional importations of critically endangered species directly from the wild. Strict conformity with CITES controls and implementation of European Association of

Zoos and Aquaria (EAZA) guidelines for the health screening of animals for transportation, combined with intensive and extensive health monitoring programs that have been promoted for many years, has been successful in establishing an animal community with excellent health status. The IUCN Species Survival Commission (SSC) provides scientific and technical advice on conservation issues related to reintroduction of species. The SSC is the largest aggregation of species specialists in the world (IUCN 2000).

As a consequence, those disease events that do occur may be characterized by their sudden peaks of incidence rather than their regular activity, and these provide ample opportunity for critical evaluation. Epizootics of yersiniosis in Rodrigues fruit bats (*Pteropus rodricensis*), of shigellosis in orangutans (*Pongo pygmaeus*), and of duck virus enteritis in Meller's ducks (*Anas melleri*), have been encountered in recent years as essentially isolated occurrences. Such events invariably prove to have been predictable based on identifiable predisposing factors such as overcrowding of captive populations and exposure to prolonged inclement weather conditions.

Where managing animal health is the focus of veterinary endeavor captive animal populations should flourish. Success can then be demonstrated by long and productive lives. The comprehensive animal records maintained by the Durrell Wildlife Conservation Trust at Jersey Zoo identify longevity records in 10 species of mammals (Mallinson and Barker 1998). Such a facility provides an excellent model for teaching and training purposes, and the trust has graduated over 1,000 students of conservation biology from more than 100 different countries at its international training center. The multidisciplinary professional environment developed by the Trust is an ideal medium for nurturing the emerging science of conservation medicine. Expanding the parameters of veterinary endeavor is crucial to the development of professional competence to meet new challenges and opportunities. Conservation medicine will benefit from the services of those who are able to address the complex environmental issues beyond the individual consequences (Boyce et al. 1992).

Gerald Durrell once famously described veterinarians, along with architects, as the most dangerous creatures to allow into the zoo. Now zoo veterinarians with vision and imagination are joining the vanguard of this new science and sharing with professional colleagues from other disciplines their unique resources of data on animal health and the dynamics of disease processes in small populations.

References

Ballou, J.D. 1993. Assessing the risks of infectious diseases in captive breeding and reintroduction programs. J Zoo Wildl Med 24:327–335.

Bielli, M., S. Lauzi, A. Pratelli, M. Martini, P. Dall'Ara, and L. Bonizzi. 1999. Pseudotuberculosis in marmosets, tamarins and Goeldi's monkeys (Callithrichidae/Callimiconidae) housed at a European zoo. J Zoo Wildl Med 30:532–536.

Boyce, W., T. Yuill, J. Homan, and D. Jessup. 1992. A role for veterinarians in wildlife health and conservation biology. JAVMA 200:435–437.

Brice, S. 1995. Screening a new world monkey colony for *Yersinia* and investigations of *Y. pseudotuberculosis* and soil. Dodo (J Wildl Preserv Trust) 31:139–147.

Hewinson, R.G., S.E.S. Rankin, B.J. Bevan, M. Field, and M.J. Woodward. 1991. Detection of *Chlamydia psittaci* from avian field samples using the PCR. Vet Rec 128:129–130.

Hutchins, M., T. Foose, and U.S. Seal. 1991. The role of veterinary medicine in endangered species conservation. J Zoo Wildl Med 22:277–281.

Lyles, A.M., and A.P. Dobson. 1993. Infectious disease and intensive management: population dynamics, threatened hosts and their parasites. J Zoo Wildl Med 24:315–326.

Mallinson, J.J.C., and P. Barker. 1998. A record of mammalian longevity at the Jersey Wildlife Preservation Trust with comparative data. Dodo (Wildl Preserv Trust) 34:8–17.

Miller, R.E. 1992. Zoo veterinarians—doctors on the ark? JAVMA 200:642–647.

Pennycott, T.W. 1999. *Salmonella, Campylobacter* and *E. coli* in wild birds. In proceedings of the seminar on zoonotic diseases of U.K. wildlife, at the British Veterinary Association Annual congress (23 Sept. 1999, Bath) pp. 19–28. BVA, Bath.

Seal, U.S. 1998. The value of veterinary medicine in wildlife conservation. Keynote address delivered at the AAZV/AAWV joint conference, Omaha, Nebr.

Teare, J.A. 1991. Medical record systems for the next century: the case for improved health care through integrated databases. J Zoo Wildl Med 22:389–391.

World Conservation Congress. 2000. Policy guidelines. Species Specialist Groups, Gland, Switzerland.

22

The Mountain Gorilla and Conservation Medicine

Michael Cranfield
Lynne Gaffikin
Jonathan Sleeman
Matthew Rooney

Certain species, especially those conservation biologists term "umbrella species," engender a broad suite of ecological values and act as an ecological proxy for habitat conservation. The mountain gorilla (*Gorilla gorilla beringei*) serves as an umbrella species within the montane forests of Nile–Zaire Divide of Central Africa. Unique challenges to the mountain gorilla dictate that a different and perhaps more concentrated approach to conservation must be employed than that used for the other two subspecies of gorilla. Subject to the risks associated with small population dynamics, the mountain gorilla numbers are estimated at 600–640 individuals, found in two equal-sized isolated populations in two protected park areas, the Virunga Mountains, including 375 km^2 in Rwanda, Democratic Republic of Congo and Uganda, and Bwindi Impenetrable Forest including 330 km^2 in Uganda (Butynski and Kalina 1998). It is postulated that these two populations have been separated for approximately 1,000 years, the length of time the landscape between these forests has been converted to intensive agricultural use. Although impossible to substantiate, historical numbers of mountain gorillas before agricultural activities began have been estimated at approximately 10,000 (C. Sholley, personal commun. 1999).

Conservation threats to mountain gorillas can be summarized as follows: habitat loss (at a rate of 0.5–1.9% yearly), incidental hunting (for both subsistence and the rapidly expanding commercial bushmeat market), intrinsic susceptibility to disease and loss of genetic fitness as a result of small population size, management challenges as a result of the political bureaucratic dimensions of three separate governments and transboundary operational difficulties for nongovernmental organizations (NGOs), and perhaps most insidious, the severe ecological imbalance brought about by the influence of humans on the gorilla-inhabited ecosystem. In a more direct fashion, people and gorillas are in greater contact with each other through pressures of expanding ecotourism and the burgeoning human population of about 300–400 people/km², with a growth rate of 3.7% annually. In addition, expanding intensive agricultural practices around the park have increased the risk of human–gorilla transmission of disease and direct or indirect degradation of gorilla habitat (Butynski amd Kalina 1998).

There are several NGOs and governmental organizations addressing the conservation of gorillas: International Gorilla Conservation Project (IGCP), Dian Fossey Gorilla Fund International, Dian Fossey Gorilla Foundation Europe, Berggorilla and Regenwald Direkthilfe, Institute for Tropical Forest Conservation, Wildlife Conservation Society, German Technical Cooperation, and CARE. All these groups are united by a common concern for gorilla survival and work cooperatively to address these unique challenges. In Democratic Republic of Congo, Uganda, and Rwanda, the three governmental organizations responsible for all decisions relating to gorillas in the respective countries are Institut Congolaise pour la Conservation de la Nature, Uganda Wildlife Authority (UWA), and Office Rwandais du Tourisme et des Parcs National. Nongovernmental organizations work to assist managers conserve gorillas through written memoranda of understanding. The Mountain Gorilla Veterinary Project (MGVP) is responsible in Rwanda and Democratic Republic of Congo for all aspects of veterinary medicine and shares this responsibility with UWA and IGCP in Uganda.

The conservation of the mountain gorilla's is inextricably linked to the health of their ecosystem, the health of humans who frequently contact gorillas, and the health of the animals themselves. As such, the mountain gorilla encapsulates the concepts of integrated health put forward in conservation medicine. First, humans and gorillas are closely related genetically: humans share over 97% genetic makeup with the Pongidae family that includes gorillas, bonobos, chimpanzees, and orangutans (Sibley and Ahlquist 1984). Given this genetic closeness, gorillas can provide an important source of comparative data on anatomy, physiology, and behavior and provide better leads to human evolution. Second, humans and gorillas share a large and ever increasing overlap in territorial space. This, coupled with the genetic similarities, results in a high potential for interspecies disease transmission. Currently, humans share with great apes more infectious diseases than with any other animal family. Finally, as mentioned above, ecotourism to view the mountain gorillas has grown in popularity, as gorilla-viewing trips are profitable enterprises. The money generated from gorilla ecotourism is one of the largest sources of foreign revenue in these African countries. Programs for visiting the mountain gorillas have been so successful that they have been heralded as a model for ecotourism elsewhere.

22.1 Historical Background

In 1902, Captain Oscar Von Beringei was the first European to observe the mountain gorilla. He and fellow explorer Dr. England spotted a group of black apes while climbing Mount Sabinyo of the Virunga Mountains. They shot two of these animals and sent them to the great German anatomist Matschu who claimed them to be a separate subspecies. This started a flurry of scientific interest throughout the world, and 54 gorillas from the Virunga area were killed in the name of science between 1902 and 1929. After shooting five mountain gorillas in 1929 for the American Museum of Natural History, Carl Ackey was so impressed with the subspecies and its habitat that he urged the Belgian Government, headed by King Albert, to make the Virunga Mountains a national park. As a result of his efforts, that same year Albert National Park was established as one of the first national parks in Africa.

Schaller (1963) initiated the first field biological study of mountain gorillas in 1959. He estimated that at that time there were 400–500 individuals in the Virunga population, plus an unknown number in the Bwindi Impenetrable Forest population. Over the ensuing two decades, approximately one third of the parkland was converted to agricultural land through an international aid program to grow *Pyrethrum*, and a market developed among local inhabitants for gorilla body parts for tourist souvenirs. By 1981, the gorilla population in the Virunga Mountains had declined to an estimated low of 260 animals, and a sustainable population of mountain gorillas had grim prospects (Harcourt 1986).

The decline in population size would likely have continued and total numbers today would certainly be lower if it had not been for the efforts of Dian Fossey, who worked in the Virungas between 1966 and 1985. The approach she used, however, characterized by the exclusion of humans and cattle from the park using tight security, focused only on a single species. What she accomplished can more accurately be labeled as preservation rather than conservation of the mountain gorilla. Her approach did not address important human socioeconomic and non-biological issues, which are critical to sustainable conservation efforts in that region; for example, external human influences outside the park, such as agricultural practices, affect wildlife inside the park.

Through increasing economic support to areas surrounding the park as a result of worldwide appeal, stronger local governmental support and donor funding, coupled with increasing activism on the part of conservationists, NGOs, and embassy and international aid agencies, efforts to save the subspecies slowly expanded to a more comprehensive conservation approach to the mountain gorilla and its habitat. This effort manifested in three major events: (1) the establishment of ecotourism in 1978, which provided a financial incentive for conserving gorillas; (2) the establishment in 1979 of an environmental education program for the people living in the vicinity of the park (which included hundreds of thousands of Rwandans); and (3) the establishment of a veterinary program in 1986 to provide emergency clinical care to gorilla individuals meeting policy criteria for medical intervention (Butynski and Kalina 1998).

Realizing the need for clinical veterinary involvement to save gorilla lives, Dian Fossey received support through the Morris Animal Foundation and its patrons to establish the Volcano Veterinary Medicine Centre, a veterinary program for gorillas in Rwanda. The center was set up to provide emergency care to severely injured or sick animals or those with human-induced injuries. Dr. James

Foster was the first wildlife veterinarian to serve at the center, beginning in 1986 (Foster 1992). The project supported a clinical veterinary program for nine years, until 1995, when it changed its name to the Mountain Gorilla Veterinary Project (MGVP). The application of clinical veterinary medicine to a wild population by creating a field base of operations was a novel concept. Many conservationists were skeptical at first that veterinary input could contribute to the long-term survival of a species. As the program evolved, the veterinary effort has expanded beyond just clinical veterinary care and now also includes preventive care approaches as well as the addition of conservation medical techniques.

22.2 Clinical Medicine

It has only been since the 1960s that veterinarians have had the technology and drugs to safely immobilize wildlife, implement successful treatments, and collect specimens. Even with technology, conservation programs are usually population oriented. Genetic studies O.E. Ryder, personal commun. 1999) have indicated that the mountain gorilla population has been reduced to such low numbers that the genetic makeup of each individual, especially females, is important to the biodiversity and long-term survival of the population (Garner and Ryder 1996). Intervention requires cautious approaches, especially when an endangered species is involved. As such, the current policy regarding when to clinically intervene in a gorilla health situation is conservative, restricted to cases deemed to be potentially life threatening to an individual or to the population, with particular emphasis on human-induced problems. A risk/benefit analysis is performed, and the local conservators, their departments, and MGVP and/or UWA veterinarians make the final decision regarding whether or not to take action. Although field veterinarians trained in primate medicine carry out interventions and treatment schemes, human medical doctors are sometimes consulted.

Due to the rugged field conditions and the need to quickly return the individual to its group before they move on, clinical medicine has been limited to basic first aid and surgery. The majority of interventions to date have been for poaching-related injuries. A recent report covering the interval 1971–95 revealed that there were 11 random years with no poacher-related injuries and 61 reports of poacher-related injuries in the other 14 years. The majority of the injuries (70%) were snare related, and 17 (28%) of them resulted in mortality. Veterinary interventions have also occurred for infectious diseases such as pneumonia.

While some interventions require only the administration of antibiotics for disease or physical restraint to remove a snare, between 1986 and 1998 there were 26 situations where anesthesia was required. Four of the 26 immobilized animals, or 15%, did not recover from anesthesia due to severe preintervention morbidity. Ketamine at 2.5 mg/kg is the anesthetic of choice because of the short duration of effect, but tiletamine/zolazepam at 1.0 mg/kg is sometimes utilized as it produces deeper and longer anesthesia with more relaxation and analgesia. Considering the compromised high-risk group being treated, interventions by trained staff appear to be a relatively safe undertaking. Immediate clinical assessment has been aided by the recent addition of a pulse oximeter (Nellcor) and a portable clinical analyzer (ISTAT) (Sleeman et al. 1998).

The MGVP continues to look for and develop novel ways of monitoring health utilizing available noninvasive samples such as feces and urine. Feces have been utilized for parasite and bacteriological studies in the past, and currently the intestinal flora of both normal and diarrheic stool of Rwanda gorillas is being intensively studied. Additionally, monitoring of fecal cortisol levels in habituated tourist groups and from wild groups will determine if stress generated by ecotourism is significant. Also, it is feasible to collect urine from habituated free-ranging gorillas for diagnostic purposes. Urine has been used to report reference values and to study endocrine function, carbohydrate metabolism, kidney and liver function, acid–base balance, urinary tract infections, neoplasia, crystals, and parasite ova (Czekala et al. 1994; Mudakikwa and Sleeman 1997).

22.3 Preventive Medicine

While clinical veterinary medicine has contributed positively over the years to the maintenance of the mountain gorilla population (Butynski and Kalina 1998), the success rate of clinical medicine applied to wild animals even in captivity is limited. This is due to gaps in existing knowledge of disease susceptibility, pathogenesis, and pharmacology, the inability to perform even moderately invasive medical procedures, and limitations in how care can be managed (e.g., continuous intravenous medications cannot be provided), due to the nature of the animal and its situation. Preventive medicine has been part of the veterinary field for many decades, and although more easily accomplished with domestic animals and captive wildlife (Fowler 1986; Franzmann 1993; Foster 1993), its application to the mountain gorilla has been expanding over the years.

22.4 Surveillance and Monitoring

Observational health monitoring, the complement of disease surveillance, involves the systematic collection and evaluation of general health data that can lead to detection of disease at even earlier stages. Types of clinical data collected as part of mountain gorilla health monitoring include abnormal discharge, breathing, fur condition, movement, and manipulative behavior as well as appetite and the presence of any injuries. Field veterinarians modify field data collection sheets to improve the relevance of clinical observations for health monitoring purposes. Once collected, this information is entered in the ARKS and MedARKS databases, produced for captive wildlife management but applicable to wild populations. Together with information collected through geographical information systems, the clinical observation data will allow for epidemiological analysis of the temporal and spatial patterns of disease. This should help define seasonal trends and high-risk groups and help evaluate outcomes of disease occurrences, allowing for improved intervention decision making. Other data are obtained through analysis of biological samples taken opportunistically during clinical interventions.

The MGVP has recently developed a universal protocol for optimally collecting samples from immobilized gorillas. While most samples are processed in the context of routine screening to yield information about the general health of the group, with newly developed field equipment, some samples can be processed immediately in the field to help the clinician with on-the-spot decisions about the individual case at hand. Samples not used for disease surveillance or health monitoring or to aid in clinical diagnosis can be stored in a variety of preservatives for future research purposes.

22.5 Serology

Analysis of existing serological data revealed evidence of an alpha herpes virus in four of seven (58%; usually younger animals) individuals tested from the Virunga mountains (Eberle 1994). This virus has similarities with, but is distinct from, human herpes simplex virus type 2. Suspicious facial macules and papules observed on adult female and infant/juvenile gorillas may represent the clinical manifestation of the virus. No mortality has been associated with this clinical presentation, but it could potentially be a serious zoonotic disease. Two other serum samples from Bwindi showed positive titers to chimpanzee cytomegalovirus and Epstein-Barr virus but were negative for influenza A + B, parainfluenza types 1–3, measles, respiratory syncytial virus, SA-8 African monkey herpes virus, herpes simplex 1 and 2, human varicella-zoster virus, simian retrovirus 1, 2, and 5, simian T-cell leukemia virus, and simian immunodeficiency virus.

22.6 Bacteriology

In a recent study of gorilla populations in Bwindi and Mgahinga National Parks, intestinal pathogens were isolated in 35% and 71% of fecal specimens, respectively. The prevalence of *Salmonella* and *Campylobacter* in habituated groups had doubled from a study conducted four years previously to a new level of 18% and 9%, respectively (Kalema 1994; Nizeyi et al. 2001). *Shigella* was isolated from wild gorillas for the first time in this study, at a prevalence of 5%. Although wild gorillas have been asymptomatic or mildly symptomatic, these are considered significant pathogens in captive collections. These pathogens are significant for their increasing prevalence and their potential to cause significant disease in the host, and as a possible source of disease for humans in close contact.

22.7 Parasitology

Mountain gorilla parasites have been the subject of many studies since before the MGVP veterinary program was initiated. Redmond (1983) completed the first study in 1978, and newer studies have yielded an expanding list of gastrointestinal parasites. Three parasites not cited in previous studies, *Trichuris trichuris, Chilomastix,* and *Endolimax nana,* have been identified by Mudakikwa and colleagues (1998). Although the reason for this could be as simple as different laboratory techniques or skill level of observers, it is equally possible that they were introduced via human contamination, since all three also are human parasites. The prevalence of helminths in mountain gorillas is higher (97–100%) than that of the eastern lowland gorilla (67%) or the western lowland gorilla (36%) (Redmond 1983; Hastings et al. 1992; Landsout-Soukate et al. 1995; Mudakikwa et al. 1998; Eilenberger 1998; Sleeman et al. 2000). Although many factors affect the prevalence of helminths, this increase could be due either to increased frequency of exposure or perhaps to a less competent immune system due to the relatively elevated population density of the mountain gorilla (MG = 1 gorilla/1.2 km², ELG = 1 gorilla/6 km², WLG = 1 gorilla/5 km⁵). Nizeyi and colleagues (1999) reported a very high prevalence (11%) of *Cryptosporidium* in the Bwindi Park, and Kalema and colleagues (1998) reported mortality due to scabies in the same area. Both of these studies would support a reduction of immune competence and possibly cross-infection from humans and/or domestic animals. Although medications for parasites are not used at the present time, they remain a potentially viable tool for clinical/preventive veterinary care.

22.8 Pathology

The MGVP has standardized the protocol for gross postmortems and has appropriate CITES and CDC permits to assure rapid transport of tissues when host country permits are issued. There have been 89 postmortems conducted on gorillas since Fossey first arrived. Detailed records exist for 42 of the postmortems, and for 25 of them there are representative fixed tissues. Preliminary results from these postmortems have been published (Lowenstine 1990; Graczyk et al. 1999; Mudakikwa et al. 2001), and the detailed study of 31 skeletal remains that revealed a high prevalence of arthritis and periapical abscesses and a low prevalence of osteomas and developmental disorders. Unfortunately, these necropsies represent only a fraction of the deaths, since dead gorillas are not always found. Investigators with varying medical backgrounds completed many of the postmortems in the early days, but their information was not always detailed. Necropsies are often performed under less than ideal field conditions, and many times tissues are not fresh.

A great deal of information can be obtained from necropsies, including diagnostic information during a disease outbreak, so that effective management decisions can be applied to remaining susceptible animals; incidental findings, which can suggest causes of significant morbidity within a population; delineating high risk subgroups within the population, allowing targeting of resources; and identifying diseases associated with poor nutrition or sanitation problems, both indicators of habitat degradation. Also, since primate diseases tend to cluster within island populations rather than by species across island populations, comparative information on the pathology of any primates in cohabitation with gorillas can be informative to their conservation (Wolfe et al. 1998).

22.9 Vaccinations

Vaccines, several of which have proven safe and effective for captive gorillas, and are technologically feasible for wild gorillas are not normally used (Hastings et al. 1991). This is due to the strict nonintervention policy. Contingency plans that include vaccination for several of the more important diseases are now being formulated, however, based on positive previous experiences with vaccination campaigns and some changes in conservation ideology. Although never definitively diagnosed, measles was implicated in a respiratory disease outbreak in 1988. Acting on the assumption that measles was the cause of morbidity, 65 gorillas (a large percentage of the individuals in the group) were vaccinated with a human-attenuated live vaccine. Subsequently, the outbreak subsided. Although the vaccine efficacy could not be proven, the experience demonstrated that it is possible to safely vaccinate a wild population of gorillas (Hasting et al. 1991).

22.10 Zoonoses

In terms of health risk to the gorillas, infectious diseases are the major concern. They are the most likely cause of premature death, pose the highest risk for affecting population viability, and are the easiest to eliminate or reduce in terms of identifying and modifying associated risk factors. Because of the genetic similarity between gorillas and humans, gorillas are susceptible to many infectious diseases transmitted from humans (anthropozoonoses). Prior to recent habituation for research and ecotourism purposes, humans and gorillas rarely, if ever, spent time in close proximity. Having been previously isolated from close human contact, the gorillas provide a potentially "naive" population for infectious organisms found in human populations, meaning they would not possess any acquired immunity to these diseases. If a human-borne organism to which they were naive were to be introduced into a gorilla population, it would likely be more virulent and disease would manifest more severely than in the general human population. Diseases of specific concern to gorillas are poliomyelitis, measles, tuberculosis, shigellosis, salmonellosis, and influenza.

22.10.1 Ecotourism

Ecotourism is an increasingly popular, nonconsumptive, sustainable use of wildlife. Many conservationists propose that ecotourism is an important natural resource management tool, allowing local authorities and the local population to derive financial benefits from the existence of the wildlife, thereby combining the goals of development and conservation. Any use of a wild species, whether consumptive or nonconsumptive, inevitably causes changes to the ecosystem (Allen and Edwards 1995). Uncontrolled ecotourism could equally destroy the flora and fauna it was designed to preserve. In addition to habitat destruction, behavioral disturbances, and other negative influences (Hartcourt 1986; Kinnard and O'Brien 1996), ecotourism also potentially poses risks to the health of wildlife, domestic animals, and human populations. As noted above, the tourism program for visiting the mountain gorillas of central Africa in their natural habitat has been heralded as a model of ecotourism (Weber 1993). It equally illustrates the dangers inherent in these types of programs and the need for veterinary and public health input into the planning, implementation, and evaluation of ecotourism programs. Mountain gorillas have been habituated to the presence of humans, allowing daily visits by tourists within five meters of the gorillas. To prevent pathogen exchange, strict regulations have been devised regarding number of visitors per day, duration of the visit, proximity to the animals, and behavior during the visit. Because of a recent study of zoonotic diseases (Homsey 1999) identifying both modes of transmission and poor compliance with the existing rules, even stricter regulations are being proposed. Established tourist rules include denying sick persons access to the gorillas; not allowing children under 15 years to visit; limiting visits to one visit to each group per day; limiting visits to eight persons and for one hour only; maintaining a distance of five meters from the gorillas; strictly forbidding touching gorillas. If a gorilla approaches, then the visitor must slowly back away; asking visitors who have to sneeze or cough to turn away from the gorillas; asking visitors who break the regulations to leave the group; burying human waste at least 30 cm into the soil; and prohibiting littering, smoking, and eating during the visit.

The situation is similar for an increasing number of free-ranging chimpanzee groups habituated for tourism in Uganda. A study was carried out using self-administered questionnaires to determine the potential risks for zoonotic disease transmission from tourists to chimpanzees during visits in Kibale National Park, Uganda (Adams et al. 1999). Key information obtained from tourists was vaccination and disease histories as well as any recent occurrence of disease. A total of 43 tourists, mostly from developed countries (93%), were interviewed between July and November 1998. Despite the fact that the vast majority of respondents were from areas now relatively free of infectious diseases, and with good access to health care, there were an alarming low percentage of up-to-date vaccinations for many diseases. For example, 84% of those surveyed were vaccinated against hepatitis A, of which only 56% were up-to-date. Among individuals who admitted that they had been previously diagnosed with a disease listed on the questionnaire, five with herpes virus infection, four with influenza, one with chicken pox, and two with tuberculosis were considered to still be infectious at the time of visitation. Importantly, there was a high prevalence of current clinical signs among the tourists, suggestive of an infectious disease. This may reflect that foreign tourists may be more prone to infectious diseases while traveling due to the stress of travel, change of diet, and exposure to novel pathogens (Wilson 1995). The most common symptoms cited by interviewees were diarrhea (53%), coughing (23%), and vomiting (13%). A large proportion of these tourists (70%) indicated that they had visited or planned to visit another group of habituated great apes during their trip to Africa. The authors concluded that the potential for human disease transmission to the chimpanzees was evident, and that this study should assist the wildlife authorities in developing appropriate tourist visit regulations to prevent disease transmission, for example, a mandatory vaccination policy, the use of face masks, mandatory washing of hands, disinfectant foot baths for all tourists prior to the visit, and adequate pit latrines for proper disposal of human waste.

22.10.2 Local Inhabitants

The local inhabitants now have increased contact with mountain gorillas habituated by ecotourism. This is partly because gorilla habituation has altered the behavior of the animals such that more time is spent foraging outside the park (M. Goldsmith, personal commun. 1999). This is also due to the burgeoning human population of host countries. For example, Uganda's human population is projected to increase from 21 million to 32 million by the year 2015 (Anonymous 1997). Surrounding the Virunga Mountains is in an area of particularly high human population density (300 people/km^2). Increases in human population will inevitably result in increased pressure to convert park areas to agricultural land and harvesting of forest products, which will further increase the potential for human contact with the gorillas.

In an effort to identify the most common, potentially zoonotic diseases, a survey was conducted of all diseases reported by local hospitals in the western districts of Uganda containing gorilla and/or chimpanzee populations in 1997. The data were accessed from the Ugandan Health Ministry's Health Monitoring Information System and from district medical offices. Diagnoses were classified according to World Health Organization guidelines, and a disease was determined to be zoonotic based on available literature (Ott-Joslin 1993). Results revealed that 2.93 million outpatient diagnoses were made in the 12 western districts studied (table 22.1). Of these, between 68.2% and 79.4% (mean for all districts = 73.2%) resulted from an infection potentially transmissible to great apes. The review identified malaria, respiratory tract infections, intestinal parasites, diarrheas (dysentery, acute and persistent), skin disorders, measles, tuberculosis, and occasionally poliomyelitis as common potentially anthropozoonotic diseases.

This kind of study provides the local health authorities, donors, and policy makers with information that should help to improve the human health in the region through appropriate public health measures such as vaccination campaigns or increasing the number of properly constructed pit latrines in the villages close to the forest. Additional studies are needed to more accurately determine the types and prevalence of zoonotic diseases specifically in the areas surrounding the mountain gorilla. The zoonotic potential of these diseases should be more fully evaluated, and measures to mitigate the impact of human contact with mountain gorillas need to be devised. The host country government's health monitoring information system and those of public health NGOs can be useful in monitoring

Table 22.1. Selected outpatient diagnoses[a] for 12 western districts of Uganda, January 1997 through December 1998

Diagnosis	Number	(% of all diagnoses)
Malaria[b]	870,866	27.9
Respiratory tract infections	636,907	22.9
Intestinal parasites	309,132	10.6
Diarrheas	160,140	6.1
Skin disorders	140,544	4.8
Measles	11,625	0.4
Tuberculosis	5,011	0.2
Poliomyelitis	2,729 (2,717 from one district)	0.1

[a]These diagnoses were considered the most likely diseases transmissible from humans to chimpanzees and gorillas of the same region.
[b]Ollomo et al. (1997) indicate that human malaria is probably not transmissible to great apes.

disease trends in the local human population and serve as a warning system or sentinel for detecting potential disease threats to the gorillas.

22.10.3 Gorilla Conservation Personnel

Gorilla personnel have the most frequent close encounters with gorillas, thus potentially confer the highest risk for human disease transmission. Despite this, no systematic approach is used to identify and monitor health and existing diseases among park personnel or field assistants. A high priority for MGVP in the immediate future is to establish an occupational health program for screening, treating, and/or vaccinating this group for diseases considered high risk to the gorillas. Clearly, such a program will also have to pay attention to nonanthropozoonotic diseases that negatively affect conservation worker (and family) health or well-being, referring them to a local health care provider as appropriate. This effort will require intensive collaboration between veterinary, human medicine, and public health sectors, and as such may represent a prime example of conservation medicine in the field.

22.11 Research Agenda

Due to the fact that tremendous gaps exist in even the most basic data necessary to effectively protect the gorillas from injury and disease, a research program has been designed as part of the MGVP to proactively seek answers to fundamental questions. The Morris Animal Foundation (MAF), through its Wildlife Scientific Advisory Committee, recently distributed its second worldwide call for research proposals. Seven studies were funded this year, totaling over US$60,000. In addition to the two previously mentioned fecal studies, other studies range in scope from the impact of rodent diseases on the gorillas to the application of modern molecular diagnostic techniques to archived material (from the 1988 respiratory outbreak) for the purposes of definitive diagnosis.

Utilizing the captive lion-tailed macaque (*Macaca silenus*) as a model, the MGVP is working on capturing and preserving live genetic material. This material will be preserved for future genetic and reproductive work on animals, for many of which there is extensive behavioral data and that may be deceased by the time advanced reproductive techniques such as cloning have been developed as a functional tool. Field protocols are being developed to culture in vitro, lymphocytes from peripheral blood and epithelial cells from biopsies collected opportunistically during interventions or at necropsy. Other protocols for gamete preservation, through either xenotransplantation or freezing, are being refined. Additionally, MAF now houses a newly constructed biological resource center in Denver, Colorado. Its mission is to apply a wide range of long-term storage techniques to biological specimens collected during medical interventions, necropsies, and routine monitoring (of feces and urine) for use by researchers from many disciplines in the future.

22.12 Conclusions

Opponents to preventive disease management of wild populations claim that the approach to wildlife conservation described in this chapter is undesirable interference in the "natural" selective forces of evolution. Yet, when one examines the human-induced pressures placed on the gorillas, one realizes that the situation is already far from "natural." The mountain gorillas have been forced to live in restricted areas of higher elevation characterized by wetter and colder climate conditions. They also are forced to experience unnatural contact with humans through ever increasing pressures of ecotourism. The era of setting aside wildlife areas without establishing a plan for managing the protected area is over in most parts of the world. There are no natural systems left on the earth either sufficiently large or free enough from human encroachment. Active management should therefore be a part of any conservation effort dealing with a highly endangered species on a fragile habitat (Franzmann 1993). Mountain gorilla conservation involves both. The future of the mountain gorilla is dubious without a medically oriented or disease-oriented program with an active research component upon which to base sound management decisions.

The MGVP has evolved over the past 14 years from a simple, fire engine veterinary practice on the side of a mountain to managing the health of a species from a clinical, preventive, public health, and research perspectives. Medical management is one of many tools that the gorilla conservation biology team needs to employ to ensure the long-term survival of the gorillas. As the MGVP continues to evolve and put into practice additional conservation medicine principles it will, by necessity, have to expand its scope to embrace other disciplines in a unified fashion.

Future objectives of MGVP include institutionalizing the surveillance of gorilla health using a health systems approach and increasing the capacity to use health data at all levels for conservation decision making. In addition, these efforts will strengthen local institutional capabilities to address health threats to mountain gorillas through the development of an international and local collaborative networks of organizations and institutions involved in gorilla research and conservation, the creation of human and institutional resources to train veterinarians in wildlife medicine, and the establishment of a health information system for mountain gorillas. This effort will involve developing a better data collection and information processing system, providing grass roots health education to all persons having contact with or influencing the gorilla habitat, collaboration with existing human health networks to ensure a two-way flow of health information and the development of coordinated strategies to reduce health risks, and developing contingency plans and using in-country institutional capacities. Mountain gorilla conservation possibly has the greatest potential for a collaborative and unified effort by veterinarians, physicians, public health officials, conservation biologists, and professionals from other disciplines to promote ecological health and well-being of a defined habitat. We believe that conservation medicine is our avenue for the conservation of mountain gorillas.

References

Adams, H.R., J. Sleeman, and J.C. New. 1999. Medical survey of tourists visiting Kibale National Park, Uganda, to determine the potential risk for disease transmission to chimpanzees (*Pan troglodytes*) from ecotourism. In Proceedings of the American Association of Zoo Veterinarians, pp. 270–271. AAZV, Columbus, Ohio.

Allen, C.M., and S.R. Edwards. 1995. The sustainable use debate: observations from IUCN. Oryx 29:92–98.

Anonymous. 1997. Statistical abstract. Ministry of Planning and Economic Development, Kampala, Republic of Uganda.

Butynski, T.M., and J. Kalina. 1998. Gorilla tourism: a critical look. *In* Milner-Gulland, E.J., and R. Mace, eds. Conservation of biological resources, pp. 294–366. Oxford University Press, New York.

Czekala, N., V.A. Lance, and M. Sutherland-Smith. 1994. Diurnal urinary corticoid excretion in the human and gorilla. Am J Primatol 34:29–34.

Eberle, R.B. 1994. Serological evidence of an alpha-herpesvirus indigenous to mountain gorillas. J Med Primatol 21:246–251.

Eilenberger, U. 1998. Der Einfluss von individuellen, gruppenspezifischen und oekologischen Faktoren auf den Endoparasitenstatus von wildlebenden oestlichen Flachlandgorillas (*Gorilla gorilla graueri*) im Kahuzi-Biega National Park von Zaire. Ein multidisziplinaerer Anatz Ph.D. Thesis, Institut für Zoo-und Wildtier Forschung. Berlin.

Foster, J.W. 1992. Mountain gorilla conservation: a study in human values. JAVMA 200: 629–633.

Foster, W.F. 1993. Health plan for the mountain gorillas of Rwanda. *In* Fowler, M., (ed.). Zoo and wildlife medicine, Current therapy 3rd ed. pp. 331–334. Saunders, Philadelphia.

Fowler, M. 1986. Zoo and wildlife medicine, 2nd ed., pp. 14–17. Saunders, Philadelphia.

Franzmann, A.W. 1993. Veterinary contributions to international wildlife management. *In* Fowler M., eds. Zoo and wild animal medicine, Current Therapy 3, pp. 42–45. Saunders, Philadelphia.

Garner, K.J. and O.A. Ryder. 1996. Mitochondrial DNA diversity in gorillas. Mole Phylogen Evol 6:39.

Graczyk, T.K., L.J. Lowenstine, and M.R. Cranfield. 1999. *Capillaria hepatica* (Nematoda) infections in human-habituated mountain gorillas (*Gorilla gorilla beringei*) of the Parc National des Volcans, Rwanda. J. Parasitol 85:168–170.

Harcourt, A.H. 1986. Gorilla conservation: anatomy of a campaign. *In* Benirsckke, E. ed. Primates: the road to self-sustaining populations. pp. 32–46. Springer, New York.

Hastings, B.E., D. Kenny, L.J. Lowenstine, and J.W. Foster. 1991. Mountain gorillas and measles: ontogeny of a wildlife vaccination. *In* Proceedings of the AAZV pp. 198–205. Calgary, Alberta.

Hastings, B.E., L.M. Gibbons, and J.E. Williams. 1992. Parasites of free-ranging mountain gorillas: survey and epidemiological factors. In Proceedings of the AAZV/AAWV joint conference, pp. 301–302. AAZV, Oakland, California.

Homsey, J. 1999. Ape tourism and human diseases: how close should we get? Report for the International Gorilla Conservation Programme regional meeting, Kampala, Uganda.

Kalema, G. 1994. Epidemiology of the intestinal parasite burden of mountain gorillas in Bwindi Impenetrable Forest National Park. Report of the Uganda Wildlife Authority, Kampala, Uganda.

Kalema, G., R.A. Koch, and I.J. Macfie. 1998. An outbreak of sarcoptic mange in free-ranging mountain gorillas (*Gorilla gorilla beringei*) in Bwindi Impenetrable Forest, Southwest Uganda. In Proceedings of the AAZV/AAWV joint conference p. 438. AAZV, Omaha, Ne.

Kinnard, M.F., and T.G. O'Brien. 1996. Ecotourism in the Tangkoko DuaSudara Nature Reserve: opening Pandora's box? Oryx 30:65–73.

Landsout-Soukate, J., C.E.G. Tutin, and M. Fernandez. 1995. Intestinal parasites of sympatric gorillas and chimpanzees in Lope Reserve, Gabon. Ann Trop Med Parasitol 89:73–79.

Lowenstine, L.J. 1990. Long distance pathology, or will the mountain gorilla fit in the diplomatic pouch? In Proceedings of the AAZV, pp. 178–185. AAZV, S. Padre Island, Tex.

Mudakikwa, A.B., and J.M. Sleeman. 1997. Urinalysis of free-living mountain gorillas (*Gorilla gorilla beringei*) normal physiological values. In Proceedings of the AAZV Conf. p. 278. AAZV, Houston, Tex.

Mudakikwa, A.B., J.M. Sleeman, J.W. Foster, L.L. Madder, and S. Patton. 1998. An indicator of human impact: gastrointestinal parasites of mountain gorillas (*Gorilla gorilla beringei*) from the Virunga Volcanoes region, Central Africa. In Proceedings of the AAZV/AAWA joint conferences, pp. 436–437. AAZV. Omaha, Nebr.

Mudakikwa, A.B., M.R. Cranfield, J.M. Sleeman, and U. Eilenbeger. 2002. Clinical medicine, preventive health monitoring and research on mountain gorillas in the Virunga Volcanoes region. *In* Sicotte, P., and K. Stewart, eds. Mountain gorillas. Cambridge University Press, Cambridge. (in press)

Nizeyi, J.B., R.B. Innocent, J. Erume, G.R.N. Kalema, M.R. Cranfield, and T.K. Graczyk. 2001. Campylobacteriosis, salmonellosis, and shigelosis in free-ranging human-habituated mountain gorillas in Uganda. J Wildl Dis 37:239–244.

Ollomo, B., S. Karch, P. Bureau, N. Elissa, A.J. Georges, and P. Millet. 1997. Lack of malaria parasite transmission between apes and humans in Gabon. A. J Trop Med Hyg 56:440–445.

Ott-Joslin, J.E. 1993. Zoonotic diseases of nonhuman primates. *In* Fowler, M.E. ed. Zoo and wild animal medicine, Current therapy 3, pp. 358–373. Saunders, Philadelphia.

Redmond, I. 1983. Karisoke parasitology research: summary of parasitology research, November 1976 to April 1978. *In* Fossey, D., Gorillas in the mist, pp. 271–286. Houghton Mifflin, Boston.

Schaller, G. 1963. The mountain gorilla: ecology and behavior. University of Chicago Press, Chicago.

Sibley, C.G., and J.E. Ahlquist. 1984. The phylogeny of the hominid primates as indicated by DNA-DNA hybridization. J Mol Evol 20:2–15.

Sleeman, J.M., K. Cameron, A.B. Mudakikwa, J.B. Nizeyi, S. Anderson, H.M. Richardson, E.J. Macfie, B. Hastings, and J.W. Foster. 1998. Field anesthesia of free ranging mountain gorillas (*Gorilla gorilla beringei*) from the Virunga Volcano region, Central Africa. In Proceedings of the AAZV/AAWV joint conference, pp. 1–4. AAZV, Omaha, Omaha, Nebr.

Sleeman, J.M., L.L. Meader, A.B. Mudakikwa, J.W. Foster, and S. Patton. 2000. Gastrointestinal parasites of mountain gorillas (*Gorilla gorilla beringei*) in the Parc National des Volcans, Rwanda. J Zoo Wildl Med 31:322–328.

Weber, W. 1993. Primate conservation and eco-tourism in Africa. *In* Potter, C., ed. The conservation of genetic resources, pp. 129–150. American Association for the Advancement of Science, Washington, D.C.

Wilson, M.E. 1995. Travel and emergence of infectious diseases. Emerg Infect Dis 1:39–46.

Wolfe, N.D., A.A. Escalante, W.B. Karesh, A. Kilbourn, A. Spielman, and A.A. Lol. 1998. Wild primate populations in emerging infectious disease research: the missing link? Emerg Infect Dis 4:149–158.

23

Linking Human and Ecosystem Health on the Amazon Frontier

Tamsyn P. Murray
James J. Kay
David Waltner-Toews
Ernesto F. Ráez-Luna

Our ability to enhance the health of ecosystems is predicated on an understanding of the interactions among ecosystem dynamics, natural resource use, and human health. Since 1996, a team of Canadian and Peruvian researchers has been developing an adaptive ecosystem approach to human health. In keeping with the standard definition of methodology used in the systems literature (Checkland and Scholes 1990), the ecosystem approach provides an interdisciplinary, holistic guide to how to investigate complex socioecological problems, drawing on and bringing together a wide variety of methods, actors, and scales of investigation. In the case study explored in this chapter, we have been particularly interested in the relationships between the structure and function of stressed ecosystems and land use strategies, natural resource use and nutritional status, and anthropogenic environmental changes and the transmission of disease. The intent of our work is to improve human health of local people through better management of their natural resources. Drawing on World Health Organization (WHO) documents and reviews of the health literature in relation to people, animals, plants, and ecosystems, we derived a working definition of health as being the capacity to achieve socially determined goals (mental, physical, and social well-being, vigor, resilience, productivity, flourishing) within a set of socioecological constraints, only one of which is disease (Sundsvall 1991; Waltner-Toews and Wall 1997; WHO 1978). This chapter describes the evolution of the ecosystem approach as it was applied in the frontier regions of the Peruvian Amazon.

23.1 Background and History of the Ecosystem Approach to Human Health

We developed the ecosystem approach in two phases, beginning with the development of a conceptual framework that brought together the most recent understanding of ecosystems as complex systems with secondary data and exploratory fieldwork, and then moving to its application in understanding and improving a problematic situation in the Peruvian Amazon. Equipped with an integrated conceptual framework for ecosystem sustainability and the tools and heuristics of systems theory, we revisited the Ucayali region in an attempt to provide new insights into the driving forces of human health. Ucayali is an area in the Peruvian Amazon where Centro Internacional de Agricultura Tropical (CIAT) and other international institutions, despite decades of research and technology development, have failed to alleviate the interrelated problems of economic stagnation, ecological degradation and deteriorating human health. A more holistic view of the region dynamics was created and has identified information gaps necessary to shed light on the key determinants of, and linkages between, ecosystem and human health, based on the existing data that had been gathered since the early 1970s.

Several fields of research were explored in developing the framework: complex systems theory (Schneider and Kay 1994); ecosystem-based management and hierarchy theory (All and Hoekstra 1992; Kay et al. 1999); post-normal science (Funtowicz and Ravetz 1994); complex adaptive systems and resilience (Gunderson et al. 1995); social and collaborative learning (Roling and Wagemakers 1998); soft systems methodology (Checkland 1981); critical systems thinking (Midgley 1992; Flood and Carson 1988); participatory action research (Greenwood and Levin 1998; Wadsworth 1998; Gijt et al. 1999); and sustainable livelihoods approach (Chambers and Conway 1992; Singh and Wanmali 1998).

23.2 The Research Setting

The Ucayali region of Eastern Peru is populated by 370,000 people and spans 100,000 square kilometers. In the 1940s a road connecting Pucallpa on the Ucayali River, a major Amazon tributary, with Lima hastened settlement from the coast and the Andes. Now, about 80% of its population lives either in Pucallpa or on the Lima road, creating agricultural production and food security challenges. Despite the natural diversity and fertility of this region, remote rural communities struggle to meet their basic needs and face a range of nutritional and health problems (INEI 1997). As a result of slash-and-burn agriculture, deforestation has steadily increased, and logging activities continue unregulated and with unknown ecological ramifications. At present, the relationship between household production, income levels, and health in the Ucayali region is complex and poorly understood. Exploitation of local resources results in diverse seasonal combinations of farming, fishing, logging, and hunting and gathering activities. It is not yet known, however, how these different resource strategies affect household health, or whether health and earnings are related—problems that have important implications for agricultural and technology development in the region. Ucayali therefore presents researchers with a very complex and dynamic set of interconnected issues. In addition, Ucayali is a benchmark site for the Consultative Group on International Agricultural Research's Eco-Regional Program and the focus of coordinated research efforts whose findings have potential application to other forest margins areas in the tropics. Therefore, it provides a valuable opportunity for the ecosystem approach to synthesize different yet interdependent dimensions within the same region.

Until the late 1990s, analyses of land use in the Ucayali region described an ecosystem dominated by small-scale cattle ranchers struggling to maintain productivity in the face of declining soil fertility. Key issues included problems of market access, technology and capital constraints, and labor scarcity. Ecological issues were confined to farm-level problems of land degradation and weed invasion. Solutions based on such an understanding tended to be technological or policy related, emphasizing economic incentives, production strategies, and access to improved varieties and/or species. Although this research provided many important insights into farm-level production and ecological constraints, it failed to link these processes with other resource sectors, actors, and regional and national policies. Despite the fact that the international research agenda in the Peruvian Amazon has slowly evolved from an agricultural focus on pastures and cattle to an integrated natural resources management approach, the evaluation of production strategies and land use technologies is confined to their contribution to income; human health and nutrition is ignored. Based on this preliminary study, we determined that (1) the Ucayali region presented us with a complex set of interacting issues and problems; (2) research and management practices designed to solve specific economic and environmental problems in a linear fashion, oblivious of context, either failed or created more problems than they solved; and (3) health (broadly defined, as above) suggested some key linkages, hitherto ignored, between people and the system in which they lived, and that might be used to suggest management strategies for general socioecological improvement.

We began to use the ecosystem approach as a guide to further research in 1999. Researchers from CIAT and the University of Guelph received funding from the International Development and Research Centre, Ottawa, for a project entitled Health, Biodiversity and Natural Resource Use on the Amazon Frontier: An Ecosystem Approach. We set out to determine the relationship between natural resource use, ecosystem dynamics, and human health. We investigated how the adaptive land use strategies of frontier communities relate to, and are synchronized with, the constantly changing floodplain. A large multidisciplinary team based in Pucallpa conducted an extensive field-based study based on the research questions raised. In summary, the field research involved household surveys, field tests, and participatory methods in eight different communities in the Ucayali region, involving a total of 345 families. To aid in differentiating the factors affecting health, we selected native and colonist communities in both the floodplain and upland forests. In addition to differences in ethnicity and ecosystem type, communities varied in their degree of access to and involvement in the market economy, the time of settlement, and dominant land use strategy (slash-and-burn agriculture, fishing, cattle ranching, and oil palm plantations). As annual flooding cycles determine the seasonal patterns in agricultural production and the availability of fish, animals, and foods gathered from local forests, fieldwork was conducted on three separate occasions: the dry season (June/July 1999), the start of the rainy season (October/November 1999), and the wet season (February/March 2000). Using these three time periods and data sets, we analyzed the cycles of food production and availability, disease outbreaks, and nutrient intake. This helped identify the critical periods when food is most scarce and disease most prevalent, thereby facilitating more targeted and effective interventions. We are currently analyzing the data, developing community action plans addressing the major health issues, and working with the Ministries of Health, Agriculture, Fisheries and Forestry to improve their policies and programs and to increase their capacity to respond to complex, interacting, environmental-agricultural-health problems.

23.3 The Conceptual Basis of the Ecosystem Approach

There are three guiding principles that form the conceptual basis of the ecosystem approach: (1) methodological pluralism, (2) multilevel investigation, and (3) local participation and action research. In this chapter, we explain these theoretical bases, how they were applied in the context of the Ucayali region, the methodological outcomes, and the resulting findings and implications. As the data analysis is still in progress, results are discussed in more general qualitative terms, with statistical verification expected within a few months.

23.4 Methodological Pluralism

No complex socioecological system can be captured using a single model or method, as no single perception is able to provide a comprehensive or adequate view of reality (Checkland 1981; Puccia and Levins 1985; Funtowicz and Ravetz 1994). Therefore, research on ecosystems necessitates a variety of forms of inquiry, multiple sources of evidence, and dialogue with persons representing different worldviews. Different methods are needed to address different forms of complexity and to answer different kinds of questions (Checkland and Scholes 1990; Midgley 1992; Holling 1995; Waltner-Toews and Wall 1997).

In order to represent these different perspectives in Ucayali, we established a multidisciplinary team with expertise in nutrition, health, anthropology, agronomy, natural resources management (NRM), fisheries, forestry, ecology, rural planning and economics. The team included relevant stakeholders from the region, local community leaders, and government professionals.

Methodologically, we combined quantitative and qualitative methods. For data on health and nutritional status, we complemented medical diagnostic tests, household surveys, anthropometry, and precise food recall with ethnographic and participatory methods exploring local diagnoses and understanding of health. For data on ecosystem dynamics and NRM, we complemented landscape level geographic information systems (GIS), spatial mapping and soil tests with in-depth household surveys and community-level participatory methods detailing the livelihood systems of families. In this way we were able to compare the results from the methods designed and driven by researchers with those that were led by the community themselves. Methods were complementary in their ability to verify results. The sequence of methods used allowed us to compensate for weaknesses in individual methods and to build knowledge systematically. For example, qualitative ethnographic techniques provided insight into local priorities and needs, yet extensive household surveys determined the extent to which these issues were common among all community members and which groups were most at risk. In addition, different methods were used to include different actors. For example, surveys involved the Ministry of Health, and mapping, timelines, and other participatory research techniques involved different members of the community. The data derived from each method was specifically targeted to the different end-users or decision-makers. For example, GIS and the regional maps generated were directed to the regional government; medical testing results, to the community health workers; and maps and drawings from participatory action research (PAR) were used in community meetings to further discussion of the community action plans.

In complementing different types of methods, we preferred the distinction between "data" and "process" approaches (Gjit 1999). The data approach stresses information extraction. Process approaches emphasize how the actual process of gathering information initiates other processes and changes existing processes within the community itself. For example, the impact of having mothers, fathers, and children view their own parasites through the microscope in the villages far surpassed the impact of the information derived from these stool samples. Parasites were no longer an abstract concept discussed only by ministry professionals; they were real and very much part of their world of poor water quality and diarrheal episodes. In each community, villagers were immediately mobilized and sought solutions to reduce water contamination and the continued transmission of parasites. Data and process approaches tap different information sources and produce different outputs and therefore should be used simultaneously.

Subsequent to 18 months of fieldwork and initial data analysis, our results reinforced the multidimensional nature of health. Ecological factors such as flooding, diversity of resources, and soil fertility affect natural resource use patterns which in turn affect nutrition and food security. Changes in vector habitats and water quality determined by flooding and land use patterns, affect cycles of disease and the dynamics of disease transmission. Economic factors such as labor supply, distance from markets, and access to credit affect income levels and access to health and educational services. Cultural differences among colonists migrating from other parts of Peru affect social cohesion in young settlements, and their lack of appropriate local knowledge undermines their ability to make use of the available abundance and diversity of resources. With further data analysis we will then be in a position to determine which of these factors, or combination of factors, are most important to human health and under what circumstances.

The inclusion of a wide range of stakeholders demonstrated that the issues facing different groups varied considerably. For example, families living in the floodplain benefit from an abundant supply of fish and animals yet are compromised by frequent parasitic infections due to poor water quality and sanitation caused by flooding. However, in the upland terraces during the wet season, the combined effects of high fish prices, scarce animal resources, and crop harvest still months away lead to predictable and seasonal food insecurity each year. Such differences demand substantive changes in intervention strategies.

23.5 Multilevel Investigation

Ecosystems exist within nested hierarchies or holarchies (Allen and Hoekstra 1992; Checkland and Scholes 1990). They are composed of smaller systems while at the same time are part of a larger whole. Similarly, a household is a part of a community, while being made up of different individuals. The different levels (household, community, region, and nation) evolve within a variety of ecological and socioeconomic contexts and constraints (Allen and Star 1982; Allen et al. 1993; Conway 1987).

Recognizing that often the determinants of individual human health may occur at levels higher within the ecosystem hierarchy, we investigated variables at four spatial scales: the individual, family, community and region/landscape. Figure 23.1 demonstrates the multilevel nature of the issues facing Ucayali. All issues are linked to and have consequence for others higher and lower within the nested hierarchy. The ecosystem approach focuses investigation on the cross-scale interactions of key variables that explain the complexity and multidimensionality of health. For example, landscape spatial mapping determined the extent to which families depended on an area larger than their farm or community for food and income. The temporal scale was seasonally determined with three field visits that captured the driest period, the start of the rains, and the height of wet season. Data on seasonal flooding levels was correlated with water quality and parasitic infections to investigate links between the hydrological cycle and disease periodicity. Comparing the nutritional status of the family with each individual determined the impact of intrahousehold food allocation and existence of gender and age inequities.

Our results demonstrate four points. (1) Patterns of natural resource use can only be understood at the landscape level. On-farm activities as well as those in the close surroundings do not capture the family livelihood strategies. Families make use of a great diversity of ecological resources located in different biotypes and parts of the region. (2) Patterns of natural resource use can only be understood if examined at different times of the year. Seasonal resource use is determined by the flooding cycles that dictate the availability of arable land, wild foods, fish

Research Variables

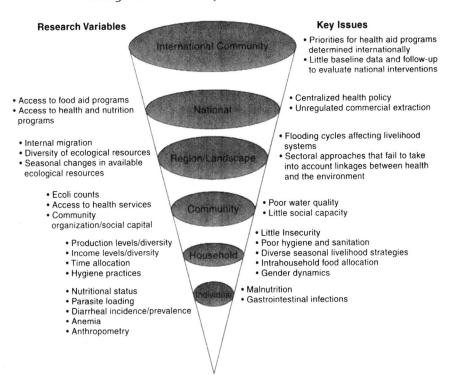

Key Issues

• Priorities for health aid programs
  determined internationally
• Little baseline data and follow-up
  to evaluate national interventions

• Access to food aid programs
• Access to health and nutrition
  programs

• Centralized health policy
• Unregulated commercial extraction

• Internal migration
• Diversity of ecological resources
• Seasonal changes in available
  ecological resources

• Flooding cycles affecting livelihood
  systems
• Sectoral approaches that fail to take
  into account linkages between health
  and the environment

• Ecoli counts
• Access to health services
• Community
  organization/social capital

• Poor water quality
• Little social capacity

• Production levels/diversity
• Income levels/diversity
• Time allocation
• Hygiene practices

• Little Insecurity
• Poor hygiene and sanitation
• Diverse seasonal livelihood strategies
• Intrahousehold food allocation
• Gender dynamics

• Nutritional status
• Parasite loading
• Diarrheal incidence/prevalence
• Anemia
• Anthropometry

• Malnutrition
• Gastrointestinal infections

Figure 23.1

and animal migrations, and access to valuable forest resources. Figures 23.2 and
23.3 portray the different cycles of disease, food availability, and income level as
perceived by the community members. The numbers indicate relative values: 1 is
low, 2 is medium, and 3 is high. This information was gathered during a community meeting where the group, using colored markers, drew the graph on a
large piece of paper. The figures demonstrate the difference between the cycles
of the upland forests versus the floodplain. For example, in the floodplain, disease (in this case diarrheal infections) is most prevalent in the dry months (June–
August) when the water quality is at its worst. In the uplands, there are similar
problems in the summer, though they also face malarial outbreaks in the wet season. In the floodplain, fish and wild animals provide sources of protein rich
foods during the summer and winter months, respectively. In January little food
is available as the flooded lands inhibit farming, hunting, and gathering and fish
have dispersed into the flooded forests. In the uplands, food shortages are in the
winter months when fish prices are high and the agricultural harvest is still
months away. Last, income in the floodplain is linked first to the sale of timber,
usually sold in April once it has all been harvested and carried along the flooded
rivers, and second, to the sale of the agricultural harvest in September. In contrast, in the uplands, income is highest in January and February, when their crops

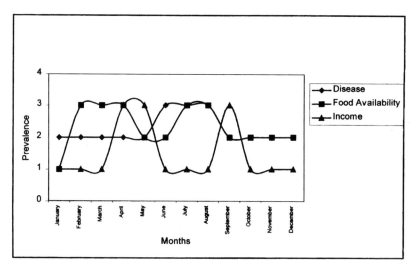

Figure 23.2

reach maturity and are sold in the Pucallpa market. People of the uplands have few other sources of income. (3) Individual nutritional status can only be understood at the landscape level. As food security and nutrition are determined primarily by the family's ability to produce, gather, or purchase food from a diversity of resources that extend across the landscape, understanding when and why there is low nutrient intake must take into account landscape-level variables. (4) Individual health can only be understood if linked to seasonal change. Patterns of

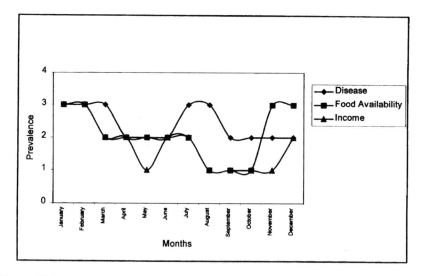

Figure 23.3

disease periodicity are linked to the environmental changes that occur with flooding. This affects vector habitats, dynamics of disease transmission, and water quality.

23.6 Local Participation and Action Research

Sustainable natural resource use is not a simple model or package to be imposed; it is more accurately described as a "process for learning" (Roling and Wagemaker 1998). As there are always multiple system goals, the trade-offs need to be negotiated among stakeholders. Research therefore requires legitimate involvement of the local community, be that via government agencies, women groups, nongovernmental organizations, or the community members themselves. Together, professional researchers and local stakeholders define the problems to be examined, generate knowledge, analyze the findings, and take action. Participatory action research recognizes its inevitable intervention in social settings. Methods are designed to explicitly investigate a situation in order to change it. Ultimately, researchers are agents of change, recognizing and acknowledging their role as catalysts in the community (Greenwood and Levin 1998; Wadsworth 1998).

In Ucayali the research questions were based on the issues identified by local stakeholders. A variety of PAR methods were used. They included community mapping, seasonal calendars, focus groups, time line, life histories, pile sorting, and key informant interviews. At the community level, the research team adapted their investigation to meet priorities identified by the community. During the process of data collection, the team and the community began to assess different options for intervention. As results became available, they were used to guide local initiatives and provide feedback for Ministry programs. There was an iterative cycle between research and action. We developed action plans that were then evaluated and the results used to refine our hypotheses and future research agenda. There was not the usual lengthy delay between research, analysis, and intervention, often carried out by different sets of actors.

At the outset it was assumed that the process from analysis to action would be completed through once. In reality, there was not one path from participation to research to action; instead, countless smaller cycles of reflection on action, learning about action, and new informed action, all fed back to one another. After 10 months in the field, several small-scale plans had already been developed, simple actions carried out, and these experiences used to focus and refine the understanding of what was really happening and what was really important to the community. Application of this adaptive and iterative research cycle led to the discovery that there were in reality several cycles existing simultaneously, moving at different speeds. The more complex research questions required a longer cycle, where rigor in methods and analysis demands years of work. Other questions that require smaller time intervals between analysis and action were critical in involving the community early on in the study. Water quality, parasite loading, and anemia tests provided a good starting point, as these tests were conducted in the field and findings were readily available. The community was quickly able to grasp the significance of the results and incorporate them into their understanding of the larger health issues. This early emphasis on action promoted greater interest and local ownership of the information, findings, and solutions. More important, as action plans were being developed early in the project, researchers were able to assess the capacity of the community as well as local organizations to manage such interventions. Social capital was lacking in most communities. Communities possessed little capacity to organize themselves and had almost no experience with cooperative action. We were therefore able to identify training needs that could address this lack of experience in collective action before embarking on the proposed interventions.

23.7 Conclusions

The application of the ecosystem approach led to several new conclusions regarding the key driving forces and relationships between human and ecosystem health in the Ucayali region. First, our research has demonstrated that the constantly changing ecology of the floodplain and upland terraces dictates the patterns of resource use as well as nutrition, food security, and disease. Engaging in resource use activities, whether it be farming, fishing, hunting, or logging, is determined by the rhythm of the rivers as they flood and later recede. Similarly, nutrient intake varies throughout the year as dietary changes reflect these resource use patterns. Disease transmission is affected by the 8–15-meter rise in river levels and the associated changes in vector habitats and water quality. The movement of people across the landscape is synchronized to the dynamics of the floodplain. Local migration for extended periods allows families to make use of a greater diversity of ecological resources that become available with the flooding cycle. To promote more sustainable resource use that is better able to meet the nutrient and health requirements of the population, interventions should therefore focus on ways to take advantage of the annual floods rather than being drowned economically by them.

Second, as people's movements mirror the dynamics of the floodplain, so too does the diversity of their livelihood strategies complement the diversity of resources available. Land use patterns that make use of a wide range of resources, that is, fish, wild animals, forest foods, and timber, as well as agriculture are more likely to provide sufficient nutrients and income, than monocultural cropping or even restriction to one type of resource use. Diverse resource strategies tend to result in more diverse sources of income and food. This diversity reduces the possibility of specific nutrient deficiencies in the daily diet, as well as reducing the vulnerability of the food supply to sudden environmental changes such as droughts and floods. On a landscape level, extensive diversified resource use patterns are more sustainable as the burden of exploitation is spread across a larger area and a greater variety of resources and is concentrated at specific times of the year. The ecosystem therefore has the opportunity to replenish itself.

Third, differentiating the spatial and temporal patterns of ecosystem dynamics and human health highlighted the ecological and cultural heterogeneity of the region. Large variations in ecology and ethnicity combined to create different patterns of resource use. Although soil quality, drainage, and flooding patterns affect the choice of crops and animals, people's knowledge, or lack thereof, of the local ecosystem influences their ability to effectively harvest the richness of the available resources. Development initiatives need to incorporate the necessary flexibility to accommodate such heterogeneity.

Fourth, the effectiveness of government health programs is undermined by centralized decision-making structures based in Lima. This results in policies disconnected from the reality of the Amazon region, and discourages the involvement of local people in the decision-making process. These structures and weak regional capacity create strong barriers to effective decentralization of knowledge and power. Lack of regional research capacity and coordination prevents meaningful dialogue between the policy and scientific communities. Consequently, policy development is largely political and education lacks the relevance and application required by decision makers. Sectoral approaches in both academia and government have prevented the integration of research, training and policy activities for health, agriculture, forestry, and fisheries. Nevertheless, problems related to local livelihood strategies, food security, biodiversity conservation, and income generation cannot be resolved without reference to the causal linkages among them, and their relationship with the annual cycles of this richly diverse river floodplain area. In order to make substantive improvements in human health in this region, these larger political and institutional challenges have to be addressed. If not, efforts at the community and individual level will continue to be thwarted from above.

Last, extensive, diversified patterns of resource use, although recommended, will eventually fail if there is no landscape-level mechanism for regulating and managing common-pool resources. On the Amazon frontier, the absence of government and nonexistent or precarious land ownership result in an institutional and political vacuum. In addition, the physical demarcation of land on the floodplain is often impossible as lands appear and disappear with the rivers' ever-changing configuration. To ensure that both the health of people and the ecosystems upon which they depend is maintained, any form of intervention must work toward establishing an institutional framework and associated political and legal mechanisms that can monitor and manage common-pool resources. Community-based management offers an alternative where the different users of the resources can come together and negotiate a long-term plan for resource use. Faced often with groups of greatly varying levels of power and influence (e.g., native communities vs. national timber companies), special care must be taken to ensure that all interests are represented. There are now several models of such community-based management in the Amazon, and currently in Ucayali the team is developing a proposal using such an approach for an important area of 2,500 km^2 and 50 communities where we find key fish habitats, endemic species, and growing conflicts between commercial fishermen, timber companies, and native residents.

This type of ecosystem-based management that incorporates the political and institutional dimensions is an inevitable outcome of a process such as the ecosystem approach that involves the relevant stakeholders and recognizes the multidimensional nature of the interaction between human and ecosystem health.

References

Allen, T.F.H., and T.W. Hoekstra. 1992. Toward a unified ecology. Columbia University Press, New York.

Allen, T.F., and T.B. Starr. 1982. Hierarchy: perspectives for ecological complexity. University of Chicago Press, Chicago.

Allen, T.F., A.W. King, B. Milne, A. Johnson, and S. Turner. 1993. The problem of scaling. Evol Trends Plants 7:3–8.

Chambers, R., and G. Conway. 1992. Sustainable rural livelihoods: practical concepts for the 21st century. Institute of Development Studies, Sussex.

Checkland, P. 1981. Systems thinking, systems practice. Wiley, Chichester.

Checkland, P., and P. Sholes. 1990. Soft systems methodology in action. Wiley, Chichester.

Conway, G.R. 1987. The properties of agroecosystems. Agricult Syst 24:95–117.

Flood, R.L., and E.R. Carson. 1988. Dealing with complexity: an introduction to the theory and practice of systems science. Plenum, New York.

Funtowicz, S., and J. Ravetz. 1994. Emergent complex systems. Futures 26:568–582.

Gijt, I. 1999. The myth of the community. Intermediate Technology Development Group, London.

Greenwood, D., and M. Levin. 1998. Introduction to action research: social research for social change. Sage, Thousand Oaks, Calif.

Gunderson, L.H., C.S. Holling, and S.S. Light. 1995. Barriers and bridges to the renewal of ecosystems and institutions. Columbia University Press, New York.

Holling, C.S. 1995. What barriers? What bridges? *In* Gunderson, L.H., C.S. Holling, and S. Light, eds. Barriers and bridges to the renewal of ecosystems and institutions, pp. 3–34. Columbia University Press, New York.

Instituto Nacional de Estadisticas y Informaticas (INEI). 1997. Población, mujer y salud: resultados de la encuesta demográfica y de salud familiar. INEI, Lima, Peru.

Kay, J., H. Regier, M. Boyle, and G. Francis. 1999. An ecosystem approach for sustainability: addressing the challenge of complexity. Futures 31:721–742.

Midgley, G. 1992. Pluralism and the legitimation of systems science. Syst Prac 5:147–172.

Puccia, J.C., and R. Levins. 1985. Qualitative modeling of complex systems. Harvard University Press, Cambridge, Mass.

Roling, N.G. and M.A. Wagemakers (eds.). 1998. Facilitating sustainable agriculture. Participatory learning and adaptive management in times of environmental uncertainty. Cambridge University Press, Cambridge.

Schneider, E.D, and J.J. Kay. 1994. Complexity and thermodynamics: towards a new ecology. Futures 24:626–647.

Singh, N., and S. Wanmali. 1998. The sustainable livelihood approach: a concept paper. Framework paper for the sustainable livelihoods workshop, 1–3 October 1998, University of Florida, Gainesville.

Sundsvall Conference on Health Promotion. 1991. Supportive environments for health: the Sundsvall statement. Health Prom Int 6:297–300.

Wadsworth, Y. 1998. What is participatory action research? Action Research International, paper 2 available on line at http//www.scu.edu.au/schools/gem/ar/ari/p-ywadsworth98. html

Waltner-Toews, D., and E. Wall. 1997. Emergent perplexity: in search of post-normal questions for community and agroecosystem health. Soc Sci Med 45:1741–1749.

World Health Organization. 1978. Declaration of Alma Ata, International conference on primary health care, 6–12 September 1978, Alma Ata, U.S.S.R.

24

Deer Tick–Transmitted Zoonoses in the Eastern United States

Sam R. Telford III

Whereas plagues were once thought to mainly affect developing countries, people now are increasingly aware of the threat within American borders. Ticks have catapulted into public awareness, to the point that if it were not for AIDS, these eight-legged creatures and their bloodsucking habits would be the focus of astounding public attention. The relatively sudden appearance of these infections seemed enigmatic at the time but is now relatively well understood. In the case of Lyme disease, babesiosis, and ehrlichiosis, reforestation in eastern North America has permitted deer to proliferate. The proliferation of deer has permitted ticks to multiply. The multiplication of these ticks has permitted Lyme disease spirochetes, babesiae, and ehrlichiae to infect people in sites where these pathogens previously had been cryptically maintained. Residential development has created new habitat, exposing human residents and their companion animals to "emerging" infectious agents.

This chapter focuses on this well-studied example of how environmental perturbation has produced a significant public health problem and documents the extent and type of evidence that allows robust testing of hypotheses of causality. Intervention methods are described that are used or proposed for reducing the risk of humans acquiring a deer tick–transmitted infection, with the suggestion that such interventions may be used as tools for probing hypotheses of causality. Finally, the lessons learned from studies of the ecology and management of Lyme disease and that might be applied to other problems in conservation medicine are summarized.

24.1 Emergent Tick-Borne Diseases: Cast of Characters

In 1975, an epidemic of oligoarticular arthritis in communities near Old Lyme, Connecticut, was described. Detailed analysis revealed a protean illness with cardiac and neurological sequelae and, in many cases, a pathognomonic rash (erythema migrans) (Steere et al. 1976). No etiological agent was identified at the time, although an infectious vector-borne disease was suspected. Virtually at the same time, an outbreak of a malarialike protozoan infection appeared on the islands off the coast of New England. The index case of infection due to *Babesia microti*, a widely distributed parasite of small rodents, was identified in 1969 from an elderly immunocompetent resident of Nantucket Island. Several other cases of "Nantucket fever" were diagnosed there by 1976, prompting an intense investigation. The piroplasms, a diverse group of blood-inhabiting protozoa (sporozoans) that include *Ba. microti*, had previously been thought to be strictly host specific. They have long been recognized as important veterinary pathogens, causing bovine babesiosis (redwater) and theileriosis (East Coast fever). Human infection was first recognized in 1957 in Yugoslavia, where a splenectomized farmer sustained a fatal infection by *Ba. bovis*. Since then, 19 cases of *Ba. divergens* or *Ba. bovis* infection have been recorded in Europe, virtually all in splenectomized patients, most of whom died (Telford and Maguire 1999).

A previously unrecognized hard tick of the genus *Ixodes*, a deer tick, was identified as the vector of Nantucket fever (Spielman et al. 1979), and this small tick (*Ixodes dammini*) was soon associated with cases of Lyme arthritis. In 1982, during a search for spotted fever rickettsiae in specimens of deer ticks that had been collected from eastern Long Island, a new spirochete was discovered. Its subsequent cultivation allowed the fulfillment of Koch's postulates, thereby establishing that *Borrelia burgdorferi* was the agent of Lyme disease (Burgdorfer 1984). Although the systematics of the deer ticks and their close relatives remain controversial (Telford 1998), the distribution of human-biting populations seems concentrated along the Atlantic Seaboard from Maine to Maryland, with a secondary group of populations centered in the upper Midwest. This distribution coincides with America's major megalopolis corridor (Boston–New York–Washington D.C.) and clearly places millions of humans at risk of acquiring infection.

Two tick-borne bacterial infections have recently emerged as threats to the public health in eastern North America. The ehrlichioses have been recognized as important veterinary infections for more than half a century, and although human ehrlichiosis was recognized as early as 1956 in Japan (Sennetsu fever), these infections attained prominence only with the description of the index case in the United States, which occurred in Arkansas in 1986. A monocytotropic organism, *E. chaffeensis*, closely related to the agent of tropical canine pancytopenia (*E. canis*), was described as the cause of human monocytic ehrlichiosis (HME). This new pathogen has infected about 750 individuals from at least 30 states, with eight fatalities (McQuiston et al. 1999). The vector for the agent of HME is the Lone Star tick, *Amblyomma americanum*. The aggressive Lone Star ticks are distributed throughout much of the mid-Atlantic, central, and southern United States. A large rural population is therefore at risk of acquiring ehrlichiosis. Since 1990, infection by a granulocytic *Ehrlichia* indistinguishable from the agent of tick-borne (pasture) fever of sheep (Gordon et al. 1932) and cattle has been described from more than 450 residents of Minnesota, Wisconsin, New York, Connecticut, Massachusetts, and other states (Bakken et al. 1994). Three cases of what is now called human granulocytic ehrlichiosis (HGE) have proven fatal. The deer tick has conclusively been demonstrated to be the main vector (Telford et al. 1996). The distribution of HGE closely parallels that of the deer tick vector and of Lyme disease.

Lyme disease is now the most frequently diagnosed vector-borne infection in the United States, with about 10,000 cases reported each year. This is considered by many to be a small fraction of the actual cases, but may be balanced by the observation that overdiagnosis appears to be frequent. In central and southern U.S. sites, a Lyme disease mimic of undescribed etiology, Master's disease, is associated with Lone Star tick bites and confounds national surveillance for Lyme disease (Armstrong et al. 2001). The public health burden of Lyme disease clearly demonstrates variation between years and between sites. Published estimates of annual incidence from prospective studies conducted in endemic American sites vary from 0% to 4%, with a median of 2.6%. Single-year studies, however, may poorly represent risk: in the two studies that were undertaken for more than five years, incidence ranged from 0% to 10% each year, probably reflecting fluctuations in vector abundance and human activity. These estimates of incidence, however, suggest that for certain communities Lyme disease may be as burdensome as other infections classified as "very common" (incidence >100/100,000 population), such as varicella or gonorrhea (Wilson 1991). The incidence of HGE and babesiosis varies among sites and between years, but either infection is approximately a fifth as common as is Lyme disease.

24.2 Environmental Correlates of Lyme Disease Risk in the United States

The confusing array of pathogens that now threaten human health in the eastern United States represents a guild of phylogenetically unrelated microbial species that utilize a common resource, coevolve, and have contemporaneously emerged as public health threats. The deer tick pathogen guild (Telford et al. 1997a) comprises *Bo. burgdorferi*, *Ba. microti*, *E. phagocytophila*, deer tick virus (a strain of Powassan, a virulent zoonotic flavivirus), and *Rickettsia helvetica* (originally thought to be a tick symbiont but now associated with perimyocarditis in humans). Although the deer tick pathogen guild is likely to have originated prior to the Ice Ages (Marshall et al. 1994), it remained cryptic because it may have been maintained in small foci with low densities of deer ticks, or by other ticks that do not feed on humans. *Borrelia burgdorferi*-like spirochetes of cottontail rabbits, for example, are maintained by *I. dentatus*, which rarely bites humans, in an independent enzootic cycle within certain northeastern U.S. sites (Telford and Spielman 1989). Such a cycle maintains spirochetes where human-biting deer ticks are not present, such as with woodrats and *I. spinipalpis* in Colorado and California

(Maupin et al. 1994). In New England, particular sites such as Martha's Vineyard appeared to be free of deer ticks earlier this century. But *Ba. microti* and *E. phagocytophila* were reported to infect voles and mice there during the 1930s (Tyzzer 1938) and probably were maintained by the now locally extinct *I. muris*, a nest-associated parasite of small rodents. This example demonstrates a main tenet for explaining many emergent diseases: natural (enzootic) cycles do not necessarily imply human risk (zoonotic) and may have existed for ages prior to becoming a public health problem. The human-biting deer tick serves a "bridge" from enzootic cycles to a zoonotic situation. (Telford et al 1997b).

Three processes appear to be responsible for the emergence of the deer tick pathogen guild as a public health burden within the last two decades: (1) the relatively recent proliferation of deer, (2) creation of habitat by the abandonment of farmland and subsequent succession to thick secondary vegetation, and (3) increased use of "woods" for human recreation or habitation. Deer, in particular, seem to serve as the proximal determinant of human risk for the deer tick–transmitted infections. Many areas with dense deer populations, however, may not be infested with deer ticks (Schulze et al. 1984; Glass et al. 1994), and thus environmental factors seem critical in explaining the current distribution and emergence of zoonotic foci. The frequency with which humans are bitten by ticks is largely a function of recreational or peridomestic exposure. The following discussion expands on the anthropogenic basis for the emergence of these infections.

24.2.1 Causal or Casual? Studies Incriminating the Main Villains

Descriptive field studies (traditionally called natural history) comprise observations of populations over time and space. Inferences about causes of ecological phenomena often are based upon statistical associations with predictor variables. Varying degrees of sophistication may be seen in the literature, ranging from crude correlations between two variables to elegant regression models. Such analyses are usually plagued by the suspicion that the observed relationship is a spurious one that is consistent with subconscious a priori conclusions (observer bias), as opposed to cause and effect. Red Sox batting averages are positively associated with OPEC oil prices by correlation analysis (Gould 1977), but no one would argue that such a relationship is causal. Accordingly, "proof" rests on the consistency with which such an association may be replicated, preferably by independent (and presumably unbiased) workers approaching the question from differing angles. I believe that this burden of proof has been adequately met by studies of the deer tick and the factors that determine its abundance.

24.2.2 Burgeoning Deer Populations, Exploding Tick Infestations

The life cycle of the deer tick is complex, spanning at least two years. Larvae emerge from eggs in July, feed during August and September, and then overwinter as fed larvae. They molt to nymphs in April and seek hosts during May and June.

Fed nymphs will molt to adults in September. These adults seek hosts October and November, as well as during warm spells through the winter and spring. If the adults do not feed by the end of May, they die. Those that have fed during the fall or spring will deposit eggs during April and May; these eggs hatch in July (Yuval and Spielman 1990). White-footed mice (*Peromyscus leucopus*) have been dogmatically proposed as the main hosts for immature deer ticks (larvae and nymphs), and the primary reservoir for the pathogen guild. These mice are the most abundant small mammal in enzootic sites and efficiently infect ticks (Donahue et al. 1987). However, larvae and nymphs ("immature" or "subadult" ticks) will infest virtually any vertebrate, including birds (Anderson 1989), and the contribution of each vertebrate to each year's cohort of ticks depends on their proportional representation in a site's biomass in that year. Because few species are stable with respect to density, varying by month and year, among habitats, and among sites, generalizations must be made with caution. In some sites, for example, *P. leucopus* may seem to feed a large proportion of immature deer ticks, but in a subsequent year of observation, when mouse populations are at a nadir or their timing of reproduction is altered, chipmunks (*Tamias striatus*) may replace them as "preferred" hosts. Over the scale of local time and space, then, the main host for immature ticks, and reservoir host for pathogens, may differ from year to year. Over many years of observation, however, and among many geographically disparate sites, mice—on average—are numerically the most important reservoir host and blood meal source for immatures.

The stability and magnitude of a deer tick infestation within a site depend upon reproduction by the adult ticks. The mortality of 2000 or so immatures (the average size of an egg mass) may be compensated for by the successful feeding, oviposition, and hatching of a single female tick's progeny. One cannot have infected ticks without small mammal reservoirs because the deer tick pathogen guild is rarely maintained by inheritance (transovarial or vertical transmission). Also, many other mammals and birds may be infected by Lyme disease spirochetes but fail to pass the infection back to the vector ("incompetent reservoirs" Telford et al. 1988). The two processes, infection and reproduction, are independent phenomena that must interact in order to maintain a cycle of transmission of any of the deer tick pathogen guild.

In parasitology, a special distinction is given the "definitive" host, in which sexual reproduction of a parasitic organism (and therefore its perpetuation) takes place. A causal analysis of the epidemic caused by the deer tick pathogen guild, then, must identify the definitive hosts for the tick. The "new" deer tick was originally thought dependent upon populations of deer because other ticks in the cosmopolitan Holarctic species complex were well known to feed mainly on medium-sized to large mammals, mainly ruminants. The only naturally occuring ruminant in coastal New England, in any numbers, was the white-tailed deer. Adult ticks seem to most frequently feed on this host, although they infest many species of medium to large mammals, including bears, coyotes, foxes, and raccoons. Much evidence supports the statement that most adult deer ticks feed on deer and therefore deer are essential for the tick's life cycle.

Deer are heavily infested with deer ticks of all stages (Piesman et al. 1987). Islands in Narragansett Bay, Rhode Island, that lack deer do not sustain deer tick populations even with alternative hosts available (Anderson et al. 1987). Deer scat pile density is strongly associated with that of immature ticks within discrete sites such as small islands (Wilson et al. 1985). The burden of immature ticks on mice is associated with the sites where deer spend most of their time, as measured by radiotelemetry (Wilson et al. 1990a). In one site, deer served as host to 95% of all feeding adult ticks compared to raccoons, opossums, and feral cats (Wilson et al. 1990b). Larval burdens on mice increased when the presence of deer was promoted by baiting the ungulates into a site that had been artificially supplemented with tons of acorns (Jones et al. 1998). Frequent sighting of deer within yards and proximity to a wildlife conservation area are associated with greater incidence of human infection (Lastavica et al. 1989). This body of literature, when taken with studies using interventions to experimentally probe the hypothesis (see discussion below), provides strong evidence that deer are the definitive hosts for the deer tick and that risk of acquiring infection is related to deer density.

On a macrogeographic scale, environmental factors seem as important as deer density in promoting deer tick infestations. Statewide analyses, such as those by geographic information systems, demonstrate that deer density by management zone is less predictive of tick density (as measured by counts from deer) than are physiographic variables (Schulze et al. 1984; Glass et al. 1994). Low elevation and acidic, well-drained soils appeared to be significantly related to deer tick density. Much work remains to be done in describing the interrelationships between deer, habitat, environment, and tick density.

24.2.3 Changes in the Landscape

Early studies of the distribution of deer ticks noted that particular physiographic regions were associated with the presence of dense tick infestations and that such habitat correlates might be as important as the presence of abundant deer (Schulze et al. 1984). The role of habitat structure in determining deer tick density is particularly important in explaining sites where deer were dense but few ticks were present. Open-grass/sparse-shrub habitat contained fewer host-seeking immature deer ticks than did high shrub areas. A simple correlation analysis demonstrated that tick burdens were greatest on mice trapped from grids with more shrub cover and woody stem densities (Adler et al. 1992). In residential sites, 67% of host-seeking ticks that were collected came from woods adjacent to homes; 22% from the ecotone; 9% from ornamental vegetation; and 2% from lawns proper (Maupin et al. 1991). In laboratory and field experiments, deer ticks are very sensitive to desiccation, requiring a microhabitat with >85% relative humidity for extended survival (Yoder and Spielman 1992; Bertrand and Wilson 1996). This body of evidence is sufficiently strong that a formal "ecological index," weighted heavily for subjective assessments of habitat structure, has been proposed to rank sites for risk (Schulze et al. 1991).

There have been vast changes in the landscape of the eastern United States within the nineteenth century (Cronon 1983). Mature forests were cleared for firewood or farmland prior to the 1900s. For example, in New York, 75% of all land was devoted to farmland by 1880. By the 1920s, successional growth had reforested approximately 60% of New England, creating edge situations favored by browsing deer. Prior to deforestation, deer density in northern Wisconsin was estimated to have been fewer than five per square mile, limited, perhaps, by large tracts of old-growth or mature forest. Deer do not thrive in mature forests. During the 1850s through the 1940s, deer density was recorded to have been as high as 35 per square mile, reflecting wildlife management and optimal habitat from reforestation and active promotion of "wildlife openings," which increase edge effects (Alverson et al. 1988). Although there are no published reports of how edge effects and habitat fragmentation might directly influence the density of tick infestations, the hypothesis that thick successional scrub promotes the survival of ticks because it has the right microhabitat might be explored. Landscape changes are directly due to the agency of humans, as is wildlife management; therefore, dense infestations of deer ticks are largely anthropogenic.

24.2.4 Habitat Use

That the landscape is increasingly dominated by human use is self-evident. Of particular relevance to the emergence of the deer tick pathogen guild in northeastern U.S. sites is the explosion of suburban residential developments, with zoning preferences that promote habitat fragmentation. Virtual ecological islands may be formed between houses, with local concentration of animals such as deer or mice. Deer adapt quickly to the presence of humans and, indeed, seem to thrive on the associated ornamental plants. Some homeowners feed deer and encourage their presence around their homes. Without natural predators, hunting is the only active control on deer populations, with car accidents, disease, and starvation passively removing animals. Hunting, however, is becoming increasingly difficult in many areas. Some communities have banned firearms discharge within their borders for social or political reasons. The main hindrance to hunting, however, is that each house effectively removes at least 1,000 square feet of habitat from use because firearms may not be safely or legally discharged within 500 feet of an inhabited dwelling or within 150 feet of a hard-surfaced road. Affluent suburbs thereby directly or indirectly promote dense deer tick infestations.

24.3 Interventions

Intervention poses a variety of logistic and planning problems. The Hippocratic mandate "first, do no harm" should serve as the basis for any action. Furthermore, objectives should be worthwhile, attainable, and sustainable. In the case of the deer tick–transmitted pathogens, intervention is worthwhile: infection is common and may burden a community economically as well as psychologically. For those who fail to respond to or delay their treatment, Lyme disease may chronically affect the quality of life. Babesiosis may kill 5% of those who become infected, despite treatment.

Attainability and sustainability of an intervention depend on the objective. For example, massive spraying of acaricides, delivered by airplanes, might kill a large proportion of ticks very rapidly, providing a gratifying and tangible effect (Uspensky 1999). But such a reduction in tick density is likely to be temporary given the two-year life cycle of the tick and the presence of overlapping cohorts that may be hidden under the exposed surface. Sustaining the reduction would require periodic respraying. Community support, however, would be unlikely because this measure is environmentally intolerable.

Gradual methods of intervention require careful planning for evaluating the effects. Baseline studies must robustly document the situation prior to and after action so that an effect may be demonstrated. Such studies must take into account the ecological relevance of the measurements. Would a single year of observation before and after an intervention provide sufficient information about effect, given the two-year life cycle of the deer tick?

24.3.1 Population-Level Interventions Against Deer Ticks

Early syntheses of the ecology of the deer tick and the origins of the epidemic pointed out the temporal correlation between the rise in the density of deer in the eastern United States and the epidemic curve for Lyme disease (Spielman et al. 1985). Deer were virtually extirpated from much of the eastern United States by the beginning of the twentieth century because of extensive habitat loss to farmland and by unregulated hunting. With the passage of the Pittman-Robertson Act in 1937, which served as the avatar of federal and state wildlife management in the United States, deer populations began increasing and spreading from the small scattered patches in which they survived, aided by the abandonment of farms and subsequent reforestation. These historical arguments, however, do not provide evidence for causality.

The large number of observational studies linking deer density with that of the deer tick, however, suggested a means of experimentally testing the hypothesis. The Great Island experiment began in 1981 on a 300-hectare tombolo on Lewis Bay, Cape Cod, Massachusetts (Wilson et al. 1988). It was estimated that between 30 and 50 deer roamed the site, roughly one square mile in area. Preintervention sampling from 1981 to 1983 provided baseline infestation levels for immature ticks on mice. Deer were removed by sharpshooters during 1983–84, and subsequently the herd has been held to fewer than six per square mile by selective shooting. The infestation levels on mice declined dramatically the year following the major reduction in deer density and has remained (as of this writing) a tenth of the preintervention magnitude. The density of other possible hosts for the adult ticks (coyotes, foxes, raccoons, skunks, opossums) was not manipulated, and despite their availability, ticks failed to propagate at high density. Eradication was not expected, but the situation of zoonotic overflow obviously had been diminished. Two cases of Lyme disease and one of babesiosis have been recorded since 1986. Prior to the intervention 16% of the population had experienced an episode of infection due to one or more of the deer tick pathogen guild. This experiment provides critical evidence for the association between deer density and risk of infection. Other studies have similarly demonstrated statistically significant reductions in deer tick density following deer reduction (Wilson and Deblinger 1993) or by the exclusion of these hosts by fencing (Daniels et al. 1993).

Deer reduction is inherently distasteful but increasingly necessary. Even if public health arguments for reduction excluded tick-borne disease, one would reasonably conclude that deer populations now need enhanced management to prevent them from exceeding either the biological or cultural carrying capacity. The latter term includes hazard from automobile collisions. Michigan, for example, recorded 56,000 car accidents with deer in 1994, and the National Safety Council estimates that the annual national burden exceeds 100 human fatalities, with 7,000 injuries (TranSafety Inc. 1997)! A healthy deer herd is said to double in size every three years, and in many sites, the limited shotgun hunting season is just enough to keep deer density from an uncontrolled trajectory.

New England communities are starting to come to terms with the foregone conclusion that a downward trend in deer populations would be a worthy objective for the long-term future. Nantucket Island, for example, had removed 250 or so deer each year during the hunting season, keeping constant the estimated population of 1,500 animals (on an island that is roughly 50 square miles in area). But increases in the prevalence of infections due to the deer tick pathogen guild stimulated community leaders to action. With the objective of starting to slowly reduce the deer herd so that their grandchildren might be less afflicted, citizens petitioned the Massachusetts Divison of Fisheries and Wildlife to expand the regular season shotgun hunt by one week. The number of deer taken each year for the past three years has exceeded 500. Although too early to demonstrate any effect on deer density or tick infestations, this approach may serve as a model for gradual control of deer herds by increased hunting pressure. Then, too, the recent invasion of New England by coyotes may help to locally reduce deer herds. Anecdotally, Naushon Island was home to nearly 300 deer in the 1970s and 1980s; coyotes were introduced in the early 1990s. A current estimate for the deer herd is 65 animals. Cause and effect, albeit not demonstrated, is suspected, based upon the lack of any other control measures undertaken on Naushon Island and absence of any epidemic or other natural acceleration of mortality.

Alternatives to deer extirpation have been proposed and are currently being tested. Proponents of the suggestion that deer populations might be controlled via contraception fail to understand that herds must first be reduced, probably to fewer than six to eight per square mile (Telford 1993), before such a measure might be used. Such an intervention assumes the availability of an effective chemoprophylactic and means to deliver it. Several steroidal or immunofertility agents appear promising in captive herd experiments, but logistic and bioethical questions remain to be answered before large-scale trials may begin.

A variety of methods designed to deliver acaricides (arthropod growth regulators or relatively harmless, pyrethroid-based toxicants; also an entomophagous fungal agent) directly to deer at bait stations are being tested. These experiments may suffer from the likelihood that coverage of individuals might be limited because of the propensity of dominant animals to exclude other members of the herd. Paradoxical effects might be observed, such as that in an ivermectin delivery trial on Monhegan Island, Maine (P.W. Rand, Maine Medical Center, personal commun. 1999). Corn dosed with ivermectin was delivered to bait stations on this island, which has no other hosts for deer ticks other than deer and feral Norway rats. Deer ingested sufficient ivermectin to render their scat bioactive in fly pupation assays, but ticks continued to infest the island, perhaps because of an explosion of Norway rats that opportunistically enjoyed the corn. (Why ticks feeding on ivermectin-fed rats failed to die remains unanswered.) Although I remain skeptical about the suggestion that mouse density locally influences that of deer ticks (Telford et al. 1999), one might expect eruptions of rodents in the sites where corn or other grain is used within bait stations, due to enhanced reproduction or a "sink" effect attracting mice from nearby areas. Finally, much research is being devoted to understanding antitick immunity with the hopes that a vaccine that inhibits tick feeding or reproduction will be developed. Such a vaccine might be delivered to remaining individuals of a reduced deer herd, along with a contraceptive agent, by hunters instructed to use their sporting skills in a novel manner.

Habitat modification would powerfully contribute to the effects of deer reduction, but in many sites it is impractical, socially unacceptable, or expensive. Fire has been historically used for managing insect infestations (Miller 1979), but the few experiments that have been conducted with deer tick populations have been inconclusive. The subconscious indoctrination of America by Smokey the Bear also sociologically hinders any use of fire to manage thick secondary growths. Other modes of brush clearing might be effective, but even reducing vegetation around summer homes meets with resistance because homeowners consider it "natural" privacy hedges.

24.3.2 Individual-Level Intervention Against Infection

Public health or "conservation medicine" approaches target population-level objectives for long-term solutions. The many effective options available at the level of the individual may diminish a community's desire to engage in interventions that might benefit future generations. Personal protection and awareness greatly decrease the risk of becoming infected by any of the deer tick pathogen guild. Use of repellants, proper clothing, and daily tick checks effectively prevent infection. The "tick check," in particular, is crucial because prompt removal of attached ticks (within 24 hours of attachment) aborts infection (Piesman et al. 1987). Communities might benefit from efforts to educate children about personal protection, with the objective of starting long-term behavioral changes.

The Lymerix vaccine has been approved by the U.S. Food and Drug Administration and is currently available for preventing Lyme disease in humans. The vaccine consists of recombinant outer surface protein A, a prominent spirochetal surface antigen. It is highly effective in laboratory experiments, and a phase III trial with 10,000 subjects suggests efficacy of 70–80%. The mode of action is novel in that the influx of antibody during the first few hours of feeding into the gut of the tick destroys spirochetes even before they are transmitted (Telford and Fikrig 1995). Although vaccination will reduce the risk of humans acquiring Lyme disease, the incidence of ehrlichiosis or babesiosis may increase because immunity is specific for *Bo. burgdorferi*. Vaccinees may fail to continue tick checks or wearing appropriate clothing because they feel protected against all tick-borne disease. Nonetheless, the announcement that Lymerix had been approved served to reduce enthusiasm for seriously considering deer reduction programs, in at least two coastal New England communities.

24.4 Lessons for Conservation Medicine

24.4.1 Public Health Versus Medicine

Medicine focuses on the individual, with the objective of improving the life of that individual within a time scale of minutes to weeks. Public health is population based, seeking to improve the quality of life for the majority, with achievement goals often based in years or decades. In public health, the greater good may dictate triage or choices of the lesser or more pragmatic evil. The deer tick–transmitted zoonoses are largely anthropogenic in origin. Undoing the ecological and behavioral changes that serve as the basis of the epidemic seems unlikely. The pragmatic evil, then, is to kill deer. Habitat modification would be equally effective but seems sociopolitically difficult and expensive. Vaccination reduces risk for acquiring Lyme disease but not the other zoonoses. Other modes of large-scale environmental intervention are at various stages of testing and are thus currently unproven.

24.4.2 Animal Versus Human Health

Whether it is useful to distinguish between animal and human infections is debatable. The fundamental elements of the transmission of pathogens are identical, as are the ecological principles that seem axiomatic in the emergence of a new public health (animal or human) hazard. Indeed, a granulocytic ehrlichia was described from meadow voles on Martha's Vineyard in the 1930s (Tyzzer 1938), and thought, at the time, to be merely a routine observation of a rodent parasite with no public health implications whatsoever. We now believe that this agent is identical to that causing HGE. This microbe parallels the history of a protozoan described as a commensal of laboratory rodents 20 years earlier (Tyzzer 1912). *Cryptosporidium* is now recognized as a zoonosis and an important cause of illness in immunocompromised individuals. One animal's arcane commensal may be another's pathogen, if one looks carefully enough. Many zoonotic "emerging diseases" of the future may turn out to be infections that are currently well recognized by veterinarians. Accordingly, veterinary medicine should serve as a sentinel for human public health.

24.4.3 Global Versus Local Perspectives

"Think globally" is a catchy pop-culture phrase but may have little application in intervening against emergent infections. The transmission of the deer tick pathogen guild has a common theme regardless of where it is studied within the eastern United States; namely, there is an obligatory cycle between a reservoir host and a specific tick vector that seems to depend on deer for reproduction. This relationship is clearly demonstrable at the local scale, but macrogeographic analyses, such as with Geographic Information Systems, fail to strongly support these observations. In addition, significant variations on the theme may be apparent between sites and between years. Although a global perspective is sometimes useful, the challenge in public health is to characterize local variation so that an intervention may be specifically planned.

24.4.4 "Sing for the Gov'nor"

Interventions that are useful in public health must be derived from an interaction among researchers, public health and government officials, and the communities that are affected. In the case of deer ticks, the Great Island experiment and Nantucket's vision of a reduced deer herd originated with and were supported by the residents of those sites. Their request for action, however, was based on information provided by research. Communication with the public, whose tax dollars ultimately fund basic research, is essential for the success of any research program or intervention (Burke 1997). Public forums and publishing in the lay press provide information for a motivated community.

References

Adler, G.H., S.R. Telford, III, M.L. Wilson, and A. Spielman. 1992. Vegetation structure influences the burden of immature *Ixodes dammini* on its main host, *Peromyscus leucopus*. Parasitology 105:105–110.

Alverson, W.S., D.M. Waller, and S.L. Solheim. 1988. Forests too deer: edge effects in northern Wisconsin. Conserv Biol 2:348–358.

Anderson, J.F. 1989. Epizootiology of *Borrelia* in *Ixodes* tick vectors and reservoir hosts. Rev Infect Dis 11:S1451–S1459.

Anderson, J.F., R.C. Johnson, L.A. Magnarelli, et al. 1987. Prevalence of *Borrelia burgdorferi* and *Babesia microti* in mice on islands inhabited by white-tailed deer. Appl Environ Microbiol 53:892–894.

Armstrong, P.M., L.R. Brunet, A. Spielman, S.R. Telford III. 2001. Risk of human Lyme disease in a Lone Star tick–infested community. Bull WHO 79:916–925.

Bakken, J.S., J.S. Dumler, S.M. Chen, et al. 1994. Human granulocytic ehrlichiosis in the upper Midwest United States: a new species emerging. JAMA 272:212–218.

Bertrand, M., M.L. Wilson. 1996. Microclimate-dependent survival of unfed adult *Ixodes scapularis* (Acari: Ixodidae) in nature: life cycle and study design implications. J Med Entomol 33:619–627.

Burgdorfer, W. 1984. Discovery of the Lyme disease spirochete and its relation to tick vectors. Yale J Biol Med. 57:515–520.

Burke, D.S. 1997. Sing for the gov'nor: a call to advocacy for tropical medicine and hygiene. Am J Trop Med Hyg 56:1–6.

Cronon, W. 1983. Changes in the land: indians, colonists, and the ecology of New England. Hill and Wang, New York.

Daniels, T.J., D. Fish, and I. Schwartz. 1993. Reduced abundance of *Ixodes scapularis* (Acari: Ixodidae) and Lyme disease risk by deer exclusion. J Med Entomol 30:1043–1049.

Donahue, J.G., J. Piesman and A. Spielman. 1987. Reservoir competence of white footed mice for Lyme disease spirochetes. Am J Trop Med Hyg 36:92–96.

Glass, G.E., F.P. Amerisinghe, J.M. Morgan, et al. 1994. Predicting *Ixodes scapularis* abundance on white-tailed deer using geographic information systems. Am J Trop Med Hyg 51:538–544.

Gordon, W.S., A. Brownlee, D.R. Wilson, et al. 1932. Tick-borne fever, a hitherto undescribed disease of sheep. J Comp Pathol Ther 45:301–304.

Gould, S.J. 1997. Ever since Darwin. Norton, New York.

Jones, C.G., R.S. Ostfeld, M.P. Richard, et al. 1998. Chain reactions linking acorns to gypsy moth outbreaks and Lyme disease risk. Science 279:1023–1026.

Lastavica, C.C., M.L. Wilson, V.P. Berardi, et al. 1989. Rapid emergence of a focal epidemic of Lyme disease in coastal Massachusetts. N Engl J Med 320:133–137.

Marshall, W.F. S.R. Telford III, P.N. Rhys, et al. 1994. Detection of *Borrelia burgdorferi* DNA in museum specimens of *Peromyscus leucopus*. J Infect Dis 170:1027–1032.

Mather, T.N., J.M.C. Ribeiro, and A. Spielman. 1987. Lyme disease and babesiosis: acaricide focused on potentially infected ticks. Am J Trop Med Hyg 36:609–614.

Maupin, G.O., D. Fish, J. Zultowsky, et al. 1991. Landscape ecology of Lyme disease in a residential area of Westchester County, New York. Am J Epidemiol 133:1105–1113.

Maupin, G.O., K.L., Gage, J. Piesman, et al. 1994. Discovery of an enzootic cycle of *Borrelia burgdorferi* in *Neotoma mexicana* and *Ixodes spinipalpis* from northern Colorado, an area where Lyme disease is nonendemic. J Infect Dis 170:636–643.

McQuiston, J.H., C.D. Paddock, R.C. Holman, and J.E. Childs. 1999. The human ehrlichioses in the United States. Emerg Infect Dis 5:635–642.

Miller, W.E. 1979. Fire as an insect management tool. Bull Entomol Soc Am 25:137–140.

Piesman, J., T.N. Mather, R.J. Sinsky, and A. Spielman. 1987. Duration of tick attachment and *Borrelia burgdorferi* transmission. J Clin Microbiol 25:557–558.

Schulze, T.L., M.F. Lakat, G.S. Bowen, et al. 1984. *Ixodes dammini* (Acari: Ixodidae) and other ixodid ticks collected from white-tailed deer in New Jersey, USA. J Med Entomol 21:741–749.

Schulze, T.L., R.C. Taylor, G.C. Taylor and E.M. Bosler. 1991. Lyme disease: a proposed ecological index to assess areas of risk in the northeastern United States. Am J Publ Health 81:714–718.

Spielman, A., C.M. Clifford, J. Piesman, et al. 1979. Human babesiosis on Nantucket Island, USA: description of the vector, *Ixodes* (Ixodes) *dammini*, n. sp. (Acarina: Ixodidae). J Med Entomol 15:218–234.

Spielman, A. M.L. Wilson, J.F. Levine, et al. 1985. Ecology of *Ixodes dammini* borne human babesiosis and Lyme disease. Annu Rev Entomol 30:439–460.

Steere, A.C., S.E. Malawista, D.R. Snydman, et al. 1976. Lyme arthritis: an epidemic of oligoarticular arthritis in children and adults in three Connecticut communities. Arthr and Rheumat 20:7–17.

Telford, S.R., III. 1993. Forum: perspectives on the environmental management of ticks and Lyme disease. *In* Ginsberg H.S., ed. Ecology and environmental management of Lyme disease, pp. 164–167. Rutgers University Press, New Brunswick, N.J.

Telford, S.R., III. 1998. The name *Ixodes dammini* epidemiologically justified. Emerg Infect Dis 4:132–133.

Telford, S.R. III, and E. Fikrig. 1995. Prospects towards a vaccine for Lyme disease. Clin Immunother 4:49–60.

Telford, S.R., III, and J.H. Maguire. 1999. Babesiosis. *In* Guerrant, R.L., D.H. Walker, and P.F. Weller, Tropical infectious diseases, vol. 1, eds. pp. 767–773 Churchill Livingstone, Philadelphia.

Telford, S.R., III, and A. Spielman 1989. Enzootic transmission of the agent of Lyme disease in rabbits. Am J Trop Med Hyg 41:482–490.

Telford, S.R., III, T.N. Mather, S.I. Moore, et al. 1988. Incompetence of deer as reservoirs of the Lyme disease spirochete. Am J Trop Med Hyg 39:105–109.

Telford, S.R., III, J.E. Dawson, P. Katavolos, et al. 1996. Perpetuation of the agent of human granulocytic ehrlichiosis in a deer tick-rodent cycle. Proc Natl Acad Sci USA 93:6209–6214.

Telford, S.R. III, J.E. Dawson, and K.C. Halupka. 1997. Emergence of tickborne diseases. Sci Med 4:24–33.

Telford, S.R. III, P.M. Armstrong, P. Katavolos, et al. 1997. A new tickborne encephalitis-like virus infecting deer ticks. Emerg Infect Dis 3:165–170.

Telford, S.R. III, A.S. Weld, H.C. Killie, and T.J. Lepore. 1999. Acorns have little to do with Lyme disease risk: mouse density and the risk of acquiring a deer tick transmitted zoonosis, Nantucket Island 1991–1998. Paper presented at the eighth international congress on Lyme borreliosis, 20–24 June 1999, Munich.

TranSafety, Inc. 1997. Deer-vehicle collisions are numerous and costly. Do countermeasures work? Web site: http//www.usroads.com/journals/rmj/9705/rm970503.htm, accessed 12 May 1997

Tyzzer, E.E. 1912. *Cryptosporidium parvum* (sp. nov.) a coccidian found in the small intestine of the common mouse. Archiv. Protistenk 26:394–412.

Tyzzer, E.E. 1938. *Cytoecetes microti*, n.g., n.sp., a parasite developing in granulocytes and infective for small rodents. Parasitology 30:242–257.

Uspensky, I. 1999. Ticks as the main target of human tickborne disease control: Russian practical experience and its lessons. J Vect Ecol 24:40–53.

Wilson, M.E. 1991. A world guide to infections. Oxford University Press, New York.

Wilson, M.L., G.H. Adler, and A. Spielman. 1985. Correlation between abundance of deer and that of the deer tick, *Ixodes dammini* (Acari:Ixodidae). Ann Entomol Soc Am 78: 172–176.

Wilson, M.L., S.R. Telford III, J. Piesman and A. Spielman. 1988. Reduced abundance of immature *Ixodes dammini* (Acari: Ixodidae) following elimination of deer. J Med Entomol 25:224–228.

Wilson, M.L., A.M. Ducey, T.S. Litwin, et al. 1990a. Microgeographic distribution of immature *Ixodes dammini* (Acari: Ixodidae) correlated with that of deer. Med Vet Entomol 4:151–160.

Wilson, M.L., T.S. Litwin, T.A. Gavin, et al. 1990b. Host dependent differences in feeding and reproduction of *Ixodes dammini* (Acari:Ixodidae). J Med Entomol 27:945–954.

Wilson, M.L., and R.D. Deblinger. Vector management to reduce the risk of Lyme disease. *In* Ginsberg, H.S., ed. Ecology and environmental management of Lyme disease, pp. 126–156. Rutgers University Press, New Brunswick, N.J.

Yoder, J., and A. Spielman. 1992. Differential capacity of larval deer ticks (*Ixodes dammini*) to imbibe water from subsaturated air. J Insect Physiol 38:863–869.

Yuval, B., and A. Spielman. 1990. Duration and regulation of the developmental cycle of *Ixodes dammini* (Acari: Ixodidae). J Med Entomol 27:196–201.

Part V

Conservation Medicine and Challenges for the Future

25

Biodiversity in Biomedical Research

Joshua P. Rosenthal
Trent Preszler

Medical research and conservation biology are operationally distinct fields, despite a great deal of overlap in their basic scientific paradigms and methods. At most universities the practitioners of these respective fields work in separate schools or departments. Furthermore, they focus on different organisms, read different journals, are funded by different organizations, and tend to conduct their research in very different environments. These differences have tended to separate researchers, their investigations, and their results in ways that are to the detriment of both.

Despite these operational divisions, some significant exchanges of theory, methodology, and findings between the biomedical and biodiversity sciences have taken place historically. For example, the fields of molecular biology and demography, as well as many of the specific techniques on which these fields depend, were originally developed in the study of humans and later applied to biodiversity sciences. Similarly, Mendelian genetics and statistical analysis of variance (ANOVA) were originally developed in horticultural and agricultural studies, respectively, and later became basic tools of biomedical science. Today, increased scientific specialization tends to obscure these exchanges and leaves a popular impression that transfer among the fields is largely an unidirectional passage of technologies from medical to biodiversity science.

Perhaps most observers would point to the model organisms *Escherichia coli*, *Rattus norvegicus*, and *Drosophila melanogaster* as the primary contributions of biodiversity to biomedical science. In fact, biodiversity contributes to biomedical science in at least four broad areas (Grifo and Rosenthal 1997). First, biological models for biomedical research have been provided by thousands of diverse species from ferns to tubeworms, from bears to sulfur bacteria, from honeybees to slime molds. Second, naturally occurring compounds found in plants, animals, and microorganisms form the basis for almost half our pharmacopoeia historically (Cragg et al. 1997; Grifo et al. 1997) and continue to be a significant source of leads for new therapeutic drugs (Demain 1998). Third, diverse organisms play a significant role in the ecology of infectious diseases, primarily as vectors and reservoirs, and as control agents for these organisms. Fourth, bacteria, algae, plants, and terrestrial and marine vertebrates, among others, play an increasingly important role as indicators of environmental quality for human health.

Our principal objective in this chapter is to identify and highlight examples of the current contributions of biodiversity to biomedical science, focusing in particular on recent research projects supported by the world's largest biomedical research funding organization, the National Institutes of Health (NIH). Our approach in this chapter is not to assess the historical importance of biodiversity but rather to characterize the body of investigation that represents its contributions at one snapshot in time. For a more comprehensive description of important events and discoveries that have emerged through the overlap of biodiversity and human health, see Grifo and Rosenthal (1997). It is our hope that this analysis of biodiversity-related biomedical research will aid the development of a new interdisciplinary field of conservation medicine by highlighting existing bridges between the respective fields and by identifying some areas that may be productive avenues for future work.

In order to characterize relationships between biodiversity and biomedical research, we have drawn upon a survey that we conducted of grants funded by NIH in 1995, 1996, and 1997. These grants provide a reasonably accurate representation of the biomedical research enterprise as a whole, primarily because NIH is the world's single largest funding organization for research on human health, providing approximately 50% of global public funding in this area. Accordingly, NIH-funded projects cover a very broad range of basic and applied research topics related to human health all over the world.

25.1 What Is Biodiversity-Related Biomedical Research?

Defining "biodiversity-related" in biomedical science is somewhat arbitrary. Biodiversity is most commonly defined as variability of life on Earth. In the broadest sense, then, biodiversity includes humans and most of their diseases. Under that definition, all biomedical research could be considered biodiversity related on some level. However, because of its genesis as an environmental concern with species loss and habitat transformation, the focus of biodiversity studies is generally on non human biota, typically in native environments.

In order to understand the diversity of relevant biomedical research projects in the NIH research portfolio, we began with a broad inclusive definition, and then progressively eliminated those studies that have only a marginal or peripheral association with biodiversity. Specifically, we searched the NIH grants database (1995–97) with a list of over 100 relevant key words and phrases. Based on its abstract, a project was initially included if it fit the following definition:

The research project describes, mimics or manipulates a naturally occurring organism(s), or biologically based compound, behavior or interaction that can exist outside the human body.

Over 10,000 projects fit this initial definition. Each of these were subsequently reconsidered individually and *excluded* if they appeared to fall under this general umbrella *only* because

1. they used a traditional or domesticated model organism such *as E. coli, Drosophila* spp., *Xenopus* spp., *Arabidopsis thaliana*, or laboratory mice, cats, dogs, or nonwild chimpanzees in the course of the study (table 25.1);
2. they examined human health effects of peripherally related environmental phenomena such as climate change, or chemical contaminants (e.g., endocrine disruptors) without significant inclusion of nonhuman biodiversity;
3. they developed chemical methodology related to the synthesis of natural products, but did not directly synthesize natural products or analogs of them; or
4. they studied the biology of a pathogenic organism or its relationship to humans without examining other biotic relationships of the pathogen (e.g., vectors or reservoirs).

The resulting set of 769 projects provides an interesting picture of biodiversity-related biomedical research supported by the NIH over a three-year period. Here we present examples and analysis of these projects. The entire database of these projects is available on request.

25.2 Strengths and Limitations of This Approach

The above-described systematic approach to categorizing research has a number of strengths and weaknesses. Perhaps its greatest strength is that it provides a relatively unambiguous and representative data set from which to draw and a logical, defined approach to inclusion. However, this is not a comprehensive picture of all relevant research for several reasons. First, as mentioned above, the

Table 25.1. Traditional biomedical model organisms

Bacteria (*Escherichia coli*)	Catfish (*Ictalurus punctatus*)
Budding yeast (*Saccharomyces cerevisae*)	Crayfish (*Orconectes* spp.)
Thale cress (*Arabidopsis thaliana*)	Sea urchin (*Arabacia incisa*)
Slime mold (*Dictyostelium discoideum*)	Squid (*Loligo* spp.)
Fungus (*Neurospora crassa*)	Mouse (*Mus musculus*)
Locust (*Locusta migratoria*)	Rat (*Rattus norvegicus*)
Fruit Fly (*Drosophila* spp.)	Cat (*Felis domesticus*)
Roundworm (*Caenorhabditis elegans*)	Dog (*Canis familiaris*)
Frog (*Xenopus laevis*)	Pig (*Sus scrofa*)
Zebrafish (*Danio rerio*)	Chimpanzee (*Pan troglodytes*)
Trout (*Salmo gairdneri*)	

NIH supports only half of the world's biomedical research. Second, that research tends to be biased toward early discovery, or basic research. Third, within the NIH pool of projects our study certainly misses many individual projects and perhaps even whole categories because (1) it is dependent on the abstract submitted by the investigators and key words attached by NIH staff, (2) our search terms were finite and reflected our own scientific knowledge biases, and (3) our criteria for inclusion in part reflect our desire to have an unambiguous and manageable data set. Defining biodiversity to exclude pathogens and traditional model organisms is obviously an arbitrary decision that has a major impact on the size and nature of this data set. With these limitations in mind, we believe the result is a useful and conservative characterization of an underappreciated dependence of biomedical research on diverse, wild, nonhuman biota.

25.3 Diverse Taxa In Medical Research

As described above, we have largely excluded from consideration of research that would only be relevant because it utilized a traditional model organism (see table 25.1 for a list of these traditional models). It is important to realize that the 20 or so species of traditional model organisms are probably utilized in over half of all biomedical research projects in the world and are a profoundly important contribution of biodiversity in the broadest sense. In the context of a discussion of wildlife conservation, they are less relevant because in almost all cases the organisms have been raised for many generations in culture. Both the biological and economic relationships with their native habitats are nearly invisible today.

In examining the abstracts of NIH projects over a three-year period, we found specific references to almost 200 other taxa, including plants, arthropods, fungi, algae, bacteria, bryozoans, sponges, tunicates, cnidarians, mollusks, birds, fish, mammals, reptiles, and amphibians. These are listed in appendix 25.1, using the level of taxonomic precision provided in the abstracts. Note that in many cases only a genus, family, or higher level identification is provided. In most cases these higher taxonomic level identifications were used to indicate the use of multiple species within that group or species for which the binomial may not have been known.

Drug discovery and development of natural products, including field collection, fermentation, chemistry, and analysis of biological activity accounted for the largest number of organisms in our list. Furthermore, the number of species identified in our data set certainly represents only a small number of those actually examined in projects that collect and bioassay large numbers of organisms from the field. Most of the plants, fungi, and algae and many marine invertebrates (e.g., sponges, tunicates) that we encountered are studied for their natural products chemistry. Arthropods are primarily examined as vectors of infectious diseases, although other uses can be found, including occasional study of their potential as natural product drug sources or as models for genetics and behavior. Mammals and birds are commonly studied as reservoir and vector organisms for infectious diseases in humans. However, because of their similarity to human physiology, a diverse group of mammals are also frequently used as model systems for human biology and behavior. Mammals (particularly rodents), fish, birds, and amphibians are also frequently studied as indicator species of environmental quality (sentinels). These categories of research are described below.

25.4 Biomedical Applications

We chose to place biodiversity-related biomedical research project categories by their relationship to the probable applications of the research findings. Focusing on the probable biomedical end of a project allows us to put the work in a context that may be useful to policy discussions. We classified projects into four general areas following, in part, findings of an earlier review of the relationships between biodiversity and human health (Grifo and Rosenthal 1997): drug discovery, biological models, disease ecology, and indicators of environmental quality. Most investigators allude to these directly in their abstracts and the rest can generally be derived without great difficulty. The largest number of projects related to drug discovery, followed by biological models and disease ecology, and we found a smaller number of projects that used biodiversity as indicators of environmental quality (table 25.2). The projects ranged from small grants to large-scale center awards. Below and in table 25.2 we provide some examples of the diversity of projects represented.

25.5 Drug Discovery From Natural Products

This general area comprised the largest number of projects that utilized biodiversity. The projects can be divided into two large classes. The first is early discovery programs that involve collection of organisms from the field, typically focused on screening the collections for therapeutic bioactivity. An example project is "Caribbean Corals (Pseudopterogoria) as a Source of New Anti-inflammatory Agents."

Table 25.2. Number of projects, by research application area, that used biodiversity as indicators of environmental quality

	1995	1996	1997
Drug discovery	120	112	126
Biological models	59	73	72
Disease ecology	50	50	60
Environmental indicators	17	12	18
Total	246	247	276

This table presents the most relevant general biomedical research areas for projects in NIH database for the three years analyzed.

The second area within drug discovery comprises a diversity of mostly lab-based analysis and development of specified natural products and their derivatives. These studies typically include elucidation of compound structure and/or aspects of medicinal chemistry, including synthesis, or pharmacology. A descriptive title of such a project is "Structure and Function of Snake Venom Glycoproteins." For other examples, see table 25.3

25.6 Models and Tools for Medical Research

The second general area of biomedical use of biodiversity we have summarized as models or tools for research. This diverse pool of studies is unified principally by the fact that they use a nontraditional model organism in the execution of their research. Projects included those focused on basic descriptive biology including taxonomy, behavior, physiology, and demography. For instance, study of the "Immune System of Bottlenose Dolphins" where these animals are exposed to environmental pollutants may provide a valuable tool to understanding human responses to pollution, and simultaneously yield information of value to dolphin conservation. Medical research models also include studies on basic synthetic chemistry associated with natural products, but without a specific therapeutic orientation. The National Center for Research Resources of the NIH (http//www.ncrr.nih.gov/) supports grants and contracts to increase the utility and availability of a variety of terrestrial and marine vertebrates for biomedical research.

Genetics and genomics encompass a large, diverse, and growing field. Of particular relevance to our analysis are the numerous projects that study the genetics of diverse species as homologous or analogous systems to understand human biology and behavior. These are quite diverse and include projects such as "Genomic Map of Honey Bee Foraging Behavior" and "Sex Determination in Reptiles—Pit Vipers and Tortoises."

The stunning achievement represented in completion of the basic map of the human genome has overshadowed parallel and smaller efforts in DNA sequencing of a number of other organisms. There are dozens of semi-coordinated efforts such as that for *Drosophila* (recently completed) zebrafish, *Dictyostelium*, and other classic model organisms. While these genome projects are not included in our database because they are, by definition, traditional model organisms, they warrant mentioning here because of their importance to biomedical research. Relatively complete genomes can also be found for several hundred viruses and several dozen archaea, bacteria, and eukaryotes. In addition, GenBank, a program of the National Center for Biotechnology Information of the NIH (http//www.ncbi.nlm.nih.gov/), compiles and analyzes genomic sequence data on most types of organisms that are submitted by scientists from around the world. At present, approximately 60,000 "species" (including many undescribed organisms at the rank of species) are represented in this database. At the present, these include approximately 324 archaea, 706 bacteria, 6445 fungi, 20803 metazoa, 18592 viridiplantae (Green plants and algae), and 4470 viruses. These efforts are not re-

Table 25.3. Examples of biodiversity-related biomedical projects by research application area

Project Number	Project Title
Drug Discovery	
1 R43 CA67557-01A1	Anticancer drugs from halophilic marine microorganisms
5 R01 CA64483-02	Biosynthesis of taxol and related compounds
3 N01 CM27704-007	Collection and taxonomy of shallow-water organisms
5 R02 AI27436-06	Coprophilous fungi—new sources of bioactive metabolites
2 R01 CA54349-04	Estrogenic plant compounds and etiology of breast cancer
1 K08 AI01307-01	Mechanisms involved in mycobacteria–host interactions
5 S06 GM08156-19(5)	Phylogenetics of moccasins (pit vipers) improving antivenin specificity
5 R37 GM13854-26	Singlet oxygen in the synthesis of bioactive molecules
1 R03 AR45071-01	Insect salivary vasoactive compounds in wound healing
1 R01 AI40641-01A1	Rational design of novel, nonneurotoxic antimalarials
Research Model	
5 R37 AI15027-18	Acylic stereoselection in natural product synthesis
5 R01 GM26166-16	Biosynthesis and metabolism of nucleosides
5 R29 CA61978-02	How didemnins inhibit protein biosynthesis
5 P51 RR00167-35	Social status and reproductive success in stump-tailed macaques
1 Z01 CP05367-11	The genetic structure of natural populations of past and present
5R01GM48665-03	Transfer RNA processing in the archaea
5 S06 GM08016-27	Color of photoreceptors in vertebrate retina and vision in estuarine fish
5 S06 GM08132-22(9)	Developmental physiology of elephant seal pups during natural, prolonged fasts
1 F32 MH11703-01	Evolution of social communication in anuran amphibians
5 P30 ES05705-07	Marine models of human diseases
Vector Biology	
5 R03 TW00240-04	Anopheles vector potential for malaria transmission
5 R29 AI34409-02	Environmental determinants of *Ixodes dammini* expansion
5 R01 AI35215-02	Evolutionary genetics of African malaria vectors
5 R01 AI34521-03	Genetics and biogeography of sand fly disease vectors
5 R01 AI11373-21	Molluscan faunas and Asian schistosomiasis
5 R01 AI16137-13	Underlying mechanisms of schistosome/snail compatibility
5 R01 AI39129-02	Epidemiology and ecology of *Vibrio cholerae* in Bangladesh
1 F32 AI09870-01	Molecular taxonomy of Lyme disease vectors
5 R01 AI36336-04	Sentinel surveillance for Four Corners hantavirus
5 R29 GM50551-04	Uptake of lipids by insect oocytes
Environmental Quality	
5 R01 GM38624-07	Causes of chemical variation in coral reef invertebrates
2P42ES04696-09	Effects-related biomarkers of toxic exposures
1P42ES07375-010001	Endocrine disrupting effects of chlorine
2 P42 ES04908-07	Microbial detoxication/degradation of hazardous wastes
1P42ES07381-010009	Sentinel species—xenobiotics, toxicity, and reproduction
1 R01 HD33563-01	Population growth, economic change, and forest degradation
5 U01 AI35894-04	New and emerging gastrointestinal pathogens
5 S06 GM08103-24(2)	Sea food poisons—toxin diversity and the ciguatera food chain
5 S06 GM08168-18	Surface antigens of endosymbiotic diatoms
5P42ES05947-060002	Assessment of metal contamination and ecological implications

flected in the data reported in the tables or appendix of this chapter. These genomic data made a major contribution to a recent phylogenetic interpretation of the still-disputed hierarchy of surviving placental mammalian orders (O'Brien et al. 1999).

25.7 Ecology and Evolution of Disease

After excluding pathogens from consideration, most relevant infectious disease research supported by the NIH during the three years analyzed relates to the biology of animals that are hosts, reservoirs, and vectors for diseases. These include over 150 field and laboratory studies of host immune systems, vector and reservoir competence, and transmission biology as well as biogeography, taxonomy, population dynamics and genetics of these animals.

In particular, there are numerous projects on the biology of mosquito and rodent vectors of parasitic, viral, and bacterial diseases. It is notable that we found very few projects that examine the relationship of habitat disruption, species invasions, climate change, or other large-scale environmental changes on the ecology of disease vectors or reservoirs, despite the importance of these phenomena for the dynamics of disease. One exception was a study on the probable effects of the Three Gorges Dam construction on several vector-borne parasitic diseases in humans, entitled, "Emerging Helminthiases in China."

25.7.1 Indicators and Modifiers of Environmental Quality

The use of diverse plants, animals, and microorganisms as "sentinels" or biomarkers is a growing field that relies on an understanding of the immunology, physiology, behavior, reproduction, and population dynamics of the organism to alert us to health hazards in an area. Hazards may be chemical, biological, or other in nature. For example, the project "Effects Related to Biomarkers of Toxic Exposures" uses a variety of mammals and plants as indicators in the analysis of Superfund toxic waste sites. This and other projects address biological degradation of environmental contaminants and the potential role of chemical pollutants as disruptors of mammalian endocrine systems. Numerous organisms, including bacteria and algae, have the capacity to alter the chemistry of environmental effluents in both positive (degradation of toxins to less harmful products) and negative (production of new toxins) ways; thus we consider these organisms to be both indicators and modifiers of environmental quality.

From a biomedical perspective, the value of indicator species derives from the tendency of many organisms, particularly those at higher levels in the food chain, to accumulate and concentrate toxic substances over time. We may thus be able to identify and quantify the effects of pollutants in other organisms before they become a threat to humans.

25.7.2 Specific Therapeutic Areas That Depend on Biodiversity

Much research on biodiversity is very basic science that does not directly affect specific disease areas. However, some trends may be discerned. Natural products drug discovery has been particularly important historically for treatment of cancer, infectious diseases, nervous system disorders, and heart disease (Cragg et al. 1997), and many projects in our sample addressed these same areas.

Medical models are integral to basic studies of genetics, physiology, chemistry, immunology, and behavior and so are relevant to almost every area of human health, including cancer, many infectious diseases, cardiovascular disorders, mental health, aging, drug and alcohol abuse, demography, and reproduction.

Infectious disease research relies on the study of diverse elements of biodiversity in ecological studies, drug discovery and numerous research models. Relevant work targets many parasitic diseases, including malaria, leishmaniasis, schistosomiasis, and trypanosomiasis, as well as numerous viral diseases such as hantavirus, dengue, encephalitis, and bacterial diseases such as Lyme disease. Further information can be gleaned by inferring therapeutic area from the institutes and centers that fund these projects (table 25.4). Across all areas the institutes whose projects most often turned up in this study were the Institutes charged

Table 25.4. Number of grants by granting institution

	1995	1996	1997
National Institute of Allergy and Infectious Diseases (NIAID)	69	70	70
National Institute of General Medical Sciences (NIGMS)	65	65	72
National Cancer Institute (NCI)	61	64	52
National Institute of Environmental Health Sciences (NIEHS)	24	11	29
Fogarty International Center	11	9	9
National Center for Research Resources (NCRR)	2	8	17
National Institute of Child Health and Human Development (NICHD)	4	4	6
National Institute of Mental Health (NIMH)	1	4	8
National Institute on Aging (NIA)	3	5	3
National Institute of Dental Research (NIDR)	3	2	1
Food and Drug Administration (FDA)	1	2	2
National Institute of Neurological Disorders and Stroke (NINDS)	1	2	2
National Institute of Health Clinical Center	0	1	1
National Heart, Lung, and Blood Institute (NHLBI)	1	0	1
National Institute of Arthritis and Musculoskeletal and Skin Diseases (NIAMSD)	0	0	1
National Institute of Diabetes and Digestive and Kidney Diseases (NIDDK)	0	0	1
National Library of Medicine (NLM)	0	0	1
Total	246	247	276

The focus of the institute and center supporting these projects in many cases provides a general indication of the therapeutic area that the project may relate to.

with research on allergy and infectious diseases, cancer, environmental health (principally toxicological exposures), and mental health, and also included those focused on aging, dental health, musculoskeletal, cardiovascular, and neurological disorders.

25.8 Significance for Conservation of Biodiversity

The need for conservation scientists, advocates, and managers to understand biomedical uses of their findings and of the organisms and habitats they study is significant for several reasons. By most any index one cares to apply—circulation rates of the major journals, availability of research funding, number of academic positions, and so forth—biologists that study humans and their afflictions vastly outnumber conservation biologists. By increasing our understanding of the overlap between these fields, we can increase public awareness and understanding of the importance of conservation, and the conservation science community can take advantage of the knowledge and resources associated with biomedical research.

Few issues are as compelling to most people as human health. People became much more interested in conservation of the weedy Pacific Yew tree and its habitat once they understand that taxol, one of our most important treatments for breast cancer, is derived from the tree. Thus, public and scientific understanding of the value of diverse organisms for human health is a potentially powerful tool for conservation policy. We are also beginning to develop models in which both the research process and the products of natural products drug discovery may contribute to conservation policy and practice (Rosenthal et al. 1999). The value of biodiversity for medicines has been widely cited, but its role in basic biomedical research, in ecology of infectious diseases, and as indicators of environmental quality has received little attention in policy discussions.

In addition to a policy role, conservation scientists should be aware of the resources and opportunities that related biomedical research endeavors may offer their work. Understanding the relevance of conservation work to medicine may help conservation scientists take advantage of substantially larger funding resources such as the NIH, the World Health Organization, and the Gates Foundation. Furthermore, the biomedical research enterprise offers a wealth of data (e.g., the public genomic sequence databases) and potential collaborators for biodiversity science, including very basic studies on inventory, ecology, and taxonomy. Growing interest in the ecology of infectious diseases and the use of indicator species signals great potential for interaction between veterinary and wildlife medicine practitioners and traditional biomedical scientists.

While much of the data that are derived from the projects described here may not contribute directly to management of biodiversity, a great deal of the genetic, physiological, and behavioral results that emerge are useful. For example, understanding the biology of malaria vectors is likely to be important for human and wildlife health. Research on the genetics of diseases in cats carried out by the NIH Laboratory on Genomic Diversity has already provided useful information for management of wild cat species (Eizirik et al. 2001).

Overharvesting of biomedically useful species is not uncommon and is frequently ignored by biomedical researchers, in part because the data necessary to evaluate this risk are not available. The biomedical research community is in need of studies on population ecology and genetics that may guide sustainable harvesting of biological materials from the wild, particularly for those species that become important sources of natural product medicines or are useful medical models.

25.9 Significance for Human Health

An analysis of the true value of biodiversity to human health is beyond the scope of this chapter. However, a few general points are worth mentioning. We know that natural products are historically the single most important source of novel molecules that generate whole new classes of drugs (Cragg et al. 1997), and we know that diverse model biological systems can lead to major breakthroughs in understanding health and disease. Furthermore, as mentioned above, understanding the biology and behavior of vector animals directly increases our ability to control some of the world's biggest killers (e.g., malaria). Thus, the impact of biodiversity on biomedical research is likely to be much greater than one might suppose looking at the number of projects in our biased subsample as a percentage of all NIH projects (<1%). Note that this does not include the database efforts of GenBank. Indeed the diversity of studies and their biomedical applications argues for a very broad influence of biodiversity on the field and is probably a better indicator of importance than the total number of projects.

One important note is that the biomedical community depends on a relatively tiny fraction of the world's biological diversity. We found references to approximately 200 taxa, and even multiplying this number by 10 to account for the higher level designations and large-scale drug screening programs, the total is a miniscule fraction of the 14 million species that are estimated to populate our planet. Clearly, in-depth understanding of the 20 or so traditional model organisms has provided incredible returns to biomedical science. However, consider that our view of the organization and diversity of life, as well as our understanding of basic metabolic pathways, has been altered dramatically in the last few years with the discovery and analysis of Arachaebacteria from extreme environments (Doolittle 1999). If biomedical science focuses too narrowly on the tiny fraction of well-described model organisms, we will certainly miss out on discoveries with profound implications for human health, and the opportunities may be lost to extinction before we realize what we have missed.

With the ever-growing impact of humans on the environment, there is a growing need to understand the ecological context of infectious disease and the geophysical and biological processes that may regulate human exposure to environmental health hazards such as toxic waste. Furthermore, the sciences of ecology, veterinary and wildlife medicine, toxicology and the biophysical disciplines of environmental health have matured a great deal in recent years, and the opportunities for biomedical science to take advantage of the biodiversity sciences are greater today than ever before.

25.10 Recent Developments, Trends, and Challenges

Natural products drug discovery continues in academia and industry, and many consider its long-term value to be enhanced by increasing scientific sophistication in synthetic, rational, and combinatorial chemistry and genomics. The astounding growth in popularity of botanical medicines over the last five years in North America and Europe caught the biomedical establishment somewhat unprepared to answer basic questions about the safety and efficacy of these diverse products. The recent establishment of the Office of Dietary Supplements at the NIH and the National Center for Complementary and Alternative Medicines is, in part, an attempt to respond to this public health information need.

Through collaborations of public and privately financed projects, the genomes of the nematode *Caenorhabditis elegans* and the fruit fly *Drosophila melanogaster* were completed in the late 1990s. As the technology increases in sophistication and ease of use, it is likely that after whole genomes of most of the traditional research organisms projects are completed, many other organisms will follow. The National Center for Biotechnology Information is a continually expanding resource that will certainly be more and more useful to conservation and biodiversity scientists in the coming years.

The observation that disease outbreaks frequently follow deforestation, dam construction, extreme weather events, and other environmental disturbances has prompted many to consider the role of ecosystem integrity in transmission of infectious diseases (Epstein et al. 1993; Molyneux 1997; Walsh et al. 1993). The need to evaluate role of large-scale ecological disturbance events on disease and develop tools for prediction of disease transmission dynamics has given rise in the late 1990s to a novel interagency grants program on the ecology of infectious diseases. The program is supported jointly by four centers and institutes of the NIH (Fogarty International Center, National Institute of Allergy and Infectious Diseases, National Institute on General Medical Sciences, and the National Institute on Environmental Health Sciences [NIEHS]) and the Bioscience Directorate of the National Science Foundation, and includes participation of three other federal agencies the (National Aeronautics and Space Agency, the U.S. Department of Agriculture/Agricultural Research Service, and the U.S. Geological Survey). The large number of applications elicited by the first announcement of the program led to a second request for applications in 2001. This effort has already led to significant findings regarding the causes of amphibian declines (Kiesecker et al. 2001)

Accumulating evidence on the sensitivity of wildlife to long-term effects of toxic waste, pesticides, radiation, and their breakdown products has broadened interest in using wildlife as indicators of environmental quality. The development of such new biomarkers is an important part of the Superfund basic research program (www.niehs.nih.gov/sbrp/home.htm) of the NIEHS.

25.11 Conclusions

We have identified a significant number of biomedical research projects that depend on diverse biota. The biomedical applications of the projects can be categorized into four general groups: drug discovery, medical models, disease ecology, and environmental indicators. The likely therapeutic areas to benefit from this work are widely distributed among diseases, including cancer, infectious diseases, central nervous systems disorders, among many others. This work depends on several hundred named biological taxa and probably includes many times that number. The organisms involved represent all three domains of life (Bacteria, Archaea, Eukarya), all four kingdoms of eukaryotes, and many major animal phyla (see appendix 25.1). Furthermore, as outlined at the beginning, our estimates of both the number and types of projects are certainly underestimates.

In the development of a new interdisciplinary field such as conservation medicine, it is critical to understand where bridges between relevant disciplines already exist. As we have shown here, recent funded biomedical research projects illustrate a diversity of such bridges. Not only is there a community of scientists already at the interface of biodiversity and human health, but there are also publicly available databases, existing avenues for research funding, and other resources to help support work at this interface. It may be tempting for pioneers in conservation medicine to define the field largely in ecological terms, focusing on the dynamics of infectious diseases and environmental exposures in wildlife and humans. However, it is important to keep in mind that humans and other animals have important long-term interests in the medicinal and research resources that biodiversity continues to provide. Such broad dependence may argue that this incipient field should also focus broadly on the interactions of biodiversity and health.

APPENDIX 25.1. Taxa identified in abstracts of NIH-funded projects in 1995, 1996, and 1997

Where possible we have added higher order designations of those for which only genus or species were identified.

Family (order)	Genus species	Common Name
Eukarya		
Plants		
Acanthaceae	—	—
Anacardiaceae	*Rhus succedanea*	—
Ancistrocladaceae	*Ancistrocladus korupensis*	—
Apiaceae	*Angelicae radix*	—
Ascelepiadaceae	*Ascelepias subverticillata*	Western whorled milkweed
Betulaceae	*Betula* spp.	Birch
Brassicaceae	—	—

(continued)

Family (order)	Genus species	Common Name
Cactaceae	—	Cacti
Carophyllidae	—	—
Celastraceae	—	—
Compositae	*Artemisia annua*	—
Compositae	*Blumea* spp.	—
Compositae	*Siegesbeckia pubescens*	—
Compositae	*Wollastonia biflora*	—
Cycadaceae	*Cycas circinalis*	—
Euphorbiaceae	—	—
Gramineae	*Festuca* spp.	Fescue
Gramineae	*Poa pratensis*	Kentucky blue grass
Gramineae	*Lolium perenne*	Perennial rye grass
Guttiferae	*Calophyllum*	—
Labiatae	*Rosmarinus officinalis*	Rosemary
Lauraceae	—	—
Leguminosae	*Medicago sativa*	Alfalfa
Leguminosae	*Astragali radix*	—
Leguminosae	*Glycine max*	Soybean
Leguminoseae	*Lathyrus sativus*	—
Liliaceae	*Allium sativum*	Garlic
Lycopodiaceae	*Lycopodium serratum*	Club moss
Malvaceae	*Gossypium* spp.	Cotton
Meliaceae	—	—
Mimosaceae	—	—
Myrtaceae	—	—
Rutaceae	*Zanthoxylum nitidum*	—
Scrophulariaceae	—	—
Simaroubaceae	*Castela* spp.	—
Solanaceae	*Solanum tuberosum*	Potato
Stemonaceae	*Croomia*	—
Stemonaceae	—	—
Taxaceae	*Taxus* spp.	Pacific yew
Theaceae	*Camellia sinensis*	Green tea
Thymelaeaceae	—	—
Xanthophyllaceae	—	—
—	*Cnidii rhizoma*	—

Fungi

Agaricaceae	*Agaricus bisporus*	Cultivated mushroom
—	—	Coprophilous fungi
—	—	Dry forest macrofungi
—	—	Endophytic fungi
—	—	Soil-associated fungi

Protists

Dasycladaceae	*Acetabularia* spp.	Algae
—	*Skeletonema costatum*	Marine diatom
—	—	Cultured microalga

(continued)

Family (order)	Genus species	Common Name

Animals

Annelids
| Glyceridae (Aciculata) | *Glycera dibranchiata* | Marine polychaete |

Arthropods
Apidae (Hymenoptera)	*Apis mellifera*	Honey bee
Archaeopsyllinae (Siphonoptera)	*Ctenocephalides felis*	Cat flea
Artemiidae (Anostraca)	*Artemia salina*	Brine shrimp
Artemiidae (Anostraca)	*Brugia malayi*	Brine shrimp
Bombycidae (Lepidoptera)	*Bombyx spp.*	Silk worm larvae
Crangonidae (Decapoda)	—	Spiny lobster
Culicidae (Diptera)	*Aedes aegypti*	—
Culicidae (Diptera)	*Aedes triseriatus*	—
Culicidae (Diptera)	*Anopheles* spp.	—
Ixodidae (Acari)	*Amblyomma americanum*	Lone star tick
Ixodidae (Acari)	*Ixodes* spp.	Tick
Ixodidae (Acari)	*Dermacentor andersoni*	Wood tick
Meloidae (Coleoptera)	—	Blister beetle
Muscidae (Diptera)	*Glossina* spp.	Tsetse flies
Psychodidae (Diptera)	*Lutzomyia* spp.	Sand fly
Reduviidae (Hemiptera)	*Rhodnius prolixus*	—
Reduviidae (Hemiptera)	—	Triatomine bugs
Sarcoptidae	*Sarcoptes scabiei*	Ectoparasitic mites
Scarabaeidae (Coleoptera)	*Tribolium confusum*	Flour beetle
Vespidae (Hymenoptera)	*Vespidae* spp.	Wasp
—	9 species	Icelandic hay mite

Bryozoans
| Bugulidae (Cheilostomata) | *Bugula neritina* | — |

Cnidarians
| Gorgoniidae (Gorgonacea) | *Pseudopterogorgia* spp. | Corals |
| — | — | Zoanthid cnidarians |

Mollusks
Aplysiidae (Anaspidea)	*Aplysia*	Marine slug
Aplysiidae (Anaspidea)	*Aplysia kurodai*	Sea hare
Conidae (Neogastropoda)	*Conus* spp.	Fish hunting sea snails
Planorbidae (Basommatophora)	*Biomphalaria glabrata*	Snails
Triculinae snails	—	—
Unionidae (Unionoida)	*Anodonta cygnea*	Fresh water mussel
—	—	Giant clams

Sponges
Adociidae (Poecilosclerida)	*Pellina triangulata*	—
Axinellidae (Halichondrida)	*Ptilocaulis spiculifer*	Caribbean sponges
Discodermiidae (Choristida)	*Discodermia dissoluta*	—
Haliclonidae (Haplosclerida)	*Reniera sarai*	—
Haliclonidae (Haploscleridia)	*Cribrochalina vasculum*	—
Myxillidae (Poecilosclerida)	*Tedania ignis*	Caribbean sponges
Phorbasidae (Poecilosclerida)	*Phorbas* spp.	—
Theonellidae (Lithistida)	*Theonella* spp.	—

(continued)

Family (order)	Genus species	Common Name
(Choristida)	—	—
(Dendroceratida)	—	—
(Hadromerida)	—	—
(Haplosclerida)	—	—
(Lithistida)	—	—
Tunicates		
Didemnidae (Aplousobranchia)	*Trididemnum solidum*	—
Perophoridae (Phlebobranchia)	*Ecteinascidia turbinata*	—
Polyclinidae (Aplousobranchia)	*Aplidium lobatum*	—
Amphibians		
Bufonidae (Anura)	*Bufo boreas*	—
Bufonidae (Anura)	—	South American bufonid toads
Discoglossidae (Anura)	*Bombina maxima*	—
Discoglossidae (Anura)	*Bombina orientalis*	—
Hylidae (Anura)	*Hyla regilla*	—
Hylidae (Anura)	*Phyllomedusa sauvagei*	—
Leptodactylidae (Anura)	*Leptodactylus rugosa*	—
Ranidae (Anura)	*Rana aurora*	—
Ranidae (Anura)	*Rana cascadae*	—
—	—	Madagascan mantellid frogs
Birds		
Corvidae (Passeriformes)	*Pitohui* spp.	—
Fringillidae (Passeriformes)	*Zonotrichia leucophrys nuttalli*	White-crowned sparrow
Muscicapidae (Passeriformes)	*Turdus migratorious*	American robins
Muscicapidae (Passeriformes)	*Sialia* spp.	Bluebirds
Muscicapidae (Passeriformes)	*Sialia sialis*	Eastern bluebirds
Psittacidae (Psittaciformes)	*Melopsittacus undulatus*	Parakeet
Sturnidae (Passeriformes)	*Sturnus vulgaris*	European starlings
Fish		
Centrarchidae (Perciformes)	*Micropterus salmoides*	Large-mouth bass
Clupeidae (Clupeiformes)	—	Estuarine fish: menhaden
Fundulidae (Cyprinodontiformes)	*Fundulus heteroclitus* (teleost)	—
Ictaluridae (Siluriformes)	*Ictalurus nebulosus*	Brown bullhead catfish
Labridae (Perciformes)	*Thalassoma duperrey*	Coral reef fish
Moronidae (Perciformes)	—	Estuarine fish: striped bass
Oryziatidae (Cyprinodontiformes)	*Oryzias laticeps*	Japanese medaka
Oryziatidae (Cyprinodontiformes)	*Oryzias latipes*	Medaka
Percidae (Perciformes)	—	Estuarine fish: yellow perch
Pomacentridae (Perciformes)	*Pomacentrus* sp.	Damselfish
Pomacentridae (Perciformes)	*Stegastes* sp.	Damselfish
Pomacentridae (Perciformes)	*Chrysiptera* sp.	Damselfish
Psettodidae (Pleuronectiformes)	—	Flounder
Rajidae (Rajiformes)	*Raja erinacea*	Little skate
Sphyraenidae (Perciformes)	*Sphyraena barracuda*	Ciguatoxic barracuda
(Tetradontiformes)	—	Puffer fish
—	—	Shark

(continued)

Family (order)	Genus species	Common Name
Mammals		
Bovidae (Artiodactyla)	*Ovis canadensis*	Bighorn sheep
Bovidae (Artiodactyla)	*Ovis* sp.	Sheep
Canidae, wild and domestic	—	—
Castoriae (Rodentia)	*Castor canadensis*	Beaver
Cebidae (Primates)	*Ateles* spp., *Alouatta* sp.	Howler monkeys
Cebidae (Primates)	*Brachyteles arachnoides*	Muriqui Monkeys
Cebidae (Primates)	*Saimiri* sp.	Squirrel monkey
Cebidae (Primates)	*Callicebus* sp.	Titi monkey
Cervidae (Artiodactyla)	*Odocoileus hemionus*	Mule deer
Cervidae (Artiodactyla)	*Odocoileus virginianus*	White-tailed deer
Dasypodidae (Xenarthra)	*Dasypus novemcinctus*	Nine-banded armadillos
Delphinidae (Cetacea)	*Tursiops truncatus*	Atlantic bottlenose dolphin
Didelphinae (Didelphimorphia)	*Didelphis virginiana*	Opossums
Felidae, wild and domestic	—	—
Hylobatidae (Primates)	*Hylobates concolor*	Gibbon
Leporidae (Lagomorpha)	*Sylvilagus* spp.	Cottontail rabbit
Muridae (Rodentia)	*Microtus* spp.	Agricultural pest rodents
Muridae (Rodentia)	*Sigmodon hispidus*	Cotton rats
Muridae (Rodentia)	*Peromyscus maniculatus*	White-footed mouse
Muridae (Rodentia)	*Neotoma* spp.	Woodrat
Phocidae (Carnivora)	*Mirounga angustirostris*	Northern elephant seal
Sciuridae (Rodentia)	*Spermophilus mexicanus*	Mexican ground squirrels
Sciuridae (Rodentia)	*Spermophilus variegatus*	Rock squirrels
Suidae (Artiodactyla)	—	Pigs
Reptiles		
Alligatoridae (Crocodilia)	*Alligator* sp.	Alligator
Elapidae (Squamata)	*Dendroaspis angusticeps*	Africa green mamba
Elapidae (Squamata)	*Bungarus multicinctus*	Banded krait
Elapidae (Squamata)	—	Cobra
Emydidae (Testudines)	*Chrysemys picta*	Painted turtle
Helodermatidae (Squamata)	*Heloderma suspectum*	Gila monster
Testudinidae (Testudines)	—	Tortoises
Viperidae (Squamata)	—	Pit vipers
Bacteria		
Actinomycetaceae (Actinomycetales)	—	Actinomycetes
Actinomycetaceae (Actinomycetales)	—	Actinomycetes from sponges
Bacillaceae (Eubacteriales)	*Bacillus thuringensis*	—
Oscillatoriaceae (Nostocales)	*Lyngbya majuscala*	Cyanophyte
Pseudomonadaceae (Pseudomonadales)	*Pseudomonas* spp.	—
Spirillaceae (Pseudomonadales)	*Vibrio* spp.	—
Streptomycetaceae (Actinomycetales)	*Streptomyces* spp.	—
Myxobacterales	*Myxobacteria* spp.	—
—	*Rhodococcus rhodnii*	—
—	—	Anaerobic and aerobic bacteria

(continued)

APPENDIX 25.1. Continued

Family (order)	Genus species	Common Name
—	—	Cyanobacteria
—	—	Cyanophytes from sponges
—	—	Cytophagalike bacteria
—	—	Proteobacteria
—	—	Thermophilic cyanobacteria
Archaea		
—	—	Deep sea vent bacteria

References

Cragg, G.M., D.J. Newman, and K.M. Snader 1997. Natural products in drug discovery and development. Nat Prod 60:52–60.

Demain, A.L. 1998. Microbial natural products: alive and well in 1998. Nat Biotechnol 16:3–4.

Doolittle, W.F. 1999. Phylogenetic classification and the universal tree. Science 284:2124–2129.

Eizirik, E., J.H. Kim, M. Menotti-Raymond, P.G. Crawshaw Jr., S.J. O'Brien, W.E. Johnson. 2001. Phylogeography, population history and conservation genetics of jaguars (*Panthera onca*, Mammalia, Felidae). Mol Ecol 10:65–79.

Epstein, P., T. Ford, R. Colwell. 1993. Marine ecosystems. Lancet 342:1216–1219.

Grifo, F., and J. Rosenthal (eds.). 1997. Biodiversity and human health. Island Press, Washington, D.C.

Grifo, F., D.J. Newman, A. Fairfield, J.T. Grupenhoff, and B. Bhattacharya. 1997. The origins of prescription drugs. *In* Grifo, F., and J. Rosenthal, eds. Biodiversity and human health, pp. 131–163. Island Press, Washington, D.C.

Kiesecker, J.M., A.R. Blaustein, and L.K. Belden. 2001. Complex causes of amphibian population declines. Nature 410:681–684.

Molyneux, D.H. 1997. Patterns of change in vector-borne diseases. Ann Trop Med Parasitol 91:827–839.

O'Brien, S.J., M. Menotti-Raymond, W.J. Murphy, W.G. Nash, J. Wienberg, R. Stanyon, N.G. Copeland, N.A. Jenkins, J.E. Womack, and G.J.A. Marshall. 1999. The promise of comparative genomics in mammals. Science 286:458–62; 479–481.

Rosenthal, J.P., D. Beck, A. Bhat, J. Biswas, L. Brady, K. Bridbord, S. Collins, G. Cragg, J. Edwards, A. Fairfield, et al. 1999. Combining high risk science with ambitious social and economic goals. Pharmaceut Biol 37(suppl):6–21.

Walsh, J.F., D.H. Molyneux, and M.H. Birley. 1993. Deforestation: effects on vector-borne disease. Parasitology 106(suppl):S55–S75.

26

Introducing Ecosystem Health into Undergraduate Medical Education

David J. Rapport

John Howard

Robert Lannigan

Robert McMurtry

Douglas L. Jones

Christopher M. Anjema

John R. Bend

The concept of health is ancient. While the notion of health intuitively may apply to the well-being of any complex system (e.g., the economy, a community, the oceans, a coral reef, the planet; Somerville 1995), for much of history its context has been limited to the physical condition of the individual (at first humans and later other biota—both plants and animals). In more recent times, the concept has won acceptance as a descriptor of the complete mental, physical, and emotional state of the individual and as a descriptor of the condition of populations (e.g., flocks, herds, human societies). The World Health Organization (WHO) in its constitution defines health as "a state of complete physical, mental and social well-being and not merely the absence of disease and infirmity" (WHO 1948). Today, the health concept has taken another leap in its formal application—being applied by ecologists, resource managers, and medical practitioners to whole ecosystems—complex organizations comprising vast numbers of species. With every shift of level, the concept of health evolves and acquires new levels of meaning (Rapport et al. 1999a).

At the level of the whole ecosystem (landscape, region), the notion of "health" is a potentially powerful integrative concept—referring to the well-functioning of complex systems comprising intimate interrelationships among humans and other species, and among life forms and their abiotic environment. A trademark of this complexity is the dependence of all life forms on functions governed by the ecosystem as a whole. Whole-system properties such as nutrient cycling, hydrological cycles, pollination, and primary productivity have a pervasive influence on all life forms that comprise a given ecosystem.

It is no secret today that the earth's ecosystems are in trouble. There is hardly a passing day when the front page of major newspapers does not contain some reference to the failing of the earth's life support systems. These failures have always been present at local scales. What is different today is that these failings are now present at regional and global scales. A recent report of the United Nations Environment Programme summed it up: the earth is in critical condition (UNEP 1999). How did it get that way? Some of the lines of causality are known—and they all point to pressures exerted by a single species: *Homo sapiens*. Ecosystem breakdown comes about by stresses, singly and interactively, as the result of human activity (Rapport and Whitford 1999). Stresses that contribute to ecosystem dysfunction at local, regional, and in some cases global scales include overharvesting, physical restructuring (e.g., building large dams that obliterate land areas and drastically alter hydrological flows), the introduction of exotic organisms, the alteration of atmospheric and stratospheric chemistry, and the release of waste residuals into air, water, and land. The consequences of ecosystem breakdown are felt in every quarter. As humans we are at risk of greater disease burdens owing to these ecological imbalances (McMichael et al. 1999; Patz et. al. 1996).

Here, we describe a new chapter in this evolution: the introduction of a program in ecosystem health as part of the development of the undergraduate curriculum in a major medical school. What motivated this development in what is otherwise a very traditional institution? What exactly is "ecosystem health" and how does it relate to the interests of medical practitioners? Why teach ecosystem health to physicians? What is the substance of the program, and how was it received by both students and trainers? These are questions addressed in this chapter. We begin with the motivation for this program.

26.1 Lessons from the University of Western Ontario

The rationale for the introduction of an ecosystem health program within the undergraduate curriculum at the University of Western Ontario in 1997 was twofold. First was the recognition that increasingly it is ecological imbalance that is at the root of many human ills. Human survivability in the twenty-first century can no longer rely solely upon the remarkable advances in the practice of western medicine (e.g., pharmaceuticals, vaccines, surgical advances, Golub 1997). Human futures are ever more dependent on achieving sustainable life support systems—that is, in restoring health to the earth's ecosystems (McMichael 1993a,b; Rapport et al. 1998a, 1999b). Second was the belief that physicians should not look at the doctor–patient relationship in isolation, but rather in the context of the physical and social environment in which both the patient and the physician exist. Traditionally, medicine emphasizes diagnosing the disease and curing it. The concept of ecosystem health emphasizes causality within the human environment. Prevention of disease in addition to treatment of disease is a natural result of this wider, holistic view of the doctor–patient relationship.

26.2 Evolving Diseases

Historical context medicine and the evolving medical curriculum have a large role to play in bringing this recognition into practice. From an ecological point of view, the history of medicine can be portrayed as a series of transitions accompanying radical shifts in human and natural ecology (Rapport et al. 1999b). The major causes of disease have shifted, and now include diseases of regional and global ecological imbalance. Some 10,000 years ago, the first major epidemiological transition occurred from a primarily hunting and gathering society to agricultural settlements. Accompanying this were a new set of risks to human health. No longer were the major health risks primarily those associated with the accidents and inherent dangers of hunting large prey. They now included a suite of infectious and parasitic diseases arising from increased contact among humans and among humans and domesticated animals. Some 7,000 years ago, the rise of city-states, crowded urban living conditions with little or no facilities for sanitation, brought about a new set of health risks. "Filth diseases"—intestinal and respiratory infections—arose from poor sanitation and maladaptive human behavior when living in close quarters. The development of major trading routes (e.g., the great silk road) between human settlements opened a further chapter in the human epidemiological history by allowing transmission of infectious diseases from one region to neighboring areas. In more recent times—in the seventeenth and eighteenth centuries—diseases such as smallpox and measles, to which Europeans had long acquired some degree of immunity, were spread rapidly to new settlement regions in which native populations, lacking such immunity, were decimated.

The late nineteenth century and much of the twentieth century appear to have offered some respite from the massive amounts of death from epidemic infectious diseases that were a central feature of human communities. Although new classes of diseases attributed to the release of human-made chemicals and ecosystem degradation are on the rise, the earlier threats from infectious diseases appeared to have subsided, particularly in the economically privileged countries. Further, the introduction of sanitary and behavioral changes proved very successful against the "filth diseases," which had accounted for very high infant mortality (Golub 1997; Last 1997, 1998). These measures were not limited only to providing proper sewage disposal and safe drinking water (Last 1998). They also included important behavioral modifications. For example, once it was understood that infected sputum might contain tuberculosis bacteria and other respiratory pathogens and thus served to spread respiratory diseases (Last 1998), spitting in public was banned, likely having a great impact. Some may recall the signs in the buses in Britain of many decades ago, "Spitting strictly prohibited"! Further changes in behavior such as covering one's nose or mouth when sneezing or coughing and washing one's hands have become common practices—also serving to reduce the spread of infectious diseases.

The tremendous strides in twentieth-century science laid the foundations for further advances in medicine: one dreaded disease after another has come under some degree of restraint as a result of scientific discovery, a process that continues to this day. The discoveries of antibiotics and vaccines put many infectious diseases at bay, although hardly eliminating them altogether as was once thought possible. (*The World Health Report 2000* in 1999 estimated that of the 55.9 million deaths worldwide, approximately 9.9 million resulted from infectious and parasitic diseases). Thus, although the dreaded diseases have not really been overcome, the likelihood of living longer even with the disease has improved. People still die of pneumonia and diabetes, although in both cases, life can be prolonged even with these diseases. In more recent times, in the United States, the rate of death from infectious diseases actually rose by 30% between 1980 and 1992 largely owing to increases in respiratory tract infections, HIV infection, and septicemia (Fineberg and Wilson 1996). Further, globalization of diseases has contributed to a significant rise in the occurrence of nonlethal infectious diseases (influenza, respiratory tract infections). The prominence of air travel almost guarantees globalization of airborne viral and bacterial illnesses in very short periods of time.

Rapid transport has also likely contributed to the leaping of continents by the West Nile virus. This pathogen, which is carried by a mosquito, has become endemic in the Nile delta region of Egypt. Infecting both birds and humans, it appeared in August 1999 in New York City. Since then dead birds (mostly crows) have been found infected by this virus in neighboring states and provinces. This may become a new mortality factor for the Western Hemisphere for birds and humans.

As we enter the twenty-first century, perhaps the most significant transition for human health lies just on the horizon: risks to human health from the degradation of the earth's ecosystems at global, regional, and local scales (Last 1998; McMichael 1993b, 1997; McMichael and Haines 1997; Rapport et al. 1998a). The consequences of the loss of ecosystem health threaten to turn back the clock on some of the major advances in human health of the twentieth century. Among the new risks to human health are climate change, with its impact on changes in distributions and virulence of disease vectors (Patz et al. 1996); reduced effectiveness of antibiotics owing to indiscriminate use in treating humans and animals; destruction of the earth's protective ultraviolet-b (UV-b) layer (the stratospheric ozone layer) and other manifestations of a world out of balance (Epstein 1995; Levins et al. 1994; Garrett 1995). Continued degradation of the earth's ecosystems have profound implications for human and animal health. It is this major transition that requires a more holistic approach to human and animal health than is now provided in medical education. It requires recognition that increasingly the determinants of individual health will be found in ecosystem health. Taking these linkages into account is the first step to preventive action and health promotion.

26.3 Ecosystem Health as a Societal Goal

Health assessments, whether at the individual, population, or ecosystem level, necessarily involve an element of subjectivity. What is "healthy" depends in part upon human values and goals. When it comes to ecosystem health, the relevant values and goals are those of the human community (Rapport et al. 1998b) whose lives and livelihoods depend on the condition of the ecosystem. Healthy ecosystems support economic, social, health, spiritual, aesthetic, and other needs of the community (Nielsen 1999).

What constitutes ecosystem health? The same question has been posed many times about human and animal health. Of course in all cases there is no one answer. However, there are some generally agreed upon criteria for assessing health at all levels. A healthy ecosystem may be defined as an ecological system that realizes its potential productivity, complexity, and resilience. Clearly a corollary to this is that the system does not display the signs of pathology, for example, impairment in ecosystem functions such as nutrient cycling, energy transformation, productivity, hydrological regulation, and the like.

Ecosystem health, both as a concept and as an emerging practice, can trace its earliest history to seminal ideas in the writings of the famous Scottish geologist James Hutton (1788), who developed the concept of the earth as an integrated system. Two decades ago, it was proposed that the fields of medicine and ecology had a lot to say to one another, in that they were both concerned with the diagnosis, prognosis, and treatment of complex biological systems (Rapport et al. 1979, 1981). The term "ecosystem medicine" was coined to describe this new field. Ecosystem medicine later evolved into principles and concepts of ecosystem health (Schaeffer et al. 1988; Rapport 1989). Conservation medicine further contributes to this thinking by providing tools to achieve ecosystem health in a conservation biology context.

Subsequently, achieving ecosystem health has become an explicit goal of many large-scale restoration projects and a slogan for new and integrative programs in national and international agencies. Yet a lively academic debate continues as to whether the notion of ecosystem health has validity (e.g., Calow 1992; Suter 1993; Wilkins 1999; Rapport et al. 1999a). Some critics have argued, for example, that ecosystems do not exist as definable bounded entities and therefore reference to ecosystem functions (which are key indicators in ecosystem health assessments) have no grounding in reality. However, widespread use of the ecosystem concept as a basis for regional environmental analysis (e.g., Chadwick et al. 1999) would refute this claim. Critics have also argued that the concept of "health" has no validity beyond the level of the individual. However, if this were true, it would delegitimize public health as we know it, in relation to human communities and populations. In so doing it would discount many of the now-recognized supraindividual influences on human health (McMichael et al. 1999). In general, those that object to the concept of ecosystem health either fail to recognize that humans are part of the ecosystem (Bormann 1996; Rapport et al. 1999a) or that ecosystems have become dysfunctional as a consequence of stress from human activity. It is the transformation of ecosystems under stress that has resulted in a reduction in the so-called ecosystem services (i.e., nutrient recycling, primary and secondary productivity, biodiversity, potable water, pollinators, soil retention, Vitousek et al. 1997). Such losses pose serious risks to the health of humans and other species and to ecologically sustainable development (Costanza et al. 1997; McMichael et al. 1999).

How is ecosystem health to be measured? Efforts to quantify ecosystem health generally rely on a set of metrics that distinguish normal system behavior from impaired behavior. Most definitions contain common elements: they focus on the vitality (productivity) of ecosystems, state of organization, and resilience (Rapport 1995c). Costanza and colleagues suggest, "An ecological system is healthy and free from 'distress syndrome' if it is stable and sustainable—that is, if it is active and maintains its organization and autonomy over time and is resilient to stress" (Costanza et al. 1992). Mageau and colleagues (1995) suggest quantitative assessments in terms of vigor (productivity), organization, and resilience. Karr (1999) suggests that simple and direct biological measures of ecosystem condition are more convincing than more elusive system properties and suffice to demonstrate the extent to which human actions have degraded living systems.

Future medical and veterinary students are likely to experience a greater interest by patients about health risks resulting from the degradation of the environment at scales ranging from local to global. Patients expect doctors and veterinarians to be conversant on these topics and to assume leadership within their communities on these issues. The experiences at the University of Western Ontario in developing a medical curriculum aimed at addressing ecosystem health have potential applications in establishing a curriculum for conservation medicine, and in preparing future physicians to be leaders in reducing health risks associated with ecosystem disturbances.

26.4 Introducing Ecosystem Health in the Curriculum of a Canadian Medical School

Ecosystem health has recently been introduced in the curricula of both veterinary and human medicine programs in Canada. This is a landmark achievement. Certainly there have also been a number of medical programs in North America that include environmental health. However, environmental health is but a small part of the considerations within an ecosystem perspective. The focus of environmental health is usually restricted to the effects of toxic substances released to the environment, in some cases occupational health hazards, and, the effects of UV-b radiation on human and animal health. This is but a small part of the scope of considerations that are germane to ecosystem health.

In undergraduate medicine, curricular components tend to not dwell on the philosophical ramifications of subject material, but rather concentrate on practical aspects. One of the constant questions in the minds of medical students is, How can this knowledge be used in the clinic? A similar question was raised by students taking the ecosystem health program offered since 1994 at the four veterinary medicine colleges in Canada (Ribble et al. 1997).

The traditional role of the physician has been the provider of diagnoses and treatment. There was an expectation of base knowledge of health and disease, but not much beyond that of their professional capacity. The conceptual realm of the physician was considered to focus solely on the patient and the disease (figure 26.1). This relationship was sufficient for many years. Viewed as a learned person, the physician was a highly respected member of society.

Recently, our society has changed and the expectations of the physician have also changed. There has been an evolution of the expectations of the general public. Greater attention is focused on the expanded role of the physicians in our

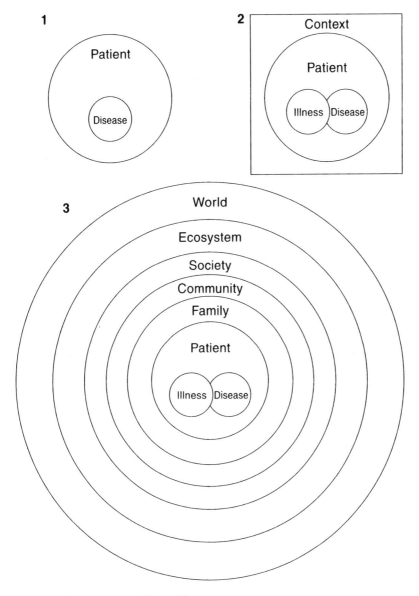

Figure 26.1 Copyright 1998 by dljones.

culture and on public accountability. In Ontario, this led to a combined set of values that were developed to guide the education and training of future physicians (Educating Future Physicians of Ontario [EFPO], www.efpo.org/uwo. Simply stated, eight values constitute what the public expects of physicians: (1) medical expert/clinical decision maker; (2) communicator (educator/humanist/healer); (3) health advocate; (4) learner/scholar; (5) collaborator; (6) gatekeeper/resource manager/steward; (7) scientist; (8) person.

These values are reflected in the more recent expectations that the role of the physician and of medical care should be "patient centered" (Stewart et al. 1995). (This is what some physicians were already doing in their practices, that now is explicitly described and recognized). The patient–centered model is a consideration of the whole patient as well as his or her context (figure 26.1). In this model, patients have diseases that they experience as an "illness experience." The illness experience is due not only to the disease process, but also to the specific physiological conditions of the patient and the context in which the patient lives (i.e., work, social structure, family, ecosystem).

Being patient centered allows a more careful consideration of the context and interaction with human health. The question is, How broad is the context? It is becoming ever more evident that the "context" is global and includes, centrally, an ecosystem perspective (Rapport et al. 1999b,c). This is not to say that the physician of the twenty-first century needs to be an expert on all aspects of the multifaceted problems facing humanity. However, this perspective suggests that physicians should be sufficiently knowledgeable to be conversant with the problems and aware of the limitations in their knowledge. It also suggests that physicians need to acknowledge the limitations of certainty about the complex interplay between humanity and the ecosystem (Costanza et al. 1997; Levins 1995; McMichael 1997). Perhaps most important, it suggests that physicians recognize the importance of the role of other professions in ecosystem health. From this perspective, ecosystem health, the study of the interaction of human health with changing environments and social responsibility, is not a standard feature of any other undergraduate medical curriculum in the world.

The evolution in thinking about "health" can be expressed as follows. Traditional medicine focuses on the questions, What disease does my patient have? and How can I fix it? The patient-centered model incorporates these questions but asks two additional questions, What are my patient's unique needs? and How can I meet those needs? The ecosystem perspective incorporates all these questions but adds the critical questions, What led to my patient coming here with this disease or illness? and How can we change things so that we can prevent other patients from developing the same disease/illness? This concept requires a significant paradigm shift from "disease fixer" to "proactive disease/illness preventer." Society has concentrated on the former since it is immediate and looks after the sufferer. Disease prevention is less concrete but of increasing importance.

The undergraduate medical curriculum at the University of Western Ontario has undergone a process of major renewal. This renewal aims to provide a curriculum that promotes the development of physicians who can well serve multiple EFPO roles. Given this mandate for the school, it was recognized that an ecosystem health program is an ideal way for students to learn these roles since it emphasizes the need for scientific knowledge acquisition, the need to communicate and to collaborate with other disciplines, the need to be an advocate for healthier patients and environments, and the obvious need to carefully manage resources. Below we describe the goals of the program and the implementation of one of its key elements, a case-based course in ecosystem health offered as an elective in the fourth year of undergraduate medicine.

26.5 Objectives

The undergraduate curriculum at Western is patient centered. The patient-centered model is similar to the traditional medical model in that there is a focus on the disease and the patient. However, the patient-centered model focuses on the patient's illness. The illness—the patient's unique experience of the disease—is a direct result of the biopsychosocial context in which the patient lives. Whereas the disease is generic to many different patients, the illness, or illness experience, is unique to the individual patient. The illness takes into consideration the whole patient (occupation, age, culture, etc.) and the context of the patient (the family, the community, the environment, etc.). The goal in the patient-centered curriculum is for the student to gain a thorough knowledge of the patient in biological and psychosocial terms, a clear understanding of the context in which the patient lives, and a clear understanding of disease entities. With this information, the student will be able to gain an understanding of the illness experience of the patient. Ecosystem health is a necessary component of this curriculum is that the students need to know the interaction of humans with the physical and psychosocial environment. However, an expansion of the ecosystem model is essential to understand the levels/layers of interplay. An additional concept is also considered in the ecosystem model—the concept of the ecosystem (Somerville 1995) or indeed the world (biosphere) as a "patient" (figure 26.1).

The ecosystem health program in medicine promotes a much needed paradigm change involving an important shift of emphasis from treatment to prevention. This sets the stage for students to recognize the value of tackling problems "upstream" before they result in disease, rather than simply, as presently, calling treatment into play "downstream" after the disease has occurred. It is very apparent to the medical educator that medical students (as most physicians) emphasize the one-patient one-doctor paradigm. To achieve "upstream" effectiveness, students need to become more globally minded. The study of ecosystem health presents the opportunity to learn and discuss the value of "upstream" prevention and interdisciplinary involvement, and to expand the learning to a new paradigm of analysis and treatment.

A course in ecosystem health could achieve the goals of a patient-centered curriculum and emphasize the need for multiple roles for physicians. It would also recognize the importance of "upstream" considerations in human health and disease. Our objective was to design and implement a course in ecosystem health that provides students with an understanding of ecosystems and how ecosystems can affect human health. We believe the course should foster the skills of systems thinking, critical appraisal, and assessment of risk, as well as developing an approach to coping with uncertainty in decision making. The course should foster appreciation of the world, social responsibility, and health advocacy and should endorse all of the EFPO values listed above. The course needs to foster holistic and global approaches to health and to go beyond environmental health and include political, economic, industrial, ethnic, and societal issues from local, regional, national, and global perspectives.

Recognizing the burgeoning amount of knowledge that medical students are expected to assimilate, it was important to define the depth of knowledge expected of the students. Equally, the course had to be relevant to the problems faced by the patients for whose care the graduating physicians would be responsible. We believe the students should have a "literal" understanding of ecosystem health, not an "expert" understanding. Students should be able to recognize the influence of the ecosystem on human health and vice versa. Students should be able to take a case consisting of a single doctor and patient and expand it to look at the regional and global ecosystems that have an influence on the interaction. Students should also be able to consider the effects of human health policy on the ecosystem. The course should also create a new way of thinking, one that recognizes that correct answers are not always present. Students must deal with uncertainty. With this in mind, the course should foster thinking about a variety of options and contingencies.

While a dedicated course on ecosystem health provided only an entry point into the curriculum, concepts of ecosystem health needed to be interspersed throughout the four years of the curriculum, with a significant presence in the final year of the program. It was expected that to enhance understanding of the components and facilitate buy-in by the students, the cases being presented should be ordered in a progressively scaled-up manner. The early cases should be chosen to have immediate and local relevance (e.g., the widespread prevalence of asthma in southwestern Ontario). Subsequent cases could be targeted to have a national or regional focus. These would be explored further for both their global significance and their local relevance.

26.6 Fourth Year Course Design

Real case studies were chosen as the most relevant method for instruction. This format emphasizes student involvement and is particularly applicable when there is no one right answer. Small group work emphasizes the need for all students to participate. This format has documented usage in other curricular offerings, as demonstrated on the Second Nature web site (www.secondnature.org accessed 10/5/01), which is devoted to medical education for sustainability. Our plan was to use a limited number of cases—five in total. The students had two two-hour small group sessions, with time between to seek out information relevant to the case. The course was set up in two-hour blocks on Mondays and Wednesdays. The case was introduced in the first hour of a Wednesday session, generally with an expert present to lead the discussion. The students then used the second hour of the session to discuss what information would be of use in order to better understand the issues and provide a starting point for further discussion for the following week (i.e., the following Monday, which was the second session for the case study). Broad-based reviews of the topics were available in the Learning Resource Centre (associated with the Faculty of Medicine and Dentistry) to act as a starting point for the student searches. Individuals could carry out the searches or several members of the group could choose to work together. The time period from Wednesday to Monday gave ample opportunity for the students to research the issues and then present their findings during the first hour of the Monday session. The second hour of the second session was devoted to a wrap-up with the expert present, where possible, to see what had been learned about the issues and what could be done to modify current practice so that sustainability might be enhanced.

While any number of cases might be developed, we tried to use topics to which students could relate on a local level in that the problem was easily identifiable in the community. Cases were also chosen so that the students would have some basic background knowledge and were therefore sufficiently confident to contribute to the discussion from the very outset. It was recognized that in some sessions, students would have tutors (discussion leaders) who were not experts. Therefore, a brief guide of topics for discussion was developed so that the students could be directed to address some of the broader issues and not just concentrate on local ones.

26.7 Case Studies

To give the flavor of course content, we provide examples of two of the cases used: one relates to the growing problem of antibiotic resistance to various classes of bacterial infection; the other, to the increasing prevalence of asthma in North America and elsewhere.

26.7.1 Antibiotic Resistance

A local issue that all students were aware of was the increasing frequency of isolates of vancomycin-resistant *Enterococcus* in the local hospitals. Originally this was going to be introduced as a formal case, but instead, the students were asked by a simple show of hands if they had been involved in managing patients that harbored vancomycin-resistant enterococci. The majority had been. This was followed by an overview of major mechanisms of antibiotic resistance, how resistance is transferred among microbes, the role of selective pressure in maintaining resistance factors in a bacterial population, and so on. As the students were in their final year, this really was a review of material with which they were already quite familiar.

Discussion then began around how big a problem antibiotic resistance was and why it was becoming more frequent. Students inevitably moved to physicians overprescribing antibiotics. The tutor expanded the discussion to make the students broaden their horizons by asking questions along the lines of the suggestions made in a tutor guide prepared for the course. As pointed out at the end of the guide, many other topics could be discussed as long as they broaden the discussion and make students think outside of the box. Students rapidly developed an appreciation of the ecosystem connections, particularly the risks to human health from antibiotic resistant pathogens resulting from widespread use of antibiotics to promote growth in and protect farm animals and aquaculture (to minimize disease in fish that are raised in high densities). As some of the resulting antibiotic-resistant organisms have infected humans, this example forces the students to think of humans as part of the ecosystem, and of human health as very much governed by what happens "out there."

The story that emerged of what happens out there is the following. The discovery of antibiotics led to the hope for the virtual elimination of bacterial disease. Initial optimism, however, has proved short-lived. For a variety of reasons including the ubiquitous applications of antibiotics, not only in medicine but also in humanconstructed ecosystems (agriculture and aquaculture), resistant strains of pathogens have increased markedly (Baquero and Blazquez 1997).

Much of the class discussion focused on why antibiotics were used on farms and in aquaculture—to produce plentiful and cheap sources of protein. We then dealt with the waste produced by thousands of animals that were excreting resistant bacteria, which were washed down to surface waters. We talked about the increased nutrient load of animal manure in the water and the production of algal blooms and the consequences of these transformations to aquatic life. One consequence is, of course, a decline in fisheries (both freshwater and marine), particularly in high-valued stocks, which cannot tolerate eutrophic (nutrient-rich) conditions. We also discussed the risk of using contaminated water in the irrigation of crops and how those crops if consumed raw (e.g., in salads) might lead to a changing epidemiology of food-borne diseases. We considered the fact that food production may be far removed from points of consumption—posing additional risks of contamination in transport, and necessitating the development of a transport infrastructure to bring food to consumers. We also touched upon international trade agreements and the issue of how to preserve foods (e.g., radiation) and the ecological consequences of using such methods to increase the shelf life of foods.

Thus, while we started with antibiotic resistance, we moved (seemingly) far away from the initial topic. In so doing, the students began to appreciate the interconnectedness of ecosystems as well as how degradation of the ecosystem could lead to human health problems.

26.7.2 Asthma

The case study on asthma was similarly constructed. Southwestern Ontario has a high prevalence and an increasing incidence of asthma. Thirty percent of children in southwestern Ontario will use asthma "puffers" to treat asthma at some-time in their childhood. The discussion can be started by either a clinical case or simply asking the question, Why?

Most frequently, the response is air pollution. Asthma, a respiratory disease that results in moderate to severe breathing difficulties, is increasing in prevalence in many parts of the world, and in Canada, the prevalence is particularly high in southwestern Ontario. This disability is thought to result from irritants that produce both allergic sensitization and bronchial hyperresponsiveness, resulting in acute or chronic inflammation, which in turn leads to chronic airflow limitation. A recent survey suggests that people in Toronto believe that poor air quality is the leading threat to their health. Only a few years ago, the same survey found that people believed diet and exercise were the major determinants of health. One possible explanation for this switch is the increasing prevalence of asthma in the Toronto area.

Although poor air quality is contributory, the linkage between air pollution and the incidence of asthma is not a strong one. In fact, the major issue is that industrial activity does not seem to be related to the prevalence of asthma. For example, the prevalence of childhood asthma in East Germany is much lower than in West Germany (Ring et al. 1999). It may cause more frequent attacks of asthma but it does not seem to be related to the numbers of people who are diagnosed with the disease in childhood. More intriguing are some pointers indicating that some natural childhood viral infections may protect against the development of asthma and even that artificial immunization might remove the protective effect by preventing disease from natural exposure. We also considered within the asthma example that some suggestions to minimize one possible cause of childhood asthma (dust mites) would be to replace carpets (upon which dust mites thrive) with wood flooring. However, this would increase the demand for hardwood and put pressure on forest ecosystems. We explored the consequences of that issue and considered that the evidence linking hardwood flooring to asthma prevention was not very strong and that this suggested remedy might be ill advised. In fact, many factors, mostly environmental, are implicated as possible causes of asthma: prevalence of dust mites in homes, shifting patterns in pollen distribution, and air pollution. Ecosystem degradation may contribute to one or more of these factors.

In both of these cases and in others we used, the students became frustrated in finding solutions. The problems often boiled down to too many people or too many expectations, and the solutions, having to make too large a population change its ways. It is important for them to appreciate, however, that these massive changes are unlikely to occur and that more manageable solutions will likely have to be implemented, as first steps toward the resolution of these problems.

26.8 The Future of Ecosystem Health in Medicine and Veterinary Medicine

Based on our experience at the University of Western Ontario, and our familiarity with programs that integrate ecological and human and animal health considerations in other institutions, we suggest that considerations of ecosystem health, climate change, and conservation medicine should gain a prominent place in the evolution of all curricula, especially in schools of human and veterinary medicine. These topics effectively integrate knowledge from ecology, ethics, conservation biology, medicine, economics, and environmental management. Our best hope for the future is in recreating integrative knowledge (Somerville and Rapport 2000) and in placing these perspectives in the educational system at all levels. Over the longer term, as knowledge expands across disciplines, ecosystem health should be incorporated throughout the university curricula.

The ecosystem health theme in one guise or another has been incorporated into the curricula of several major schools of public health, including Harvard and Johns Hopkins universities. It is also part of the Diploma Program in Health and the Environment at McMaster University (Canada). Ecosystem health is also part of the curricula in schools of veterinary medicine both in the United States, for example, at the University of Illinois, and in Canada at the four schools of veterinary medicine.

Despite the best efforts of medical educators to introduce new subject matter into medical school curricula, the degree to which the material is accepted by faculty and students alike depends upon how prominently the material is evaluated in national qualifying examinations. While this does fly in the face of good educational practices where curriculum content drives the evaluation, when faced with large amounts of subject matter to learn, students necessarily devote more time to subjects well represented in the evaluation process. In fairness to examination boards, developing questions that provide a valid measure of students' understanding of a subject is a lengthy process, frequently requiring extensive testing before reliable questions can be incorporated into national exams. This creates a lag time between new materials appearing in curricula and evaluation of such material at the national level. The converse is also true in that examining boards are unlikely to spend time developing evaluations for material that is not well represented in the majority of medical school curricula. Many aspects of preventive approaches to disease management are given only cursory treatment in medical schools, so there is no pressure on the national examining board to respond to evaluative material. The end result is that new subjects, no matter how important they may be, take time to be broadly incorporated into mainstream medical education.

It is imperative to foster concepts such as ecosystem health, ecosystem medicine, and conservation medicine—all of which integrate knowledge from the life sciences, natural sciences, and social sciences. This we believe will emerge as one of the essential "coping strategies" for human futures in the twenty-first century.

References

Baquero, F., and J. Blazquez. 1997. Evolution of antibiotic resistance TREE 12:482–487.
Bormann, F.H., 1996. Ecology: a personal history. Annu Rev Energy Environ 21:1–29.

Calow, P. 1992. Can ecosystems be healthy? Critical considerations of concepts. J Aquat Ecosys Health 1:1–15.

Chadwick, O.A., L.A. Derry, P.M. Vitousek, B.J. Huebert, and L.O. Hedin. 1999. Changing sources of nutrients during four million years of ecosystem development. Nature 397: 491–497.

Costanza, R., B.G. Norton, and B.D. Haskell (eds.). 1992. Ecosystem health: new goals for environmental management. Island Press, Washington, D.C.

Costanza, R., R. d'Argre, R. de Groot, S. Farber, M. Grasso, B. Hannon, S. Naeem, K. Limburg, J. Paruelo, J.R.V. O'Neill, R. Raskin, P. Sutton, and M. van den Belt. 1997. The value of the world's ecosystem services and natural capital. Nature 387: 253–260.

Epstein, P.R., 1995. Emerging diseases and ecosystem instabilities: new threats to public health. Am J Publ Health 168:172.

Fineberg, H.V., and M.E. Wilson. 1996. Social vulnerability and death by infection. N Eng J Med 334:859–860.

Garrett, L. 1995. The coming plague: newly emerging diseases in a world out of balance. Penguin Books, New York.

Golub, E.S. 1997. The limits of medicine: how science shapes our hope for the cure. University of Chicago Press, Chicago.

Hutton, J. 1788. Theory of the earth; or, An investigation of laws observable in the composition, dissolution and restoration of land upon the globe. Trans R Soc Edinb 1: 209–304.

Karr, J.R. 1999. Defining and measuring river health. Freshw Biol 41:221–234.

Last, J.M. 1997. Human health in a changing world. *In* Last, J.M., ed. Public health and human ecology, 2nd ed., pp. 395–425. Appleton and Lange, Stamford, Conn.

Last, J.M., 1998. Human-induced ecological determinants of infectious disease. Ecosys Health 4:83–91.

Levins, R. 1995. Preparing for uncertainty. Ecosys Health 1:47–57.

Levins, R.T. Awebuch, U. Brinkmann, I. Eckardt, P. Epstein, N. Makhoul, C. Albuquerque de Possas, C. Puccia, A. Spielman, and M.E. Wilson. 1994. The emergence of new diseases. Am Sci 82:52–60.

Mageau, M.T., R. Costanza, and R.E. Ulanowicz. 1995. The development and initial testing of a quantitative assessment of ecosystem health. Ecosys Health 1:201–213.

McMichael, A.J. 1993a. Global environmental change and human population health: a conceptual and scientific challenge for epidemiology. Int J Epidemiol 22:1–8.

McMichael, A.J. 1993b. Planetary overload: global environmental change and the health of the human species. Cambridge University Press, Cambridge.

McMichael, A.J. 1997. Global environmental change and human health: impact assessment, population vulnerability, research priorities. Ecosys Health 3:200–210.

McMichael, A.J., and A. Haines. 1997. Global climate change: the potential effects on health. Br Med J 315:805–808.

McMichael, A.J. B. Bolin, R. Costanza, G.C. Daily, C. Folke, K. Lindahl-Kiessling, E. Lilndgren, and B. Niklasson. 1999. Globalization and the sustainability of health: an ecological perspective. Bioscience 49:205–210.

Nielsen, N.O. 1999. The meaning of health. Ecosys Health 5:65–66.

Patz, J.A., P.R. Epstein, T.A. Burke, and J.M. Balbus. 1996. Global climate change and emerging infectious disease. JAMA 275:217–223.

Rapport, D.J. 1995. Ecosystem health: an emerging transdisciplinary science. *In* Rapport, D.J., C. Gaudet, and P. Calow, eds. Evaluating and monitoring the health of large scale ecosystems, pp. 5–31. Springer, Heidelberg.

Rapport, D.J., and W. Whitford. 1999. How ecosystems respond to stress. Common responses of aquatic and arid systems. Bioscience 49:193–202.

Rapport, D.J., C. Thorpe, and H.A. Regier. 1979. Ecosystem medicine. Bull Ecol Soc Am 60:180–182.

Rapport, D.J., H.A. Regier, and C. Thorpe. 1981. Diagnosis, prognosis and treatment of ecosystems under stress. *In* Barrett, G.W., and R. Fosenberg, eds. Stress effects on natural ecosystems, pp. 269–280. Wiley, New York.

Rapport, D.J., R. Costanza, P. Epstein, C. Gaudet, and R. Levins, eds. 1998a. Ecosystem health. Blackwell, Malden, Mass.

Rapport, D.J., R. Costanza, and A.J. McMichael. 1998b. Assessing ecosystem health: challenges at the interface of social, natural, and health sciences. TREE 13:397–402.

Rapport, D.J., G. Bohm, D. Buckingham, J. Cairns Jr., R. Costanza, J.R. Karr, H.A.M. de Kruijf, R. Levins, A.J. McMichael, N.O. Nielsen and W.G. Whitford. 1999a. Ecosystem health: the concept, the ISEH, and the important tasks ahead. Ecosys Health 5: 82–90.

Rapport, D.J., N. Christensen, J.R. Karr, and G.P. Patil. 1999b. The centrality of ecosystem health in achieving sustainability in the 21st century: concepts and new approaches to environmental management. Trans R Soc Canada 9:3–40.

Rapport, D.J., A.J. McMichael, and R. Costanza. 1999c. Assessing ecosystem health: a reply to Wilkins. TREE 14:69–70.

Ribble, C., B. Hunter, M. Larvière, D. Bélanger, G. Wobeser, P.Y. Daoust, T. Leighton, D. Waltner-Toews, J. Davidson, E. Spangler, and O. Nielsen. 1997. Ecosystem health as a clinical rotation for senior students in Canadian veterinary schools. Can Vet J 38: 485–490.

Ring, J., U. Kramer, T. Schafer, D. Abeck, D. Vieluf, and H. Behrendt. 1999. Environmental risk factors for respiratory and skin atopy: results from epidemiological studies in former East and West Germany. Int Arch Allergy Appl Immunol 118:403–407.

Schaeffer, D.J., E.E. Henricks, and H.W. Kerster. 1988. Ecosystem health: 1. Measuring ecosystem health. Environ Manage 12:445–455.

Somerville, M.A., 1995. Planet as patient. Ecosys Health 1:61–72.

Somerville, M.A., and D.J. Rapport (eds.). 2000. Transdisciplinarity: re-creating integrated knowledge. EOLSS Publishers, Oxford.

Stewart, M., J. Belle Brown, W.W. Weston, I.R. McWhinney, C.L. McWilliam, and T.R. Freeman. 1995. Patient-centered medicine: transforming the clinical method. Sage, Thousand Oaks, Calif.

Suter, G.W. 1993. A critique of ecosystem health concepts and indexes. Environ Toxicol Chem 112:1533–1539.

United Nations Environment Programme (UNEP). 1999. Global Environment Outlook 2000. Web site: http//www.grida.no/geo2000/index.htm, accessed 10/4/01.

Vitousek, P.J.M., H.A. Mooney, J. Lubchenco, and J.M. Melillo. 1997. Human domination of Earth's ecosystems. Science 277:464–499.

Wilkins, D.A., 1999. Assessing ecosystem health. A critique. TREE 14:69.

World Health Organization (WHO). 1948. Text of the constitution of the World Health Organization, Off Rec WHO 2:100.

World Health Organization (WHO). 1999. The world health report 2000. Health systems: improving performance. Web site: http//www.who.int/whr/, accessed 10/4/01.

27

Ecotourism
Unforeseen Effects on Health

Mary E. Wilson

The desire to find regions untouched by human development leads travelers to explore increasingly remote areas and brings them in contact with relatively isolated human and animal populations. Current materials, equipment, and methods of travel allow individuals to rapidly penetrate physical and vegetation barriers that would have stopped, or markedly slowed, even intrepid explorers in the past. The time to reach remote areas is often short, and travelers may sample remote areas for a brief time, sometimes a few days or less. The process of travel typically involves multiple places and populations, and all may be affected directly or indirectly by the travel of humans.

Travelers form an important bridge to diverse geographic areas and increase the mingling of genetic material from multiple species as they also transform the landscape (Wilson 1995a). This chapter focuses on three themes: movement, mixing, and pressures as they relate to changing patterns of infectious diseases. A number of examples will illustrate the complex interactions that influence infectious diseases.

Movement involves all species, but the volume, speed, and reach of human travel today is unprecedented. Movement of other species is closely linked to human traffic as humans transport other species with them or move them for commerce. Humans also displace animals through land development and by altering habitats (Wilson 1995b). Although some animals migrate, sometimes over long distances, they typically follow paths or routes that may remain remarkably constant over many generations. Human activities, including importation of exotic animals, move animals out of traditional habitats and migration pathways.

Travel by the majority of humans tends to follow set paths, roads, and routes. Modern technology is changing the range of destinations, making it easier for travelers to stray from well-worn paths and to reach new destinations. Technologic advances also permit travelers to enter and survive in more extreme environments. Travelers can be dropped into areas (e.g., via helicopter) that are inaccessible by road or foot. They can reach remote areas where populations of humans and other animals may have had limited contact with other populations.

Today, it is estimated that more than 1.4 million persons cross international borders on commercial flights every day. According to the World Tourism Organization more than 43 million tourists visited the United States in 1995—and the volume of within-country and international travel continues to increase. In the 1990s, more than 5,000 airports had regularly scheduled international flights. At the same time that an increasing percentage (now estimated at 50%) of the world population lives in urban centers (including megacities), technology and resources allow travelers to reach more remote areas. The world has more and faster links than ever before in history. Travelers can now reach most places on the planet within 24–48 hours, a period shorter than the incubation period for many infectious diseases. The ease and speed of travel make more frequent trips possible. Grubler and Nakicenovic (1991) estimated that spatial mobility of the average human has increased more than 1,000-fold over the 200-year period from 1800 to 2000.

Schafer and Victor (1997) note a predictable relation between income and spending on transport and, on that basis, speculate about future travel. Although, in the future, humans may spend a relatively constant average fraction of daily time on travel, they will travel greater distances because a larger fraction of travel will be by high-speed transport.

Although travel for exploration and recreation is increasing, so is travel for business, education, and humanitarian reasons. Movement also includes population displacements because of war and political conflicts, socioeconomic reasons, and environmental changes, including extreme weather events and catastrophes such as earthquakes and volcanoes (Wilson 1997).

The globalization of trade brings all manner of materials and species into areas where they are not naturally found. An interest in the remote and exotic has fueled an increase in trade in exotic animals. Although this is not new, the volume and variety of species being brought into the United States and going into homes and communities are unprecedented. For example, in 1996 animal imports through Miami alone included >7,000 mammals, >1 million reptiles, and 110 amphibians (S. Ostrowski, Division of Quarantine, CDC, personal commun 2000).

The process of movement disturbs the environment in multiple ways. Roads and railroads fragment habitats, which can lead to species loss. Roads and bridges also breach physical barriers such as mountains, rivers, and deserts and can serve as new conduits for movement of biologic life. Such physical barriers traditionally have served to limit natural movement and hence affect the distribution of certain arthropod vectors and intermediate or reservoir hosts that could carry pathogenic microbes or other parasitic species (Wilson 1991). Roads, railroads, and the vehicles that pass on them carry humans, animals, seeds, and a range of biologic life quickly and repeatedly into new regions. Travel and trade are often the source of unwelcome introductions of invasive species of plants and animals that alter local ecosystems (Crosby 1972). Introductions, whether planned or an unintentional consequence of travel, may irreversibly change the local landscape and ecology (Vermeij 1991).

Even persons going to remote destinations typically use heavily traveled routes for a portion of the trip. The process of travel links buses, trains, airplanes, terminals, and boats and juxtaposes people from multiple regions and varied destinations. The vehicle of transport becomes a habitat where persons from diverse regions often spend hours to days in close proximity sharing recirculated air. Commercial jets, for example, recirculate and filter air. On modern planes, the systems provide about 20 air exchanges per hour during flight, though airflow is decreased during takeoffs and landings (WHO 1998). Humidity on aircraft is low, typically 10–20%, and can dry mucous membranes, potentially increasing susceptibility to respiratory infections. Transmission of airborne infections on aircraft has been documented on many occasions. In one example (Moser et al. 1979), a jet with 54 persons aboard was delayed from takeoff for three hours. During that period the ventilation system was not working. Over the next three days, 72% of those on board the aircraft became ill with an influenzalike illness. Influenza A/Texas/77 (H3N2) was isolated. A person on board the aircraft was in the early stages of influenza, and the virus spread easily in this closed environment. More recently, a person with cavitary multidrug-resistant tuberculosis during a long flight on a Boeing 747 transmitted infection to 4 of 13 (30.8%) persons seated within two rows and to 2 of 55 (3.6%) persons seated elsewhere within the same section of the aircraft (Kenyon et al. 1996). Transmission of measles (probable) and smallpox has also been reported on aircraft (WHO 1998).

Ships are important vehicles for travel, trade, and recreation. As of 1993, about 80% of the world trade volume was transported by ship (Committee on Ships' Ballast Operations 1996). Cargo ships must use ballast (from Middle English, meaning "useless load") to help maintain the stability of the ship as cargo is added and unloaded and as fuel is used. Although rocks, sand, and other "useless" materials were used in the past, for the last 100 years ships have increasingly used water as ballast. This eliminates the time-consuming loading process and avoids shifts in solid materials that could destabilize the ship. Ballast water is typically taken on and discharged in multiple locations. It is a veritable biological soup that includes bacteria, viruses, fungi, and basically all plants and animals less than 1 cm in size that are adjacent to the boat when the water is drawn in for ballast. A study of ships traveling between Japan and Coos Bay, Oregon, found at least 367 different taxa in the ballast water (Carleton and Geller 1993). In the 1990s, it was estimated that ballast water may transport >3,000 species of animals and plants per day around the world. The volume of ballast is huge and on larger tankers may be 200,000 cubic meters. Material in the ballast may be hours to months old and may include accumulated life forms from around the world. When ballast is discharged near coasts, coastal currents may disperse introduced species to other sites. Species that are controlled by predators in their native ranges may proliferate unchecked in a new site.

Although most of the life forms introduced in ballast water do not become established in a new environment, some do (e.g., zebra mussel) and have caused extensive damage. Ballast can also carry potential human pathogens. Researchers were able to identify *Vibrio cholerae* (cause of human cholera) in the ballast and bilge of cargo ships in the Gulf of Mexico whose recent ports of call had been in Latin America (McCarthy and Khambaty 1994). The strains identified were identical to those causing epidemic cholera in Latin America.

Cruise ships have provided an increasingly popular form of travel and entertainment. In 1995 about five million persons in the United States traveled on cruise ships. This is expected to reach seven million in the year 2000. Globally, cruise ships have the capacity to carry 47 million passengers per year. Cruise ships are growing in size; one ship may carry 2,000–3,000 passengers. Cruise ships can be considered floating islands and are ideally also suited for the spread and dispersal of infections (Minooee and Rickman 1999). Persons from multiple regions gather on ships and spend a period of several days together sharing the same environment. The ship may pick up and drop off passengers and crew at multiple stops along the way, often in several different countries. Cruises have become especially popular among the elderly and frail—so a large portion of the passengers may be particularly susceptible to infections. For example, on one cruise ship (traveling between Montreal and New York) studied by the Centers for Disease Control and Prevention (CDC) because of an outbreak of influenza, 77.4% of the 1,448 passengers were 65 years of age or older and 26.2% had chronic medical problems (CDC 1997). At the end of the cruise, passengers disperse to many locations, often carrying with them new microbiologic baggage, which can be introduced into their home communities. The trip home typically involves one or multiple airplane flights, where travelers mingle with other passengers from all over the world.

Outbreaks of gastrointestinal infections have been especially common on cruise ships and have included cyclosporiasis, *Escherichia coli* diarrhea, Norwalk virus gastroenteritis, typhoid fever, shigellosis, salmonellosis, and staphylococcal food poisoning, among others. Outbreaks of legionnaire's disease and rubella have also been documented on cruise ships. Among the most important is influenza because of its ease of spread and its potential for being introduced widely to other populations.

Influenza is a particularly instructive example that pulls together many themes of this chapter. Caused by an RNA virus that has a remarkable capacity to change rapidly by genetic shift and drift, influenza is highly transmissible from person to person and can cause disease that ranges from trivial to fatal. Because it is common and familiar, its impact is often underestimated and underappreciated. In the United States alone, it causes 10,000 to 40,000 or more deaths per year. Animals are also infected with influenza viruses (Geraci et al. 1982) and are an important source of viral genetic material that reassorts or combines with influenza viruses that infect humans (Wright et al. 1992). Pig respiratory epithelial cells have receptors that allow attachment of both avian and human influenza viruses. Avian influenza viruses are shed in respiratory secretions and feces of birds. Infected ducks, for example, shed virus for at least 30 days. Influenza virus from the feces of waterfowl can be recovered from surface water. Avian species develop infection that ranges from asymptomatic to lethal. Avian influenza has caused major outbreaks in poultry farms.

Influenza virus can undergo genetic mutations in hemagglutinin or neuraminidase (antigens on the surface of the virus) that can lead to epidemics. Much less commonly, a completely new hemagglutinin or neuraminidase emerges—with the new genetic material coming from animals. This genetic shift typically leads to pandemics. Until 1995, only three of the 15 influenza hemagglutinins that had been identified were known to cause infections in humans. Birds have all 15 identified hemagglutinins and nine neuraminidases. New influenza viruses often emerge from southern China, a region characterized by a large, densely settled human population and abundant pigs and ducks living in close proximity to humans.

Until events in Hong Kong in 1997, scientists thought that avian influenza posed no direct threat to humans. In 1997, after causing influenza outbreaks on chicken farms, avian influenza (H5N1) spread to humans (Claas et al. 1998). Eighteen human cases were confirmed, six of them fatal. Infection was concentrated in children and young adults, unlike the pattern in most outbreaks where morbidity and death are most common in older adults. The virus recovered from humans was identical to that found in birds (Subbarao et al. 1998). Epidemiological studies suggested that there had been multiple independent introductions of the influenza virus into the human population from birds, but that very limited person-to-person spread occurred. At the time of the human cases, there were estimated to be 300–600 live bird markets in Hong Kong, where mixing of different avian species (ducks, chickens, pheasants, pigeons, wild birds) was possible. When the Hong Kong live bird markets were studied, 10% or more of birds were found to be shedding H5N1, in multiple avian species (geese, chickens, ducks). The birds (more than one million) were killed, and no additional human cases of H5N1 have been documented. In 1999, human infection with H9N2, another avian influenza strain widespread in Asia, was also documented for the first time in humans, at a time of enhanced surveillance (Peiris et al. 1999).

The events in Hong Kong have led to heightened global surveillance for influenza in humans and animals. There was reason to be concerned about the events in Hong Kong, a densely populated city with extensive links to the rest of the world. In 1993, there were an estimated 41.4 million passenger movements (boat, train, car, airplane) to and from Hong Kong.

In recent years, multiple large outbreaks of influenza have been documented on cruise ships (as well as on smaller boats, barges, and in other settings). In the summer of 1998, 40,000 tourists and tourism workers were affected during influenza A outbreaks in Alaska and the Yukon territories (CDC 1999). Many were initially infected during land travel, but transmission continued on cruise ships. Air-conditioned buses and other places with shared air may have been sites of transmission. A cruise ship with 1,500 or more passengers who dispersed widely after a trip potentially spread the influenza virus widely.

Influenza typically occurs during the winter months in temperate regions. Increasingly, summertime outbreaks are being documented. Travel to tropical regions (where influenza transmission occurs year round) or to the opposite hemisphere may be the source of some infections. Creation of new habitats may be another reason. Influenza is transmitted most effectively where there is indoor crowding and low relative humidity (Schaffer et al. 1976). The creation of air-conditioned spaces where people congregate during hot weather may provide a good environment for influenza transmission.

Travelers need food, water, and shelter and often request or demand a level of comfort, sanitation, and space that is unavailable to the local population. Because travelers bring money into a region, vast by comparison to local standards, the focus of the local economy may change to attend the needs of travelers. Even when travelers are sensitive to environmental issues and try to minimize their impact, they change the places they visit. Traveler interests and needs may lead to improved sanitation, a better water supply, and an increase in cash flow—but may also mean new roads, buildings, and changes in the local landscape and social structure.

Travel is typically a loop with many points of connection and interactions with multiple populations along the way. Concern about infections related to travel often focuses only on disease risks to the person traveling. Migrating humans, while at risk for new infections, also aid in the global dispersal of infections. Humans carry their microbiologic flora and other baggage when they travel and may add to it, exchange it, or leave some of it along the way. Humans may pass on or pick up infections at any point in the travel and not just at the destination. Populations and ecosystems potentially affected include human populations (en route, destination, home community), animal populations (including arthropods), and plants.

Isolated human populations have been especially vulnerable to infections that are new to them. In past centuries, explorers carried infections such as smallpox and measles that decimated immunologically naive local populations (Crosby 1972). In November 1918, the steamship *Talune* from Auckland, New Zealand, docked on the islands of Western Samoa. Members of the crew carried influenza, which spread to the local population, killing 8,000 people over the next two months. One fifth of the people on the islands died during the epidemic (Kolata 1999). An epidemic with high attack rates and high case fatality rate (15.1%) occurred in a remote jungle highland population of Irian Jaya, Indonesia, in 1996 (Corwin et al 1998). The epidemic was notable for high mortality among the young adults, similar to that seen in the 1918–19 influenza pandemic. Serologic studies confirmed recent influenza A infection in the population. Investigators were unable to prove with certainty that influenza A was the cause of the epidemic. Historically, influenza outbreaks have been especially severe in remote populations. In this outbreak, the population may have had little immunity to influenza because of limited contact with other populations. In addition, they may have been relatively compromised because of poor nutrition and high background rates of other infections, including malaria (Corwin et al. 1998).

International travelers could be a source of infections for animals. The U.S. Department of Agriculture (USDA) is especially concerned about the health of animals that are economically important (e.g., livestock and poultry) and has recently reviewed the risks for International Office of Epizootics (OIE) List A diseases (Centers for Epidemiology and Animal Health 1998). Microbes that might threaten animals can potentially be carried by humans by two main methods: mechanically (e.g., on footwear in contaminated soil on shoes, on dirty clothing) or biologically (by infecting humans, who then transmit infection). The USDA report concluded that two diseases (Newcastle disease and swine vesicular disease) posed high risk for mechanical transmission. Biologic transmission was considered a risk (low) for only two infections, avian influenza and Newcastle disease.

Humans have been a source of measles epizootics in nonhuman primates, presumably via airborne spread (Acha and Szyfres 1987). Attack rates have been high in captive populations. Spread is most likely to occur in closed spaces. Of note, some of the measles vaccines used to protect humans (live, attenuated virus vaccine) can cause clinical measles and death in marmosets and owl monkeys.

Many examples exist of microbes that have the capacity to cross species barriers and cause vastly different clinical outcomes in different species. Several viruses, such as the hantaviruses, can be carried by rodents with little or no apparent adverse effect, yet cause disease with mortality of up to 50% when they enter humans. A herpesvirus that usually causes latent infection in African elephants appears to cause fatal hemorrhagic infection in young Asian monkeys (Richman et al. 1999). Herpes B virus (ceropithecine herpesvirus I) causes infection in monkeys similar to herpes simplex infections in humans. In healthy monkeys, herpes B virus establishes latent infection in sensory nerve ganglia and can reactivate at times of stress or with immune compromise (Weigler et al. 1993). Infections in humans, while rare, are usually fatal if untreated. Serologic studies of primate handlers have found no evidence of asymptomatic infection in humans (Freifeld et al. 1995).

Travel to remote and relatively unexplored habitats can be expected to uncover yet other microbes with the capacity to infect and to harm humans. The greatest concern must be for infections that are casually transmissible from person to person. However, even a virus such as HIV, which may have entered the human population from primates and which requires sexual, blood, or other close contact, has been remarkably successful in spreading throughout the world.

Infected humans or animals can introduce disease-causing microbes into a new region. Travelers viremic (virus in the blood) with dengue virus, on many occasions, have carried the virus into areas that are already populated with mosquitoes competent to transmit dengue (Wilson 1991). Outbreaks and epidemics of dengue have followed.

In 1999 an outbreak of West Nile virus encephalitis, a mosquito-transmitted flavivirus (Hubalik and Haluzka 1999), occurred for the first time ever in the Western Hemisphere. Using reverse-transcriptase polymerase chain reaction, researchers cloned viral material from human brain tissue. They assembled a genome sequence and compared it with genetic sequences from West Nile isolates from other geographic regions. Jia et al. (1999) found that the West Nile virus isolated in New York in 1999 most closely resembled viruses from the Middle East (one from Israel and one from Egypt). Although the precise mechanism by which the virus reached New York has not been defined, the most likely suspects include a viremic person traveling from areas where West Nile virus is endemic, and importation of infected birds. Competent mosquito vectors exist in the United States, and the virus has been recovered from birds and mosquitoes outside of New York. Because birds can migrate, researchers will follow and study the birds along their migratory paths. Surveillance systems will monitor birds and mosquito vectors to determine whether the virus has become established in the northeastern United States and whether and where it has spread.

At the same time that movement and mixing of various species is occurring at an unprecedented level, there are also increasing pressures for genetic change in microbes. Use of antimicrobials in humans, animals, and agriculture has been associated with rise in antibiotic resistance. Increasing resistance is seen among many classes of organisms (e.g., bacteria, viruses, protozoa, fungi) and among a number of major human pathogens (*Streptococcus pneumoniae, Staphylococcus aureus, Mycobacterium tuberculosis*, malaria, *Salmonella typhi*). Although resistance may emerge locally, global movement of people, animals, and foods is contributing to the expansion of the problem (Ayliffe 1997). *Campylobacter jejuni* is a common cause of acute gastroenteritis in the United States and elsewhere. When treatment is needed, erythromycin or a quinolone antimicrobial (such as ciprofloxacin) is commonly used. In Minnesota, the percentage of quinolone-resistant *C. jejuni* increased from 1.3% in 1992 to >10% in 1998. Of note, most of the increase in resistance was due to infections acquired during travel. Mexico was the most common place of travel for patients with quinolone-resistant *Campylobacter*. The sale of fluoroquinolones for use in poultry in Mexico increased approximately fourfold between 1993 and 1997 (Smith et al. 1999).

Molecular markers now make it possible to track the spread of specific clones or strains of microbes. Humans are the reservoir hosts for *Streptococcus pneumoniae*, the most common cause of bacterial pneumonia. Resistance of *S. pneumoniae* to penicillin and many other antibiotics is making treatment more difficult and more costly. Studies have documented the spread of a multiresistant clone of serotype 6B *S. pneumoniae* from Spain to Iceland in the late 1980s (Soares et al. 1993) and of a multiresistant clone of serotype 23F from Spain to Cleveland (Munoz et al. 1991). In most instances, multiresistant clones persist and spread once they have been introduced into a geographic area—presumably by a traveler who was infected or colonized with the resistant bacteria. The abrupt appearance of a clone of multidrug-resistant *Staphylococcus aureus* in Portugal (and subsequent spread) was thought to be due to the intercontinental transfer of the strain from Brazil to Portugal (Aires de Sousa et al. 1998).

The size, density, and location of human populations can influence infectious diseases (Horton 1996). More people live on the earth now than ever before in history, and more live in urban areas, many in densely packed megacities, some with populations exceeding 10 million. Population growth and increase in urban populations are occurring most rapidly in low-latitude areas. Typically these are also areas that are most at risk for outbreaks because of poor sanitation, lack of clean water, and a warm climate that provides the appropriate milieu for the survival and transmission of a greater array of infections. These megacities are linked to the world through travelers, especially those who visit or pass through via an airplane flight.

Large populations provide abundant hosts for viral or other microbial replication. A study found that the number of lineages of dengue virus (the cause of mosquito-transmitted dengue fever) has been increasing roughly in parallel with the increasing size of the human population over the last two centuries (Zanotto et al. 1996; Holland 1996). This provides the potential for more viral evolution and expanded chances for the appearance of more virulent strains.

Animal husbandry practices have changed. Very large populations of genetically similar animals kept in restricted spaces have become increasingly common. Modern chicken farms, for example, may house 100,000 or more chickens. Animal populations may also be moved. The outbreak of Nipah virus encephalitis, initially recognized in Malaysia (Chua et al. 1999), extended into Singapore, appearing in workers with contact with pigs imported from Malaysia (Paton et al. 1999). Nipah virus, a previously undescribed paramyxovirus related to the Hendra virus (which caused fatal infections in humans and horses in Australia), may have entered the pig population from its reservoir host. Preliminary studies suggest that fruit bats (also known as flying foxes) may be the natural reservoir host for the virus. Once the virus entered the pig population, because of crowding of pigs in farms and movement of infected pigs, the infection was able to spread to other farms. In Malaysia, of 889 farms tested, 50 had infected pigs (and all the pigs on those farms were culled). Pigs developed a febrile respiratory illness with a barking cough (called "one mile cough" because it could be heard from a long distance). Humans typically develop encephalitis. Illness in humans is severe, with about 40% fatality rate. As of 1999, 285 human cases had been reported.

These many examples remind us of the interconnectedness of the world today and the close associations between human and animal populations. The study of disease emergence must look at ecosystems, evolutionary biology, and populations of all species (Wilson 1999).

References

Acha, P.H., and B. Szyfres 1987. Zoonoses and communicable diseases common to man and animals, 2nd ed., pp. 402–405. Scientific Publication No. 503. Pan American Health Organization, Washington, D.C.

Aires de Sousa, M., I.S. Sanches, M.L Ferro, J.J. Vaz, Z. Saraiva, T. Tendiero, J. Serra, and H.D. Lencastre. 1998. Intercontinental spread of a multidrug-resistant methicillin-resistant *Staphylococcus aureus* clone. J Clin Microbiol 36:2590–2596.

Ayliffe, G.A.J. 1997. The progressive intercontinental spread of methicillin-resistant *Staphylococcus aureus*. Clin Infect Dis 24(suppl):S74–S79.

Carlton, J.T., and K.B. Geller. 1993. Ecological roulette: the global transport of non-indigenous marine organisms. Science 261:78–82.

Centers for Disease Control and Prevention (CDC). 1997. Update: influenza activity—United States, 1997–98 season. Morbid Mortal Wkly Rep 46:1094–8.

Centers for Disease Control and Prevention (CDC). 1999. Outbreak of influenza A infection among travelers—Alaska and the Yukon Territory, May–June 1999. Morbid Mortal Wkly Rep 48:545–546, 555.

Centers of Epidemiology and Animal Health. 1998 The potential for international travelers to transmit foreign animal diseases to US livestock and poultry. Centers for Epidemiology and Animal Health, USDA:APHIS:VS, Fort Collins, Colo.

Chua, K.B., K.J. Goh, K.T. Wong, A. Kamarulzaman, P.S.K. Tan, T.G. Ksiazek, S.R. Zaki, G. Paul, S.K. Lam, and C.T. Tan. 1999. Fatal encephalitis due to Nipah virus among pig-farmers in Malaysia. Lancet 354:1257–1259.

Claas, E.C.J., A.D.M.E. Osterhaus, R. van Beek, J.C. De Jong, G.F. Rimmelzwaan, D.A. Senne, S. Krauss, K.F. Shortridge, and R.G. Webster. 1998. Human influenza A H5N1 virus related to a highly pathogenic avian influenza virus. Lancet 351:472–477.

Committee on Ships' Ballast Operations, Marine Board, Commission on Engineering and Technical Systems, National Research Council. (1996) Stemming the tide: controlling introductions of nonindigenous species by ships' ballast water. National Academy Press, Washington, D.C.

Corwin, A.L., C.H. Simanjuntak, G. Ingkokusumo, N. Sukri, R.P. Larasati, B. Subianto, H.Z. Muslim, E. Burni, K. Laras, M.P. Putri, C. Hayes, and N. Cox. 1998. Impact of epidemic influenza A-like acute respiratory illness in a remote jungle highland population in Irian Jaya, Indonesia. Clin Infect Dis 26:880–888.

Crosby, A. 1972. The Columbian exchange. Greenwood Press, Westport, Conn.

Freifeld, A.G., J. Hilliard, J. Southers, M. Murray, B. Savarese, J.M. Schmitt, and S.E. Straus. 1995. A controlled seroprevalence survey of primate handlers for evidence of asymptomatic herpes B virus infection. J Infect Dis 171:1031–1034.

Geraci, J.R., D.J.S. Aubin, I.K. Barker, R.G. Webster, V.S. Hinshaw, W.J. Bean, H.L. Ruhnke, J.H. Prescott, G. Early, A.S. Baker, S. Madoff, and R.T. Schooley. 1982. Mass mortality of harbor seals: pneumonia associated with influenza A virus. Science 215:1129–1131.

Grubler, A., and N. Nakicenovic. 1991. Evolution of transport systems: past and future. Research Report RR-91-008. International Institute for Applied Systems Analysis, Laxenburg, Vienna.

Holland, J.J. 1996. Evolving virus plagues. Proc Natl Acad Sci USA 93:545–546.

Horton, R. 1996. The infected metropolis. Lancet 347:134–135.

Hubalik, Z., and J. Haluzka. 1999. West Nile fever—a reemerging mosquito-borne viral disease in Europe. Emerg Infect Dis 5:643–650.

Jia, X-Y., T. Briese, I. Jordan, A. Rambaut, H.C. Chi, J.S. Mackenzie, R.A. Hall, J. Scherret, and W.I. Lipkin. 1999. Genetic analysis of West Nile New York 1999 encephalitis virus. Lancet 354:1971–1972.

Kenyon, T.A., S.E. Valway, W.W. Ihle, I.M. Onorato, and K.G. Castro. 1996. Transmission of multidrug-resistant *Mycobacterium* tuberculosis during a long airplane flight. N Engl J Med 67:1097–1100.

Kolata, G. 1999. Flu: the story of the great influenza pandemic of 1918 and the search for the virus that caused it, p. 294. Farrar, Straus, and Giroux, New York.

McCarthy, S.A., and F.M. Khambaty. 1994. International dissemination of epidemic *Vibrio cholerae* by cargo ship ballast and other nonpotable waters. Appl Environ Microbiol 60:2597–2601.

Minooee, A., and L.S. Rickman. 1999. Infectious diseases on cruise ships. Clin Infect Dis 29:737–744.

Moser, M.R., T.R. Bender, H.S. Margolis, G.R. Noble, A.P. Kendal, and D.G. Ritter. 1979. An outbreak of influenza aboard a commercial airliner. Am J Epidemiol 110:1–6.

Munoz, R., T.J. Coffey, M. Daniels, C.G. Dowson, G. Laible, J. Casal, R. Hakenbeck, M. Jacobs, J.M. Musser, B.G. Spratt, and A. Tomasz. 1991. Intercontinental spread of a multiresistant clone of serotype 23F *Streptococcus pneumoniae*. J Infect Dis 164:302–306.

Paton, N.I., Y.S. Leo, S.R. Zaki, A.P. Auchus, K.E. Lee, A.E. Ling, S.K. Chem, B. Ang, P.E. Rollin, T. Umapathi, I. Sng, C.C. Lee, E. Lim, and T.G. Ksiazek. 1999. Outbreak of Nipah-virus infection among abattoir workers in Singapore. Lancet 354:1253–1256.

Peiris, M., K.Y. Yuen, C.W. Leung, K.H. Chan, P.L.S. Ip, R.W.M. Lai, W.K. Orr, and K.F. Shortridge. 1999. Human infection with influenza H9N2. Lancet 354:916–917.

Richman, L.K., R.J. Montali., R.L. Garber, M.A. Kennedy, J. Lehnhardt, T. Hildebrandt, D. Schmitt, D. Hardy, D.J. Alcendor, and G.S. Hayward. 1999. Novel endotheliotropic herpesviruses fatal for Asian and African elephants. Science 283:1171–1176.

Schafer, A., and D. Victor. 1997. The past and future of global mobility. Sci Am, 277: 58–61.

Schaffer, F.L., M.E. Soergel, and C.D. Straube. 1976. Survival of airborne influenza virus: effects of propagating host, relative humidity, and composition of spray fluids. Arch Virol 51:263–273.

Smith, K.E., J.M. Besser, C.W. Hedberg, F.T. Leano, J.B. Bender, J.H. Wicklund, B.P. Johnson, K.A. Moore, and M.T. Osterholm. 1999. Quinolone-resistant *Campylobacter jejuni* infections in Minnesota, 1992–98. N Engl J Med 340:1525–1532.

Soares, S., K.G. Kristinsson, J.M. Musser, and A. Tomasz. 1993. Evidence for the introduction of a multiresistant clone of serotype 6B *Streptococcus pneumoniae* from Spain to Iceland in the late 1980s. J Infect Dis 168:158–163.

Subbarao, K., A. Klimov, J. Katz, H. Renery, W. Lim, H. Hall, M. Perdue, D. Swayne, C. Bender, J. Huang, M. Hemphill, T. Rowe, M. Shaw, X. Xu, K. Fukuda, and N. Cox. 1998. Characterization of an avian influenza A (H5N1) virus isolated from a child with a fatal respiratory illness. Science 279:393–396.

Vermeij, G.J. 1991. When biota's meet: understanding biotic interchange. Science 253: 1099–1104.

Weigler, B.M., D.W. Hird, J.K. Hilliard, N.W. Lerche, J.A. Roberts, and L.M. Scott. 1993. Epidemiology of cercopithecine herpesvirus 1 (B virus) infection and shedding in a large breeding cohort of rhesus monkeys. J Infect Dis 167:257–263.

Wilson, M.E. 1991. A world guide to infections: diseases, distribution, diagnosis. Oxford University Press, New York.

Wilson, M.E. 1995a. Travel and the emergence of infectious diseases. Emerg Infect Dis 1:39–45.

Wilson, M.E. 1995b. Infectious diseases: an ecological perspective. Br Med J 311:1681–1684.

Wilson, M.E. 1997. Population movements and emerging diseases. J Trav Med 4:183–186.

Wilson, M.E. 1999. Emerging infections and disease emergence. Emerg Infect Dis 5:308–309.

World Health Organization (WHO). 1998. Tuberculosis and air travel: guidelines for prevention and control. WHO/TB/98.256. WHO, Geneva.

Wright, S.M., Y. Kawaoka, G.B. Sharp, D.A. Senne, and R.G. Webster. 1992. Interspecies transmission and reassortment of influenza A viruses in pigs and turkeys in the United States. Am J Epidemiol 136:486–497.

Zanotto, P.M. A., E.A. Gould, G.F. Gao, P.H. Harvey, and E.C. Holmes. 1996. Population dynamics of flaviviruses revealed by molecular phylogenies. Proc Natl Acad Sci USA 93:548–553.

28

Global Ecological Integrity, Global Change, and Public Health

Colin L. Soskolne
Roberto Bertollini

Ecological degradation, precipitating large-scale ecosystem collapses, likely will have an impact on human health (McMichael et al. 1999). Bearing the principle of sustainable development in mind, a pilot workshop on "Global Ecological Integrity and Human Health" took place in 1998. It was convened at the World Health Organization (WHO) European Center for Environment and Health, Rome Division, Italy, 3–4 December (Soskolne and Bertollini 1999). An initial review of the available scientific evidence and philosophical considerations was undertaken to link global ecological integrity with the sustainability of human health and well-being. "Sustainable development" is development that meets the needs of the present without compromising the ability of future generations to meet their own needs (Brundtland Commission Report 1987).

28.1 Global Ecological Integrity

The term "global ecological integrity" (global EI) is an umbrella concept that includes the following criteria: the ecosystem must retain its ability to deal with outside interference and, if necessary, regenerate itself; the systems' integrity reaches a peak when the optimum capacity for the greatest number of possible ongoing development options, within its time/location, is reached; and it should retain the ability to continue its ongoing change and development, unconstrained by human interruptions, past or present (Lemons et al. 1997; Westra 1994, 1998; Westra and Lemons 1995; Pimentel et al. 2000). More familiar terms include "ecological health and ecosystem health." Conversely, the terms ecological disintegrity, ecosystem impoverishment, and ecological degradation are used to convey the opposite of a state of integrity and health. Biological impoverishment underlies each of the latter terms (Karr and Chu 1999). Three basic points with respect to the relationship between EI and human health follow.

(1) Human population and individual health ultimately depend on the integrity of ecosystems and the ecosphere (i.e., without an environment capable of supporting life, no population and, hence, no health can exist).

(2) Healthy populations can exist in local environments that have lost their EI—such as most urban regions—only if healthy ecosystems exist elsewhere to support them. This is a function of technology and trade and is a feature of human culture that uniquely distinguishes humans from other animal species dependent on their local environments. That is, human health can be maintained by healthy ecosystems (or, at least, productive ones) elsewhere. Under this modus operandi, the local population imposes its ecological footprint (i.e., the "mark" left on the earth through the depletion of ecological capital beyond a local population's political boundaries) on the global commons (i.e., the planet's resources available to support the world's population) and on other regions or countries (Rees and Wackernagel 1996; Wackernagel and Rees 1996; Rees 1996, 1997; Pimentel et al. 2000). This interregional dependency can obscure the connection between people and their health with the health of ecosystems.

(3) It is unlikely that increasing "footprints" are sustainable in the long run. The concepts of global commons, environmental health, ecological health, ecosystem health, ecological integrity and disintegrity, and the like, generally relate to the conditions of the biosphere that support life. Indicators have been developed by agencies such as the World Wildlife Fund (WWF) (WWF 1998) as standard measures that provide a sense of the health of ecological life support systems. The ecological footprint of nations (Wackernagel and Rees 1996), the index of biotic integrity (IBI; Karr and Chu 1999), the measure of mean functional integrity (MMFI; Loucks et al. 1999), and the findings by those metrics reported in the 1998 WWF report all point to humanity exceeding the global carrying capacity and to associated collapses in ecological life support systems.

All of these concepts and measures are related to public health by virtue of the link between the sustainability of human health as a function of the sustained health of ecological life support systems.

28.2 Feasible Adaptations in the Face of Progressive Declines

Public health consequences of the ecological nonsustainability of current global trends in the short, medium, and long term depend on whether we have a managed decline or a catastrophic decline. In the short term, we would probably see "classical" environmental effects associated with air pollution, toxics, flooding, and famines. In the medium term, we might see resource depletion, which could result in civil strife and even war. Longer term possibilities are more uncertain and could include extensive resource depletion leading to large-scale famine and societal disruption.

What adaptations are feasible in the face of progressive declines? The global community has three options for adaptation. First, we could accept the status quo in environmental trends, trusting (and hoping) that humans will be able to adapt to these changes (making no mention of the ability of other biological systems to adapt). Second, we could implement superficial remedies through patchy approaches to currently perceived problems. Third, we could proactively promote substantive policy reforms requiring major paradigm shifts in socioeconomic policies designed to restore, maintain, and protect the ecosystems upon which human and other life forms completely depend.

Some are criticized for their belief that humans will simply adapt to global climate change or other human-induced ecological change (Daily 1997; Korten 1998). This confidence may be comforting, but just what does adaptation mean? There may be danger if "to adapt" is taken to mean that we should merely adjust to the inevitable (e.g., by applying sunscreen, moving our towns and cities, developing new crop types). By itself, such reactive adaptation is foolhardy and perhaps even immoral as a policy directive because of its shortsighted approach; it will probably not likely be effective in the long run, and detracts from the more serious causes of ongoing ecological degradation.

There are evolutionary biologists who claim that fears about ecological degradation and the collapse of ecological life support systems are unfounded. The considered response of this position is that ecological concern is based on the lack of appreciation for the rate at which changes in life support systems are indeed taking place (Lubchenco 1998). The time frame of the evolutionary biologist tends to be many thousands of years. The evolutionary perspective, requiring millennia for a recovery to establish itself in support of life as we know it today, thus could indeed be correct. However, the time frame of concern from the public health–related discipline perspective is for current and future generations (i.e., decades and perhaps a century or two), not thousands or even millions of years into the future.

Presently, adaptation to ecological disintegrity has meant high levels of disease and unemployment among most of the world poor, as can be seen in many of the cities in the developing world. The passive option benefits a minority of wealthy people and seems to be embraced by more economic policy makers in developed countries each year (Funtowicz and Raretz 1994; Korten 1998; UNDP 1998).

Superficial measures can be defined generically as those that treat the symptoms without addressing their causes. They often provide short-term benefits mortgaged against longer term harms, for successive generations of children considered in future time frames. Thus, short-term benefits could be seen as being mortgaged against the well-being of future generations. One example of this type of remedy is voluntary (or mandatory) transfer payments from beneficiaries of resource "mining" (e.g., Canadian government) to those who suffer from the effects (direct consequences) of the resource/ecological collapse (e.g., fishermen who suffered the collapse of the Newfoundland codfish industry; Westra 1998).

While simple adjustments to shifting conditions may actually be necessary in the short to medium term, effective longer term social adaptation may well require changes in values and behaviors. These changes would be designed to avoid further climate or ecological change with a view to ultimately reversing the dangerous trends that we already see taking place in many of the world's ecological life support systems. Thus, adaptation measures should be used only in the short term, as emergency measures, and should not be incorporated into medium- and longer term strategies that may result in lulling the public into a false sense of security.

In the longer term, it is paramount that population pressures and critical resource consumption be reduced so as to prevent global change or collapse. Longer term adaptations would include lifestyle/equity changes, using both grass roots and policy directives. As a start, we need to recognize the global commons; that is, we need to foster a better understanding of the dynamic interrelations between local behaviors and global economics. Socioeconomic policy reform would involve finding ways to invest in social/human capital so that lowered levels of natural capital are more equitably redistributed. This has the added benefit of promoting health and preventing disease (e.g., Kerala, India, and Mondragon cooperatives in Spain; Ratcliffe 1978; Alexander 1997) through reduced disparities within societies (Wilkinson 1996; Schrecker 1999; Sen 1999). Policy and behavioral changes would need to be framed in positive rather than negative terms. Finding solutions that make changes attractive rather than burdensome (i.e., win–win arguments), such as creating urban transportation systems that are less damaging and that promote physical exercise, will be a crucial challenge. Perhaps promoting the benefits for children would be a strong motivator.

Attempting to answer these questions could lead to a new longer term vision and goal for public health and public health agencies. This activity would require new research projects with a focus on 25–50-year trends in the relationships between population health indicators and local/global ecological change (or loss of integrity). Specific case studies could be developed to test particular hypotheses and provide models for data collection and analysis. An analogous project design might be the Long-term Ecological Research Network (LERN) of projects in the United States. The LERN program provides a basis for integrating the monitoring of disease events and ecological parameters over the long term as well as case studies. This approach has been named "ecological epidemiology" (Epstein et al. 1998).

28.3 Feasible Adaptations in the Face of Sudden Declines

Public health action in anticipation of a nonsustainable environment will require both urgent responses such as famine and disaster relief and longer term strategies such as raising public awareness, data collection, measures, and indicators, as well as mobilizing action and policy formulation. The latter two issues are not addressed here, but are discussed elsewhere (Soskolne and Bertollini 1999).

28.4 Raising Public Awareness

Given the high level of uncertainty coupled with possible very serious consequences, some sort of massive, democratic mobilization is required. There is, on the one hand, a real danger of demagoguery (which will backfire) and, on the other, bureaucratic wrangling (which will result in too slow a reaction to avert catastrophe). There are no easy solutions.

To raise public awareness medium- and long-term scenarios need to be identified, and then talented movie/TV/video communicators could write and produce material. For example, a movie depicting paradigm shift options with their potential to mitigate the effects of global ecological change might serve to promote public dialogue about policy options. A movie concerning adaptations probably necessary in the wake of continuing environmental degradation and diminishing EI would be useful for creative discourse on enlightened policies in support of paradigm shifts. This material should be developed to specifically target each social sector (e.g., various socioeconomic, gender, cultural, religious, and age groups). Movie development could be aided by cooperation with educators.

The WHO could consider expanding its annual *State of Global Health Report* to not only include sensitive new indicators reflecting the state of the world's life support systems for human health, but also periodically devote a special focus report to the analysis of such data. This could be helpful for uncovering both data and indicator strengths and weaknesses. In this way, data needs would be identified, and more sensitive indicators developed in relation to the most appropriate health outcomes (OECD 1998; Rapport et al. 1998; Sieswerda 1999; Sieswerda et al. 2001). Progress in coping with global health problems depends on our ability to monitor the positive and negative effects of global change on both personal and population health (not only country-by-country tables and statistics, but also by latitude, longitude, and altitude, presenting indicators and differences). Emerging and reemerging diseases, such as malaria, cholera, dengue fever and plague, should be tracked, and scenarios based on alternative policies should be developed.

National and international agencies must agree on a standardized set of variables to monitor diseases or disease risk defined broadly as the converse of the WHO definition of health. Standardized international reporting methods across national borders are needed as the data are provided to WHO (1992, 1997) for the proposed expansion of the *State of Global Health Report*. Similar agreements for necessary (including biological) indicators must also be made.

28.5 Data Collection, Measures, and Indicators

Every effort must be made to remove barriers to the furnishing, gathering, verification, archival, and retrieval of data if the needed data and resulting information are to become available. Collection of data that would guide the charting of ecological impoverishment may be the clearest and most immediately feasible goal. Collation and reporting such data should be elaborated in relation to the many inventories (e.g., those from WHO, WRI, UNEP, UNDP, WWF, OECD, and World Bank) that show declining resource stocks (i.e., impoverishment), especially on a per capita basis as well as on a resource ratio or rate basis. This information is central to dealing with the poverty issues that WHO is pursuing and that many studies show are the dominant underlying drivers of disease.

There are two strategies for implementing data collection/evaluation/analysis and modeling: (1) assembling the trend analyses that may indicate progressive degradation or integrity of ecological systems (especially mountain regions) and (2) documenting case studies that show emerging health problems associated with areas where ecological systems are impoverished (i.e., the 1998 floods in China and Bangladesh; unstable hillsides in Honduras, China, and Peru; *Pfiesteria* in Chesapeake Bay and North Carolina [Burkholder and Glasgow 1997; Epstein et al. 1998; Russell 1998]; morbillivirus in the North Sea seals). For example, while studies show that nitrogen- and phosphorus-enriched coastal waters are facilitating harmful algal blooms in some coastal regions of the world, these results need to be given more emphasis in an international human health context. Single-nation reports tend not to provide a complete picture.

Public health agencies worldwide, while continuing to collect data on events such as infectious disease outbreaks, should also consider gathering additional data on their socioecological contexts. Indeed, contrasting similar communities without such outbreaks, and using these as controls (as a grouped-incident case-control study at the outbreak level), could help to identify upstream ecological disintegrity features associated with the outbreaks.

Furthermore, ecological risk assessment of chemicals and environmental interventions should be undertaken; that is, the ecological ramifications and implications for the spread of pests and pathogens should be examined. The full ecological and health implications of the complete cycle of fossil fuels could be useful.

In addition, public health agencies should identify and then promote additional good measures of health (self-perceived health is probably much better than many scientists believe) and select a variety of communities to monitor prospectively, collecting both health and socioecological data. Such studies need to be at a scale of meaningful ecological integration (watersheds, bioregions, ecodistricts), rather than in a form determined by convenient geopolitical boundaries. Family planning, a central component in all modern public health policies, is crucial to issues of population growth. Data on and case studies of family planning programs that have been both successful and unsuccessful need to be used as models for what does and does not work in developing future programs.

Consumption indices should be developed and used. Sensitive measures/indicators of the early effects from the erosion of EI (outcomes) also need to be identified. Such sources as the Global Terrestrial (and Oceanic/Atmospheric) Observations System data (WMO and FAO 1997) can be used to track the loss of natural ecological capital, including, for example, tracking the disappearance of forests and changes in freshwater streams.

The ecological footprint measures the energy equivalent required to support a concentration of human actions (Wackernagel and Rees 1996); the IBI measures the biota (i.e., the living system's or the life support system's condition; Karr and Chu 1999) in the places where the people with the footprint are living. At the same time, it can be used to measure the biological condition of places that are being influenced by those footprints from elsewhere. This single measure (i.e., the ecological footprint) is an energetic or thermodynamic view of the magnitude of human effects, while the IBI is a direct biological measure of the biological effects of those activities measured at a variety of sites. In short, the IBI is a measure of the effects and consequences of footprints in nonhuman biological terms.

While the metrics of WWF (1998) complement the ecological footprint analysis, they are narrow in the framing of biological effect and broad in the geographic area that is assessed. Broader and more integrative views of the condition of living systems and better diagnosis of the probable causes of degradation and their variation from place to place are needed. These require use of measures of biological condition combining large-scale and narrow measures of biological effects with more detailed and comprehensive biological evaluation such as the IBI. In essence, more than one human disease and more than aggregate counts of endangered species need to be tracked over broad areas to understand the effects of local ecological, explicitly biological, degradation. Consistent with the WHO definition of health, the loss of languages, loss of community, and loss of other quality of life characteristics beyond conventional disease measures should be included.

The database should include time-series data on health outcomes using newly developed standardized methods. Public health statistics (e.g., population growth and density, urbanization, resources available for and population benefiting from sanitary sewers and potable water, their association with indicators such as algae blooms, temperature increases, and heat deaths) correlate trends in equity. The extent to which nations and national health depend upon imported carrying capacity should be monitored. An example is the effects expected from climate change and other global changes. These effects would not simply be an increasing frequency of violent storms as a consequence of climate change, but also increased damage relating to loss of local forest cover and integrity, soil erosion, and related predisposing factors of infrastructure that would compound the extent of the damage. To determine such effects, public health agencies would need to develop measures, analyses, and projection capabilities for health based on the "drawdown" of integrity and natural capital.

Public health priorities for data collection and analyses could be established, based on three levels of increasing complexity: (1) comprehensive indicators (with no consideration of a causal relation to health), (2) indicators for which there is a *tentative* suggestion of a linkage to health, and (3) development of tools (i.e., models) for projecting global EI change and its implications for health.

It is most important that reports of the data acknowledge that this line of inquiry is in a very early stage with respect to recognizing and starting to capture information on the linkages between ecological damage and health outcomes. In particular, the global transport of pollutants and their health effects present a more recent challenge. We need to share more of the information that already is, or could easily be, captured through other agencies. Developing new paradigms and methods for dealing with the data, including establishing linkages, analyzing, modeling, and predicting, needs to be a priority. The present paradigm of epidemiology tends to focus on the immediate past, with some reasonable ability to forecast to the immediate future. Ecological disintegrity (or impoverishment) and global climate change arguments should succeed in shifting the focus of epidemiology toward scenario-based risk assessments with the necessary information needed for forecasts (initially imprecise) coming from multiple disciplines. This will require the work of many other disciplines. Current limits of epidemiologic methodology necessarily direct this focus toward specific diseases at the level of individuals (disease, not health) and ignore the complex issues involved with projecting outcomes for global public health.

Even within the constraints of presently accepted methodology, it is possible to shift emphasis: instead of the traditional emphasis on the best-fitting statistical model, research should focus on outliers—those points that lie well outside the fitted curve—in that it is within those points that answers to EI problems may lie (e.g., Kerala, India, which produces 1/65th of the U.S. gross domestic product, yet is among the first-world countries in standards of health and literacy; see Alexander 1997; Ratcliffe 1978).

28.6 Conclusions

Current patterns of human consumption worldwide are unsustainable according to a number of measures, for example, the ecological footprints of nations, the IBI, the MMFI, and the WWF analyses. Major disparities in resource consumption among nations are evident. Indeed, from nation to nation, and even within industrialized nations, there are great disparities between those who reap the benefits from the consumption of natural capital and those whose livelihood and health are affected by the degradation. A fuller understanding is needed of the linkages, both proximate and distant, between human health and EI, as well as its converse, human disease and ecological disintegrity. The examination of case studies would be useful.

Global life support systems are inextricably tied to EI. Issues relating EI to human health and the survival of human life on Earth should therefore become fully part of the public health agenda. Improved cooperation among the various United Nations agencies would make the attainment of sustainable goals more likely. If ecological life support systems indeed should fail, the extent of morbidity and mortality could be more extreme than any human calamity known before.

Maintenance of current levels of public health will become increasingly difficult, if not impossible, in the face of declining resources and collapsing natural life support systems. Disregarding concerns about EI may be considered equivalent to mortgaging the well-being of future generations against the greed of present generations, measured in terms of current trends in drawing down natural capital through overconsumption, population growth, and the abuse and/or inequitable use of technology. Compromising the ability of future generations to meet their needs is contrary to the principle of sustainable development as articulated in the Brundtland Commission Report (1987).

Consequently, the issue of declining global EI and its consequences for public health should be included on the public health agenda. The WHO and other public health agencies will need to develop tools to monitor and evaluate ongoing changes in ecological life support systems, their subsequent health effects, and the effects of interventions (including empirical evidence and predictive modeling, both of which require advances in methodology). The current paradigms of economic and ecological analysis will also need to be reconsidered. Worldwide application of the precautionary principle is encouraged.

In addition, mutual or social learning is a prerequisite to any new paradigms proposed. Appropriately targeted and effective messages are urgently needed for informing both the public and policy makers of the underlying issues, and of the consequences of adhering to current paradigms. The challenge lies in developing messages, jointly with community groups and other stakeholders, that will be both credible and able to be assimilated, and that will result in timely actions in support of agreed-upon paradigm shifts.

Public health measures promoting health and longevity contribute directly to population growth and cannot be disconnected from population control policies. In this context, public health practices need to be linked more strongly with professionals in the areas of education, economics, sociology, fertility, and population.

One practical method for making an operative response to declining global EI would be to improve measures of environmental monitoring (e.g., an index such as the "ecological footprint of nations" complemented by the IBI). Then, public health professionals and decision makers would have a concrete set of metrics for identifying actions/targets to be addressed at the local level. Future regional successes could be assessed in relation to such metrics.

Declines in global EI could result in mass migrations, extreme weather events (e.g., floods, droughts, famines, hurricanes, cyclones, heat waves, ice storms, tornadoes), epidemics, wars, social disruptions, and skirmishes over access to remaining resources. Physicians trained in emergency care, mental health, and a broad range of other clinical and surgical specialties will be needed. They will also contribute data for monitoring the effectiveness of their interventions in relation to their frontline experiences.

The single greatest adaptation will be for professional groups to transcend each of their traditional boundaries and utilize interdisciplinary approaches to human health that integrate the natural, social, and health sciences in a humanities context. Emergent concepts and methods are the hallmark of the transdisciplinary effort. Such innovative adaptation by the health-related professions will be needed if they are to maintain capacity for serving society's best interests by providing the rational basis for informing policy.

References

Alexander, W.M. 1997. Exceptional Kerala, efficient and sustainable human behavior. *In* James, V., ed. Capacity building in developing nations. Praeger, New York.

Brundtland Commission Report. 1987. Our common future. United Nations World Commission on Environment and Development. Oxford University Press, New York.

Burkholder, J.M., and H.B. Glasgow Jr. 1997. *Pfiesteria piscicida* and other *Pfesteria*-like dinoflagellates: behavior, impacts and environmental controls. Limnol Oceanogr 42: 1052–1075.

Daily, G.C., ed. 1997. Nature's services: societal dependence on natural ecosystems. Island Press, Washington, D.C.

Epstein, P., B. Sherman, E. Spanger-Siegfried, A. Langston, S. Prasad, B. McKay. 1998. Marine ecosystems: emerging diseases as indicators of change. In Health of the oceans from Labrador to Venezuela—Year of the Ocean special report. Health Ecological and Economic Dimensions of Global Change Program, The Center for Health and the Global Environment, Harvard Medical School, Boston.

Funtowicz, S.O., and J.R. Ravetz. 1994. Emergent complex systems. Futures 26:568–582.

Karr, J.R., and E.W. Chu. 1998. Restoring life in running waters: better biological monitoring. Island Press, Washington, D.C.

Korten, D.C. 1998. Do corporations rule the world? And does it matter? Org Environ 11: 389–398.

Lemons, J., L. Westra, and R. Goodland, eds. 1997. Ecological sustainability and integrity: concepts and approaches. Kluwer, Dordrecht.

Loucks, O., O.H. Ereksen, J.W. Bol, R.F. Gorman, P.C. Johnson, and T.C. Krehbiel. 1999. Sustainability perspectives for resources and business. Lewis Publishers, Boca Raton, Fla.

Lubchenco, J. 1998. Entering the century of the environment: a new social contract for science. Science 279:491–497.

McMichael, A.J., B. Bolin, R. Costanza, G.C. Daily, C. Folke, K. Lindahl-Kiessling, E. Lindgren, and B. Niklasson. 1999. Globalization and the sustainability of human health: an ecological perspective. Bioscience 49:205–210.

Organisation for Economic Co-operation and Development (OECD). 1998. Environmental indicators: towards sustainable development. OECD, Paris.

Pimentel, D., L. Westra, and R.F. Noss, eds. 2000. Ecological integrity: integrating environment, conservation, and health. Island Press, Washington, D.C.

Rapport, D.J., R. Costanza, and A.J. McMichael. 1998. Assessing ecosystem health. TREE 13:397–402.

Ratcliffe, J. 1978. Social justice and the demographic transition: lessons from India's Kerala State. Int J Health Serv 8:1.

Rees, W.E. 1996. Revisiting carrying capacity: area-based indicators of sustainability. Pop Environ 7:195–215.

Rees, W.E. 1997. Is "sustainable city" an oxymoron? Local Environ 2:303–310.

Rees, W.E., and M. Wackernagel. 1996. Urban ecological footprints: why cities cannot be sustainable (and why they are a key to sustainability). EIA Rev 16:223–248.

Russell, D. 1998. Underwater epidemic. Amicus J 20:28–33.

Schrecker, T. 1999. Money matters: incomes tell a story about environmental dangers and human health. Alt J 25:12–18.

Sen, A. 1999. Health in development. Keynote address. Presented at the 52nd world health assembly (agenda Item 4, A52/DIV/9), 18 May, Geneva.

Sieswerda, L.E. 1999. Towards measuring the impact of ecological disintegrity on human health. Masters thesis. University of Alberta, Edmonton.

Sieswerda, L.E., C.L. Soskolne, S.C. Newman, D. Schopflocher, and K.E. Smoyer. 2001. Towards measuring the impact of ecological disintegrity on human health. Epidemiology 12:28–32

Soskolne, C.L., and R. Bertollini. 1999. Global ecological integrity and "sustainable development": cornerstones of public health. Web site: http//www.who.it/Emissues/Globaleco/globaleco.htm, accessed July 1999.

United Nations Development Programme (UNDP). 1998. Human development report 1998. Oxford University Press, New York.

Wackernagel, M., and W. Rees. 1996. Our ecological footprint: reducing human impact on the earth. New Society Publishers, New Haven, Conn.

Westra, L. 1994. An environmental proposal for ethics: the principle of integrity. Rowman and Littlefield, Lanham, Md.

Westra, L. 1998. Living in integrity: a global ethic to restore a fragmented earth. Rowman and Littlefield, Lanham, Md.

Westra, L., and J. Lemons, eds. 1995. Perspectives on ecological integrity. Kluwer, Dordrecht.

Wilkinson, R.G. 1996. Unhealthy societies: the afflictions of inequality. Routledge, London.

World Health Organization (WHO). 1992. Our planet, our health. Report of the WHO Commission on Health and Environment. WHO, Geneva.

World Health Organization (WHO). 1997. Health and environment in sustainable development: five years after the Earth summit. WHO, Geneva.

World Meteorological Organization (WMO) and Food and Agricultural Organization (FAO). 1997. Global terrestrial observing system: GCOS/GTOS plan for terrestrial climate-related observations (version 2.0). WMO/FAO, Rome, Italy.

World Wide Fund for Nature (WWFN). 1998. Living planet report: over-consumption is driving the rapid decline of the world's natural environments. WWFN, Washington, D.C.

29

Wildlife Health and Environmental Security
New Challenges and Opportunities

Jamie K. Reaser

Edward J. Gentz

Edward E. Clark, Jr.

29.1 Environmental Security

Environmental security is the concept that social (and thus political and economic) stability affects, and is affected by, the abundance and distribution of natural resources. Poor human health, mass human migrations, and border zone conflicts are often symptoms of environmental insecurity, when human populations perceive that they are not able to acquire natural resources at an adequate level to sustain themselves (Dabelko 1998; Kennedy 1998; U.S. Department of State 1997). When the environment is adversely stressed, the ability of natural biological processes to maintain and renew the structure and function of ecosystems, the ultimate source of natural resources, is compromised (Daily 1997; Spellerberg 1996). We face ever-increasing threats to environmental security in this millennium: the demand for natural resources will intensify as the human population tops 7.5 billion by 2015 (OECD 1997), the climate continues to warm (IPCC 1996; Gore 1992), and advanced technologies will enable resource extraction more extensively and at faster rates than at any time in history.

Mortality is a crude measure of stress. Yet, the information typically used to gauge environmental condition, and thus environmental security, derives from the assessment of trends in biodiversity, especially the decline and decimation of certain species (Halvorson and Davis 1996; Soulé 1986; Ehrlich and Ehrlich 1981). However, by the time a decline in a population is perceived, the factors that led to the decline are often indeterminable, population recovery is costly if not impossible, and the repercussions of the decline have had a broad negative impact throughout the ecosystem (Spellerberg 1996; Soulé 1986). Because acute and chronic stress may each contribute to detectable illness before death, the assessment of wildlife health can provide a timely means to gauge environmental quality (Daszak et al. 2000) and thus effectively manage for environmental security.

Recent surges in disease outbreaks thoughout a diversity of taxonomic groups and ecological systems (e.g., Epstein et al. 1998; Morell 1999) have scientists asking the following questions:

- Is a general decline in environmental quality compromising immune systems and making organisms more susceptible to typically benign microbes (Epstein et al. 1998; Carey et al. 1999)?
- Are climatic shifts in the environment enabling microbes to increase in virulence, range, and/or diversity (Colwell 1997; Kennedy et al. 1998)?
- Are increases in our technological ability to transport people and products farther and faster than at any time in the history of the biosphere facilitating the introduction of microbes to novel environments and hosts (Morell 1999)?
- Are two or more of these processes operating concurrently to lead to epidemics in wildlife, humans, and livestock (Daszak et al. 2000)?

The answers to all of these questions will have profound implications for the management of environmental security. Furthermore, as human society further encroaches into the remaining wilderness, environmental security will require wildlife disease monitoring in order to control the potential transmission of pathogens between wildlife and domestic animals (e.g., Aguirre et al. 1995; Irby and Knight 1998), zoonotic transmission from wildlife to humans (e.g., Weiss and Wrangham 1999), and anthropozoonotic transmission from humans to wildlife (e.g., Mudakikwa et al. 1998).

Recent epizootics of disease decimating coral reefs and amphibian populations have illustrated the connection between wildlife disease and environmental degradation, drawing the attention of policy makers and the general public alike, particularly when wildlife health, human health, and sustainable development issues intersect (e.g., Epstein et al. 1998; Daszak et al. 2000). Similarly, the interconnectedness of environmental disruption and potentially devastating effects of agents such as *Pfiesteria*, Ebola, and simian immunodeficiency virus (SIV) warrant attention. Now is the time to raise the profile of conservation medicine among the policy-making community, develop well-integrated wildlife health monitoring programs, and create effective lines of communication between people in the conjoined fields of conservation medicine and environmental policy that will lead to timely, well-informed decision making.

The purpose of this chapter is to highlight the issue of wildlife disease in the context of environmental policy making. This is a broad topic, and the challenges and opportunities for addressing wildlife disease through the environmental policy process are numerous. This chapter therefore describes only the tip of the iceberg. Professionals in the fields of conservation medicine and environmental policy have different informational requirements. We hope that this chapter serves the needs of both. Those in the field of conservation medicine will benefit from suggestions of how to approach the policy process, and policy makers will gain a basic understanding of how wildlife disease is relevant to environmental security.

29.2 Case Studies

The most rapid way to raise the profile of conservation medicine and to illustrate to policy makers the connection between wildlife disease and environmental security is to draw attention to the role wildlife disease is already playing in high-profile issues in environmental conservation, human health, and sustainable development. Here we provide several case studies of such high-profile issues; each could be a starting point for policy dialogues on wildlife disease and environmental security. This approach will be most productive if case studies are chosen that illustrate issues that policy makers and the public have already decided are of priority concern, thus deserving of time and money.

29.2.1 Coral Reefs

The coral reef is the most complex, species-rich, and productive marine ecosystem (Bryant et al. 1998; Sebens 1994; Stafford-Deitsch 1993), with one estimate (Reaka-Kudla 1996) that coral reefs have about one million species, with only 10% described. The benefits derived from coral reefs are both immediate and long term, making them a priority for conservation and a major resource for sustainable development (Bryant et al. 1998; Roberts et al. 1998; Baskin 1997; Fenical 1996; Carte 1996; Jameson et al. 1995). As one of the first systems to show signs of ecological stress from global warming (IPCC 1998) on a worldwide scale (Reaser et al. 2000; Wilkinson et al. 1999), coral reefs are valuable ecological indicators (Hayes and Goreau 1991).

Coral reefs face numerous threats, many of them human induced and well documented (Bryant et al. 1998; Peters 1997). Persistent chemical and nutrient pressures (largely from coastal zone erosion), compounded by climate changes and globalization, may be accelerating the evolution, virulence, and spread of pathogens in reef-building corals (Epstein et al. 1998). The number, severity, and extent of coral diseases have been expanding rapidly in the last decade, and the affect of disease on coral reef communities is often devastating (e.g., Richardson et al. 1998; Hayes and Goreau 1998; Goreau et al. 1998; Peters 1997; Richardson 1996; Gladfelter 1992). Coral colonies stressed by infectious disease may be less resistant to the stresses of climate change and other localized stressors, and the converse may also be true (Williams and Bunkley-Williams 1990; Hayes and Goreau 1998; Reaser et al. 2000).

29.2.2 Amphibian Declines

Amphibians warrant substantial conservation attention. They are considered valuable indicators of environmental quality and have multiple functional roles in aquatic and terrestrial ecosystems (Blaustein and Wake 1990; Stebbins and Cohen 1995; Reaser 1996, 2000). Furthermore, amphibians provide significant cultural and economic value to human society (Grenard 1994; Stebbins and Cohen 1995; Reaser and Galindo-Leal 1999). In the ongoing overall biodiversity crisis, many amphibian populations have been declining and undergoing range reductions (reviewed in Blaustein and Wake 1990, 1995; Stebbins and Cohen 1995; Reaser 1996).

A wide diversity of microbes are commonly associated with amphibians (e.g., Gibbs et al. 1966; Carr et al. 1976; Brodkin et al. 1992; Blaustein et al. 1994), and they are susceptible to infection by a wide variety of pathogens, especially when stressed. Scientists investigating declines of amphibians in relatively remote, undisturbed regions have frequently pointed to fungi (e.g., Berger et al., 1998), viruses (e.g., Laurance et al. 1996), or bacteria (e.g., Worthylake and Hovingh 1989) as the proximal cause of death. In some of these cases, introduced fish have been implicated as vectors (Blaustein et al. 1994; Laurance et al. 1996). Chytridiomycosis, a fungal disease, appears to be capable of causing amphibian population declines and species extinctions in previously pristine sites, possibly due to human-associated introduction of the pathogen to naive populations (Daszak et al. 1999).

29.2.3 Pfiesteria *and Marine Coastal Ecology*

Fish and shellfish that feed on blooms of toxic, dinoflagellate algae species become poisonous to humans. Epidemics of ciguatera fish poisoning and paralytic, neurotoxic, diarrheic, and amnesic shellfish poisoning are increasingly common around the world (Morris 1999). *Pfiesteria piscicida,* a single-celled dinoflagellate, poses a significant threat to fisheries and human health in the Mid-Altantic region of the United States. *Pfiesteria* is associated with waterways that have been enriched with nitrogen and phosporus, often due to pollution with animal waste products originating from swine and poultry operations (Burkholder 1999). Since the late 1980s, the pathogen has been implicated in substantial fish disease and mortality in North Carolina, Maryland, and Virginia (Burkholder 1999). Human health concerns arose when laboratory workers investigating *Pfiesteria* and commercial fisherman in infected areas began to experience unusual clinical symptoms, including memory loss, possibly as a result of exposure to the dinoflagellates' toxin(s). Laboratory research continues in a Biohazard III facility, although the disease itself remains enigmatic. Unquestionably, water quality, fish health, and human health are inextricably linked; two thirds of all Americans live within 50 miles of a coastline (Burkholder 1999).

29.2.4 *Ebola Virus*

In 1967, 31 laboratory workers in Germany were infected with Ebola virus after being exposed to green monkeys imported from Africa. Seven of the workers died (Peters and LeDuc 1999). In the 1970s, Ebola virus caused highly virulent epidemics of hemorrhagic fever in Sudan and the Democratic Republic of Congo. It surprised the U.S. quarantine and health officials when it first appeared in a Reston, Virginia, research facility in monkeys imported from the Philippines and most recently in a Texas quarantine facility in 1996. Although primate import and quarantine policies have been strengthened in the United States and other countries in response to these outbreaks (DeMarcus et al. 1999), the virus will continue to be a threat until the hosts and pathways for infection have been definitively identified. Current evidence suggests that the wild reservoir for Ebola virus is a rare African species, or one that has limited contact with humans. It is also possible that virus may not be readily passed between human and nonhuman primates under most circumstances, making transmission difficult to observe. Human infections have occurred in Gabon from exposure to wild chimpanzees butchered for meat that were infected with Ebola virus, and in the Ivory Coast from the necropsy of wild, infected chimpanzees. Interestingly, antibodies to Ebola are significantly higher in some African hunter-gatherer tribes than in subsistence farmers (Monath 1999).

29.2.5 Simian AIDS

Human acquired immunodeficiency syndrome (AIDS) is caused by either of two distinct human immunodeficiency viruses, HIV-1 and HIV-2, both of which have primate reservoirs (Gao et al. 1999). Simian AIDS and human AIDS are strikingly similar (Geretti 1999), although simian AIDS has a more rapid course. A simian immunodeficiency virus (SIV) indistinguishable from HIV-2 is found in wild sooty mangabey monkeys, in the area around the epicenter of the HIV-2 epidemic, where they are both hunted for food and kept as pets. Recently, HIV-1 has been shown to have arisen from an SIV associated with chimpanzees. In addition to sooty mangabeys, African green monkeys, Sykes monkeys, and mandrills have all been found to possess their own SIV strains. As in the case of Ebola, HIV may be difficult to eradicate until the use of primates for bushmeat (Weiss and Wrangham 1999), pets, and research subjects is abolished.

29.3 Policy Objectives

Some policy makers are beginning to take actions that show they recognize that wildlife diseases pose a significant threat to environmental security and that conservation medicine requires new resources. For example, Canada recently established the Canadian Science Center for Human and Animal Health, which will be administered jointly by Health Canada and the Canadian Food Inspection Agency (Mervis 1999). As part of its national action plan, the U.S. Coral Reef Task Force has called for the creation of a coral bleaching and disease center (U.S. Coral Reef Task Force 2000).

Each of these programs provides an opportunity to establish better baseline data and develop interdisciplinary research programs to address gaps in current knowledge. Unfortunately, such programs are the exception, not the rule. Wildlife health, as a critical aspect of environmental health and an issue for policy consideration, is seldom incorporated into regulatory or management processes. As a result, wildlife and the services wildlife provides remain vulnerable worldwide. Rectifying this situation will require a strategic approach. Unless new programs are coordinated through a comprehensive policy strategy, the transfer of research findings into practical measures to address the causes and consequences of wildlife disease will be slow and largely fortuitous.

Here we recommend a series of broad policy objectives to increase the profile of conservation medicine and to enhance our ability to effectively detect, classify, monitor, and control wildlife diseases. Achieving success in these efforts will depend on the cooperation of the federal, state, and local governments, international organizations, and the private sector, as well as the public health, medical, and veterinary communities. Substantial financial support is needed to put these priorities into practice. Because the causes and consequences of wildlife disease may be global and long term in scale, governments around the world need to work together to make available the funds that will enable these important initiatives.

29.3.1 National Level

1. *Raise the profile of the issue and increase the number of practitioners.* Wildlife disease has, until recently, had a low profile in the biological and veterinary science communities. As a result, there are few practitioners qualified to address this problem, now increasingly complex and global in scale. We need to encourage expansion of the field of conservation medicine by advising scientific and medical institutions to expand training in wildlife disease. Academic degree programs in fields such as biology or ecology need to deepen the emphasis on wildlife disease and the understanding of the physiology of individual organisms within ecological systems. By enhancing the capability of the veterinary profession to detect disease in wild animals before that disease reaches epidemic proportions and populations decline, we will have new opportunities to prevent or mitigate the effects of problems that presently affect wildlife on a broad scale. We can further enhance the field by urging that wildlife disease be given greater emphasis in fellowship programs and on veterinary certification and recertification examinations. Ultimately, support for the field needs to come in the form of new career opportunities, in both the public and private sectors, for wildlife disease specialists.

2. *Strengthen the wildlife disease surveillance and response domestically, at federal, state, and local levels and at ports of entry into the United States.* Due to increases in trade and human transport, pathogens can be rapidly relocated around the world, quickly causing epidemics of disease to which we are ill prepared to respond. Monitoring and reporting systems for the surveillance, detection, and response to wildlife disease must become more effective and employed as tools to prevent epidemics rather than merely document them. Monitoring methods need to be developed through which the health of *individual* organisms within the ecosystem is more closely monitored. This will require improved diagnostic support resources, such as public and private laboratories, and an expanded pool of wildlife health professionals. These improved surveillance techniques need to be applied in the field and at the "hubs" of trade and transport,

such as ports of entry, captive wildlife inspection stations, and quarantine facilities. If these monitoring programs are standardized, the results could be reported to and compiled by a centralized (or at least networked) information management center with the capacity to rapidly analyze and disseminate the data, enabling a more timely and effective response to disease events. In most cases, the regulations, procedures, and resources at ports of entry and other relevant facilities will need to be reviewed and updated in order to meet higher standards of operation.

3. *Expand and establish the authority of relevant U.S. government agencies to contribute to a global wildlife disease prevention, surveillance, and response network.* Wildlife disease must be given a higher priority by the U.S. government. Experience indicates that waiting until a wildlife disease event occurs is too little too late to prevent, contain, or mitigate most epidemics. A wildlife disease task force should be created with the authority and mandate to develop a proactive strategy to address wildlife disease and to coordinate efforts within the federal government, as well as between the federal government and other sectors. Ideally, this task force would establish a mechanism to monitor wildlife health, identify potential disease situations, and swiftly mount a coordinated response when disease outbreaks occur. This would require the task force to identify and establish mechanisms to ensure the responsible agencies have the authority, emergency procurement powers, and resources to respond to large-scale wildlife disease outbreaks. There are many models for such a task force within federal government and individual states, and some, such as the U.S. task forces on coral reefs, on emerging infectious disease, and on amphibian declines and deformities, already consider wildlife disease as a program element (table 29.1). In order to have the full cooperation of the federal and state agencies and institutions with the resources and expertise needed to address this issue, the wildlife disease task force will require the authority and support from the highest levels of the government.

Table 29.1. U.S. task forces with relevance to wildlife disease

Name	Lead Agencies	Establishment
National Invasive Species Council	U.S. Departments of Interior, Agriculture, and Commerce	Executive Order 13112, February 1999
U.S. Coral Reef Task Force	U.S. Departments of Interior and Commerce	Executive Order 13089, June 1998
U.S. Task Force on Emerging Infectious Disease	Center for Disease Control and White House Office of Science and Technology Policy	Committee on International Science, Engineering and Technology (CISET) of the National Science and Technology Council (NSTC)
U.S. Task Force on Amphibian Declines and Deformities	U.S. Department of Interior	June 1998

29.3.2 International Level

1. *Establish a global wildlife disease surveillance and rapid response system and strengthen the ability of other countries to detect, monitor, and control wildlife disease.* This system should be based on a standardized set of protocols, operated through a series of regional hubs and linked by modern communication technologies. Because few people have training in wildlife epidemiology, capacity building will need to be a top priority of this global program. Zoological and wildlife parks should be the primary focus of such activities; their personnel are most in need of this training, and their facilities can typically support training courses. There are a variety of surveillance programs that could serve as models for a global wildlife disease surveillance and rapid response system. Some of these are very technical programs run by professionals, while others manage largely anecdotal information collected by volunteers. A few examples include the various programs of the Centers for Disease Control and Prevention, Pro-Med, and the Global Coral Reef Monitoring Network.

2. *Seek ways to strengthen bi- and multilateral policy frameworks.* Existing bi- and multilateral agreements could be modified in order to strengthen their coverage of wildlife disease issues. These include science and technology agreements and the environmental side agreements to trade pacts. This might be accomplished most successfully in language relating to "invasive alien species." Invasive alien species, which include many pathogens and pathogen hosts, are a by-product of globalization. Because the world trading system is so extensive and complex, controlling trade-mediated epidemics will be a formidable challenge (Bright 1998). Every week, about one million people move between the industrial and developing worlds; every day, about two million people cross an international border (IOM 1997). Furthermore, people who come in contact with wildlife pathogens (including biologists), and thus might transmit them, are increasingly traveling long distances and into remote regions. An expert consultation under the Convention of Biological Diversity could be convened in conjunction with the Office International des Epizooties in order to review the existing policy frameworks, standards, and guidelines and make recommendations for how they might best be utilized in the wildlife disease context. Special emphasis will need to be given to policies and standards regulating the quarantine of wildlife crossing international borders, as well as the trade and transport of plants, animal, and microbial materials.

The United States can encourage other nations and international organizations to assign higher priority to wildlife disease issues by (1) raising wildlife disease in bilateral, regional, and multilateral discussions and (2) negotiating cooperative agreements with other nations to promote the establishment of a global wildlife disease monitoring and action network.

3. *Review and modify sustainable development and humanitarian assistance programs.* Sustainable development and humanitarian assistance programs rarely consider wildlife disease in their programs of operation. The United States should develop guidelines for "best practice" goals of wildlife disease prevention and

control and review their assistance programs, including both direct aid and loan guarantees, for compatibility with these guidelines, modifying them as needed.

29.3.3 Crosscutting

1. *Promote public awareness of the causes and consequences of wildlife disease.* Policy makers need to use their resources to make the public more aware of the causes and consequences of wildlife disease. The implementation of policies that reduce the emission of green house and ozone-depleting gases, pesticide contaminants, habitat degradation, and the direct exploitation of wildlife could help reduce the emergence and spread of wildlife disease. Many public constituencies are in frequent contact with wildlife and thus are well suited to contribute to programs that provide early warnings of and rapid response to wildlife disease outbreaks. These include wildlife rehabilitators, humane societies, park and refuge rangers, naturalists, and bird watchers. Government programs could be developed to address wildlife disease by supporting the exchange of information and public–private partnerships among these diverse groups.

2. *Strengthen research activities to improve the diagnosis, treatment, and prevention and improve the understanding of the ecology of wildlife disease.* The research priorities for both government and private institutions need to include analyses of the physiology and viability of animals and their associated pathogens within the ecosystems of which they are a part. Special emphasis should be placed on investigations of the response of various organisms to environmental factors, whether pathogenic or toxic. Included in these research efforts should be the investigation of improved disease prevention, detection, and mitigation methods. These tools and technologies need to be effectively disseminated through training initiatives directed at both the policy and field level of wildlife agencies, academic degree programs, and other relevant institutions.

3. *Ensure the availability of diagnostic tests and treatments needed to combat wildlife disease.* In order to prevent and mitigate the impacts of wildlife disease, we need the capability to identify environmental contaminants, isolate pathogens, evaluate the progression of wildlife disease episodes, and administer effective treatments. Thus, we need to develop regional inventories and networks of resources for combating wildlife disease, enhance the laboratory and diagnostic resources available throughout these networks, and explore steps to strengthen surveillance and response capabilities of internationally.

29.4 Conclusions

In order to maintain environmental security, the United States and other nations must now take strategic action to reduce the risks posed by wildlife disease. Wildlife disease events may become more frequent and severe as the human pressures on natural resources grows and the climate continues to warm, exposing wildlife to an increasingly hostile environment. In order to maintain environmental security, significant attention needs to be given to the identification and monitoring of wildlife disease, research on the projected and realized effects of global climate change on pathogens, and measures to reduced the localized anthropogenic factors that already threaten the environment. Even wildlife granted well-enforced legal protection in sanctuaries, or managed for sustainable use, is threatened by disease.

In order for policies, regulations, and management strategies to afford true protection and the sustainable use of natural systems, wildlife health must be given equal status with benchmarks such as air quality and water quality. This will require a significant investment of resources, a more comprehensive assessment of environmental threats, and an integrated, interdisciplinary approach to policy making. Just as the silenced canary warned coal miners of impending danger, so too can wildlife health be a sentinel of environmental and human condition. We have the capability to heed warnings. We must employ the wisdom and determination: while a coal miner could depart a gaseous mine, we have no place to go.

References

Aguirre, A.A., E.E. Starkey, and D.E. Hansen. 1995. Wildlife diseases in national park ecosystems. Wildl Soc Bull 23:415–419.
Baskin, Y. 1997. The work of nature: how the diversity of life sustains us. Island Press, New York.
Berger, L., R. Speare, P. Daszak, D.E. Green, A.A. Cunningham, C.L. Goggin, R. Slocombe, M.A. Ragan, A.D. Hyatt, K.R. McDonald, H.B. Hines, K.R. Lips, G. Marantelli, and H. Parkes. 1998. Chytridiomycosis causes amphibian mortality associated with population declines in the rainforests of Australia and Central America. Proc Nat Acad Sci USA 95:9031–9036.
Blaustein, A.R., and D.B. Wake. 1990. Declining amphibian populations: a global phenomenon? TREE 5:203.
Blaustein, A.R., and D.B. Wake. 1995. The puzzle of declining amphibian populations. Sci Am 272:52–57.
Blaustein, A.R., D.G., Hokit, R.K. O'Hara, and R.A. Holt. 1994. Pathogenic fungus contributes to amphibian losses in the Pacific Northwest. Biol Conserv. 67:251–254.
Bright, C. 1998. Life out of bounds: bioinvasion in a borderless world. Norton, New York.
Brodkin, M.A., M.P. Simon, A.N. DeSante, and K.J. Boyer. 1992. Response of *Rana pipiens* to graded doses of the bacterium *Pseudomonas aeruginosa*. J Herpetol 26: 490–495.
Bryant, D., L. Burke, J. McManus, and M. Spalding. 1998. Reefs at risk: a map-based indicator of threats to the world's coral reefs. World Resources Institute, Washington, D.C.
Burkholder, J.M. 1999. The lurking perils of *Pfiesteria*. Sci Am 281:42–49.
Carey, C., N. Cohen, and L. Rollins-Smith. 1999. Amphibian declines: an immunological perspective. Dev Comp Immunol 23:459–472.
Carr, A.H., R.L. Amborski, D.D. Culley Jr., and G.F. Amborski. 1976. Aerobic bacteria in the intestinal tracts of bullfrogs (*Rana catesbeiana*) maintained at low temperatures. Herpetologica 32:239–244.
Carte, B.K. 1996. Biomedical potential of marine natural products. Bioscience 46:271–286.
Colwell, R.R. 1997. Microbial biodiversity and biotechnology. *In* Reaka-Kudla, M.L., D.E.

Wilson, and E.O. Wilson, eds. Biodiversity II: understanding and protecting our biological resources, pp. 279–299. National Academy of Sciences, Washington, D.C.

Dabelko, G.D., ed. 1998. Environmental change and security project report. Issue 4. Woodrow Wilson Center, Smithsonian Institution, Washington, D.C.

Daily, G.C., ed. 1997. Nature's services: societal dependence on natural ecosystems. Island Press, Washington, D.C.

Daszak, P., L. Berger, A.A. Cunningham, A.D. Hyatt, D.E. Green, and R. Speare. 1999. Emerging infectious diseases and amphibian population declines. Emerg Infect Dis 5: 735–748.

Daszak, P., A.A. Cunningham, and A.D. Hyatt. 2000. Emerging infectious diseases of wildlife—threats to biodiversity and human health. Science 287:443–449.

DeMarcus, T.A., M.A. Tipple, and S.R. Ostrowski. 1999. U.S. policy for disease control among imported nonhuman primates. J Infect Dis 179(suppl. 1):S281–S282.

Ehrlich, P.R. and A.H. Ehrlich. 1981. Extinction: the causes and consequences of the disappearnce of species. RandomHouse, New York.

Epstein, P., B. Sherman, E. Spanger-Siegfried, A. Langston, S. Prasad, and B. McJay. 1998. Marine ecosystems: emerging disease as indicators of change. Health Ecological and Economic Dimensions of Global Change Program, Durham, N.H.

Fenical, W. 1996. Marine biodiversity and the medicine cabinet: the status of new drugs from marine organisms. Oceanography 9:23–27.

Gao, F., E. Bailes, D.L. Robertson, Y. Chen, C.M. Rodenburg, S.F. Michaels, L.B. Cummins, L.O. Arthur, M. Peeters, G.M. Shaw, P.M. Sharp, and B.H. Hahn. 1999. Origin of HIV-1 in the chimpanzee *Pan troglodytes troglodytes*. Nature 397:436–441.

Geretti, A.M. 1999. Simian immunodeficiency virus as a model of human HIV disease. Rev Med Virol 9:57–67.

Gibbs, E.L., T.J. Gibbs, and P.C. Van Dyck. 1966. *Rana pipiens*: health and disease. Lab Anim Care 16:142–160.

Gladfelter, W. 1992. White band disease in *Acropora palmata*: implications for the structure and growth of shallow reefs. Bull Mar Sci 32:639–643.

Gore, A. 1992. Earth in the balance: ecology and the human spirit. Houghton Mifflin, New York.

Goreau, T.J., J. Cervino, M. Goreau, R. Hayes, L. Richardson, G. Smith, K. DeMeyer, I. Nagelkerken, J. Garzon-Ferrera, D. Gil, G. Garrison, E. Williams, L. Bunkely-Williams, C. Quirolo, L. Patterson, J. Porter, and K. Porter. 1998. Rapid spread of diseases in Caribbean coral reefs. Rev Biol Trop 46:157–172.

Grenard, S. 1994. Medical herpetology. Reptile and Amphibian Magazine, Pottsville, PA.

Halvorson, W.L., and G.E. Davis. 1996. Science and ecosystem management in the national parks. University of Arizona Press, Tucson.

Hayes, R.L., and T.J. Goreau. 1991. Tropical coral reef ecosystems as a harbinger of global warming. World Resource Rev 3:306–322.

Hayes, R.L., and N.I. Goreau. 1998. The significance of emerging disease in the tropical coral reef ecosystem. Rev Biol Trop 46:173–185.

Institute of Medicine (IOM), Board of International Health. 1997. America's vital interest in global health. National Academy Press, Washington, D.C.

Intergovernmental Panel on Climate Change (IPCC). 1998. The regional impacts of climate change: an assessment of vulnerability. Watson, R.T., M.C. Zinyowera, and R.H. Moss, eds. Cambridge University Press, New York.

Intergovernmental Panel on Climate Change (IPCC). 1996. Climate change 1995: the science of climate change. Houghton, J.T., L.G. Meira Filho, B.A. Callender, N. Harris, A. Kattenberg, and K. Maskell, eds. Cambridge University Press, Cambridge.

Jameson, S.C., J.W. McManus, and M.D. Spalding. 1995. State of the reefs: regional and global perspectives. U.S. Department of State, Washington, D.C.

Kennedy, D., D. Holloway, E. Weinthal, W. Falcon, P. Ehrlich, R. Naylor, M. May, S. Scheider, S. Fetter, and J. Choi. 1998. Environmental quality and regional conflict. Carnegie Commission on Preventing Deadly Conflict, New York.

Irby, L., and J. Knight, eds. 1998. International symposium on bison ecology and management in North America. Montana State University, Bozeman.

Laurance, W.F., K.R. McDonald, and R. Speare. 1996. Epidemic disease and the catastrophic decline of Australian rainforest frogs. Conserv Biol 10:406–413.

Mervis, J. 1999. Canada dedicates new human, animal labs. Science 284:1902.

Monath, T.P. 1999. Ecology of Marburg and Ebola viruses: speculations and directions for future research. J Infect Dis 179(suppl 1):S127–S138.

Morrell, V. 1999. Are pathogens felling frogs? Science 284:728–731.

Morris, J.G., Jr., 1999. *Pfiesteria*, "the cell from hell," and other toxic algal nightmares. Clin Infect Dis 28:1191–1198.

Mudakikwa, A.B., J. Sleeman, J.W. Foster, L.L. Meader, and S. Patton. 1998. An indicator of human impact: gastrointestinal parasites of mountain gorillas (*Gorilla gorilla beringei*) from the Virunga Volcanoes region, Central Africa. *In* Proceedings of the AAZV/AAWV joint conference, pp. 436–437. AAZV, Omaha, Ne.

Organization for Economic Cooperation and Development (OECD). 1997. Shaping the 21st century: the contribution of development co-operation. Paper presented at the 34th high level meeting of the Development Assistance Committee, 6–7 May 1996, Paris.

Peters, C.J., and J.W. LeDuc. 1999. An introduction to Ebola: the virus and the disease. J Infect Dis(suppl 1):ix–xvi.

Peters, R.L. 1997. Diseases of coral reefs organisms. *In* Birkland, C., ed. Life and death of coral reefs, pp. 114–136. Chapman and Hall, New York.

Reaka-Kudla, M.L. 1996. The global biodiversity of coral reefs: a comparison with rainforests. In Reaka-Kudla, M.L., D.E. Wilson, and E.O. Wilson, eds., Biodiversity II: understanding and protecting our natural resources, pp. 83–108. National Academy Press, Washington, D.C.

Reaser, J.K. 1996. The elucidation of amphibian declines. Amphib Rept Conserv 1:4–9.

Reaser, J.K. 2000. Amphibian declines: an issue overview. U.S. Taskforce on Amphibian Declines and Deformities, Washington, D.C.

Reaser, J.K. and C. Galindo-Leal. 1999. Vanishing frogs: Mesoamerica and the Caribbean. U.S. Federal Task Force on Amphibian Declines and Defomities (TADD) and IUCN Declining Amphibian Population Task Force (DAPTF), Washington, D.C.

Reaser, J.K., R. Pomerance, and P.O. Thomas. 2000. Coral bleaching and global climate change: scientific findings and policy recommendations. Conserv Biol 14:1500–1511.

Richardson, L. 1996. Horizontal and vertical migration patterns *of Phoridium coralyticum* and *Beggiatoa* spp. associated with black band disease of corals. Microb Ecol 32: 323–335.

Richardson, L., R.B. Aronson, W.M. Goldberg, G.W. Smith, B. Ritchie, K.G. Kura, S.L. Miller, E.C. Peters, and J.C. Hales. 1998. Coral disease outbreak on reefs of the Florida Keys: plague type II. Nature 392:557–558.

Roberts, C.M., J. Hawkins, F.W. Schueler, A.E. Strong, and D.E. McAllister. 1998. The distribution of coral reef fish biodiversity: the climate-biodiversity connection. Paper presented at the fourth session of the Conference of the Parties of the United Nations Framework Convention on Climate Change, 2–13 November 1998, Buenos Aires.

Sebens, K.P. 1994. Biodiversity of coral reefs: what are we losing and why? Am Zool 34: 115–133.

Soulé, M.E., ed. 1986. Conservation biology: the science of scarcity and diversity. Sinauer, Sunderland, Mass.

Spellerberg, I.F., ed. 1996. Conservation biology. Longman, Essex.

Stafford-Deitsch, J. 1993. Reef: a safari through the coral world. Sierra Club Books, San Francisco.

Stebbins, R.C., and N.W. Cohen. 1995. A natural history of amphibians. Princeton University Press, Princeton, N.J.

U.S. Coral Reef Task Force. 2000. The national action plan to conserve coral reefs. U.S. Coral Reef Task Force, Washington, D.C.

U.S. Department of State. 1997. Environmental diplomacy: the environment and U.S. foreign policy. State Department Publication 10470. Bureau of Oceans and International Environmental and Scientific Affairs, Washington, D.C.

Weiss, R.A., and R.W. Wrangham. 1999. From *Pan* to pandemic. Nature 397:385–386.

Wilkinson, C., O. Linden, H. Cesar, G. Hodgson, J. Rubens, and A.E. Strong. 1999. Ecological and socioeconomic impacts of 1988 coral mortality in the Indian Ocean: an ENSO impact and a warning of future change? Ambio 28:198–206.

Williams, E.H., Jr., and L. Bunkley-Williams. 1990. The worldwide coral reef bleaching cycle and related sources of coral mortality. Atoll Res Bull 335:1–71.

Worthylake, K.M., and P. Hovingh. 1989. Mass mortality of salamanders (*Ambystoma tigrinum*) by bacteria (*Acinetobacter*) in an oligotrophic seepage mountain lake. Great Basin Nat 1989:364–372.

Index